D1276516

NCRP REPORT No. 58

Second Edition

A HANDBOOK OF RADIOACTIVITY MEASUREMENTS PROCEDURES

With Nuclear Data for Some Biomedically Important Radionuclides, Reevaluated between August, 1983 and April, 1984

**Recommendations of the
NATIONAL COUNCIL ON RADIATION PROTECTION AND MEASUREMENTS**

Issued February 1, 1985
Second Reprinting May 15, 1989
Third Reprinting February 28, 1994

National Council on Radiation Protection and Measurements
7910 WOODMONT AVENUE / BETHESDA, MD. 20814

Library of Congress Cataloging in Publication Data
National Council on Radiation Protection and Measurements.
Main entry under title:

A Handbook of radioactivity measurements procedures.

(NCRP report ; no. 58)
"Recommendations of the National Council on Radiation Protection and Measurements."
"February 1, 1985"
Bibliography: p.
Includes index.
1. Radioactivity—Measurement. 2. Radioactivity—Instruments. I. National Council on Radiation Protection and Measurements. II. Series.
QC795.42.H36   1985   539.7′7   84-25423

ISBN 0-913392-71-5

Preface to the First Edition

In 1961 the NCRP issued NCRP Report No. 28, *A Manual of Radioactivity Procedures*, which was published as the National Bureau of Standards Handbook 80. The report came to be one of the definitive works in the area of radioactivity measurements. As the development of new techniques, procedures, and equipment made parts of the Report obsolescent, the NCRP recognized the need to update and extend the document. The report consisted essentially of three parts: Part One being "Radioactivity Standardization Procedures"; Part Two, "Measurement of Radioactivity for Clinical and Biological Purposes"; and Part Three, "Disposal of Radioactive Material". Since the disposal of radioactive material constitutes a subject in itself, the NCRP, in preparing the revision of Report No. 28, decided to concentrate on the first two parts of the Report.

Because of the projected bulk of the revised report, it was decided to divide the task of preparation into two parts. NCRP Scientific Committee 18A was given the task of drafting the material on general measurement and standardization procedures; and Scientific Committee 18B was charged with drafting the material on those measurement procedures specifically concerned with medical and biological applications. This report is the result of Committee 18A's work. As such, it follows, to a high degree, the general format of the original report. Thus, the physics of detectors is discussed in Section 2.0 and then the uses to which these detectors may be put are discussed in those sections dealing with direct and indirect measurements. At the same time, each of the sections has been developed in much more detail than previously and it is therefore inevitable that some duplication occurs. In fact, some duplication has been deliberately included so that individuals may utilize given sections of the Report without having to refer back to earlier sections where similar material may have been covered. In almost every case, it is believed that the repetition will not only emphasize the suitability of a detector for a particular application, but will also add detail to the information already given.

In a report of this length it is inevitable that the supply of unique symbols would be exhausted. Where, however, there may have been some duplications in the use of symbols, every effort has been made to keep them consistent within the short section in which they appear.

Acknowledgments are made to Dr. Ayres who acted as secretary to the Committee, and to Dr. Hoppes who, while making no specific written contribution, served invaluably by spending many hours in valuable discussion with the Chairman of the Scientific Committee during the course of editing and clarifying the original text. The assistance of Miss P. A. Mullen in the work of coordinating the references is also gratefully acknowledged.

The Council has noted the adoption by the 15th General Conference of Weights and Measures of special names for some units of the Système International d'Unités (SI) used in the field of ionizing radiation. The gray (symbol Gy) has been adopted as the special name for the SI unit of *absorbed dose, absorbed dose index, kerma,* and *specific energy imparted.* The becquerel (symbol Bq) has been adopted as the special name for the SI unit of *activity* (of a radionuclide). One gray equals one joule per kilogram; and one becquerel is equal to one second to the power of minus one. Since the transition from the special units currently employed—rad and curie—to the new special names is expected to take some time, the Council has determined to continue, for the time being, the use of rad and curie. To convert from one set of units to the other, the following relationships pertain:

$$1 \text{ rad} = 0.01 \text{ Jkg}^{-1} = 0.01 \text{ Gy},$$

$$1 \text{ curie} = 3.7 \times 10^{10} \text{ s}^{-1} = 3.7 \times 10^{10} \text{ Bq (exactly)}.$$

Serving on Scientific Committee 18A during the preparation of this report were:

The Council wishes to express its appreciation to the members and consultants for the time and effort they devoted to the preparation of this report.

WARREN K. SINCLAIR
President, NCRP

Bethesda, Maryland
February 15, 1978

Preface to the Second Edition

By 1983 supplies of NCRP Report No. 58, *A Handbook of Radioactivity Measurements Procedures*, which was issued in 1978, were exhausted. In view of the progress which continues to be made in the field of radionuclide metrology it was decided to reissue it, not just by reprinting, but as a substantially revised edition, that now constitutes the present text.

Many of the individuals who participated in the earlier work of Scientific Committee 18A and a few others have contributed to this revision. The chairman of the Committee has asked that the contribution of several new pages of extensively revised text by A. P. Baerg, R. L. Heath, J. G. V. Taylor, and J. A. B. Gibson, mostly to Sections 2, 3 and 4 be especially acknowledged. He also wishes to thank R. Collé, E. H. Eisenhower, H. H. Ku with K. Eberhardt, D. D. Hoppes and R. Loevinger for countless and lively discussions of as many drafts of the new Sections 8.3 and 8.4 before a consensus was almost achieved; and to acknowledge very valuable discussions with B. M. Coursey and A. Grau Malonda when writing the new sub-section 4.3.9. Very grateful acknowledgement is due to M. J. Martin for organizing a special printing of nuclear data for selected radionuclides for this edition from the ENSD File of the Oak Ridge Nuclear Data Project. Contributions ranging from suggestions to references and discussions have been welcomed from W. C. Eckleman, T. G. Hobbs, J. M. R. Hutchison, B. Kahn, J. R. Noyce, C. T. Peng, L. A. Slaback, B. N. Taylor and M. P. Unterweger.

Especial thanks are due to D. D. Hoppes for very many cogent and fruitful discussions that were grossly intrusive on his time that was so graciously spared; and to Commander J. A. Spahn (U.S.N., Ret.) who substituted at short notice for C. J. Maletskos to represent the NCRP Secretariat and who has given weeks to meticulous and devoted editing of this new edition.

The NCRP wishes to acknowledge the deep debt of gratitude owed to Dr. Wilfrid B. Mann for the extensive effort he devoted to the scientific editing of this revised second edition and for his dedication to the never ending task of updating important technical information in this edition of the report.

Warren K. Sinclair
President, NCRP

Bethesda, Maryland
September 12, 1984

Contents

1. Introduction

1.1 General

Radioactivity is the term used to describe those spontaneous, energy-emitting, atomic transitions that involve changes in state of the nuclei of atoms. The energy released in such exoergic transformations is emitted in the form of electromagnetic or corpuscular radiations. The radiations interact, in varying degrees, with matter as they traverse it.

Radioactivity was discovered in 1896 by A. H. Becquerel, who was investigating the fluorescence of a double sulphate of uranium and potassium, using a photographic plate. He found that the plate was affected by certain *radiations actives*, irrespective of whether or not the salt was caused to fluoresce. Marie Curie, who coined the word *radioactivité*, investigated this property in a number of minerals containing uranium, which she found to be more active, atom for atom, than uranium itself. Postulating an element that was unknown except for its radioactivity, she and Pierre Curie set about the isolation of this hypothetical element, thereby applying, for the first time, techniques that now comprise the field of radiochemistry. In 1898 they discovered polonium, and, in collaboration with G. Bémont, radium. E. Rutherford showed that two types of radiation were emitted by uranium, namely the alpha rays, that were completely stopped by a thin sheet of paper, and the beta rays, that were much more penetrating. In 1900, even more highly penetrating radiations, the gamma rays, were discovered by P. Villard. Subsequently, the alpha and beta rays were shown to be ionized helium-4 atoms and electrons, respectively, and the gamma rays to be electromagnetic in nature, i.e. energetic photons.

In due course, three separate generic families of naturally occurring radioactive materials were discovered, namely the uranium, thorium, and actinium series. In each of these, one radionuclide—uranium-238, thorium-232, and uranium-235 (known originally as actinium)—in undergoing radioactive change, gives rise to a series of between twelve and eighteen radioactive progeny, usually referred to as *daughters*. The atomic mass number of each member of each series is obtained by substituting $n = 51, 52 \ldots 59$ in $(4n + 2)$ (uranium), $n = 52, 53 \ldots 58$ in $4n$ (thorium), and $n = 51, 52 \ldots 58$ in $(4n + 3)$ (actinium).

Following the discovery of the transuranic elements, a fourth such family of radioactive elements was discovered, namely the neptunium series of which the atomic-mass numbers are given by substituting n = 52, 53 ... 60 in (4n + 1). The common step of 4n in each series represents, of course, the loss of an alpha particle, the loss of a beta particle from the nucleus leaving the atomic-mass number unchanged.

The early history of radioactivity, and the underlying principles of its detection and measurements, are treated in such textbooks as *Radiations from Radioactive Substances* by E. Rutherford, J. Chadwick, and C. D. Ellis (1930); *The Atomic Nucleus* by R. D. Evans (1955); *Nuclear and Radiochemistry* by G. Friedlander, J. W. Kennedy, and J. M. Miller (1964); *Alpha-, Beta-, and Gamma-ray Spectroscopy*, edited by K. Siegbahn (1965) and *Radiation Detection and Measurement* by G. F. Knoll (1979). A brief and more elementary treatise, that may be useful for those involved as technicians in nuclear-medicine measurements, is *Radioactivity and Its Measurement* by W. B. Mann, R. L. Ayres, and S. B. Garfinkel (1980).

The then state of the art in radioactive metrology was the subject of an International Summer School on Radionuclide Metrology held at Herceg Novi, Yugoslavia in 1972. The proceedings, published a year later, (Herceg Novi, 1973) are a very valuable contribution to the literature of the subject.

In 1932 the neutron was discovered by James Chadwick and, in 1933, artificial radioactivity by Fréderic and Irène Joliot-Curie. Thereafter, in the late 1930's and in the 1940's many artificially produced radioactive elements were discovered by many investigators using nuclear reactions produced in the cyclotron of Ernest Lawrence and the nuclear reactor developed by Enrico Fermi and his colleagues. In discussing such nuclear reactions the following designations are used: n for neutron, p for proton, d for deuteron, α for alpha particle, and γ for gamma radiation, and the format, (n,γ), for example, meaning capture of a neutron by a nucleus and emission of a prompt gamma ray. The reactions giving rise to artificial radioactivity in cyclotrons and other particle acclerators are many, including (p,n), (p,γ), (d,n), (d,p), (He^{++},p), and so on. In many cases accelerated ions of higher atomic-number elements have also been used to produce nuclear reactions that result in radioactive elements. Among these, delayed proton emitters have also been observed. In reactors, in addition to the many radioactive fission products produced in the nuclear fission process, a vast quantity of radioactive materials has been produced by (n,γ), $(n,2n)$, (n,p), (n,α), and other reactions. With high-energy accelerators producing photons of many millions of electron volts of energy, prolific photonuclear reactions occur, corresponding to ($h\nu$,xn) and ($h\nu$,xp) where $h\nu$ (product of Planck's constant and photon frequency) represents the photon, and x (an integer representing the number of prompt neutrons and protons emitted) can be as high as ten. In the very high-energy accelerators, mesons can also be produced.

This report will be concerned with the measurement of radioactivity in general, but specifically it will deal with the vast number of different radioactive materials that have become available in the last three decades, from nuclear reactors and particle accelerators, for applications in medicine, scientific research, and industry. It will also be concerned with low-level radioactivity measurements for the monitoring of radioactivity in environmental media, such as air and water, in connection with the control of radioactive effluents associated with the production of nuclear power or the use of radionuclides.

It would be inappropriate to enlarge here upon this brief historical introduction. Some familiarity on the part of the reader with any of the textbooks referred to already, or their equivalent, must be assumed.

1.2   Radiations Emitted from Radioactive Atoms

In addition to the alpha particles, beta particles, and gamma rays already mentioned, energy from radioactive atomic transformations can be emitted as protons, neutrons, neutrinos, internal bremsstrahlung, conversion electrons, x rays, and Auger electrons. The beta particles can be either negative or positive electrons (β^- and β^+), β^- from neutron-rich and β^+ from neutron-deficient nuclei. The emission of a β^+ from the nucleus is always simultaneously accompanied by the emission of a neutrino, and that of the β^- by an antineutrino. An alternative to β^+ decay is capture of an atomic electron from the same atom with emission of a neutrino. The sharing of the available energy for decay between the beta particle and the neutrino accounts for the continuous beta-particle spectra; the sum of the energies of the beta particle and neutrino in a given transition is always constant, being *equivalent* to the mass difference ($E = mc^2$) between the parent and daughter atoms, less such energy as may be emitted in the form of gamma rays (or conversion electrons, x rays, or Auger electrons) by the daughter *atoms* in their transitions from excited levels to the ground level. Thus, ^{60}Co transforms (or "disintegrates" or "decays") mainly to an excited energy level of ^{60}Ni by the emission of a β^- particle and antineutrino with a total energy of 0.3179 million electron volts (MeV). The ^{60}Ni nucleus is left in an excited energy level, 2.5057

MeV above its ground level and decays with the prompt emission of two gamma rays, one with energy equal to 1.1732 MeV followed by another of 1.3325 MeV. Nuclear, or radioactive, transitions are conventionally depicted by means of a "decay scheme" such as those shown in Figure 1, representing the modes of decay of ^{64}Cu, ^{90}Sr-^{90}Y, ^{187}Re, and ^{210}Po. In the case of copper-64, the atomic mass of this isotope of copper is greater than the atomic masses of both its neighbors in atomic number, Z, namely nickel-64 and zinc-64, so that transitions to both these "isobars" are energetically permissible.

The case of ^{187}Re is unusual and is of interest since it illustrates the fact that the energy balance in radioactive decay involves the *whole* atom. In the transition of ^{187}Re to ^{187}Os (Fig. 1d) a total of 2.6 keV of energy is carried away by the β^- particle and antineutrino. Since, however, the atomic-electron binding energy of osmium is greater than that of rhenium by 15.3 keV, it follows that the mass of the rhenium nucleus is less than the mass of an osmium nucleus and an electron by the mass equivalent of 12.7 keV (i.e., 15.3-2.6 keV). A ^{187}Re nucleus stripped of all of its electrons would therefore be stable against β^- decay.

Crasemann (1973) has discussed the influence of atomic electrons on the modes of nuclear decay. In the case of ^{7}Be, an electron capturer in which the atomic binding energies are comparable with chemical binding energies, the value of the decay constant (Section 1.3) can be very slightly modified by both chemical composition and pressure (see, for example, Hensley *et al.*, 1973).

In alpha-particle decay, no simultaneous radiation, comparable to the neutrino, is emitted from the parent nucleus and hence groups of alpha particles are always homogeneous in energy. Again, gamma rays may be emitted promptly from the daughter nucleus in its decay to the ground level.

If the daughter nucleus does not decay promptly to its ground level, it may exist in a metastable state for a considerable time and exhibit radioactivity, so to speak, in its own right. Thus, ^{99}Mo decays to a 0.143-MeV level of ^{99}Tc which exists long enough, several hours, for the technetium to be chemically separated as a useful radioactive species, isolated from its parent, ^{99}Mo. The delayed transition of an excited daughter nucleus to a lower-energy level of the same nucleus is called an *isomeric transition* and such nuclear isomerism (as distinct from chemical isomerism) is denoted by the addition of the letter m (for metastable) after the atomic-mass number for the given nuclear species, e.g., ^{99m}Tc. In such cases where an isomeric species decays to a radioactive daughter, the letter g (for ground) is sometimes used to

Fig. 1. Typical decay-scheme representations for (a) ^{210}Po, (b) ^{64}Cu, (c) ^{90}Sr-^{90}Y, and (d) ^{187}Re. Energies are expressed in millions of electron volts (MeV). ^{210}Po normally decays to the ground state of ^{206}Pb with the emission of a 5.3045-MeV alpha particle. Occasionally, however, the decay is to an excited level of ^{206}Pb with the emission of a 4.524-MeV alpha particle, followed by a 0.8031-MeV gamma ray. The difference between the energy balances in the two branches, α_1 and α_2, is due to the energy associated with the recoil of the ^{206}Pb nucleus (see Section A.1.2.1. of Appendix A). In the case of ^{64}Cu, its atomic mass is greater than those of ^{64}Ni and ^{64}Zn by 0.0018008 and 0.0006168 atomic-mass units, respectively. From the Einstein equation, $E = mc^2$, 1 atomic-mass unit is found to be equivalent to 931 MeV. Hence the energy available for the transition of ^{64}Cu to ^{64}Ni or ^{64}Zn would be respectively 1.6765 and 0.5742 MeV. The transition to ^{64}Ni can occur either by emission of a positive electron ($\beta+$) or by electron capture (ϵ). The vertical line appearing in the β^+ branch represents the rest energy of two electrons, 1.022 MeV, because of the loss from the system, in the process of decay, of a positive electron and an excess atomic electron.

Most such decay schemes indicate gamma-ray abundances, or probabilities per decay, in percent, either before or after the gamma-ray transition energy. Thus, in Fig. 1a the 0.8031-MeV transition is shown as 0.001%. Where more than one gamma ray originates from any one energy level, all abundances would be shown.

More recently evaluated nuclear-decay data are available from the Oak Ridge Nuclear Data Project, as indicated in Section A.1.1 of Appendix A.

denote the ground state of the daughter nucleus, e.g., ^{81g}Kr is the radioactive daughter of ^{81m}Kr.

The energy associated with the electromagnetic decay of an excited state is not always emitted as gamma radiation; it may be transferred by a competing radiationless transition directly to a bound electron from a shell of the same atom. This process is known as *internal conversion*, a misnomer dating from early workers who erroneously assumed the existence of an intermediate gamma ray. The (internal) conversion electrons are ejected in monoenergetic groups having energies equal to the transition energy less the binding energies of the various electron shells.

For a given transition, the ratio of the number of decays proceeding by internal conversion to the number proceeding by gamma-ray emission is called the *internal-conversion coefficient* denoted by $\alpha(= e/\gamma)$. Similarly, coefficients for internal conversion from specified shells are written, e.g., $\alpha_K(= e_K/\gamma)$ so that $\alpha = \sum_i \alpha_i (i = K, L_I, L_{II}, L_{III}, M_I \ldots)$. If the transition energy is large compared to the K-shell binding energy, α_K is the largest term; if the transition energy is less than the i-shell binding energy, α_i is necessarily zero. As a rule α is small for prompt, high-energy transitions in low-Z nuclei and is larger for delayed or low-energy transitions or transitions in high-Z nuclei. Internal-conversion coefficients for transitions of interest are included in compilations of decay schemes such as the *Table of Isotopes* (Lederer *et al.*, 1967). Conversion-electron probabilities per decay for selected radionuclides are given in Appendix A.

The emission of radiation as a conversion electron is a transition involving the *whole* atom, and this process emphasizes again how important it is to consider the interactions of the entire atom in radioactive decay. In the decay of both ^{90m}Nb and ^{99m}Tc, there are highly converted transitions of relatively low energy, of about 2.2 keV, which are again sensitive to the chemical or physical environment of the atoms. Experiments have shown that the decay constants of each of these isomers can be modified both by chemical composition (Cooper *et al.*, 1965; Crasemann, 1973) and pressure (Mazaki *et al.*, 1972).

Either an electron-capture or conversion-electron event creates a vacancy in an atomic shell of the daughter atom. The filling of this vacancy gives rise either to an x ray, characteristic of the daughter atom, or to one or more electrons (Auger electrons) from its outer shells. In the filling of atomic-shell vacancies the ratio of x rays emitted to the total number of vacancies to be filled in any given shell is called the fluorescence yield and is usually represented by ω (ω_K, ω_L, and so on). The emission of an Auger electron instead of an x ray is analogous to the emission of a conversion electron instead of a gamma ray, in that the excess energy is transferred directly to the electrons. Values for fluorescence yields and Auger-transition probabilities are available in a review article by Bambynek *et al.* (1972).

Electron capture and β^+ decay are alternative processes, the latter being possible only if the atomic mass of the parent, M_A^Z, is greater than the atomic mass of the daughter, M_A^{Z-1}, by more than two electron rest masses, $2m_e$, i.e.

$$M_A^Z > M_A^{Z-1} + 2m_e \qquad (1.1)$$

The difference of $2m_e$ has nothing to do, as is sometimes asserted, with the creation of an electron-positron pair, but is simply due to the fact that the nucleus of atom M_A^Z loses one electron mass by decay (β^+) and one atomic electron is lost because of the decrease in atomic number from Z to $Z-1$ (Mann *et al.*, 1980, Chap. 5).

If, however,

$$M_A^{Z-1} < M_A^Z < M_A^{Z-1} + 2m_e \qquad (1.2)$$

the transition from M_A^Z to M_A^Z can still occur by the process of electron capture, excess energy being carried away by the neutrino. The electron captured by the transforming nucleus originates in the K shell or outer electronic shells of the decaying atom, with the number of extra-nuclear electrons balancing the new nuclear charge. For each type of capture (e.g., K, L, or M), monoenergetic neutrinos are emitted.

Values for the energies and probabilities of emission per decay, or "intensity," of Auger electrons and x rays arising from K or L capture are given in Appendix A. The effects of other possible transitions within and between atomic shells are discussed by Bambynek *et al.* (1972).

It must be emphasized that the emission of gamma rays or conversion electrons from the daughter atom *follows* the transformation of the parent by intervals of time varying from fractions of a picosecond to tens of thousands of seconds. The definition of isomerism, or what time interval we choose to say has to elapse to call a transition isomeric rather than prompt, is largely a subjective judgment conditioned to some extent by our instrumental capability for measuring short intervals of time.

Finally, the parent atoms may decay either to the ground state or to excited energy levels of the daughter atoms.

For a given radioactive *nuclide* (or atomic species), the proportions of transitions to the different nuclear energy levels of the daughter atom are given by the *branching ratios*. Such ratios for selected radionuclides, together with other decay data which are important for both their assay and use, are given in Appendix A.

For a more detailed treatment of the subject matter of this section, reference should be made to the textbooks mentioned in Section 1.1.

1.3 The Half Life, Decay Constant, and Activity of A Radioactive Element

In 1903, Rutherford and Soddy, in a paper entitled "Radioactive Change," published in the *Philosophical Magazine*, proposed "the law of radioactive change." They had observed, in every case that they had investigated, that the activity of a radionuclide decreased in geometric progression with time. This meant that if the activity decreased from any initial value to half that value in a given time, then in a further equal interval of time the activity would decrease by half again, or to one-quarter of the initial activity. In 1904, Rutherford introduced the term *half life* to denote the time required for the activity of a radioactive element to decay from any given initial value to half that value. Rutherford and Soddy also stated the law that is appropriate to a geometric decrease of activity with time, namely that $N_t/N_0 = e^{-\lambda t}$, where N_0 is the number of radioactive atoms present at $t = 0$, N_t is the number after a further time t, and λ is a constant. This constant they called the "radioactive constant" but we now usually call it the *decay constant*. By differentiation they obtained $dN_t/dt = -\lambda N_t$ and concluded that, "The proportional amount of radioactive matter that changes in unit time is a constant." The fractional rate of decrease in the number of radioactive atoms is thus clearly equal to λ, which can be expressed as $(1/N_t)(dN_t/dt)$. (Also see Appendix B2.)

From the radioactive decay law, the time, $T_{1/2}$, required for a given number of radioactive atoms of a single radionuclide to decrease to half is very simply derived. Thus, if $N_t = N_0/2$ then $(N_0/2)/N_0 = e^{-\lambda T_{1/2}}$ whence the half life, $T_{1/2} = \ln 2/\lambda$.

For a given number of atoms N of a given radionuclide, unfed by any parent, λN is the expected number of decays per unit time at a given time, but only if we regard N as being a continuous function of t can we say that the decay rate is equal to dN/dt. But, by simple observation, it is known that the *measured* normalized decay rate $(dN/dt)/N$ is not constant but fluctuates over a range of values (see Appendix B). In other words, radioactivity is a random or stochastic phenomenon, and dN/dt, known as activity from the earliest days of radioactivity, only has meaning when we consider the average of a very large number of decays. For this reason, activity has been defined in the following rather precise manner, namely, the activity of an amount of a nuclide in a specified energy state at a given time is the expectation value, at that time, of the number of spontaneous nuclear transitions in unit time from that energy state. By this definition, zero activity would be equivalent to stability of the nuclide.

To define activity as dN/dt can be confusing as, for example, in parent-daughter relationships where dN/dt for the daughters can be positive, negative, or zero.

The dimension of activity is reciprocal time, and activity is usually expressed as the number of radioactive events per second. Such radioactive events are comprised of any atomic transitions that involve a change in nuclear state, and therefore include isomeric transitions.

A commonly used unit of activity that is accepted as a unit outside the Système International (SI) is the *curie* (Ci) (see BIPM, 1973; NBS, 1977). One curie is equal to exactly 3.7×10^{10} radioactive events per second. Multiples and sub-multiples of the curie are the megacurie (MCi), millicurie (mCi), microcurie (μCi), and so on. The *becquerel* (Bq) has recently been adopted as the SI unit of activity, being equal to one radioactive event per second. Other quantities frequently used in radioactivity are *activity concentration* (activity per unit mass or volume of sample, *e.g.*, $s^{-1}\ mg^{-1}$, $Ci\ ml^{-1}$, $Bq\ mg^{-1}$) and *specific activity* (activity per unit mass of element, radioisotope of which is in question, *e.g.*, $Bq\ g^{-1}$). (ICRU, 1963)

As has been noted in Section 1.2, the definition of isomerism is a subjective one, and depends on whether the half life of the excited state of the daughter nucleus is long enough for the daughter to be separated from the parent and exist as a radioactive substance in its own right. Thus, one might say that the nearly instantaneous decay of the 2.506-MeV energy level of ^{60}Ni, following ^{60}Co decay, is not considered to be radioactivity in its own right, as distinct from, say, the 0.1426-MeV level of ^{99m}Tc, decaying with a half life of 6.02 hours. (The half life of the 1.,3325-MeV energy level of ^{60}Ni is about 0.7 picosecond.) On the other hand, ^{137m}Ba, with a half life of only 2.55 minutes, is often more closely identified with its radioactive parent, ^{137}Cs.

Care must be taken, in certifying the activity of a radioactive substance, to state very specifically what substance is under consideration. Thus, in the case of the two β^- emitters, ^{90}Sr and its daughter ^{90}Y, the two in equilibrium have approximately double the activity of freshly separated ^{90}Sr. As the ^{90}Y grows back into equilibrium after separation, its activity increases until, in approximately 300 hours, a state of equilibrium is again approached (e.g., in 5 half lives the ^{90}Y activity is 97 percent of the ^{90}Sr activity). It is therefore imperative, when dealing with an equilibrium mixture, to state whether or not the reported value is the *sum of the activities* of the two radionuclides in equilibrium, or the activity of the parent.

It is important to note that, in the condition of equilibrium between any radioactive parent that decays only to a short-lived daughter, the activity of the daughter is greater than the activity of the parent. It can be shown (see, for example, Mann *et al.*, 1980) that, if A represents the parent and B the daughter, and λ_A and λ_B their respective decay constants, then, at a sufficiently long time, t, after separation

$$dN_B/dt = [\lambda_B/(\lambda_B - \lambda_A)](dN_A/dt). \qquad (1.3)$$

If $\lambda_B \gg \lambda_A$, then the activities of parent and daughter tend to become equal, and we have the condition of *secular* equilibrium. When, however, λ_A and λ_B are comparable in magnitude, the activity of B is significantly different from that of A, in the proportion of $\lambda_B/(\lambda_B - \lambda_A)$. This kind of equilibrium is known as *transient*.

In the case of ^{140}La ($T_{1/2} = 40$h) in transient equilibrium with its parent ^{140}Ba ($T_{1/2} = 12.8$d), the activity of the former is 15 percent higher than that of the latter (Figure 2).

When $\lambda_B < \lambda_A$, no state of equilibrium can exist.

1.4 The Interaction of Radiation with Matter

As mentioned in Section 1.1, the radiations emitted in radioactive transformations interact with matter, and it is in this manner, whether

Fig. 2. Growth and decay of ^{140}La ($T_{1/2} = 40$ h) from freshly separated ^{140}Ba ($T_{1/2} = 12.8$ d) and the decay of the ^{140}Ba as a function of time. The ^{140}La activity is a maximum of $T_m = 6.57$ d. (From Mann *et al.*, 1980).

it be in the interaction with a photographic emulsion or in the ionization of atoms in the surrounding air, that radioactivity can be detected and the amount of radioactivity, or the quantity of radioactive material, can be measured.

The earliest measurements of radioactivity were based on the fact that radiation, as it traverses air, ionizes the gaseous molecules present. Under the influence of an electric field, the electrons and the negative and positive ions migrate respectively to and from regions of higher positive potential. An ionization current is thus established. In the earliest experiments, simple parallel-plate ionization chambers and gold-leaf electroscopes were used to detect and to measure radioactivity. Thus, Pierre and Marie Curie measured radioactivity in terms of units of 10^{-11} ampere of ionization current flowing across a charged parallel-plate condenser, with the plates horizontal, when powdered radioactive minerals were placed on the lower plate. For those who may be interested in the details of these earlier, yet very elegant experiments, reference is made to Mann et al. (1980). Reference to this and to Evans (1955), to Friedlander et al. (1964), and to other texts, should be made for details of the interactions of alpha and beta particles, and of photons, with matter.

1.4.1 Interactions of Charged Particles with Matter

When charged particles traverse air, they undergo elastic and in-elastic collisions with the atoms and molecules, the elastic producing deflections and the inelastic producing positive ions and electrons, usually referred to as ion pairs. An average of nearly 35 eV of energy is expended for each ion pair formed, so that a 3.5-MeV alpha particle will, for example, produce some 10^5 ion pairs in air before expending all its energy and coming to rest. The production of ion pairs by charged particles is by Coulomb interactions. The alpha particle, being some 7000 times heavier than an electron, will "plow" its way through a gas and only rarely, when it comes close to an atomic nucleus, will it be deflected from its linear trajectory. Beta particles are, on the other hand, of the same order of mass (except at high velocity) as the atomic electrons removed by ionization. The path of a beta particle as seen, say, in a cloud chamber, is therefore meandering and circuitous by comparison with that of an alpha particle. The responses of ionization chambers, Geiger-Müller counters, and proportional counters, are all based upon the principle of gas ionization, and the physics of these detectors will be described in later sections of this report.

Charged particles also interact with photographic emulsions, with scintillating fluids and solids, such as anthracene and plastic fluors, or with semiconductor detectors. The scintillation of zinc-sulfide crystals when bombarded with alpha particles was one of the earliest methods used by Rutherford and his collaborators in the detection of these radiations. Energetic electrons, beta particles, Auger electrons, and alpha particles also cause solid and liquid fluorescent materials to emit light by imparting energy to electrons of the atoms comprising these fluors. The photons so emitted can be detected by sensitive electron-multiplier phototubes to provide quantitative information on the radioactivity present. Semiconductor detectors, to be more fully described in subsequent sections of this report, act as solid-state ionization chambers.

The energy carried by charged-particle radiations, when completely absorbed by matter, appears as an equivalent amount of heat which can be measured in a sensitive calorimeter. With the type of calorimeter described by Jordan and Blanke (1967), Jordan (1973) has achieved such sensitivity as to be able to observe, in a period of from 2 to 4 hours and with a sample of about 3 kilocuries, the 12.3-year half life of tritium.

1.4.2 *Interactions of Energetic Photons with Matter*

The term "energetic" in this section's title is meant to refer to energies above, say, one kiloelectron volt, or in other words, to x rays and gamma rays, rather than to visible or ultraviolet photons. By way of reference, the energy of the "D line" of sodium corresponds to an approximately 2-eV transition of the atomic electrons involved.

The x and gamma rays transfer energy to matter chiefly by three extranuclear processes, namely (i) the *photoelectric effect*, which is important for energies below 1 MeV; (ii) the *Compton effect*, which is important over a wide range of energies; and (iii) *electron-positron-pair production*, which occurs at gamma-ray energies above 1.022 MeV.

In the *photoelectric effect*, the whole energy of the photon is transferred to a bound electron of another atom, which electron leaves its parent atom with an energy equal to that of the interacting photon, less the binding energy of the K, L, M, etc., shell from which it is ejected. Electrons originating in the same shell are therefore mono-energetic for incident monoenergetic photons.

In the *Compton effect*, the photon is scattered and behaves like a particle, with conservation of energy, $h\nu$, and momentum, $h\nu/c$, where

h is Planck's constant, ν is the photon frequency, and c is the speed of light. The scattering equations are derived in the textbooks referred to in Section 1.1, but, essentially, energy is imparted to a free (or loosely bound) electron with a consequent loss in frequency and energy of the scattered photon.

In *pair production*, an energetic gamma ray produces a negatron-positron pair (usually referred to as an electron-positron pair). The rest energy of the electron, derived from the Einstein relation $E = mc^2$, is equal to 0.511 MeV. A photon of energy greater than 1.022 MeV can therefore produce an electron-positron pair. To conserve energy and momentum, such a process must occur within the Coulomb field of a nucleus and any energy in excess of 1.022 MeV is imparted to the recoiling particles. The positron subsequently annihilates with an electron with the production (usually) of two 0.511-MeV photons. The processes of pair production and annihilation normally occur within the resolving time of a detector, and, for the low-energy gamma rays occurring in radioactive decay, the annihilation photons propagate essentially at 180° to each other.

The detection of gamma rays is made possible by the interactions, with matter, of the electrons produced by the three processes enumerated above. These electrons ionize gases, and consequently, ionization chambers, Geiger-Müller, and proportional counters can be used to detect the photo-electrons, scattered electrons, and electron pairs. In addition, gamma rays produce such electrons within the volumes of solid detection devices, such as sodium-iodide, germanium, and silicon crystals. In the case of sodium-iodide crystals, the secondary electrons produce light pulses that can be detected by phototubes, while the germanium and silicon crystals act as solid-state ionization chambers. The physics of such detectors are dealt with in detail in Sections 2.4 and 2.5 of this report.

If gamma rays or x rays are completely absorbed by matter, the electrons produced, if also absorbed, give an equivalent amount of heat that can again be measured by sensitive calorimeters. Such measurements are more often made, however, to determine the absorbed dose rate from therapeutic gamma-ray sources and are not the concern of this report. (See, for example, ICRU, 1970.)

Linear attenuation coefficients for collimated monochromatic-photon beams as a function of energy are shown in Figures 3 and 4 for aluminum and lead, based on the data of Grodstein (1957), Hubbell (1969), and Hubbell *et al.* (1975). For more recent mass attenuation coefficients and linear absorption coefficients over the range 1 keV to 20 MeV, see Hubbell, 1977, 1982. In Figures 3 and 4 the total attenuation coefficients, together with the photoelectric, Rayleigh (coher-

Fig. 3. Linear attenuation coefficients as a function of photon energy for a collimated, parallel beam of monochromatic photons in aluminum. The mass attenuation coefficient can be derived by dividing the ordinates by 2.70 g cm⁻³, the density of aluminum. The "Compton (a) and (b)" curves are those pertaining to free and bound electrons, respectively. (These curves are derived from data given in Hubbell [1969] and Hubbell et al. [1975], kindly provided by J. H. Hubbell.)

ent), Compton (incoherent), and pair-production components, are shown. The attenuation coefficient is, effectively, the probability, per unit path length, of a photon being either absorbed or scattered. Experimental observations of the removal of photons from narrow collimated beams give results that closely approximate the attenuation coefficients shown in Figures 3 and 4, and the attenuation coefficient is then synonymous with what is still sometimes called the absorption coefficient. These values, divided by the density, give the mass attenuation coefficient. The interactions of energetic photons with matter are discussed in more detail in Section 4.2.2.

1.4.3 *Other Interactions*

Other radiations emitted from the nucleus in radioactive decay, such as protons and neutrons, can be detected by ionization and other

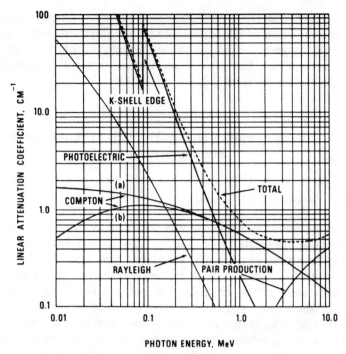

Fig. 4. Linear attenuation coefficients as a function of photon energy for a collimated, parallel beam of monochromatic photons in lead. The mass attenuation coefficient can be derived by dividing the ordinates by 11.34 g cm^{-3}, the density of lead. The "Compton (a) and (b)" curves are those pertaining to free and bound electrons, respectively. (These curves are derived from data given in Hubbell [1969] and Hubbell *et al.* [1975], kindly provided by J. H. Hubbell.)

methods, for example, by observing the etched tracks of protons and alpha particles in crystals, glass, and plastics (see Section 6.1.2.5). Neutrinos, which are particles of zero mass and zero charge and, therefore, have practically no interaction with matter, must be detected by more sophisticated means, but are of no immediate concern to the measurement of radioactivity. When electrons are accelerated or decelerated they also emit electromagnetic radiation known as *bremsstrahlung*. The beta particles or electrons, which are retarded by matter, emit photons having a range of energy from zero up to the maximum energy of the beta particles or electrons being stopped. Many of the photons from a typical x-ray tube are produced by the bremsstrahlung process. Bremsstrahlung photons have also been used in the assay of tritium (Curtis, 1972). *Internal bremsstrahlung* photons are emitted from transforming nuclei, but these are rarely used to measure radioactivity, because of their very low abundance.

High-energy charged particles can also be detected by observing the Čerenkov radiation which is generated when they traverse a medium with a velocity greater than the velocity of light in that medium. This again has, however, but limited application to the measurement of radioactivity (see Sections 2.6, 4.5.1, 6.1.3.3).

1.5 Fundamental or Direct Methods of Radioactivity Standardization

1.5.1 *Introduction; Terminology*

Direct methods for the measurement of radioactivity are sometimes referred to as "absolute," and standards so produced are often designated as "absolute" or "primary" standards. The word "absolute" bears the connotation, however, of being without error. Only the defined standards, such as the kilogram, are without error. An indirect or relative radioactivity standard (see Section 1.6) is "absolute" in the sense that it is measured in terms of the reciprocal second. Such a standard is an "absolute measure" but, like all measurements made in relation to the defined standards, is not devoid of error. Nowadays the term "absolute" is often used merely to indicate that a measurement is direct, or that the measurement of an activity is independent of any other measured activity.

The terms "primary," "secondary," "tertiary," etc. can also be misleading in that an international "secondary" (as in the case of radium) could be a national "primary" standard. Or a national secondary or tertiary standard may be a state, or provincial, or hospital primary, and so on. Where possible, therefore, it would seem to be preferable to refer to the *international* standards, or to *national* standards (McNish, 1958 and 1960) or to the *working, laboratory,* or *calibration* standards of hospital X or university Y. In many cases, a national laboratory may issue a *national* standard based on a relative measurement in, for example, a calibrated ionization chamber, while a university or commercial laboratory may produce an "in-house" standard by a direct method of measurement such as coincidence counting.

It is, however, more important to describe the method of preparation and measurement, whether *direct* or *indirect* (or *relative*), and to characterize a radioactivity standard fully with regard to purity and estimated uncertainty in an accompanying statement or certificate. In the last resort, the method of production of the standard is of less importance than a realistic appraisal of the uncertainty of its value.

1.5.2 *Direct Methods of Measurement*

There are, broadly speaking, four methods of direct calibration in which the activity of a radioactive source can be determined from the count-rate data alone, without reference to any supplementary data determined from previous experiments, apart from the need to have a prior knowledge of the mode of decay of the radioactive substance itself.

The most powerful of these direct methods is probably that of coincidence counting (Geiger and Werner, 1924; Dunworth, 1940; NCRP, 1961; Herceg Novi, 1973). This method is applicable to all radionuclides decaying by the emission of two or more coincident radiations. Thus, where a radionuclide emits an alpha or beta particle, or an x ray (as a result of electron capture), followed promptly by a gamma ray, the method of coincidence counting (Sections 3.2, 3.3, and 3.4) can be used to determine the efficiency of the detectors of the two coincident radiations, and thence to calculate the activity of the radionuclide in question. In the case of ^{60}Co, the activity may be determined either by beta-gamma coincidence counting of the ^{60}Co beta particles and the ^{60}Ni gamma rays, *or* by gamma-gamma coincidence counting of the two ^{60}Ni gamma rays in cascade. For more complex decay and, generally for the highest accuracy, efficiency variation techniques are now used (Section 3.2).

Modifications of the method of coincidence counting are those of anti-coincidence, correlation counting and selective sampling (Sections 3.2.3 and 3.2.4). These are particularly useful where a radionuclide decays to a delayed excited state of a daughter nucleus.

For charged-particle emitters the direct measurement of activity can be achieved by placing the source on a very thin source mount within a windowless detector subtending an angle of 4π steradians around the source (Section 3.5). Here the geometrical efficiency is unity, and all corrections for scattering and absorption outside the source are eliminated. For very thin, or "mass-less," sources of alpha or beta particles of not too low an energy, practically 100-percent detection efficiency can be achieved in the case of 4π gas proportional counters. Internal gas counting of radionuclides in gaseous form (Section 3.6) and liquid-scintillation counting of energetic alpha- and beta-particle emitters (Sections 2.4.5 and 3.8) are also forms of 4π counting in which the total detection efficiency closely approaches 100 percent.

One of the oldest methods of direct standardization of activity is that of defined-solid-angle counting where the defined solid angle is 1π steradians or less. It has been used for measuring alpha-particle, beta-particle, and photon activities, but is now considered unsuitable for direct beta-particle standardization, because of the large errors introduced by scattering. Bambynek (1967) states, however, that for alpha particles with energies of several MeV, and for photons in the range of energy of 1 to 80 keV, accuracies of within ±0.1 percent for alpha particles and ±1 percent for photons are attainable. His criterion for the successful use of the defined-solid-angle method is that the radiations shall be "lightly scattered but heavily absorbed."

Low- and medium-geometry defined-solid-angle counters have been used by Spernol and Lerch (1965), Bambynek *et al.*, (1966) and Bambynek (1967) to standardize alpha-particle and low-energy photon sources. A very successful $1\pi\alpha$ counter has been described by Robinson, H. (1960) and a modified $1\pi\alpha$ counter (based on Robinson's design) is in use at the National Bureau of Standards (Hutchinson *et al.*, 1968). Counting in a 1π geometry, or less, normal to the surface of the thick source support, eliminates errors due to multiple alpha-particle scattering, because such scattered alpha particles are emitted at angles of less than 5° to the plane of the surface of the source support (Walker, 1965). (See Section 3.7.7.)

The fourth direct method is applicable to radionuclides that decay with the emission of at least two photons in coincidence with no transitions of the parent nucleus to the ground state of the daughter. This method requires a photon detector of relatively high intrinsic and geometrical efficiency. In such circumstances the two photons will sum in the detector so that not only two photopeaks corresponding to the energies of each of the photons will be produced, but also a sum-peak corresponding to the summing of the energies totally deposited by two coincident photons in the detector (Section 3.3). The count rates under these three peaks are usually measured by the analysis of the spectrum of pulses from the detector using a multichannel analyzer. From the total gamma-ray count rate, and the count rates in the three photopeaks, the activity may be calculated with no other measurement information, apart from certain relatively minor corrections for counting losses (Section 2.7) inherent in the nature of the electronic circuitry involved.

1.5.3. *"Near-Direct" Methods of Measurement*

In these methods the disintegration rate, or activity, is not measured directly but is calculated from a direct measurement of some property of the decay of the radionuclide in question, such as rate of energy emission, together with previously measured decay data, in this case the average energy emitted per disintegration. Some "near-direct" methods are briefly described in the subsections below. Table 1 gives conversion factors for deriving disintegration rates from the various quantities measured in different units.

1.5.3.1 *Weighing of a Radioactive Element of Known Isotopic Abundance.* Historically, weighing is the oldest method of radioactivity standardization, having first been applied in the case of radium in 1911. This radium, which was used to determine the atomic weight of radium-226, was considered to be isotopically pure. Weighing, or its equivalent, such as measuring the pressure, volume, and temperature of a gas, may also be used to determine the activities of other radioactive nuclides, provided that the isotopic abundance of the radionuclide in question and its decay constant are known. Where a radionuclide is mixed with inactive isotopes, such as ^{63}Ni in nickel, the relative abundances may be determined by mass spectrometry (Section 3.10). The disintegration rate, $-dN/dt$ can then be determined from the radioactive decay law $-dN/dt = \lambda N$, by using the measured value of N and the known decay constant λ.

1.5.3.2 *Calorimetry.* Particularly suitable for short-range particle emitters (such as alpha or low-energy beta) is the calorimetric method in which the rate of energy emission from the radioactive substance is measured. The conversion to activity requires a knowledge of the average energy per disintegration. (See Section 3.9.)

1.5.3.3 *Ionization Measurements.* If the average energy expended by radiation in forming one ion pair in a gas is known, the total energy of the radiation from a source can be determined from ionization-current measurements. The calculation of the activity of the source then requires a knowledge of the average radiation energy emitted per disintegration for the nuclide in question. (See Section 3.11.)

TABLE 1—*Conversion of various measures of radioactivity into nuclear transitions per second*

Quantity of radioactive nuclide expressed as	Transitions per second Bq	Remarks
1 curie	3.7×10^{10}	Defined as exactly 3.7×10^{10} by international agreement. Symbol for the "curie" is Ci.
1 gram of pure radionuclide	$\dfrac{6.022 \times 10^{23}}{\text{at. wt.}}\lambda$	λ decay constant (s^{-1}), atomic weight (mass) (unified scale, i.e., carbon-12)
1 watt total rate of energy emission	$\dfrac{6.24 \times 10^{12}}{E}$	E average energy emitted per disintegration (MeV).
1 roentgen/hour at 1 cm (from gamma-ray-emitting point source).	$3.7 \times 10^7 \, \Gamma_\delta$	If exposure-rate constant Γ_δ is in R cm^2 mCi^{-1} h^{-1}. See Eq. 3.35 *et seq.*

1.5.3.4 *Loss-of-Charge Method.* For charged-particle emitters the loss-of-charge method allows the number of such particles emitted to be calculated from a measurement of the charge transfer. Since the method is rather insensitive, it is not recommended for practical use. (See Section 3.12.)

1.5.3.5 *Activity from Production Data.* The activity of a radionuclide produced in a particular irradiation may be calculated approximately from irradiation, cross-section, and decay-scheme data. (See Section 3.13.)

1.6 Indirect or Relative Methods of Radioactivity Standardization

In the national standardizing laboratories, the calibration data for many radionuclides (particularly a short-lived radionuclide) are usually preserved by using a standardized source to calibrate a reference instrument such as an ionization chamber. This instrument can then be used to calibrate further samples of the same radionuclide to give derived or "secondary" standards at any later date, or to check the consistency of subsequent "absolute," or direct, standardizations. The stability of the instrument is usually checked by measuring a reference source of some long-lived material, such as radium, whenever a calibration is made. One may either take the activity of the sample to be proportional to the response of the instrument (after its stability has been checked), or to the ratio of the response to that produced by the reference source (in which case variations in the sensitivity of the instrument are immaterial as long as they effect the responses to both sources equally).

Methods of relative standardization are described and discussed in Section 4 of this report.

2. Physics of Some Radiation Detectors

2.1 Ionization Chambers

An ionization chamber is an instrument in which an electric field is applied across a volume of gas. Often its geometry is cylindrical, with a cylindrical cathode enclosing the volume of gas and an axial insulated rod anode. The anode is also frequently in the form of a hollow cylinder into which a photon-emitting radioactive source can be inserted for calibration.

Charged particles moving through matter undergo inelastic collisions with atoms or molecules. In a gas, these particles produce positive ions and electrons which, in the absence of an electric field, will recombine. In some gases the electrons may become attached to neutral molecules to form negative ions. When an electric field is applied to the gas, the ions drift along the lines of force to produce an ionization current. Under usual conditions, electrons drift at speeds of the order of 10^6 cm s^{-1}. The drift velocity of ions is many orders of magnitude less. Collisions with gas molecules prevent the ions from attaining high velocities. Some values of W_{air}, the average energy expended in the creation of an ion pair in air, are shown in Table 2. As can be seen, this quantity does not depend markedly on the energy, and is approximately the same for electrons and alpha particles, when measured in the same gas (see also Valentine and Curran, 1958; Klots, 1968; and ICRU, 1979a).

For applications and further details of the ionization process see Section 4.4 and Corson and Wilson (1948); Wilkinson, D. H. (1950); Staub (1953); Price (1964); Taylor, D. (1962); Curran and Wilson (1965); and Smith (1965).

2.1.1 *Current Ionization Chamber*

When an ionization chamber is in a radiation field, the measured ionization current first increases with increasing voltage and then levels off. Figure 5 shows the so-called "saturation" curve. In general, the voltage required to attain saturation current for any chamber will

TABLE 2—*Average energy expended in the production of an ion pair in air*

Radiation	W_{air}	Reference
	eV/ion pair	
^{3}H β	33.9	Jesse and Sadauskis, 1955
^{35}S β	33.7	Bay *et al.*, 1957
^{35}S β	33.6	Gross *et al.*, 1957a
^{63}Ni β	34.0	Jesse and Sadauskis, 1955
2-MeV x rays	33.9	Weiss and Bernstein, 1955
^{210}Po α	35.5	Jesse and Sadauskis, 1953
^{210}Po α	35.0	Bortner and Hurst, 1954

Fig. 5. Saturation curve (ionization current *versus* collecting voltage) for a typical ionization chamber.

depend on the rate at which ionization is being produced and on the rate of recombination. At saturation, the ionization current, I, is related to the number of ion pairs produced per unit time, n, by the equation $I = ne$, where e is the electronic charge. The chamber, therefore, measures the integrated effect of a large number of ionizing events. The time constant of the current-detecting device is generally made long to suppress statistical fluctuations, and whether the electrons are collected as free electrons (as, for example, in argon), or attached to slow-moving molecules as ions (as, for example, in oxygen), is of no importance. However, the probability of recombination is lower in a gas in which negative ions are not formed, and saturation can be reached with a lower field strength.

When the ionization current is sufficiently large, it can be measured with a microammeter. Generally, however, it is necessary to use more sensitive methods, for instance, by allowing the charge to collect on a capacitor and measuring the rate of change of voltage on the capacitor with an electrometer. The number of ion pairs formed per unit path length is a function of the density of the gas in an ionization chamber. Therefore, in unsealed chambers, corrections of the ionization current to standard temperature and pressure may have to be made when the range of the primary radiation is greater than the dimensions of the chamber. With the pressure ionization chambers now often in use, and which mostly contain argon at a pressure of some 20 atmospheres, this problem does not arise.

2.1.2 Pulse Ionization Chamber

An incident particle depositing all its energy in the gas of an ionization chamber will give rise to a number of ion pairs which is proportional to the energy deposited. An externally applied voltage will cause a pulse of current to flow between the anode and cathode of the chamber. If the voltage drop due to the resulting ionization current is measured across a high-megohm resistance, R, in parallel with the applied voltage and the capacitance of the chamber and measuring system is C, a voltage-time dependence is found, such as is shown in the lower curve of Figure 6, which is representative of a time constant of, typically, $RC = 1$ μs. When the time constant is increased, the voltage rises as shown in the upper curve, for $RC = \infty$, of Figure 6. Here, a fast rise (to point a) at the beginning of the pulse, due to the rapidly moving electrons, is followed by a slower increase (to point b) due to the movement of the positive ions. When it is feasible to use a gas that allows fast electron collection, a differentiating time constant of only a few microseconds is usually chosen to prevent pulse pile-up at high counting rates. If the differentiating time constant is longer than the time for electron collection but much shorter than for collection of the positive ions, the pulse shape is determined by the movement of the electrons. The remaining positive ions induce a

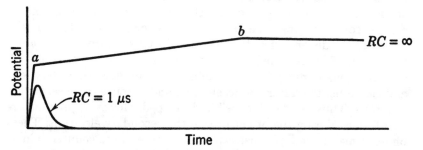

Fig. 6. Voltage pulse on collecting electrode of pulse ionization chamber. (From Halliday, 1955.)

charge in the electron-collecting electrode and reduce the pulse amplitude, the magnitude of this charge depending on the location of the initial ionization. Therefore, in a parallel-plate chamber, the pulse amplitude will depend on where the ionizing particle traversed the chamber. This undesirable effect can be overcome by placing a grid in the chamber (e.g., Frisch grid chamber) to screen the collecting electrode from the positive ions.

The use of a pulse ionization chamber is limited at low particle energies by the inherent noise of the associated amplifier. This noise is equivalent to the arrival on the ionization-chamber anode of about 200 electrons, which might represent an energy of about 7 keV dissipated by a particle traversing the chamber. For further details, see, for example, Price (1964) and Mann et al. (1980).

2.2 Proportional Counters

When the electric field strength at the center electrode of a pulse ionization chamber is increased above a certain level, the size of the output pulse from the chamber starts to increase but is still proportional to the initial ionization. A device that is operated in such a fashion is called a proportional counter. The great advantage of the proportional counter is that it allows one to detect a very low initial ionization, even down to a single ion pair.

For a counter with cylindrical geometry, the field strength E_r, at radius r from the axis is given by

$$E_r = \frac{V_0}{r \ln(b/a)} \qquad (2.1)$$

where a is the radius of the inner electrode, b is the radius of the outer cylinder (for $a < r < b$), and V_0 is the voltage applied to the counter. In a proportional counter the anode is usually a fine wire of diameter about 0.001 to 0.010 inch, so that the radius r can be small, and E_r can be correspondingly large. The field strength close to the wire increases rapidly and when the electrons drifting toward the wire enter the region of high-intensity electric field close to the wire, they will acquire enough energy between collisions to produce secondary ions and electrons; the latter are accelerated and produce more secondary electrons. Thus, an "avalanche" is developed. However, photons are also formed during the production of the secondary electrons. These photons may release photoelectrons anywhere in the counter volume or walls, depending on the nature of the filling gas.

The following rather simple model will serve to illustrate the operation of the proportional counter. Consider such a counter containing a pure gas filling of, say, argon. Suppose that $\bar{m}$ is the average electron multiplication in the avalanche so that each of n electrons produced in the primary ionizing event gives rise, in its acceleration toward the anode wire, to an average total of $\bar{m}$ electrons in the avalanche. As a result of the inelastic collisions between the accelerated electrons and gas atoms, photons will be emitted and these, in turn, can interact with the walls (in the case of a rare gas) to produce further photoelectrons. If P_e is the probability that such a photoelectron be formed for every electron in the avalanche, then the n electrons produced by the primary ionizing event will produce in the initial avalanche, or Townsend discharge, in the high-intensity electric field near the wire, an average of $n\bar{m}$ electrons by collision and $n\bar{m}P_e$ photoelectrons elsewhere in the counter. However, the $n\bar{m}P_e$ photoelectrons can now also be accelerated to the anode producing $n\bar{m}^2P_e$ more electrons by collision and $n\bar{m}^2P_e^2$ photoelectrons, which in turn form more generations of electrons by collision.

Thus, the final number of electrons collected at the anode wire as a result of each primary event is

$$N = n\bar{m} + n\bar{m}^2P_e + n\bar{m}^3P_e^2 + \cdots. \tag{2.2}$$

But, in a proportional counter, $\bar{m}P_e$ is much less than one, so that multiplication is essentially by electron avalanches, and the series converges giving

$$N = n\,\frac{\bar{m}}{1 - \bar{m}P_e}, \tag{2.3}$$

and the total number of electrons in the avalanche is proportional to the number of primary ion pairs, n. This is the essential property of a proportional counter, but it must be emphasized that $\bar{m}$ is an average number, and the actual number of electrons formed from one generation to the next will fluctuate statistically. $M = N/n$ is called the multiplication factor.

Multiplication factors have been worked out theoretically (Rose and Korff, 1941). Figure 7 shows experimental values for M as a function of applied voltage for argon. A comparison of theory and experiment has been made recently by Hendricks (1972).

2.2.1 Shape of the Output Pulse

In contrast to the pulse ionization chamber where the main part of the pulse is determined by the rapid movement of electrons, the pulse

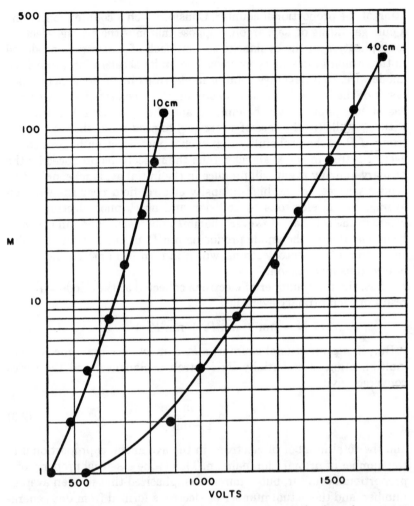

Fig. 7. Multiplication factor, M, as a function of applied voltage for argon at pressures of 10 and 40 cmHg (1.33×10^3 Pa = 1 cmHg), radius of center wire = 0.013 cm, radius of outer electrode = 1.105 cm (Staub, 1953).

shape measured on the wire of a proportional counter is essentially determined by the movement of the positive ions. Since, in a proportional counter, most of the ionization develops very close to the wire, the electrons travel only short distances and their effect on the pulse rise is rather small. Figure 8 shows a typical pulse shape. A delay, of time t_1, elapses before the primary electrons reach the multiplication region. This time t_1 is determined by the location of the initial ionization in the counter and (depending on its construction) may be

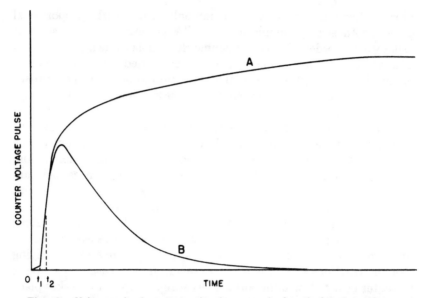

Fig. 8. Voltage pulse from proportional counter. As described in the text, curve A represents the undifferentiated and curve B the differentiated responses. (From Rossi, 1952.)

of the order of 1 μs. This delay time, which is variable, should not be overlooked when coincidence experiments are carried out.

Initially the pulse rises very fast (curve A), first as a result of the movement of the electrons (t_1 to t_2), then of the positive ions traveling in the strong field near the wire. As the positive ions move into the region of weaker field, the rise is slower and ceases when the ions reach the outer cylinder. This, for an average counter, takes about 200 μs.

Curve A shows the shape of the voltage pulse for no loss of charge from the anode wire (i.e., for an infinitely long time constant). In practice, in order to provide for time resolution, suitable RC differentiation is used in the output to give the type of pulse shown by curve B. This differentiation is accomplished by placing a capacitance, C, in series with the signal, followed by a high resistance, R, to ground. The charge from the detector will be exponentially dissipated by this circuit; the time required for the charge to be reduced to $1/e$ of its value being RC, known as the time constant of the circuit. When RC is small compared with the time between ionizing events, the charges collected during individual events (or the corresponding voltages) may be measured by means of appropriate pulse circuitry. Differentiation times as short as 0.1 μs have been used. Low-noise charge-sensitive preamplifiers are now more commonly used with proportional counters. In such preamplifiers the differentiating time constant is usually of the order of 1 ms, but double RC or delay-line differentiation of less than 1 μs in the main amplifier, combined with measurement of the time at which the bipolar pulse changes polarity (i.e., crosses zero voltage), can provide excellent time resolution over a wide range of pulse amplitudes.

As all the pulses will have the same general shape (except when the ionizing event takes place very close to the wire), the height of the differentiated pulse will be proportional to the initial ionization. However, ionization and the formation of the electron cascade are statistical processes and a distribution in the height of the output pulses is observed, which broadens when the initial ionization decreases. (For further details, see Rossi, 1952; Wilkinson, D. H., 1950; and Bambynek, 1973.) A further spread can also be caused by a non-uniform center wire and by end-effects, both of which cause varying field distributions along the wire. However, in most cases the shape of the outer electrode is unimportant, because multiplication takes place very close to the anode wire, where the electric field gradient is highest (Eq. 2.1). Asymmetric shapes of the outer electrode may result, however, in significant variation in electron-collection times from different regions of the sensitive volume of the counter.

2.2.2 *Choice of Counting Gas*

A proportional counter should be filled with a gas in which the negative charge is carried by electrons. However, some pure, electronegative gases such as hydrogen, nitrogen, and the noble gases exhibit a rapid change of multiplication factor, M, with applied voltage due to the generation of photoelectrons from the cathode walls. These photoelectrons can arise from photons, generated in the Townsend avalanche near the anode wire. Such photoelectrons give rise to *spurious pulses* that occur within 10 or 15 microseconds for most counters of typical dimensions, and operating pressures and voltages (see Campion, 1973). Longer delayed pulses, called *after-pulses*, arise from the neutralization of the positive ions close to the cathode wall with the emission of a photoelectron from the cathode wall (see BIPM, 1976). The addition of small proportions of gases with complex molecules that absorb photons without ionization increases the stability of the counter. Methane-argon mixtures are in common use and multiplication factors up to 10^5 are possible. Polyatomic gases, such as methane, can also be used very satisfactorily without any admixture.

2.2.3 *Characteristics of Proportional Counters*

The proportional counter can be used to determine the energy of particles and photons if they are totally absorbed in the counting gas. For low-energy x and gamma radiation, almost complete absorption can be achieved with a suitable filling gas (Figure 9), or by operating at elevated pressure (Allen, 1957; Baerg, 1973b; Legrand *et al.*, 1973; Bambynek, 1973).

Proportional counters also have wide application to particle counting without energy determination. In common practice, the source itself is located inside the counter to avoid window absorption, and the counter is operated at atmospheric pressure, generally with methane or methane-argon mixtures flowing continuously (flow counter). Counter pulses due to a beta-particle spectrum vary in size, by a factor of 1,000 or more when the counter is operated at elevated pressure. Therefore, the associated amplifiers must have good overload characteristics, which may, for example, be achieved with a low-noise charge-sensitive preamplifier and a doubly differentiating main amplifier (Baerg *et al.*, 1967).

A characteristic that is used to measure the performance of a gaseous multiplication counter in the proportional region is the so-called plateau of the counter. As the applied voltage is increased, the increasing gas gain permits the triggering of the circuit discriminator by lower-energy primary ionizing events in the counter. When the applied

Fig. 9. Percentage of quanta absorbed in 50 mg cm^{-2} of argon, krypton, and xenon (i.e., 5-cm path at 5 atmospheres in argon, at 2.5 atmospheres in krypton, or at 1.5 atmospheres in xenon). The break in the xenon curve is at the K edge, i.e., when a photon has sufficient energy to remove an electron from the K shell. (From Curran, 1950.)

voltage is such that nearly all primary events are counted, a plot of counting rate against applied voltage becomes nearly horizontal, and the flatness of the curve in this region, the plateau, is a measure of the ability of the counter to detect essentially all primary ionizing events. Any further increase in voltage will cause only a small increase in counting rate. An increase in the effective sensitive volume of the counter, or the onset of spurious pulses, can also cause the plateau to have a positive slope. The use of low gas gains, as permitted by low-noise amplifiers even for particle energies of a few keV, may be important in avoiding possible spurious pulses (Campion, 1973; Williams and Smith, 1973; BIPM, 1976), which are, however, usually negligible ($<10^{-4}$) when the counter is operated near the middle of the plateau. The applications of proportional counting to specific metrological problems are discussed in Section 3.5.6.

2.3 Geiger-Müller Counters

If the voltage on a proportional counter is increased beyond the high-voltage end of the plateau, $\overline{m}P_e$ (Eq. 2.2) also increases and approaches unity, and an electrical discharge is propagated along the anode wire. The product $\overline{m}P_e$ represents the number of photoelectrons formed by the emission and subsequent absorption of photons in the walls of the counter when the gas is a pure electronegative gas. The emission of photons will be isotropic and their number, which was less than one per secondary electron formed by collision in the proportional counter, is now very considerable. In the region of intense electric field near the wire, the Townsend avalanches caused by the electrons produced in the primary ionizing event will now produce intense photon emission with consequent photoelectric absorption throughout the counter, including the region of intense electric field around the anode wire. Under these conditions a discharge takes place, which spreads all along the wire forming a positive-ion "sheath," and gives rise to an output pulse, the size of which is independent of the amount of the initial ionization and which may amount to several volts. Less additional amplification is therefore required than for proportional counters, and the discharge can be initiated by a primary event producing as little as one ion pair. The discharge will spread along the wire at a velocity as great as 10^7 cm s^{-1}.

This mode of gas-ionization-detector operation is said to occur in the Geiger region and the detector itself is known as a Geiger-Müller counter.

Unless the discharge is "quenched" it may be self-sustaining or multiple pulses may occur, due to the neutralization of the positive ions at the cathode and consequent release of photoelectrons from the cathode after a delay of several hundred microseconds. Such quenching may be achieved externally, by electronic means, or internally, by adding a suitable polyatomic gas. A gas filling that has been frequently used is argon at a pressure of about 90 mmHg (12 kPa) with ethyl alcohol vapor, or other organic gas, at a pressure of about 10 mmHg (1.3 kPa). Alcohol has a lower ionization potential than argon, so that the ions moving toward the cathode will, after a few collisions, consist only of alcohol ions. In contrast to argon ions, alcohol ions dissociate and do not produce photoelectrons when neutralized at the cathode. Therefore, multiple pulsing is avoided. However, the alcohol ions dissociate upon being neutralized and when the supply of alcohol is exhausted, after some 10^8 counts, poor plateau characteristics result. Halogen vapors also have the same quenching effect (Liebson and Friedman, 1948; Friedman, 1949). As the halogen molecular ions do not remain dissociated, these counters have very long lives. The operating voltage is also considerably lowered. Argon, with the addition of small quantities of xenon, nitrogen, and oxygen, has also been found to be a satisfactory filling for Geiger-Müller counters (Shore, 1949; Collinson et al., 1960).

After an ionizing event, the sheath of positive ions along the counter wire effectively lowers the wire potential and makes the counter inoperative until the ions have drifted to the cathode. The inoperative time is known as the dead time, and in the Geiger-Müller counter dead times are long, ranging from 10 to 500 μs. Thus, this type of counter does not allow of such high counting rates as does the proportional counter. It is also inferior because the slopes of the Geiger-Müller-counter plateau are steeper than those of proportional counters, values of less than 1 percent variation in counting rate per hundred volts being uncommon unless an artificial dead time is used to reduce spurious counts. When it is necessary to establish an actual counting rate, it is difficult to know at which point on the plateau to count. This is not so serious with a proportional counter where a plateau slope as low as 0.1 percent, counting-rate change per hundred volts, can be achieved. Therefore, for standardization work, or for any other precise counting, preference should be given to the proportional counter. However, the efficiency of either a proportional or Geiger-Müller counter may be determined, in the case of beta- and photon-emitting nuclides, by coincidence counting (Section 3.2). In the case of length-compensated internal gas counters, used for the assay of radioactive nuclides in the gaseous form, a plot of the difference in the count rates, at the same voltage, for two cylindrical counters, having the same diameters but different lengths, will give "difference plateaux" that are very flat, both in the Geiger-Müller and proportional regions (Section 3.6).

2.4 Scintillation Detectors

2.4.1 *Introduction*

It has long been known that many substances emit visible light when exposed to nuclear radiation, and this was the basis of Sir William Crookes' spinthariscope which utilized a ZnS screen to permit the observation of single alpha particles emitted in the decay of natural radioactivity. In 1944, Curran and Baker first substituted an electron-multiplier phototube for the human eye, creating the first scintillation detector as we know it today. Kallmann (1947) used naphthalene, an organic crystal, as a scintillation medium that is transparent to the fluorescent radiation emitted by the interaction of the charged particles within the scintillator. It was subsequently observed that the quantity of light emitted was proportional to the energy of the absorbed particle, and this established the scintillation process as a valid means for energy spectrometry of charged particles emitted in radioactive decay. The discovery of the scintillation properties of thallium-activated sodium-iodide crystals, NaI(Tl), by Hofstadter (1948), and the possibility of their use as gamma-ray spectrometers, led to the development of pulse-amplitude spectrometry as a method for the measurement of charged-particle and photon energies and intensities associated with radioactive decay. In order to achieve the level of sophistication required to utilize the potential of the scintillation detector as an energy-measurement device, it was necessary to develop an entire family of electronic circuitry that was unavailable at the time. The linear pulse amplifier and multichannel pulse-height analyzer were developed to utilize the energy-measurement potential of the scintillation detector. The general development of the scintillation spectrometer is outlined in the text *Alpha-, Beta-, and Gamma-Ray Spectroscopy* (Siegbahn, 1965). The characteristic energy-sensitive response of scintillators has been applied to develop techniques for gamma-ray spectrometry that are presently employed extensively for the quantitative assay of radioactivity and the identification of radionuclides.

NaI(Tl) detectors (frequently 3″ × 3″ or 2″ × 2″ crystals with or without sample wells) are used in almost every laboratory where radioactivity is measured (see Section 4.2).

NaI(Tl) crystals can be hot-pressed or extruded into large or intricate shapes which would be economically prohibitive if made from single crystals. The result is a strong polycrystalline material with no change in scintillation efficiency or transparency (Nestor, 1983). (Scintillator rods 75-cm long formed by this process are used in hundreds of airport-security luggage-scanning x-ray systems giving a clear image at a film-safe dose of 0.005 mR per package.)

Advances in medical computed tomography (CT) in the 1970's spurred increased interest and consequent improvements in high-atomic-number scintillators; especially bismuth germanate (BGO), $Bi_4Ge_3O_{12}$, (Nestor and Huang, 1975) and cadmium tungstate, $CdWO_4$ (Nestor, 1983). Compared with NaI these have about twice the density with 4.5 (BGO) and 3.5 ($CdWO_4$) times higher linear absorption coefficients at 150 keV. Their low afterglow intensities compared with NaI(Tl) make them particularly suitable for applications such as CT that involve high and rapidly changing counting rates.

BGO has largely replaced NaI(Tl) in positron CT imaging where its high density and absorption coefficient allow the use of smaller but more efficient detectors to give better spatial resolution with lower patient doses (Thompson et al., 1979). In addition, its fluorescence decay time [300 ns vs. 230 ns for NaI(Tl)] is compatible with short coincidence resolving times (Moszynski et al., 1981). For x-ray CT applications $CdWO_4$ has an advantage over BGO in that its higher light output and longer emission wavelengths are better matched to silicon-avalanche photodiodes used in place of conventional photo-tubes (Farukhi, 1983; Reiff et al., 1983). The relatively long (5 μs) fluorescence decay time of $CdWO_4$ is not a drawback in systems recording currents rather than pulses because CT scan rates are typically in the millisecond range. On the other hand, the very low afterglow of $CdWO_4$ [< 0.005% residual light output after 3 ms, 60 to 1000 times less than NaI(Tl)] is a significant advantage at these scan rates.

BGO is also valuable as a detector of high-energy gamma rays in nuclear physics research (Häusser et al., 1983) and has found applications in gamma-ray borehole logging (Stromswold, 1981). At the time of writing (1984), the higher cost, lower light output and poorer energy resolution of BGO compared with NaI(Tl) continue to make BGO a less attractive choice for routine x-ray and gamma-ray measurements. For more details see Heath, 1983.

2.4.2 The Scintillation Process

In order to understand the modern scintillation detector and the concepts of pulse-height analysis, let us consider the basic processes involved. The operation of a scintillation detector may be described by a sequence of four essential processes: (1) the interaction of charged particles or photons with the detector, resulting in the production of electron hole-pairs; (2) the conversion of the energy of the charge carriers into a proportional àmount of light by the scintillator; (3) the conversion of the light emitted by the scintillator into photoelectrons at the photocathode of a phototube; and (4) the multiplication (amplification) of the initial number of photoelectrons into a measurable electrical current pulse. The result of these processes, operating in sequence, is to produce at the output of the electron-multiplier phototube an electrical pulse whose amplitude is proportional to the energy deposited in the scintillator by the incoming charged particle or photon. The fact that all of these processes occur in a very short interval of time (of the order of a microsecond) requires that the amplitude of individual electrical pulses be measured if one is to relate the output of the detector to the energy of the radiation being detected. A typical NaI(Tl) scintillation detector is shown in Figure 10 to illustrate the sequence of events following detection of a photon. In the following section, and Section 4.2, the principles of photon spectrometry will be presented together with the application of this technique to the measurement of radioactivity.

A charged particle (alpha or beta) incident upon the detector, or an electron excited in the detector following an interaction with an incoming photon, will dissipate all of its energy within the scintillator, provided that the dimensions of the scintillator are greater than the range of the particle. The scintillation process results in a portion of this energy being emitted as photons in the visible or near ultraviolet region of the electromagnetic spectrum. The number of photons, p, of average energy $h\nu$ produced by the absorption of energy E within the scintillator, may be written:

$$p = \frac{E}{h\nu} \epsilon_\nu. \tag{2.4}$$

The quantity ϵ_ν is referred to as the intrinsic efficiency of the scintillator. It is nearly independent of the energy of the ionizing particle for most inorganic scintillators.

Many materials have been observed to emit light when excited by ionizing radiation, including both inorganic and organic chemical compounds. In organic materials such as anthracene and stilbene, this luminescence is a molecular property; thus organic compounds will fluoresce in any physical form—crystalline solid, liquid, or vapor phase. Inorganic phosphors are mainly limited to alkali halides in crystalline form. In general, the scintillation process in crystalline inorganic materials requires the presence of impurity atoms within

Fig. 10. Illustration of the various processes that contribute to the output from an electron-multiplier phototube by the interaction of gamma rays from a point source with a NaI(TI) scintillation detector. (From Heath, 1964.)

the crystal lattice. Thallium is employed as an activator in NaI and CsI, being introduced as an impurity when the crystal is grown from the molten state. Bismuth germanate (BGO) and cadmium tungstate ($CdWO_4$) are intrinsic scintillators and require no activators.

Following the excitation of a scintillator by an individual event, the time dependence of photon emission appears to follow an exponential law of the type

$$p(t) = constant\ e^{-t/\tau}, \qquad (2.5)$$

where $p(t)$ is the total number of photons emitted at time t. The decay time τ is characteristic of the scintillator and determines the rise time of pulses produced at the output of the phototube. Decay times, to 1/e, are of the order of 1 μs or less for inorganic scintillators. Decay times for most organic scintillators, plastic or liquid, vary from 0.2 to 10 ns. Anthracene crystal has a decay time of about 30 ns, and naphthalene 75 ns.

In addition to the time response, a major consideration in the performance of a phosphor as a scintillation detector is the absorption of the light emitted by the scintillation process in traversing the scintillator material. NaI(Tl), for example, exhibits almost no measurable attenuation of the light traversing the crystal; thus the amplitude of the light pulse emitted from a large crystal detector fabricated from this material is independent of the point of origin of the light within the volume of the scintillator. This has a significant bearing upon the performance of the device as a photon-energy spectrometer. Organic phosphors, in general, exhibit appreciable attenuation. Relevant characteristics of some of the more commonly used scintillators are given in Tables 3 and 4.

2.4.3 Conversion of Fluorescent Radiation to Electrons (Phototube)

To increase the collection efficiency of the light emitted by the scintillator (as shown in Figure 10), it is surrounded by a thin reflector of aluminum oxide. Of the photons produced in the scintillator, the number striking the cathode surface of the phototube will depend upon the geometry and the optical properties of the system, including those of the reflector and any light guides between the scintillator and the cathode of the phototube. For good scintillators and reasonable geometries, the optical efficiency, ϵ_L, can approach unity. Photons falling on the cathode surface of the phototube are converted into electrons by the photoelectric process with an efficiency ϵ_C. Thus, the

TABLE 3—*Inorganic scintillators*

Scintillator (activator in parentheses)	Density (g cm⁻³)	Emission wavelength (nm)	Refractive index	Light yield (anthracene = 1.00)	Decay time[a] (μs)	Remarks
ZnS (Ag)	4.1	450	2.4	$(2.0)^b$	$(>1)^b$	Only very small crystals, transparency poor.
CdS (Ag)	4.8	760	2.5	$(2.0)^b$	$(>1)^b$	Only small crystals, yellow.
NaI (Tl)	3.67	410	1.7	2.0	0.25	Excellent crystals, hygroscopic.
CsI (Tl)	4.51	565	1.79	1.5	1.1	Excellent crystals, not hygroscopic.
KI (Tl)	3.13	410	1.68	0.8	>1	Excellent crystals, not hygroscopic.
NaCl (Ag)	2.17	245–385	1.54	1.15	>1	Excellent crystals.
LiI (Eu)	3.49	470	1.96	0.7	0.94	Hygroscopic.
LiF (AgCl)			1.39	0.05		
CsBr (Tl)	4.44		1.70	2.0	>1	
CaWO₄	6.06	430	1.92	1.0	>1	Small crystals, transparency good.
Bi₄Ge₃O₁₂ (BGO)	7.13	480	2.15	0.2	0.3	Not hygroscopic, very low afterglow.
CdWO₄	7.90	540	2.3	0.8	5	Not hygroscopic, very low afterglow.

[a] Time for light pulse to fall to 1/e of initial value.
[b] The light yield and the decay time of the zinc-sulfide and cadmium-sulfide phosphors depend strongly on former treatment (quenching) and intensity of excitation. To a lesser degree this is also the case for other inorganic phosphors.

TABLE 4—*Organic scintillators*

Scintillator	Density	Emission wavelength	Refractive index	Light yield	Decay time	Remarks
	g cm⁻³	nm			ns	
Anthracene	1.25	445	—	1.00	32	Large crystals, not quite clear.
Quaterphenyl	—	438	—	0.85	8	Pure crystals difficult to synthesize.
Stilbene	0.97	410	—	0.73	6	Good crystals readily obtainable.
Diphenyloxazole	—	—	—	0.78	—	
Diphenylbutadiene	—	461	—	0.67	8	
Diphenylanthracene	0.997	480	—	0.65	—	
Terphenyl (para)	1.23	415	—	0.55	12	Good crystals, readily obtainable.
Diphenylacetylene	—	390	—	0.26–0.92	7	
Phenanthrene	1.03	435	1.66	0.46	10	Clear crystals difficult to obtain.
Naphthalene	1.15	345	1.58	0.15	75	Good crystals easy to obtain.
Chloroanthracene	—	—	—	0.03	—	

number of photoelectrons, P, produced by p photons generated in the scintillator may be expressed by the following relationship

$$P = \epsilon_L \epsilon_C p. \tag{2.6}$$

Typically ϵ_L may be as large as 95 percent while ϵ_C usually lies between 10 and 20 percent. In order to make ϵ_C as large as possible, the phototube is designed so that the spectral response of the photocathode material closely matches the photon spectrum emitted by the scintillator material being used. It is of interest to combine the quantities presented above to obtain the number of photoelectrons to be expected for a given amount of energy expended in the scintillator, i.e.

$$P = \frac{\epsilon_\nu \epsilon_L \epsilon_C}{h\nu} E. \tag{2.7}$$

The quantity $\epsilon_\nu \epsilon_L \epsilon_C / h\nu$ provides a measure of the efficiency of a given scintillation detector. For a high-quality scintillation detector the energy required to release one photoelectron (E/P) from the photocathode is between 300 and 2,000 eV. This may be compared with 30 eV required to produce an ion pair in a typical gas counter.

Recently large-area silicon-avalanche photodiodes have been used in place of phototubes in space and medical instrumentation where the bulk, bias-power requirements, or fragility of phototubes are a drawback (Reiff et al., 1983). Future developments may lead to the replacement of phototubes by semiconductor devices in more routine applications.

2.4.4   Output-Pulse Processing

As outlined above, the electron-multiplier phototube embodies two functions that are essential to the performance of a scintillation detector: (1) the detection and conversion of the light emitted by the phosphor into a pulse of electrons; and (2) the amplification of this minute amount of current to a level that can be suitably handled in conventional electronic-amplifier circuits. The amplification of the initial number of photoelectrons emitted from the photocathode of the tube is accomplished by a process of electron multiplication. The initial photoelectrons, as illustrated in Figure 10, are accelerated toward the first of a series of metal plates which are termed *dynodes*. The dynode surface is coated with a material (typically cesium or a bimetallic compound) that has the property of emitting secondary electrons when struck by an accelerated electron, the number emitted being a function of the coating material and accelerating voltage, and

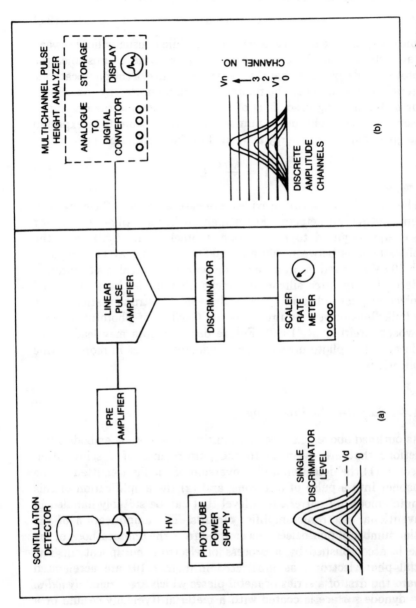

Fig. 11. Block diagram of typical scintillation-counter systems: (a) for integral count-rate measurements for photon energies above those corresponding to a discriminator voltage V_d; (b) for "pulse-height-spectrometer" measurements at different photon energies corresponding to discriminator voltage intervals $V_1, V_2, \cdots V_n$.

typically varying from 3 to 5. These secondary electrons are then accelerated toward a second dynode and the process is repeated. With 10 or more secondary emission stages, the electron-multiplier structure can achieve a current gain of up to 10^7 or 10^8. At this level the electron pulse arriving at the final collector of the phototube is sufficient in amplitude to be handled effectively by conventional electronic circuitry. It is important to note that the electron-multiplier phototube used as a current pulse amplifier is extremely fast and linear, and has minimal influence on the speed of scintillation detectors.

A typical scintillation-detector system is illustrated in Figure 11. In this figure two different output configurations of the detector system are presented to illustrate different applications of scintillation detectors. For simple applications, where the detector is to be employed for integral photon or particle counting, the system comprises the detector, phototube voltage supply, a pulse amplifier (with preamplifier), discriminator to eliminate electronic noise from the output, and a scaler or count-rate meter for indication of counting rate. If differential energy measurement of charged particles or photons is required, then the output of the linear pulse amplifier is fed to a multichannel pulse-height analyzer (PHA) which provides a means of measuring the amplitude of each pulse from the detector, and presents the energy spectrum of the radiation being measured in the form of a differential pulse-amplitude distribution. A typical pulse-amplitude distribution obtained with a NaI(Tl) detector is shown in Figure 12. This spectrum, resulting from the detection of gamma rays emitted in the decay of ^{137}Cs, exhibits a sharp peak that results from the delayed emission from ^{137m}Ba of monoenergetic gamma rays having energies of 0.6616 MeV. This unique energy-sensitive characteristic of NaI(Tl) provides the basis for using pulse-height spectrometry to measure the energy and intensity of radiation emitted in the decay of radionuclides. A description of pulse-amplitude spectrometry and its application in the measurement of radioactivity is presented in Section 4.2.

The development of the scintillation detector in its many forms has generated a variety of detection devices that have application in the measurement and characterization of all types of radiation emittted in the decay of radioactive nuclei. Although all scintillating phosphors can be utilized for the detection of radiation, the properties of different fluorescent media may be optimized for a particular application. For example, the detection of high-energy gamma radiation with optimum efficiency requires a detector with high density and preferably with high atomic number. For this purpose, NaI (Tl), BGO, CsI(Tl) and $CdWO_4$ offer the best choices. On the other hand, the measurement of low-energy beta-emitting radionuclides such as ^{3}H and ^{14}C can best

Fig. 12. A typical pulse-height spectrum obtained with a NaI(Tl) spectrometer, illustrating the energy response of inorganic scintillators. The scale of the abscissa is 1 keV per channel. (From Heath, 1964.)

be achieved by incorporating the radioactivity into a liquid scintillator. Tables 3 and 4 provide a partial listing of solid scintillators normally employed for a selection of radiation-measurements applications. The unique capabilities of the liquid-scintillation method have found many uses in biomedical applications, and have led to the development of a sophisticated technology. This specialized measurement technique is described below and in Section 4.3. (Also see Heath, 1983.)

2.4.5 *Liquid-Scintillation Counting*

2.4.5.1 *Counting.* The wide application of liquid-scintillation counting derives from its particular usefulness for measurement of low-energy beta-emitting radionuclides such as ^{3}H and ^{14}C. Several monographs and reviews treat various aspects of liquid scintillation (Birks, 1964; Horrocks, 1966b; Horrocks, 1974; Bransome, 1970; Horrocks and Peng, 1971; Dyer, 1971; Crook et al., 1972; Steyn, 1973a; Crook and Johnson, 1974, 1977, 1978; Noujaim et al., 1976; Peng, 1977; Simonnet and Oria, 1977; and BIPM, 1980). Liquid scintillators are normally composed of one or more fluorescent solutes in an organic solvent. In the preparation of the sample for liquid-scintillation counting, the radioactive material is introduced into, and thoroughly mixed with, the liquid scintillator. It is generally accepted that radiation energy is expended in the ionization and excitation of the solvent. This energy is subsequently transferred to the solute (or fluor) and is re-emitted as photons in the violet and ultraviolet. These photons, having an average wavelength characteristic of the solute, can then be detected at the photocathode of one or more phototubes, although in the general application of the method a "wavelength shifter" may also be used to produce an output pulse of photons having wavelengths compatible with the optimum sensitivity of the photocathode. Following amplification in the phototube, the signal may be routed to additional amplifiers and finally recorded with a suitable counter such as a single-channel or multichannel analyzer.

At high beta-particle energies, the magnitude of each pulse produced in a liquid-scintillation-counter system is proportional to the energy deposited in the scintillator, and both the amplitude and the shape of the pulse will be dependent on the nature and linear energy transfer (LET) by the ionizing radiation. For example, the light output due to an alpha particle will be only about one-tenth that produced by a beta particle of the same energy. This difference arises because of the higher specific ionization along the alpha-particle path. Where high-energy beta particles produce excited and ionized solvent molecules which are relatively isolated from each other, alpha and low-energy beta particles produce them in close proximity in regions of high ionization density in which energy is dissipated in nonradiative transitions. This process is referred to by Birks (1964) as ionization quenching.

In order to evaluate the effectiveness of the various energy-transfer steps in the liquid-scintillation process Birks (1964, 1970) has developed an expression relating the pulse amplitude at the anode to the energy deposited in the scintillator. In an abbreviated form of his earlier equation, Birks (1975) defines a "scintillation figure of merit,"

$$F = sfqm, \tag{2.8}$$

which is useful for comparing the performance of different scintillators using the same phototube and optical arrangement. The terms on the right hand side of Equation 2.8 are best defined by considering the intermediate steps in the liquid-scintillation process.

(i) In the first step, the energy of the charged particle is deposited in the solvent, and s, the relative scintillation efficiency, is a measure of the efficiency, relative, usually, to that of toluene, for the transfer of energy from the incident radiation to those excited levels of the solvent molecules from which energy can be transferred to the fluor in solution [see Table 5 and Birks (1975)].

(ii) In the second step, energy is transferred to the solute and f, the probability of this process occurring, is dependent both on the nature of the solvent and the concentration of the solute.

(iii) The third step is emission of light by the solute. The "quantum efficiency," q, of the solute is the ratio of the number of light photons emitted to the number of solute molecules in excited states.

(iv) The last term in Equation 2.8 is called the "spectral matching factor," m. This parameter, which varies from 0 to 1, indicates the degree of overlap of the emission spectrum of the solute and the absorption spectrum of the photocathode.

Thus the first three terms in F are dependent upon the choice of scintillator while m depends on the scintillator-phototube combination. Values for some of these parameters are given in Table 6. Examples of the use of the "scintillation figure of merit" are given in Birks (1975).

Table 5 shows solvent scintillation efficiencies relative to that of toluene. It is evident that all good solvents have an aromatic structure. Dioxane is included in the table because of its significance as a solvent for water and aqueous solutions. Dioxane alone has a low relative scintillation efficiency, but with the addition of naphthalene the efficiency becomes comparable with that of benzene.

Many scintillation liquids contain an additional component generally referred to as a secondary solute, or secondary scintillator. The secondary scintillator was originally used to match the spectral response of the phototube to that of the light emitted by the scintillator. Therefore, the term *wavelength shifter* was often used. This matching is no longer essential because many scintillators exhibit m values of 95 to 97 percent. However, secondary scintillators are still used to improve the performance of liquid-scintillation systems, particularly in quenched systems.

2.4.5.2 *Quenching.* In liquid scintillation, the production of light is

TABLE 5—*Scintillation efficiencies relative to toluene*, s, *of solutions containing 3 g l^{-1} of PPO*[a]

Solvent	Chemical structure	s
p-Xylene	CH_3—⬡—CH_3	1.12
m-Xylene	CH_3—⬡ (CH₃)	1.09
Xylene (mixture of isomers)		1.07
Phenylcyclohexane	⬡—⬡	1.02
Toluene	⬡—CH_3	1.00
o-Xylene	⬡ CH_3 CH_3	0.98
Ethylbenzene	⬡—CH_2—CH_3	0.96
Benzene	⬡	0.85
Anisole	⬡—O—CH_3	0.83
Mesitylene	CH_3—⬡ CH_3 CH_3	0.82
Dioxane	O⬡O	0.20
Dioxane + naphthalene (~6%)	O⬡O + ⬡⬡	0.80

[a] From Birks (1964). For toluene, s is of the order of 3 percent.

sometimes inhibited, or the final light "signal" is "quenched" in the energy or light-transfer processes in the liquid. The details of the quenching processes, in particular chemical-structural effects on quenching, are not completely understood.

TABLE 6—*Properties of some organic fluors (solutes) commonly used in liquid-scintillation counting: relative fluorescent quantum efficiencies (q), spectral matching factors (m), fluorescent life times (t), average fluorescent emission wavelengths (λ), and solubilities*

Solute	Chemical structure	q*	m^b	t*	λ*	Solubility^c (Toluene)
				ns	nm	g l⁻¹
TP p-Terphenyl		0.93	0.97	1.0	342	8.6 (20 °C)
PPO 2,5-Diphenyloxazole		1.00	0.97	1.4	375	414 (20 °C)
POPOP 1,4-Di-(2-(5-phenyloxazolyl))-benzene		0.93	0.83	1.5	415	2.2 (20 °C)
M₂-POPOP 1,4-Di-(2-(4-methyl-5-phenyloxazolyl))-benzene		0.93	0.78	1.7	427	3.9 (20 °C)
PBD 2-Phenyl-5-(4-biphenylyl)-1,3,4-oxadiazole		0.83	0.96	1.1	375	12 (20 °C)
tert-Butyl-PBD 2-(4'-t-Butylphenyl)-5-(4''-biphenylyl)-1,3,4-oxadiazole		0.85	0.95	1.2	385	119 (20 °C)

bis-MSB
p-bis-(*o*-Methylstyryl)-benzene

0.94	—	1.2	425	5.5 (25 °C)

 Unless otherwise stated, the values shown for q, *t*, and λ are taken from Berlman (1971). All of these values were obtained for dilute solutions of the solutes in cyclohexane. Fluorescent quantum yields are given relative to an assumed value of 1.0 for 9,10-diphenyl anthracene.

 Spectral matching factors, m = $\int_0^\infty I(\lambda)\eta(\lambda)d\lambda / \int_0^\infty I(\lambda)d\lambda$, were taken from Birks (1975). The values of $I(\lambda)$, the emission spectrum, were obtained from dilute solutions of the solutes in toluene and values for $\eta(\lambda)$, the phototube response, for bialkali phototubes.

 Solubilities are given here only for comparative purposes, since most data have been taken from suppliers' literature and wide variations were noted for several compounds. Suggested optimum concentrations for many of these solutes may be found in Birks (1964) and Horrocks (1974).

For practical applications three classes of quenchers may be distinguished, namely, diluting, chemical, and color quenchers. The first class consists of substances that dilute the mixture and thus reduce the probabilities of absorption of ionizing radiation by the solvent, and of the various energy transfer processes that follow the absorption of radiation. Aliphatic hydrocarbons and alcohols fall into this category.

Chemical quenchers interfere in the scintillation process by deexcitation of the solvent. If this interference is small, these quenchers are termed mild quenchers. Examples of mild quenchers are R—Cl, R—NH$_2$, R—CH = CH—R, R—SR, RCOOH (R representing an alkyl group). Compounds that strongly interfere with the process are termed strong quenchers, such as R—Br, R—SH, R—CO—R, RI, and R—NO$_2$. The distinction between these two types is, however, arbitrary. In particular, molecules with complex chemical structure often have unpredictable quenching properties. The most common quenching agent is oxygen which is often removed by bubbling an inert gas, such as nitrogen or argon, through the scintillator solution.

Color quenchers are materials that reduce the light transmission through the sample and thus lessen the probability of transfer of photons to the cathode of the phototube. Obviously, there are compounds that are both color and chemical quenchers.

An excellent review of the use of liquid scintillators in radionuclide metrology has been given by Vatin (1980) in BIPM Monograph No. 3 (1980), in which monograph references to just over 300 publications in the field are listed.

2.4.6 Gas-Scintillation Counting

2.4.6.1 *Gas-Scintillation Counters.* Perhaps the most common effect observed due to the passage of ionizing radiation through gases is the emission of light. Light may be readily observed, for example, from electric discharges or from an intense beam of charged particles passing through a gas. As ionizing radiation passes through a gas, ion pairs and electronic excitations are produced in the molecules and atoms. The electronic excitation energy may then be given up either by nonradiative collision or by photon emission.

A detector utilizing a gas as a scintillator was first constructed by Grun and Schopper (1951). It consisted of a chamber containing argon gas at atmospheric pressure with a small nitrogen impurity. The light output was measured by a phototube attached to a quartz window.

Grun and Schopper also found that the light yield from argon was about 100 times greater than that from the polyatomic molecular gases such as nitrogen, hydrogen, oxygen, and carbon dioxide. Subsequent investigations, typically by Northrup and Nobles (1956) and Nobles (1956), indicated that a noble gas of high purity made the most suitable scintillator, with xenon giving the highest light yield. It was demonstrated that the noble-gas light yield was a maximum in the visible to far-ultraviolet regions of the spectrum and that the light output was nearly proportional to the energy deposited over a wide range of stopping powers and incident radiation energies. The scintillation decay times were found to be less than 10^{-8} second and adjustable by pressure. The addition of even trace amounts of polyatomic-molecular gases reduced the light output and its linearity with deposited energy and lengthened the decay times. A comprehensive and detailed review of the scintillation process in gases is given by Birks (1964).

Gas-scintillation detectors have been successfully applied to alpha-particle and fission-fragment detection (Sayres and Wu, 1957). Gas-scintillation counters utilizing neutron-induced reactions have been successfully used for neutron cross-section measurements (Sayres and Wu, 1957; Engelke, 1960).

2.4.6.2 *Gas-Proportional-Scintillation Counters.* An intriguing development of the gas-scintillation counter employs a chamber resembling that of a gas proportional counter, replete with electrodes for the application of an electric field (Koch, 1958). When the electrons resulting from the passage of ionizing radiation are accelerated in the electric field, which is maintained below that required for charge multiplication, the resultant atomic excitations increase the light output by factors of 100 to 1000 (Policarpo et al., 1968), and thus reduce the detection limit to a few hundred electron volts. Such a counter (Policarpo et al., 1972), filled with xenon, has an energy resolution better, by almost a factor of two, than that of a conventional gas proportional counter, in which the resolution additionally involves the variance due to charge multiplication.

Studies of different gases and gas mixtures (always involving at least one noble gas) have led to a better understanding of the light production (Thiess and Miley, 1974) and of the significance of long rise times, different pressures, and wavelength shifting for better matching of the light output to the response of the phototube. The role of the electric-field distribution and counter layout was investigated (Palmer, 1975; Conde et al., 1975) in attempts to increase the detector size and achieve better resolution.

The continued development of the gas-proportional-scintillation detector may lead to its general use where large window areas and good resolution are required at low energies.

2.5 Semiconductor Detectors

2.5.1 *Introduction*

The development of radiation detectors fabricated from single crystals of semiconducting material has, in recent years, revolutionized the detection of x rays and gamma rays. These devices, which are now available in reasonably large volumes, have the advantage of the high detection efficiency for gamma radiation afforded by NaI(Tl) scintillation detectors, but exhibit superior energy resolution. Pulse-amplitude analyzing systems employing detectors fabricated from single crystals of silicon and germanium are extensively utilized for elemental and radionuclidic analysis. The application to x- and gamma-ray spectrometry and the techniques of pulse-height analysis are described in Section 4.2.

These devices have the properties of solid-state ionization chambers. If a voltage is impressed across the detector, electrons that have been raised to the conduction bands of the solid material by excitation, and the "holes" left vacant in the valence band, are free to move and may be collected at the appropriate electrode of the device.

The development of the semiconductor radiation detectors has paralleled the development of scintillation detectors for the detection and energy measurement of charged particles and photons. McKay (1951) reported the detection of alpha particles with a semiconductor-junction diode. This was followed by the development of p-n *diffused-junction* counters for x-ray and charged-particle energy measurement. In 1958, Walter and Dabbs (1958) obtained marked improvement in energy resolution by operating germanium surface-barrier junctions at liquid-nitrogen temperatures. Even then, however, it was only possible to collect charge carriers from very thin layers of material, and consequently the counters had very limited sensitivity to high-energy gamma radiation. Pell (1960) demonstrated that lithium ions could be drifted through large volumes of very pure silicon or germanium to compensate for the presence of residual p-type impurities and thus increase the volume of the depleted region. With this technique, lithium ions may be used to create a p-i-n structure such as is shown in Figure 13.

Typical bias voltages of 500 to 4000 V are applied across the depleted region to assure collection of charge and optimum rise-time response.

Fig. 13. Illustration of the operation of a Ge(Li) detector. The term, n⁺, signifies an excess of electron donors. (From Heath, 1969.)

At saturated drift velocity, the time required for a charge carrier to travel a distance of 1 cm is approximately 0.1 μs. In large-volume devices the rise time will also depend on the location of the primary event in the detector, resulting in a variation in pulse shape, limiting the time resolution of the device.

When used as a detector of gamma rays, the interaction of a gamma ray in the detector produces a primary excited electron that, in turn, excites a cascade of secondary electrons in the process of dissipating its energy within the detector volume. In this manner, the energy of the primary electron is expended in the production of electron-hole pairs which are then collected. The number of electrons collected will be proportional to the energy of the primary electron, and hence related to the energy of the detected photon.

Semiconductor crystals that are free from impurities or structural defects (which can act as traps for secondary electrons) can be used directly for the fabrication of detectors. During the early period of development of these detectors, sufficiently pure material was not available and it was necessary to compensate for the impurities by drifting lithium ions through the detector material at elevated temperature. As the mobility of lithium ions within the crystal lattice is not negligible at room temperature, such detectors must be maintained at cryogenic temperatures throughout their lifetime. A description of early work on the application of lithium-ion drifted germanium (Ge(Li)) detectors for gamma-ray-energy spectrometry is given in Webb and Williams (1963), Tavendale (1964), and Goulding (1966). More recent reviews of semiconductor fabrication techniques have

Fig. 14. Schematics of cylindrical annular and semi-annular types of Ge(Li) and HPGe detectors.

been given by Goulding (1978), and Trammell (1978). Due to difficulties in obtaining semiconductor material with sufficient purity, it has not been possible to achieve complete collection of charge at depths exceeding 1.5 to 2 cm. To overcome this limitation, Tavendale and Ewan (1963) proposed a coaxial geometry. In these devices the lithium is drifted from the outer surface of the crystal, leaving a p-type core. This type of structure provides a large sensitive volume and has been widely used in the fabrication of Ge(Li) detectors for gamma-ray spectrometry. Active volumes in excess of 150 cm^3 can be obtained. Geometries employed in the fabrication of coaxial Ge detectors are illustrated in Figure 14.

Single crystals of germanium are now produced with sufficient purity to make high-quality detectors without lithium drifting to compensate for residual acceptors (Pehl, 1977). Planar detectors with areas of 20 cm^2 and volumes of 30 cm^3 and coaxial detectors with volumes greater than 150 cm^3 are now available. Such detectors, which are sometimes termed "intrinsic" but more properly "high-purity" (HPGe),[1] have many advantages. As lithium-ion mobility is not a problem they need

[1] The meaning of the term "intrinsic" as applied to semiconductors has unfortunately changed with time. Originally (as defined by Webster's Dictionary in this context) "intrinsic" meant that the concentration of charge carriers was characteristic of the material itself in the absence of impurity. Thus, in *absolutely* pure germanium every electron elevated to the conduction band leaves a hole in the valence band and the number of electrons, N_e, equals the number of holes, N_h, created in the valence band. Presently "intrinsic" often means that the absolute number of acceptor sites, N_A, minus the number of donor sites N_D, *i.e.*, $|N_A - N_D|$, is less than approximately 10^9 cm^{-3}, and that $N_A + N_D$ is of the order of 10^{10} cm^{-3} or less. In an *absolutely* pure crystal $N_e - N_h = 0$. It is therefore, preferable, in general, not to call "high-purity" material "intrinsic," but simply to refer to "drifted," "compensated" or "high-purity" detectors. (See Knoll, 1979, or Mann *et al.*, 1980.)

not be kept cold at all times. They can be assembled at room temperature, warmed up deliberately or accidentally without damage, shipped without liquid nitrogen and, if necessary, annealed by the user to repair radiation damage. In addition, manufacturing time is a few days in contrast to the weeks or months required to drift lithium through a large detector. The consequent savings eventually overtook the higher cost of purification and it has been reported that the last Ge(Li) detector manufactured in the United States was delivered early in 1983 (Hubbard, 1983).

It should be stressed that HPGe as well as Ge(Li) detectors must be operated near liquid-nitrogen temperatures to eliminate thermally generated electrons from the conduction band. With the cryostat at 77 K, the balance between the thermal conductivity of the detector mount and thermal radiation from the vacuum enclosure holds the detector between 80 and 85 K. In systems employing a cooled-FET preamplifier, the input-stage FET (field-effect transistor) is mounted on the cryostat to keep its temperature in the range 100 to 140 K.

The operating characteristics of Ge(Li) and HPGe detectors are similar and "Ge detector" can mean either type. Discussions in this report referring to Ge(Li) detectors nearly always apply equally to modern HPGe detectors.

Detector-grade HPGe is not quite "intrinsic"; the very low residual concentration of acceptors or donors ($<10^{-10}$ cm^{-3} or $< 1/10^{12}$ Ge atoms) leaves it slightly p-type or n-type. The use of n-type HPGe (with a thin p-type outer contact) instead of the more usual p-type HPGe makes possible detectors sensitive to low-energy x-rays down to 3 keV (Raudorf et al., 1979; ORTEC, 1983). As is indicated in Figure 14, the electrode configuration of the n-type HPGe is the reverse of that for the closed-end p-type HPGe detector; a core is removed and the thick ($\approx 300 \ \mu$m) lithium n-contact is diffused on the inner surface. The outer p-type contact forming the p-n junction is made by ion-implantation of a boron layer only 0.3-μm thick. Thus the inactive surface layer is eliminated and the low-energy detection limit is determined principally by the cryostat vacuum enclosure which can be fitted with a thin beryllium window.

A useful bonus with the reversed-electrode geometry in coaxial detectors is a greatly reduced sensitivity to radiation damage, particularly from fast neutrons, an important consideration in accelerator-target rooms. This comes about because the principal damaging effect of radiation is the production of hole traps. With the reversed electrodes the holes are collected at the outside contact and in cylindrical geometry more material is closer to the outside so most holes have a shorter distance to travel and a smaller chance of being trapped. Also, compared with a p-type detector, for a given bias voltage the space-charge distribution results in a higher field in the outer regions which further reduces trapping (Raudorf et al., 1979). As a result of these two effects coaxial n-type HPGe detectors can sustain a greater than 30-fold more radiation damage than p-type detectors before their performance is significantly degraded (Pehl et al., 1979).

The necessity of cooling Ge detectors limits their use where space or liquid nitrogen supply are a problem, e.g., around nuclear reactors, in industry, in space and in nuclear medicine. Needs in these fields have motivated several research groups to develop semiconductor γ-ray detectors which can be operated from room temperature to $\approx 50°C$ (e.g., medical probes must operate at 37°C). The operating temperature of a semiconductor detector is limited by the requirement that negligible numbers of thermally excited electrons cross the band gap, E_g, from the valence to conduction bands. In practice, for room-temperature operation this means $E_g > 1$ eV, preferably in the range ≈ 1.4 to 2.4 eV (cf $E_g = 1.12$ eV for Si and 0.68 eV for Ge).

The most promising semiconductors for room temperature nuclear-radiation detectors are HgI_2 ($E_g = 2.13$ eV) and CdTe ($E_g = 1.47$ eV), with GaAs ($E_g = 1.43$ eV) running a not-too-close third, but gaining. To-date, however, defects due to residual impurities and small departures from stoichiometry have prevented good charge collection and limited spectroscopy-grade HgI_2 and CdTe detectors to volumes of ≈ 100 mm^3 as compared to the ≈ 100 cm^3 available in Ge. These small detectors by reason of their high atomic numbers, compared with Ge, have very high detection efficiencies per unit volume, particularly for x rays. They have found many applications where small size is a virtue rather than a drawback, especially as medical probes, x-ray detectors and personal radiation monitors ("chirpers") (Zanio, 1978; Wolf et al., 1982; Steidley et al., 1980). CdTe radiation monitors respond to thermal neutrons as well as to x and γ rays (Cornet et al., 1975). At present (1984) there is little prospect for large HgI_2 or CdTe detectors with γ-ray efficiencies and resolutions approaching those of available Ge detectors, so they are unlikely to be widely used for routine radionuclide measurements in the foreseeable future. For the state of the art at the time of writing see Siffert (1983) for CdTe, and Squillante et al. (1983) for HgI_2 detectors.

Planar Si detectors are widely used for room-temperature charged-particle detection and α-particle spectrometry (see Section 6.1.2.4) while Si(Li) detectors operated at 77 K give the best resolution (≈ 150-eV FWHM at 6 keV) for low-energy x rays (see Section 6.1.4.4.1). Lithium ions are much less mobile in Si than in Ge, and Si(Li), unlike Ge(Li) detectors, may be stored at room temperature.

2.5.2 *Energy Resolution*

As previously indicated, the major application of these devices is for x- and gamma-ray energy spectrometry. The development of low-noise pulse amplifiers that employ field-effect transistors operated in the liquid-nitrogen range of temperature (Smith and Cline, 1966) has made it possible to achieve energy resolutions approaching the statistical limits imposed by the variance in the number of electron-hole pairs created in the detector. Figure 15 presents a pulse-height spectrum of the gamma radiation emitted in the decay of ^{205}Bi, that was obtained with a large-volume Ge(Li) detector. To indicate the superior energy-resolving capability of the semiconductor detector, the pulse-height response of a 3″ × 3″ NaI(Tl) scintillation detector is shown for comparison.

The observed experimental width of a monoenergetic spectral peak resulting from the interaction of a gamma ray in the intrinsic region of a semiconductor detector is determined by a number of factors. The most important of these are: (1) variations in the number of electron-hole pairs; (2) noise contribution from the detector, such as leakage current or charge-collection problems; (3) electronic-noise contribution from the input stages of the pulse-amplifier system; (4) the performance of the pulse-shaping filter networks in the pulse-amplifier system; (5) the accuracy of the analogue-to-digital converter used to measure the amplitude of the detector pulse; and (6) the stability of the electronic system. Of these, the basic limitation in resolution is imposed by variations in the production of charge pairs within the detector. In germanium, on the average, 2.98 eV of energy are required to produce an electron-hole pair. This may be compared with the 30 eV of energy required to produce one ion pair in most of the proportional-counting gases. Following the production of a high-energy electron in the detector, energy is lost either by the production of electron-hole pairs or by processes that give rise to lattice vibrations. Because the sharing of energy between these two competing modes is statistical in nature, we must speak of an average number of electron-hole pairs resulting from the loss of a given amount of energy in the detector.

It is customary to express the experimental resolution of a detector in the following manner:

$$R = 2.355(FE_\gamma \bar{E}_p)^{1/2}, \tag{2.9}$$

where R is the measured width of a peak in keV at a point equivalent to one-half the maximum amplitude (FWHM), E_γ is the energy of the gamma ray in keV, $\bar{E}_p$ is the average energy required to create an electron-hole pair, and F is the so-called Fano factor (Fano, 1947),

Fig. 15. Pulse-height spectrum of gamma rays emitted in the decay of ²⁰⁶Bi obtained with a Ge(Li) detector. The response of a 3" × 3" cylindrical NaI(Tl) detector is shown to illustrate the improved energy resolution of semiconductor detectors. "SE" and "DE" indicate single- and double-escape peaks.

which is related to the fractional amount of the total energy absorbed
that results in the production of electron-hole pairs. To illustrate the
performance of semiconductor detectors, Figure 16 shows the resolution measured as a function of photon energy for a medium-sized
Ge(Li) detector used for general-purpose gamma-ray spectrometry. At
the lowest energies, the resolution is generally limited by the noise
performance of the electronic system. At higher energies, the resolution can be impaired by charge-collection problems in the detectors
and instabilities of the electronic systems. For comparison, the predicted width from Eq. 2.9 is shown for a germanium device with an
assumed value for the Fano factor of 0.13.

2.5.3 Characterizing Detector Performance

In order to provide a basis for evaluation of the performance of
semiconductor detectors in gamma-ray spectrometry, some conven-
tions have been established which are presently in use by commercial

Fig. 16. The measured energy resolution, expressed as the full width at half
maximum of the full-energy gamma-ray peak (FWHM), as a function of gamma-ray
energy for a Ge(Li) detector. The Fano Factor is chosen so that R vs. E_γ (Eq. 2.9) forms
a lower limit to the measured curve.

manufacturers of germanium detectors. Performance characteristics usually include: energy resolution (FWHM), peak shape (usually referring to asymmetric "tailing" which is described in Section 4.2.8), and detector efficiency. Resolution is usually quoted for the 1.332-MeV gamma ray from the decay of ^{60}Co together with a FWHM value for electronic noise. In the case of thin planar devices to be used for x-ray or low-energy gamma-ray measurements, a value may be quoted for the 22-keV Ag x ray. The usual method for specifying the severity of low-side peak tailing is to quote the full width at one-tenth maximum (FW0.1M) for the peak, in addition to the more conventional full width at half maximum (FWHM). The ratio of FW0.1M/FWHM for a Gaussian peak is 1.823.

The "peak-to-Compton ratio" is sometimes quoted as one feature of Ge semiconductor-detector spectra, and is defined as the ratio of the count in the highest photopeak channel to the count in a typical channel just below that of the Compton edge. It is conventionally quoted for the 1.332-MeV gamma rays from the decay of ^{60}Co, and is a measure of the combined effects of detector energy resolution and the fraction of total detector response appearing in the full-energy peak in the pulse-height spectrum. Typical values of the ratio for Ge detectors vary from 10 to 60. Detector efficiency is usually quoted by specifying the photopeak efficiency relative to that of a standard NaI(Tl) scintillation crystal. Values are provided for the 1.332-MeV gamma ray, measured for a source-detector spacing of 25 cm. Values for typical large coaxial devices range from 5 to 50%.

Several performance standards on the use of solid-state semi-conductor-detector systems have been published by the American National Standards Institute (ANSI, 1971, 1975b, 1977, 1978a, c, 1980b, 1982a, b).

2.6 Čerenkov Counting

When a charged particle passes through a medium, it causes local polarization along its path. When the polarized molecules return to their respective original states, they emit electromagnetic radiation. The propagation of the electromagnetic radiation occurs with a phase velocity of c/n where c is the speed of light in $vacuo$ and n is the refractive index of the medium. If the velocity of the charged particle v is smaller than c/n, the electromagnetic radiation will interfere destructively within the medium. When $v > c/n$, coherent directional electromagnetic radiation results, which is called Čerenkov radiation. The threshold for production is, therefore, $v = c/n$.

TABLE 7—*Čerenkov photon yields as a function of electron energy in water for selected spectral regions*[a]

Electron Energy	Number of photons/spectral region in nm						
	250–300	300–350	350–400	400–450	450–500	500–550	550–600
keV							
275	0.0032	0.0023	0.0017	0.0013	0.0011	0.0009	0.0007
300	0.0041	0.0029	0.0022	0.0017	0.0013	0.0011	0.0009
350	0.273	0.194	0.146	0.113	0.091	0.074	0.062
400	0.723	0.513	0.385	0.300	0.240	0.196	0.163
450	1.37	0.975	0.731	0.569	0.455	0.373	0.309
500	2.22	1.58	1.18	0.922	0.737	0.604	0.501
550	3.26	2.31	1.73	1.35	1.08	0.884	0.734
600	4.45	3.16	2.37	1.85	1.48	1.21	1.00
650	5.80	4.12	3.09	2.41	1.92	1.58	1.31
700	7.59	5.39	4.04	3.15	2.51	2.06	1.71
750	8.94	6.35	4.76	3.71	2.96	2.43	2.01
800	10.7	7.59	5.69	4.43	3.54	2.90	2.41
850	12.6	8.92	6.69	5.21	4.16	3.41	2.83
900	14.6	10.3	7.75	6.04	4.83	3.95	3.28
950	16.6	11.8	8.87	6.91	5.52	4.52	3.75
1000	18.9	13.4	10.0	7.82	6.25	5.12	4.25
1200	28.5	20.3	15.2	11.8	9.46	7.75	6.43
1400	39.6	28.1	21.1	16.4	13.1	10.7	8.92
1600	51.7	36.7	27.5	21.4	17.1	14.0	11.6
1800	64.8	46.0	34.5	26.9	21.5	17.6	14.6
2000	78.8	56.0	42.0	32.7	26.1	21.4	17.8
2200	93.6	66.5	49.8	38.8	31.0	25.4	21.1
2400	109	77.6	58.2	45.3	36.2	29.7	24.6
2600	125	89.1	66.8	52.0	41.6	34.1	28.3
2800	142	101	75.8	59.0	47.2	38.7	32.1
3000	160	114	85.1	66.3	53.0	43.4	36.0
3200	178	126	94.7	73.8	58.9	48.3	40.1
3400	196	140	105	81.5	65.1	53.3	44.3
3600	215	153	115	89.3	71.4	58.5	48.6
3800	235	167	125	97.5	77.9	63.8	53.0
4000	255	181	136	106	84.5	69.3	57.5

[a] From Ross (1970).

Čerenkov radiation is emitted mainly in the higher-frequency range of the visible spectrum, in a narrow cone around the direction of propagation of the incident charged particle.

Table 7 shows photon yields for electrons in water, for various regions of the electromagnetic spectrum, according to Ross (1970). The threshold is about 260 keV. Figure 17 shows that the threshold electron energy decreases as the refractive index increases. Unfortunately, there are practical limitations in using high-refractive-index solutions, and thus water remains the most commonly used solvent for Čerenkov counting.

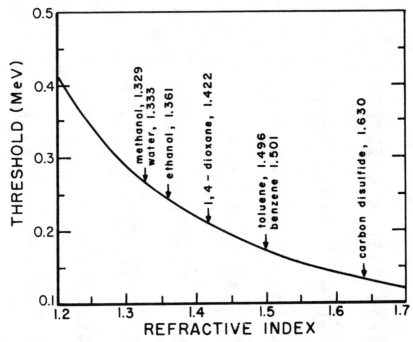

Fig. 17. Electron threshold energy for Čerenkov emission as a function of refractive index of several solvents. Refractive indices are for a wavelength of 589 nm (Na D lines) at 20 °C. (From Parker and Elrick, 1970.)

The use of Čerenkov radiation for the indirect measurement of radioactivity is discussed in Section 4.5.1 and in Jelley (1958).

2.7 Dead-Time, Decay, and Background Corrections

2.7.1 *Introduction*

The counting rates obtained with any of the variety of detectors available will require corrections for dead-time losses and backgrounds as well as decay during the counting interval, if this is significant, and, as a rule, decay for the elapsed time between the beginning of the counting interval and some chosen reference time. While other corrections, depending on the detector, may also be required, those just mentioned are common to all and are treated in the following subsections.

2.7.2 *Dead-Time Corrections*

2.7.2.1 *Introduction.* For any counting system consisting of a detector with its associated electronic equipment there exists a minimum time interval, τ, by which two consecutive events must be separated for both to be recorded. Events arriving during this dead time are lost and the fraction of events lost increases with increasing counting rate. In this section the terms dead time and dead-time losses are used instead of the alternatives, resolving time and coincidence losses, to avoid confusion with coincidence-counting terminology. (See Section 3.2.)

Dead times can be of two types depending on the system response to events arriving *during* a dead time (Figure 18a). If the dead time is increased by such input events, so that an interval τ must elapse

Fig. 18.   Nonextendable and extendable dead times. (a) A sequence of input events (top row) is shown with the corresponding outputs from a nonextendable dead time (middle row) and an extendable dead time (bottom row). With the nonextendable dead time, events 3 and 6 are lost but 4 survives because it is separated by more than τ from 2, the last preceding event to produce an output. With the extendable dead time, event 4 is lost also because it is separated by less than τ from 3 which extended the dead time initiated by 2. Event 6 extended the dead time initiated by 5 but not by enough to block the following event. (b) The observed output rates as functions of the input rate for the two types of dead time are shown as plots of $n\tau$ vs $N\tau$. (This is equivalent to plotting n vs N both in units of τ^{-1}.) Note that with an extendable dead time, the output rate n_1 could be observed for two input rates, N_1 or the many times larger N_2.

following the last input event before another output event can be produced, it is said to be extendable (or paralyzable, cumulative, or updating). If the dead time is not increased by input events arriving while the system is dead, the dead time is said to be nonextendable (or nonparalyzable or noncumulative).[2] The responses of real systems lie between these limiting cases, but, as we shall see, it is usually desirable to design systems to have as nearly nonextendable dead times as possible.

If the input events occur at random times at a mean rate N, the time intervals between events are described by the Poisson distribution (see Section 7.1.1 and Appendix B.3.2) and the probability that no events occur in an interval τ is $e^{-N\tau}$. Then the expected output rate, n, for a system with an extendable dead time, τ, is

$$n = Ne^{-N\tau}. \tag{2.10}$$

Equation 2.10 cannot be solved for N in terms of n (i.e., in closed form), but the approximation

$$N \approx n/[1 - n\tau - (n\tau)^2/2] \tag{2.11}$$

is good to better than 0.1 percent for losses ≤ 10 percent. For somewhat larger losses, or higher accuracies, the approximate N from Eq. 2.11 may be substituted in the exponent of Eq. 2.10, and N obtained to the required accuracy by iteration. The output rate, Eq. 2.10, reaches a maximum of $(e\tau)^{-1}$ for an input rate of τ^{-1}, decreases for higher input rates, and approaches zero, with the system almost completely paralyzed, for input rates $\gg \tau^{-1}$. Because the input rate is a double-valued function of the output rate, there is danger with extendable dead times of wrongly assigning the lower input rate to an unexpectedly strong source (Figure 18b).

If a system with a nonextendable dead time τ gives an average output counting rate n, then the system is dead for a fraction $n\tau$ and "live" for a fraction $1 - n\tau$ of the counting interval, t. If the input events arrive randomly at an average rate N that does not vary significantly during t, then N (the dead-time-corrected counting rate) is simply the observed rate divided by the live-time fraction, or

$$N = n/(1 - n\tau). \tag{2.12}$$

Conversely, if the input rate is known, the expected output rate given by Eq. 2.12 is

$$n = N/(1 + N\tau). \tag{2.13}$$

[2] The two types have also been called Type I and Type II dead times but this terminology is best avoided because different authors (e.g., Feller, 1948; Evans, 1955) do not agree as to which is which.

Equations 2.12 and 2.13 give the exact expectations for the usual case in which the beginning of the counting interval, t, is chosen at random ("equilibrium process"). Expectations given by some authors for the cases in which t commenced either with an event, i.e., at the beginning of a dead time ("ordinary process"), or commenced only if the system is "live" ("shifted" or "free-counter process"), differ, respectively, by terms of the order of τ/t or $N\tau^2/t$ (Müller, 1974) that are several orders of magnitude smaller than the statistical uncertainties of practical measurements. A system with a nonextendable dead time is never paralyzed; the output rate asymptotically approaches τ^{-1} for input rates $\gg \tau^{-1}$ (Figure 18b).

Equation 2.12 suggests a simple method for comparing dead times. If pulses from a detector are fed to two systems with dead times τ_1 and τ_2, then

$$N = n_1/(1 - n_1\tau_1) = n_2/(1 - n_2\tau_2),$$

and

$$\tau_2 = \tau_1 + (1/n_2 - 1/n_1). \qquad (2.14)$$

Thus if τ_1 is known, τ_2 can be found directly, or, if τ_1 is known to be nonextendable and rate independent while τ_2 is suspect, τ_2 can be quickly checked over a range of N.

If the half life of the sample activity is so short that decay during the counting interval, t, is significant, the losses are not constant and must be integrated over t. The resultant formulae are given in Section 2.7.3.

2.7.2.2 *Statistics of Dead-Time Distorted Processes.* A property of the Poisson distribution that is applicable to radioactive events is that the variance is equal to the mean (Section 7.1.1 and Appendix B.3.5). Therefore, for a hypothetical system with zero dead time, the estimated variance of the observed total counts in interval t is equal to the observed total counts:

$$\hat{\sigma}_{nt}^2 = nt, \text{ or } \hat{\sigma}_{nt} = \sqrt{nt}. \qquad (2.15)$$

This simple relation does not hold when the dead time is not zero because the intervals between observed counts no longer constitute a simple Poisson distribution; all intervals $<\tau$ have been removed. For a nonextendable dead time the standard deviation of the *observed count* is less than for a Poisson process by a factor $(1 - n\tau)$ or

$$\hat{\sigma}_{nt} = (1 - n\tau)\sqrt{nt}. \qquad (2.16)$$

The standard deviation of the *corrected count* is greater than for a Poisson process by a factor $\sqrt{1 + N\tau}$, or

$$\hat{\sigma}_{Nt} = \sqrt{(1 + N\tau)Nt}. \tag{2.17}$$

These results (Feller, 1948; Vincent, 1973) should be kept in mind when assessing statistical uncertainties if the losses are large. Note, however, that the *relative* standard deviation of the *corrected rate* is still given by the usual expression

$$\hat{\sigma}_N/N = 1/\sqrt{nt}. \tag{2.18}$$

Baerg (1973a) has shown that Eq. 2.18 holds whether the data are taken by preset time, preset live time, or preset count, and, if background is negligible, even if the source decays significantly during t. For moderate losses, Eq. 2.18 also holds to a good approximation with extendable dead times (Vincent, 1973). Cohn (1966) has prepared graphs showing relative standard deviations as functions of counting losses for both types of dead time.

2.7.2.3 *Dead Times of Practical Systems.* For accurate measurements over a range of counting rates involving significant dead-time losses, one component of the system ideally should have a dead time that is stable, nonextendable, independent of pulse rate and amplitude, and longer than the maximum dead time in any other stage. If no electronic component (e.g., discriminator, single-channel analyzer, or scaler) meets these requirements, it may be necessary to insert a special dead-time-determining stage into the system. This dead time becomes the effective system dead time, the principle being that accurately determinable corrections for dead-time losses are preferred to somewhat smaller ones that are poorly defined.

Except for Geiger-Müller counters for which the intrinsic dead times range from several tens to a few hundreds of microseconds, the dead times that are imposed on systems employing proportional, scintillation, or solid-state detectors are determined by the time constants of the associated electronics and are usually a few microseconds. Some methods for measuring these are described in the following section.

Commercial liquid-scintillation-counting (LSC) systems generally have fast electronics combined with inherently fast scintillators. Because LSC is widely used to assay large numbers of relatively low-activity sources and dead times are usually ≤ 1 μs, it is common practice to neglect the dead time corrections. If, however, accuracies approaching 1 percent are required for sources giving count rates of $\geq 10^4$ s^{-1}, rate-dependent effects should be investigated for the channel settings of interest (see Sections 3.8 and 4.3).

The methods of measuring dead times and correcting for losses applicable to single-detector systems apply to the independent singles channels of coincidence-counting arrangements, but the corrections to the observed coincidence counts are more complex and are treated in Section 3.2.

Most multichannel gamma-ray spectrometers have variable, amplitude-dependent dead times, and commercial models usually compensate the dead-time losses during each run by the use of automatic live-timing devices, as described in Section 4.2.4.4. Provided the input rate does not vary significantly over the counting interval, these are accurate to the extent that the contributions to dead-time losses from the detector and all the circuits preceding the analyzer are negligible. This may not be true for high-resolution Ge(Li) spectrometers employing long amplifier time constants, base-line restoration, and pile-up rejection circuits. The pile-up rejectors are essentially extendable dead-time devices and may contribute substantially to the system losses at high input rates. A bibliography containing some 170 references dealing with pile-up effects has been published by the International Bureau of Weights and Measures (BIPM, 1981a).

2.7.2.4 *Measuring Dead Times.* The choice of method for measuring dead times depends on the accuracy required, the time and equipment available, and the type of system. A careful investigation of the counting-loss characteristics of a new system can be time-consuming; routine tests to check that the dead time has not changed take only a few minutes. Some methods are outlined below.

2.7.2.4.1 *Oscilloscope Method.* High-rate output pulses from the dead-time-determining stage are viewed with a triggered oscilloscope and the minimum time between successive pulses estimated. By varying such parameters as source strength, gain, and discrimination level, any gross change of dead time with counting rate or instrument setting can be observed. Although the minimum pulse separation might be read to, say, 100 ns on the screen, the error in the dead time could be larger. For reasons discussed below (Section 2.7.2.5), it is impossible to achieve a perfectly nonextendable overall system dead time, and the average effective dead time will be larger than the minimum.

2.7.2.4.2 *Double-Pulse Method.* A train of suitably shaped pulse pairs from a double-pulse generator is fed into the preamplifier test input, the pair separation is reduced until the output rate is halved, and the separation read from an oscilloscope or from the pulser setting if it is accurately calibrated. As with the method given in Section 2.7.2.4.1, this method is rapid, but does not test for dead-time extendability. Nor does it simulate the random amplitudes or random arrival times of detector pulses. A triple pulse generator, if available, can be used to test for extendability.

2.7.2.4.3 *Two-Source Method.* Two sources, (1) and (2), of about equal activity sufficient to give several percent losses are counted in this sequence; (1), (1 + 2), (2) with observed rates n_1, n_{12}, and n_2, respectively, and background rate n_b. Then, if the dead time, τ, is nonextendable and independent of the counting rate,

$$\tau = \{1 - [1 - (\Delta - n_b)q/p^2]^{\frac{1}{2}}\}p/q, \qquad (2.19)$$

where

$$\Delta = n_1 + n_2 - n_{12},$$

$$p = n_1n_2 - n_bn_{12},$$

and

$$q = n_1n_2n_{12} + n_b(n_1n_2 - n_1n_{12} - n_2n_{12}).$$

Diethorn (1974) derives Eq. 2.19 and discusses various approximations that neglect the background rate. If the background can be neglected, then, for $n_b = 0$, Eq. 2.19 becomes

$$\tau = [1 - (1 - \Delta n_{12}/n_1n_2)^{\frac{1}{2}}]/n_{12}. \qquad (2.20)$$

The background may be included to a first order approximation by putting $\Delta - n_b$ for Δ in Eq. 2.20 (Kohman, 1945). Then Eq. 2.20 will usually be adequate for $n_b/n_1 \approx n_b/n_2 \lesssim 10^{-3}$.

The two-source method has the advantage of giving the effective overall dead time by using real pulse distributions. Several precautions must be taken to get correct results. Since Δ is a small difference between large numbers subject to statistical fluctuations the individual rates must be measured with high precision. For this reason the entire system must be extremely stable with respect, for example, to rate-dependent gain shifts. The presence of one source must not affect the input counting rate from the other source; to equalize scattering effects an identical blank should replace the absent source when each is counted alone. Rate independence is important because if $n_1 \approx n_2$, if the dead time is τ at rate n_1 (or n_2), and if the dead time is $\tau + \delta\tau$ at rate n_{12}, the method gives an apparent dead time of $\approx \tau + 2\delta\tau$. To test for rate dependence two or three pairs of sources covering the counting-rate range of interest may be used. More sophisticated multi-source methods have been described (Kohman, 1945; Thomas, 1963).

Approximations to Eq. 2.20 widely cited in textbooks, but not recommended for accurate work, are

$$\tau = \Delta/(n_{12}^2 - n_1^2 - n_2^2) \quad \text{or} \quad (\Delta - n_b)/(n_{12}^2 - n_1^2 - n_2^2) \qquad (2.21)$$

and

$$\tau = \Delta/2n_1n_2 \quad \text{or} \quad (\Delta - n_b)/2n_1n_2.$$

Of these, the last is preferred and may often be adequate for less critical work, or when only small dead-time corrections are anticipated. For an extendable dead time an exact solution for τ cannot be obtained in closed form but, neglecting background,

$$\tau \approx (n_{12}/2n_1n_2) \ln[(n_1 + n_2)/n_{12}] \qquad (2.22)$$

to better than 0.5 percent if $n_{12}\,\tau \leq 0.15$ and if $n_1 \approx n_2$ within a factor of two. (Eq. 2.22 is exact if $n_1 = n_2$ and $n_b = 0$.)

2.7.2.4.4 *Source-Pulser Method.* This is similar to the method given in Section 2.7.2.4.3 except that one source is replaced by a train of suitably shaped periodic pulses fed into the preamplifier test input. If the observed rates for the source, source plus pulser, and pulser are n_s, n_{sp}, and n_p then, if the dead time is nonextendable and rate independent,

$$\tau = [1 - (1 - \Delta/n_p)^{1/2}]/n_s \ , \qquad (2.23)$$

where $\Delta = n_s + n_p - n_{sp}$, and the background is included in n_s. Equation 2.23 is almost exact for $n_p\tau \lesssim 0.30$, a condition that should always obtain for practical measurements. Müller (1976a, 1976b) has derived the corrections applicable to Eq. 2.23 for higher pulser rates. This method has many advantages over that described in Section 2.7.2.4.3 (Baerg, 1965), and is recommended for general use. In particular, any source may be used, the restrictions on τ with respect to extendability and rate dependence are less critical than with two sources, there are no background terms, and, since the pulser output is not subject to statistical fluctuations, a given precision can be reached in about one-third the counting time. As with the two-source method, an exact solution for an extendable τ cannot be obtained in closed form, but the approximation

$$\tau \approx pe^{-p}/n_s \ , \qquad (2.24)$$

where

$$p = (q^2 + 4q - 1)^{1/2} - (q + 1)$$

and

$$q = n_p/(n_{sp} - n_s) \ ,$$

is good to better than 0.3 percent for $n_s\tau \leq 0.20$ and a wide range of pulser rates; i.e., the error in the approximation in Eq. 2.24 is independent of n_p.

2.7.2.4.5 *Two-Pulser Method.* This is another variant of the two-source method (Müller, 1973). If n_1 and n_2 are the rates of the two pulsers, then $\tau = \Delta/2n_1n_2$ exactly. The pulsers must be extremely stable, have incommensurable rates (i.e., the ratio of the rates must not be expressable as a ratio of integers), and be perfectly isolated (i.e., must never "lock-in"). The method has the disadvantage, common to the double-pulse method (Section 2.7.2.4.2), that no random pulse distributions are used. It is most useful for the fast, accurate testing of electronic dead-time circuits, especially for extendability.

2.7.2.4.6 *Decaying-Source Method.* A source of a pure activity with a conveniently short, accurately known, half life is followed (e.g., ^{128}I, $T_{1/2} = 24.99$ m; ^{56}Mn, $T_{1/2} = 2.578$ h). The decay plotted on semilogarithmic paper is compared with a straight line with a slope corresponding to the known half life. Corrections are taken directly from the graph. This method is relatively time consuming and imprecise but it has several advantages. It does not matter if the dead time is extendable, nonextendable, or rate-dependent; the appropriate correction is read directly, without calculation, from data obtained under normal operating conditions. The method is particularly suitable for use with Geiger-Müller counters and may be used at higher rates for liquid-scintillation counting if the source-pulser method cannot be used. If the dead time is, in fact, nonextendable and rate-independent, τ can be found from the ratio of slope to intercept of a plot of $ne^{\lambda T}$ vs. n (Martin, 1961), where T is the elapsed time.

2.7.2.4.7 *Proportional-Sources Method.* This method requires two or more sources for which the activity ratios are accurately known and the intrinsic (rate-independent) efficiencies are equal (Stevenson, 1966). The first condition may be met by careful quantitative source preparation and the second may obtain for, *e.g.*, liquid-scintillation counting or integral gamma-ray counting in a well crystal, but not, as a rule, for solid-source beta-particle counting. For a nonextendable dead time, negligible background, and two sources with an activity ratio $A_1/A_2 = R$,

$$\tau = (R - n_1/n_2)/n_1(R - 1). \tag{2.25}$$

Typically $2 < R < 10$; a series of sources may be prepared to cover a wide range of counting rates.

For an extendable dead time, the approximate solution

$$\tau \approx \ln(Rn_2/n_1)/n_2[R - 1 + \ln(Rn_2/n_1)] \tag{2.26}$$

is good to better than 0.5 percent for $n_2\tau \leq 0.10$ for a wide range of $R > 1$: i.e., the accuracy of Eq. 2.26 depends only on the losses for the *weaker* source.

Proportional sources or a decaying source as in Section 2.7.2.4.6 may also be used to test for non-linear rate dependence in detectors operated in current-measuring or charge-integrating modes, such as ionization chambers.

2.7.2.4.8 *Reference-Peak Method.* Dead-time losses in multichannel gamma-ray spectrometers are often too complex to be calculated from measured system parameters. To monitor these losses directly, reference pulses having excellent amplitude stability and exactly the same shape as the detector pulses are injected at the preamplifier test input. The pulse amplitude is chosen so that the corresponding peak will appear in the spectrum well-removed from any gamma-ray peaks, e.g., beyond the highest energy gamma ray. The pulse rate should be comparable to the rates of the principal gamma-ray peaks but should not exceed 10 percent of the total counting rate if the pulses are periodic. This restriction is unnecessary if a random pulser is used. The total number of reference pulses presented to the preamplifier during the counting interval is recorded separately.

After computer analysis of the peak areas in the spectrum, the ratio of the recorded reference-pulse rate to the computed reference-pulse peak area is used to convert the gamma-ray peak areas to gamma-ray rates. Provided all the radionuclides in the sample have half lives long compared to the counting interval and there are no amplitude-dependent losses, this method corrects for all rate-dependent losses, whether from analyzer dead time, pile-up (including random summing) or pile-up-rejector dead time, or base-line shift, or base-line restorer dead time. It also compensates for any bias in the computer method of determining the peak areas. It does not, of course, correct for true coincidence summing, which is rate independent. For details see, e.g., Bolotin *et al.* (1970) or Wiernik (1971a, 1971b).

A similar principle has been used to monitor rate-dependent losses in medical scintillation (Anger) cameras. Here, again, the effective system dead time may be extendable and rate-dependent, effects that make accurate corrections difficult to calculate (Budinger, 1974). A reference source is placed in a peripheral position of the camera collimator outside the region of interest. The effect of all loss mechanisms can be compensated digitally, frame by frame, by having the system computer normalize all intensities to a standard value for the reference source (Freedman *et al.*, 1972; Adams *et al.*, 1974).

In systems using multichannel gamma-ray spectrometers, the reference peak method is not readily applicable to spectra distorted by amplitude-dependent (i.e., energy-dependent) losses. Such distortions are not common but may be introduced by, for example, pulsed-optical-feedback techniques in some low-noise field-effect-transistor (FET) preamplifiers because the detector pulse that initiates an FET bias restoration, or "reset," is lost, and this initiating pulse is more likely to be a large one than a small one. Losses of this type from a reference peak relative to losses from individual gamma-ray peaks depend on the peak separation and there is no direct way, when a single reference peak is used, of obtaining undistorted, dead-time-corrected spectra. If the source is a mixture of short- and long-lived activities with the shorter half lives comparable to, or shorter than, the counting interval, there is no simple method of making accurate dead-time corrections, or even corrections that give the true relative intensities of the component activities. Electronic methods of providing either continuous automatic compensation of losses, or constant losses, have been proposed for applications involving rapidly decaying multi-component samples, e.g., activation analysis (Harms, 1967; De Bruin *et al.*, 1974). Monte Carlo studies have shown that the Harms method should give correct results automatically for a wide range of multi-component, time-dependent spectra (Masters and East, 1970). Harms' approach has been extended to include compensation for losses (including pile-up) in the circuits preceding the multichannel analyzer (Westphal, 1979) and incorporated into systems designed to handle high and variable counting rates (Allsworth *et al.*, 1982; Nuclear Data Inc., 1984).

2.7.2.5 *Dead Times In Series.* As a rule, it is impracticable to have the dead-time-determining stage immediately at the detector output before any pulse-shaping takes place. If not, the overall system dead time will always be somewhat extendable because any circuit has some dead time and it is impossible to achieve a perfectly nonextendable dead time with dead times in series, unless the first is the longest (Lewis, 1948; Damon and Winters, 1954; Müller, 1973). When the longest dead time ends, a preceding stage may be dead from an event that arrived after this stage recovered, but before the end of the longer dead time (Figure 19). Thus, the average dead time of the system suffers rate-dependent extension. In this way, pulse pile-up and base-line shift in early stages may extend the effective dead time at high input rates. Therefore, rate dependence must be studied carefully before high-accuracy measurements can be made when the losses exceed 10 percent or so. At low input rates these effects are negligible. At 2-percent losses, Eq. 2.11 differs from Eq. 2.12 by only 0.02 percent, but at 10-percent losses, the difference is ≈ 0.5 percent. The differences for a partially extendable dead time would be much less. Practically, and sometimes even conceptually, it is difficult to distinguish between (a) rate-dependence of the dead time caused by dead-time extendability or actual dead-time variations and (b) an apparent rate-dependence caused by gain shifts, discriminator shifts, base-line shifts, pulse pile-up at high rates, and similar effects that are amplitude-dependent as well. Single-channel analyzers can show apparently anomalous effects at high rates, because pile-up may shift events both into and out of

Fig. 19. Nonextendable dead times in series. A high-rate sequence of input events (top row) is shown with the corresponding output from a single dead time, τ (second row). Events 2, 5, and 7 are lost. The bottom two sequences show the effect of a preceding, shorter dead time, τ_1, followed by the principal dead time, τ_2; the irregular termination of τ_1 indicates that τ_1 may be somewhat variable, e.g., pulse-amplitude dependent. One effect of τ_1 has been to increase the losses by removing event 3 from the output of τ_2. In the sequence of events 4 to 7, however, the effect of τ_1 on the output of τ_2 has been to remove event 6 and restore event 7 for no additional net loss of counts. Thus, the effect of τ_1 is second order, complex, and nonlinearly rate dependent. Note, however, that the sequences illustrated correspond to a situation in which dead-time losses exceed 40 percent.

the window region. If discriminators or single-channel analyzers are set to select a small fraction of the total input events, the input rates will be many times the observed output rates and the series-dead-time effect may be exaggerated.

2.7.2.6 *Further Reading.* More complete discussions of dead-time phenomena have been given by several authors, e.g., Evans (1955), chapter 28; Stevenson (1966), chapter VII; Vincent (1973), chapter 4; Müller (1973). A bibliography of some 580 references dealing with the measurement of dead times, the corresponding corrections, and closely related subjects has been prepared by the International Bureau of Weights and Measures (BIPM, 1981b).

2.7.3 *Background and Decay Corrections*

In the absence of decay over the counting interval, the background rate n_b is subtracted from the observed counting rate after correction for dead-time losses. For a nonextendable dead time

$$N = \frac{n}{1 - n\tau} - n_b . \tag{2.27}$$

As background count rates are normally low, dead-time corrections to the observed background count rate may nearly always be neglected. As a rule, when the ratio of source-to-background count rates is small the dead-time correction is small. Statistical considerations governing the optimum division of available time between sample and background measurements are discussed in Section 7.1.2. The origins of the background, and methods of minimizing it, are discussed in Sections 6.1.1.5, 6.1.4.2.2 and 6.1.4.3.3.

In the absence of significant background and dead time, but with significant decay during the counting interval, it can readily be shown that the counting rate N_0 at time t_0, the beginning of the counting interval, is given by

$$N_0 = n\, \frac{\lambda t}{1 - e^{-\lambda t}}, \tag{2.28}$$

where t is the duration of the counting interval and λ the decay constant. Alternatively, if the average counting rate is assigned to the midpoint of the counting interval, the influence of decay during that interval is much reduced. The observed rate exceeds the midpoint rate by only 0.5 percent for an interval of half a half life or by only 0.02 percent for a counting interval of one tenth a half life. For greater accuracy or longer counting intervals, the decay-corrected counting rate at the midpoint, N_m, is given by

$$N_m = n\lambda t e^{-\lambda t/2}/(1 - e^{-\lambda t}) = n(\lambda t/2)/\sinh(\lambda t/2). \tag{2.29}$$

A third possibility is to specify the time, t_a, at which the rate is equal to the observed average rate, viz.,

$$t_a = t_0 + \lambda^{-1}\ln[\lambda t/(1 - e^{-\lambda t})] \tag{2.30}$$

which is earlier than the midpoint $t_0 + t/2$, by 1.5 percent of t for t equal to half a half life. Eq. 2.30, however, is not usually as convenient as either of the two preceding expressions if a large amount of counting data must subsequently be corrected to a common reference time.

Exact expressions treating the combined effects of dead time, decay, and background have been given by Axton and Ryves (1963). An approximation, also given by these authors, has the form,

$$N_0 = \left(\frac{n}{1 - n\tau} - n_b\right)\left\{\frac{\lambda t}{1 - e^{-\lambda t}}\right\}\left(1 + \frac{n\tau(\lambda t)^2}{12}\right). \tag{2.31}$$

This expression is accurate to 0.1 percent for the condition that $\lambda t < 2$ and $n\tau < 0.26$. Clearly N_m may be obtained to the same approximation by multiplying the right side of Eq. 2.31 by $e^{-\lambda t/2}$. When the combined dead-time and decay corrections are such that the last term of Eq. 2.31 can be neglected, N, the dead-time- and background-corrected rate, may be substituted for the observed rate, n, in Eqs. 2.28 and 2.29; e.g., if $\lambda t = 0.10$ and $n\tau = 0.10$, then $n\tau(\lambda t)^2/12 < 10^{-4}$. The effect of dead time, background, and decay on the variances of the corrected rates has been discussed by Baerg (1973a) and a detailed analysis including extendable dead times can be found in Vincent (1973).

Finally, the rate N_0 (or N_m) may be converted to the rate N_R at the reference time by the well-known relation derived in Section 1.3, viz.,

$$N_R = N_0 e^{-\lambda T} = N_0 e^{-(\ln 2)T/T_{1/2}}. \tag{2.32}$$

The elapsed time, T, from the beginning of the counting interval (or, for N_m, the midpoint) to the reference time, is negative if the reference time is earlier than the time of counting.

It is important to estimate the contribution of the decay correction to the systematic uncertainty in N_R; it will depend on the uncertainty in $T_{1/2}$, the sensitivity of the detection system to any radioactive impurities in the sample, and the size of the correction. For pure samples of radionuclides with accurately known half lives, decay corrections for periods of a half life or less may not increase the error significantly. On the other hand, corrections for long decay periods based on inaccurate half life values or applied to impure samples can lead to large errors. Only in unusually favorable circumstances can decay corrections for more than five or six half lives be made with any degree of confidence.

The simple decay correction, Eq. 2.32, can be applied to combined parent and daughter activities only when they are in equilibrium as discussed in Section 1.3. When the half lives of parent and daughter are not very different, e.g., ^{95}Zr-^{95}Nb, equilibrium may not be reached during the useful life of the sample. The more complex equations describing the growth and decay of serial activities not in equilibrium were treated by Bateman (1910) and are covered in standard textbooks, e.g., Evans (1955), chapter 15, or Friedlander et al. (1964), chapter 3.

3. Fundamental or Direct Measurements of Activity in Radioactive Decay

3.1 General

The question arises, what is a fundamental or direct measurement of activity in radioactive decay. Possibly the simplest answer is that it is based only on count-rate measurements of the source in question. Such a measurement invokes only the standard of time, and is never based on any other radioactivity standard.

The whole subject of radioactivity is, however, not just a question of measuring so many nuclear transitions in a given period of time. Implicit in most measurement results is a fairly detailed knowlege of the modes of decay of the radionuclide under investigation. In sum peak, or gamma-gamma, and in conventional (non-efficiency-extrapolation) electron-photon coincidence counting, as examples, it is necessary to know the probability per decay of the relevant radiations, before the source disintegration rate can be deduced from the counting data, and with calorimetric methods, the average energy or energy per decay is required to compute activity. Only within the limitation that the decay-scheme data, of the kind given in Appendix A.3, are sufficiently reliable, can such measurements be regarded as direct or fundamental.

On the other hand, even seemingly direct measurements with 4π proportional counters, defined-solid-angle detectors, liquid-scintillation counters, and internal gas counters may be subject to various corrections such as those for source self-absorption, counter wall effects, detector threshold level, and the energy distribution of the radiation being measured. Where it is not possible to make these corrections to a sufficient degree of accuracy, these measurements, too, cannot properly be regarded as fundamental.

Methods using $4\pi\beta\text{-}\gamma$ coincidence counting have been most thoroughly studied and developed to provide, for a wide range of nuclear species, measurements that are independent of the precise values of nuclear-decay-scheme parameters and free of requirements for corrections other than those for decay, dead time, resolving time, and backgrounds. These methods, based on efficiency extrapolation, must be regarded as the most reliable of the fundamental or direct methods. It should be noted, however, that no one method is applicable to all nuclear species. The choice of method is finally determined by the nature of the nuclear decay, and whether the attainable accuracy of measurement is considered to be acceptable.

A calibration method might also be described as direct, or "absolute", if it is "self-contained." Thus the efficiency-tracing method, referenced in Sections 3.2.9 and 3.5.6.2, can combine the measurement of the tracer and the traced into one experiment; if a short-lived tracer is used that decays in a sufficiently short time to a small fraction of the traced activity, then this method is certainly "self-contained".

In this section, methods are described for nuclear species and conditions under which the measurements are not relative to a previously measured activity of another source of the same radionuclide, or to the activities of other radionuclides with similar radiations.

Some might consider methods requiring auxiliary data such as internal-conversion coefficients as "non-absolute", but neither are they relative insofar as they do not involve comparison with a previously calibrated standard, or an instrument calibrated with that standard. Moreover, most activity measurements depend on some prior knowledge regarding their mode of decay—whether it be by particulate emission, or electromagnetic, or both. And the method of calibration of a standard, and its continued usefulness depend very much on a knowledge of the half life of its activity.

Thus the dividing line between direct and indirect methods can be tenuous.

3.2 Beta-Gamma Coincidence Counting

3.2.1 *General*

The measurement of activity by the coincidence method can be applied to any nuclear decay involving two or more distinguishable radiations occurring in prompt succession.[3] Two detectors are usually required, each of which responds, ideally, to one only of the two types

---

[3] Such coincidence methods, to measure source disintegration rates, cannot be based on coincidences between primary radiation and associated secondary radiation generated outside the atom, e.g., beta particles and external bremsstrahlung and gamma rays and scintillator photons.

of radiation. A third counting channel, derived from a coincidence circuit, records those events from the two detectors that occur in coincidence. Then, for simple two-stage beta-gamma decay of a source with a true disintegration rate N_0, and with beta- and gamma-detector efficiencies ϵ_β, ϵ_γ, the channel counting rates would be:

$$N_\beta = N_0\epsilon_\beta, \qquad N_\gamma = N_0\epsilon_\gamma, \quad \text{and} \quad N_c = N_0\epsilon_\beta\epsilon_\gamma. \qquad (3.1)$$

The value of N_0 is then simply $N_\beta N_\gamma/N_c$ and no knowledge of detector efficiencies is required, but, in this simple case, these latter clearly emerge from the data, with ϵ_β equal to N_β/N_0 or N_c/N_γ, and ϵ_γ equal to N_γ/N_0 or N_c/N_β.

Various modifications of this simple result are required to take account of (1) dead time, resolving time, and backgrounds, the effects of which are count-rate dependent and (2) decay-scheme effects that affect detection efficiencies. These modifications will be considered in the following subsections in the context of beta-gamma decay, but the treatment may be taken as equally applicable to other forms of decay.

A $4\pi\beta$-γ-coincidence system, shown in the block schematic of Figure 20, is the most widely used in standards laboratories. Commercially available amplifiers and other electronic components may be used to assemble an acceptable system but the detector assemblies are usually of individual design, some even including automatic sample-changing facilities (e.g., Garfinkel *et al.*, 1973a; Figure 21). Two NaI(Tl) detectors (e.g., 7.6 × 7.6 cm) are commonly used for higher efficiency. The dead-time circuitry in these systems is usually of the nonextending type since it is fairly well understood theoretically and the count-rate-dependent effects are minimal. The decay-scheme effects, on the other hand, are much reduced when a $4\pi\beta$ proportional counter is used (Steyn and Haasbroek, 1958; Campion, 1959) and this type of counter also conforms well to general theoretical requirements for complex decay (Baerg, 1966).

The measurement of the counting rates for the separate beta and gamma channels is, in practice, usually made in the conventional manner appropriate for a single detector. The data are, however, usually collected simultaneously and the counting is controlled by a single timer. The determination of coincidences between events recorded in the beta and gamma channels is not so simple and several methods are described briefly in the following three sections.

3.2.2 *Coincidence Counting*

The most widely used method of detecting coincident beta and gamma events depends essentially on summing the two series of

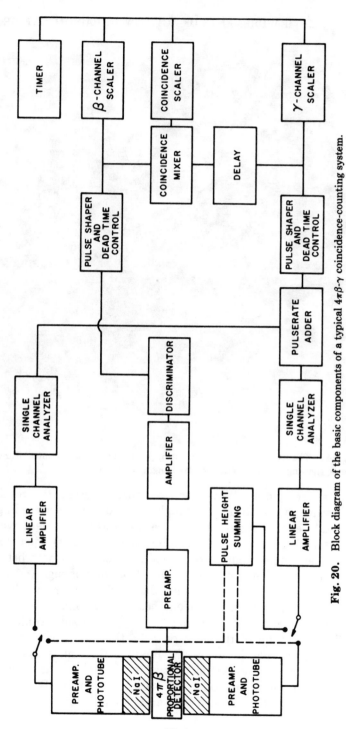

Fig. 20. Block diagram of the basic components of a typical $4\pi\beta$-γ coincidence-counting system.

77

Fig. 21. The $4\pi\beta$-γ coincidence-counting assembly at the National Bureau of Standards, showing the automatic sample-changing mechanism.

shaped, square pulses. Whenever two pulses overlap, the resultant pulse is twice the amplitude of a single one and thus, by exceeding the threshold level of a trigger circuit, is recorded as a coincident event.

The resolving time, τ_r, of the coincidence circuit is equal to the width of the square input pulses, usually set equal for both beta and gamma channels. This pulse width must be long enough ($\sim$1 μs) to avoid the loss of coincidence counts due to fluctuations in the time delays in the arrivals of coincident beta and gamma pulses at the coincidence mixer circuit. This time "jitter" is due to electronic effects as well as variation of the electron collection time in proportional counters. Errors may be incurred (Gandy, 1961) unless the relative delays between beta and gamma pulses are adjusted to a net zero mean delay (Williams and Campion, 1965; Munzenmayer and Baerg, 1969).

Chance overlap of unrelated beta and gamma pulses leads to accidental coincidences. Their rate of occurrence is directly dependent on the pulse widths, which therefore should be kept to a minimum, consistent with avoiding losses due to time jitter.

When the half life for gamma decay of the excited state of the daughter nucleus is comparable to, or longer than the selected minimum pulse width or resolving time of about 1 μs, serious loss of coincidence events will occur. In principle, it is still possible to use direct methods of counting the coincidences if the pulse widths are increased. For example, in the electron-capture decay of ^{85}Sr to the 1-μs delayed, 514-keV state of ^{85}Rb, excessive loss of coincidence events could be avoided by increasing the resolving time to, say, 10 μs or more. However, the correction for accidental coincidences would be increased correspondingly. It is, therefore, considered preferable to use either anticoincidence techniques or correlation counting methods for cases of delayed gamma-ray emission, and to reserve direct coincidence counting methods for decays in which the parent-daughter radiations occur in prompt succession ($< \sim 10^{-7}$ s).

3.2.3 AntiCoincidence Counting

With anticoincidence counting, those gamma events not correlated with detected beta events are counted and the coincidence rate can then be obtained from the difference between the total and uncorrelated rates. The approach leads to important advantages in eliminating instrumental effects and dealing with delayed nuclear states. Corrections to observed counting rates can be significantly reduced, simplified or eliminated, depending on which of several methods is used:

1) *Anticoincidence counting* as introduced and envisaged by Bryant (1962, 1967) (see also Chylinski *et al.*, 1972) offers the appealing advantage over coincidence counting that accidentals are completely excluded and that only corrections for dead times and backgrounds are required.

In the application of Bryant's anticoincidence method to prompt β-γ decay, only those gamma signals that precede or follow signals in the beta channel by some minimum time T, are accepted as anticoincident gamma events. This is achieved by delaying the gamma signals by a fixed time T, and closing an anticoincidence gamma-channel gate for a fixed time $2T$ after each beta signal.

Using the method for beta decay followed by delayed gamma emission (Bryant, 1967), it is only necessary that, for each beta signal, the anticoincident gamma-channel gate remain closed for a further period τ beyond the time $2T$ required for prompt β-γ decay. By making $T + \tau$ equal to m half lives of the delayed gamma radiation, the probability of recording unwanted "coincidence" events may be reduced to any desired practicable limit, 2^{-m}.

In the treatment given by Bryant (1962, 1967) the individual beta- and gamma-channel counting rates are corrected for deadtime losses and backgrounds in the usual way but correction for the anticoincidence channel is not so well understood. He gives the approximate expression

$$N_A = \frac{N_A'(1 + N_\beta\tau_\beta)}{\exp[-N_\beta(2T + \tau)] - N_A'\tau_\gamma(1 + N_\beta\tau_\beta)} \quad (3.2)$$

where N_A, N_β are the dead-time corrected anticoincident and beta-channel counting rates and where primes indicate uncorrected rates, while τ_β and τ_γ are, respectively, the beta- and gamma-channel nonextending dead times.

2) *Live-timed anticoincidence counting* with extending dead-time circuitry (Baerg *et al.*, 1976) eliminates essentially all corrections due to instrumental effects. The method consists in first delaying the gamma signals, relative to the beta events, by an amount greater than any intrinsic β-γ time jitter and beta-pulse delay. Then both beta- and gamma-logic signals are combined so that either type of pulse will trigger a dead-time generating circuit of the extending kind. This extending dead time, τ, is adjustable and made to have a value longer than any overall β-γ jitter and delay (*e.g.*, $\tau \geq 2 \mu s$). The output pulses from this circuit are then applied as common gating signals to close simultaneously a livetiming clock channel and the beta and the gamma channels. A slight delay ($\approx$5ns) in the gating allows the original beta or gamma pulse which initiated the extending dead time, to proceed through its counting channel to be recorded before the gates are closed.

This arrangement ensures that only gamma events not correlated with any beta channel events can be recorded in the gamma channel. This follows from a basic characteristic of an extending dead time, namely that no other pulse (beta or gamma) can have been detected in the interval τ preceding the observed pulse. It may also be stated that no beta-correlated gamma events can be recorded because the delay of the gamma signal ensures that the gamma channel would already be blocked by a gating signal initiated by a correlated beta event.

The live timer, gated by the same pulses which control the beta and gamma counting channels, measures cumulatively the time intervals of the combined β-γ series. The cumulated count in each channel divided by the cumulated live time gives the correct counting rate for that channel. In contrast with coincidence counting, no corrections are needed for dead-time and resolving-time effects, nor do β-γ time-jitter effects arise. Normal background and decay corrections only are required, and, delayed gamma emitters (*e.g.*, ^{85}Sr) may be treated simply by increasing the value of the adjustable extending dead time, τ.

3) *"Selective sampling"* is an interesting method recently introduced by Müller (1981). It is also based on the use of an extending dead time τ (*e.g.*, $\tau = 25$ μs), which is inserted artificially into the sequence of detected beta events. From the basic characteristics of an extending dead time, whenever a beta pulse is registered, it is assured that no other event has been detected in the beta channel within the time interval τ, which preceded the observed beta pulse. Therefore any gamma event registered in this time zone is uncorrelated, and cannot have a partner in the beta channel, at least not for prompt β-γ decay.

The basic experimental arrangement offered by Müller for recording the gamma events which precede a registered beta event involves use of a multichannel analyzer operated in time mode. The gamma pulses are first delayed (*e.g.*, ~60 μs) relative to the beta events and then these gamma-pulse arrival times are accumulated by the analyzer which is started by a registered beta event. This procedure of "selective sampling" is repeated many times to yield the arrival-time distribution for the gamma events. This distribution shows two main regions of interest. First, a region of comparatively high gamma count rates per channel in which all detected gamma events are recorded. Secondly, there is a region, corresponding to the 25-μs period, equal to the extending dead time, preceding the initiating beta event, in which only uncorrelated gamma events are registered.

From the mean number of gamma count rates per channel for the two regions of the arrival-time distribution, one can derive an experimental ratio, R_γ, of uncorrelated-to-total count rates for the gamma channel. The coincidence-to-total ratio, N_c/N_γ, is, by difference, $1-R_\gamma$ and again all problems associated with coincidence-circuit resolving times are absent, including delay-matching, time-jitter and random-coincidence effects. In the case of delayed gamma emission (*e.g.*, ^{85}Sr) the transition zone in the arrival-time distribution is merely changed in shape to conform with the decay function for the delayed state.

For evaluation of the source disintegration rate, the observed beta-channel count rate, as well as the gamma rates determined from the distribution on the multichannel analyzer, still need correction for backgrounds, decay and dead-time losses. Müller also notes that the slowness of commercial multichannel analyzers for this application can be circumvented by using an intermediate data storage technique.

3.2.4 *Correlation Counting*[4]

The underlying principle of correlation counting is that there is a correlation between parent-daughter decays, irrespective of whether the daughter decay is prompt or not. The method also uses the fact that, in a Poisson distribution, the variance is equal to the mean. Determination of the covariance between the two series of parent-daughter decays, measured separately in many equal and simultaneous short intervals of time by two detector systems, provides a measure of coincident or time-correlated events, without the use of a coincidence circuit and, hence, without the limitation of its resolving time. This was first pointed out in 1955 in a Russian publication (Goldansky *et al.*, 1955). The linguistic barrier was not, however, crossed until 1964 (Friedländer, 1964), and the first definitive experiments using these so-called correlation techniques were performed by Williams and Sara (1968). These experiments showed that the coincidence and correlation-counting methods gave good agreement in measuring the activity of ^{60}Co sources, although the dead-time corrections and imprecision were somewhat larger, for the same source measured for equal times, in the correlation method. The power of the correlation method, like that of anti-coincidence counting, lies in the measurement of activity where there is a significant time delay between the parent and daughter decays. The method has also been applied to the measurement of half-lives that are too long for measurement by the delayed-coincidence method (Petroff and Doggett, 1956; Foglio Para and Mandelli Bettoni, 1970), and has been used to detect the presence and number of spurious pulses in a detector consequent upon a real pulse (Campion, 1973; BIPM, 1976).

---

[4] The NCRP is indebted to J. W. Müller of the Bureau International des Poids et Mesures for making available a preprint of his report (Müller, 1975) on which much of this section is based.

Only a brief summary of the theory of correlation counting will be given in this report, and the interested reader is referred to the literature references, given in this section, for further details.

Consider the general case of prompt intermediate states with different radiations, from parent and daughter nuclides, detected by two separate counters. The series of detected events, p and d, are stochastic and the distribution of each series remains the same as time progresses. The two series are measured separately for identical time intervals and their covariance is given by

$$\text{cov}(p, d) = \text{E}[(p - \langle p \rangle)(d - \langle d \rangle)] = \langle pd \rangle - \langle p \rangle \langle d \rangle, \quad (3.3)$$

where E(f), or $\langle f \rangle$, denotes expected value, and $\langle p \rangle$ and $\langle d \rangle$ are the expected values of p and d in a given time t, which is usually quite short. (The expected, or expectation, value is the average value of *all possible* values weighted by their respective probabilities.)

In the case of a radionuclide emitting beta particles and *prompt* gamma radiation we can say that, in the given time interval t,

$$p = b + c \quad \text{and} \quad d = g + c, \quad (3.4)$$

where b and g are the number of uncorrelated pulses detected in the beta and gamma channels, respectively, and c is the number of true coincidences in each interval t. For simplicity, dead-time corrections will be ignored. Then b, g, and c are independent, and one obtains

$$\text{cov}(p, d) = \text{E}[(b + c)(g + c)] - \langle b + c \rangle \langle g + c \rangle$$
$$= \langle c^2 \rangle - \langle c \rangle^2 = \text{var}(c). \quad (3.4)$$

As the coincidences have a Poisson distribution, $\text{var}(c) = \langle c \rangle$ (Appendix B.3.5) and, therefore,

$$\text{cov}(p,d) = \langle c \rangle. \quad (3.6)$$

An experimental value of the $\text{cov}(p,d)$ can be equated to the average value $\bar{c} = \epsilon_p \epsilon_d N_0 t$, where N_0 is the activity of the source, and $\epsilon_p \, (= \bar{p}/N_0 t)$ and $\epsilon_d (= \bar{d}/N_0 t)$ are the respective detection efficiencies of the detectors of the parent and daughter radiations, and $\bar{p}$ and $\bar{d}$ are the experimental average values. Therefore

$$N_0 t = \overline{pd}/\bar{c}. \quad (3.7)$$

The experimental value for the covariance is obtained from n readings (p_i, d_i), each of duration t, by using the expression

$$\text{cov}(p,d) = \frac{1}{n-1}\left[\sum_1^n p_i d_i - \frac{1}{n}\left(\sum_1^n p_i\right)\left(\sum_1^n d_i\right)\right]. \quad (3.8)$$

For zero dead time, it may be shown (*e.g.*, Williams and Sara, 1968; Williams *et al.*, 1973) that

$$\frac{\text{Var}(c)}{\bar{c}^2} = \frac{1}{n\bar{c}} + \frac{1 + \epsilon_p\epsilon_d}{n\epsilon_p\epsilon_d}. \tag{3.9}$$

It is of interest to note that the first term on the right-hand side of this equation is merely the Var $(C)/C^2$ for the ordinary coincidence method where $\bar{C}$ is equal to $n\bar{c}$, the total average number of coincidences for a total counting period nt. It is, therefore, apparent that, for equal total coincidence counts, the correlation method, because of the second term on the right-hand side of Eq. 3.9, cannot be as precise as the coincidence method.

It is also of interest to note that the second term on the right-hand side of Eq. 3.9 reduces to $(1/\epsilon_p\epsilon_d + 1)n^{-1}$. It is, therefore, important for ϵ_p and ϵ_d to be large in order to achieve higher relative precision in the determination of $\bar{c}$.

Using the relationships $\epsilon_p = \bar{p}/N_0t$, $\epsilon_d = \bar{d}/N_0t$, and $\bar{c} = \bar{pd}/N_0t$, Eq. 3.9 reduces to

$$\frac{\text{Var}(c)}{\bar{c}^2} = \frac{1}{n}\left(\frac{\bar{pd}}{\bar{c}^2} + 1\right) + \frac{1}{T(\bar{c}/t)}, \tag{3.10}$$

where T, equal to nt, is the total counting time and $\bar{c}/t$ is the mean coincidence count rate. As the mean coincidence count rate and the ratio of the mean count rate $\bar{pd}/\bar{c}^2$ do not depend on the time interval t, the relative standard deviation of c, namely $\sigma(c)/\bar{c}$, is made smaller by making both n and T large. Therefore, at reasonable count rates, the duration, t, of the individual counting times is kept short, and the number, n, of runs is kept high. Thus, in practice, t may be 100 μs and n may be 10^6.

In general, the conventional method of coincidence counting for the standardization of radionuclides having long daughter lifetimes is not to be recommended because the coincidence resolving time must be increased to include a reasonable proportion of the delayed coincidences, but this increase will also increase the relative size of the correction necessary for accidental coincidences.

Accidental coincidences do not occur in the correlation method, so that, in the case of delayed daughter decays, the time interval t can be increased to include larger fractions of the daughter decays. Thus, in the measurement of the activity of ^{75}Se, which, in one branch, feeds a 17.5-ms 304-keV intermediate state in ^{75}As, Lewis *et al.* (1973) set t equal to 45ms. In such cases, the covariance equation to be identified with the coincidence count becomes

$$\text{Cov}(p,d) = \epsilon_p \epsilon_d N_0 t \left(1 - \frac{1 - e^{-\lambda t}}{\lambda t}\right), \tag{3.11}$$

where λ is the decay constant of the intermediate state.

3.2.5 Count-Rate-Dependent Corrections

The count-rate-dependent corrections for the combined effects of dead time and resolving time in coincidence counting have been rather thoroughly studied. The corrections for dead time are not yet so well understood for correlation counting or for the earlier applications to anti-coincidence counting. These corrections do not arise with live-timed anti-coincidence methods. Here the correction formulae used for the more familiar coincidence counting are outlined.

The count rates observed for the beta and gamma channels are corrected for non-extending dead time in the well-known manner, namely

$$N'_\beta = \frac{N''_\beta}{1 - N''_\beta \tau} \quad \text{and} \quad N'_\gamma = \frac{N''_\gamma}{1 - N''_\gamma \tau}, \tag{3.12}$$

where single primes indicate corrected rates, still including backgrounds, while double primes indicate observed rates and τ is the dead time, usually set equal for both beta and gamma channels. The corrections which are to include resolving-time effects in the coincidence channel are more complicated (Campion, 1959; Gandy, 1961, 1962; Bryant, 1963). Although various formulae in current use are of comparable accuracy, the one given by Bryant is widely used and has the form,
N'_c

$$= \frac{(N''_c - 2\tau_r N''_\beta N''_\gamma)[1 - (N''_\beta + N''_\gamma)\tau/2]}{(1 - N''_\beta \tau)(1 - N''_\gamma \tau)[1 - (N''_\beta + N''_\gamma)(\tau + 2\tau_r)/2 + N''_c \tau]}, \tag{3.13}$$

where τ_r is the coincidence resolving time. Recent detailed analyses by Cox and Isham (1977) and Smith (1978), for conditions which can be closely approximated in practice, offer an exact but more elaborate solution for coincidence and correlation counting. This may be needed for high accuracy or when counting rates are very high or dead times long.

Background rates may be subtracted directly from the count-rates calculated from equations such as 3.12 and 3.13, provided source decay during the measurement is negligible. The corrections are more complicated when decay is significant, as is the decay correction itself as a result of the dead-time effect (Axton and Ryves, 1963). The decay correction for live-timed anti-coincidence counting with extending dead time is no more complicated (Baerg et al., 1976). It should be noted that dead time, decay, and backgrounds also influence the estimate of variance of the corrected counting rates (Baerg, 1973a).

3.2.6 Efficiency-Dependent Effects

With any practical counting system at least one of the detectors, usually the beta detector, is not exclusively sensitive to one type of radiation. Contributions to the beta-counting rate also arise from conversion electrons, and the simple expressions of Eq. 3.1 must be further modified for complex beta-gamma decay (Steyn and Haasbroek, 1958; Campion, 1959). For simple two-stage beta-gamma decay the expressions for the counting rates, already corrected for background and rate-dependent effects, become

$$N_\beta = N_0 \left[\epsilon_\beta + (1 - \epsilon_\beta) \left(\frac{\alpha \epsilon_{ce} + \epsilon_{\beta\gamma}}{1 + \alpha} \right) \right]$$

$$N_\gamma = N_0 \epsilon_\gamma / (1 + \alpha) \tag{3.14}$$

$$N_c = N_0 [\epsilon_\beta \epsilon_\gamma / (1 + \alpha) + (1 - \epsilon_\beta) \epsilon_c],$$

where ϵ_{ce}, $\epsilon_{\beta\gamma}$ are respectively the beta-counter detection efficiencies for conversion electrons and gamma rays and α is the conversion coefficient, while the second term in the expression for N_c represents the probability of observing additional coincidences that are not derived from coincident beta-gamma events, background, or accidental coincidences. The probability of observing such additional coincidences is the product of two probabilities, namely (i) that of the beta event being undetected; and (ii) that, ϵ_c, of simultaneous events occurring in both detectors, due, for example, in simple beta-gamma decay, to Compton scattering from one detector to another. An effect of this kind does not arise with anti-coincidence counting where the rate, Y, is simply

$$Y = N_0 (1 - \epsilon_\beta) \frac{\epsilon_\gamma}{1 + \alpha}.$$

Then the rate for coincidences may be obtained by difference,

$$N_c = N_\gamma - Y.$$

The considerable advantage to using a $4\pi\beta$ counter is that N_0 in Eqs. 3.14 reduces to $N_\beta N_\gamma / N_c$ when the beta-detector efficiency approaches unity. Additional terms due to complex branching are of a similar efficiency-dependent character (see next subsection) so that small corrections can sometimes be estimated to an acceptable accuracy by using available nuclear constants, and measured (Merritt and Taylor, 1967b) or calculated (Urquhart, 1967) values for $\epsilon_{\beta\gamma}$. Other effects include beta-gamma angular correlations (Hayward, 1961), non-uniform source efficiency (Putman, 1953), beta sensitivity of the gamma detector and efficiency instabilities (Campion, 1959). These may, however, usually be ignored when a $4\pi\beta$ proportional counter is used, along with some care in selecting gamma-channel energy response.

While the $4\pi\beta$ proportional counter is the most widely used, and is the only type leading to valid fundamental measurements for many radionuclides, especially those with complex decay, there are cases for which low-efficiency beta-particle detectors may be advantageously used. One example is that of ^{129}I, which decays by β-emission ($E_{\beta\text{max}} = 0.150$ MeV) to the 39.58-keV level in ^{129}Xe. This in turn decays with the emission of 39.58-keV gamma rays and a preponderance of K-shell conversion electrons, thus giving rise to xenon x rays of energies between 29 and 34 keV. Therefore these K x rays also signal transitions from the 39.58-keV state in ^{129}Xe. Mann and Hutchinson (1976) have used as the beta-particle and photon detectors, respectively, thin (0.019-inch) plastic-scintillator and thin ($\sim\frac{1}{16}$-inch) NaI(Tl)-crystal phototube assemblies, the latter with a 0.005-inch beryllium window that, together with the intervening air, is thick enough to absorb all beta particles. Sufficient absorber material is also placed between the source and beta-particle detector to absorb all conversion electrons. Because the gamma rays and x rays have energies well below that of $E_{\beta\,\text{max}}$, those few photons that do interact in the thin plastic scintillator are excluded by pulse-height discrimination. Under these special conditions the measurement is a simple one of determining the beta-particle, photon, and coincidence counting rates, inserting these quantities in Eqs. 3.12 and 3.13, and solving for $N_0 = N'_\beta N'_\gamma / N'_c$.

3.2.7 Complex Beta-Gamma Decay

When, as is often the case, the nuclear decay proceeds through two or more alternative beta branches, simple equations like 3.1 are not adequate. Instead, for coincidence measurements on a nuclide with n beta branches, each of which may be followed by prompt gamma emission, sums of equations like 3.14 are needed to describe the observed counting rates. Thus

$$N_\beta = N_0 \sum a_r \left[\epsilon_{\beta_r} + (1 - \epsilon_{\beta_r}) \left(\frac{\alpha \epsilon_{ce} + \epsilon_{\beta\gamma}}{1 + \alpha} \right)_r \right],$$

$$N_\gamma = N_0 \sum a_r \frac{\epsilon_{\gamma_r}}{1 + \alpha_r}, \tag{3.15}$$

$$N_c = N_0 \sum a_r \left[\frac{\epsilon_{\beta_r} \epsilon_{\gamma_r}}{1 + \alpha_r} + (1 - \epsilon_{\beta_r}) \epsilon_{c_r} \right],$$

and

$$Y = N_0 \sum a_r \left[(1 - \epsilon_{\beta_r}) \left(\frac{\epsilon_{\gamma_r}}{1 + \alpha_r} \right) \right],$$

for anti-coincidence gamma counting. In these equations the summations extend over the n branches, and a_r is the fractional branching for the r-th branch.

The majority of quantities on the right of Eqs. 3.15 are not usually known with sufficient accuracy to extract a reliable value for N_0 directly from the observed counting rates on the left. In the general case, it is necessary to choose detectors and experimental conditions to provide additional information to solve these equations. Methods have been developed that do not require direct knowledge of nuclear decay-scheme constants or estimates of branch counting efficiencies. However, only an outline of the theoretical analyses leading to these methods will be given here. The reader interested in applying them should refer to the literature on the subject (*e.g.* Baerg, 1973c, and other references cited there).

The primary requirements for reducing Eqs. 3.15 to a practical formula are that the beta detector is one of 4π geometry and that the beta detection efficiencies for the various branches are functionally related. Such an interdependence derives from the fact that all branches are subjected to the same physical conditions responsible for counting losses. Then, as an example, the beta efficiency of the r-th branch, given as a function g_r (usually a polynomial) of that for the s-th branch, is

$$\epsilon_{\beta_r} = 1 - g_r(1 - \epsilon_{\beta_s}) \to 1, \qquad \text{as} \quad \epsilon_{\beta_s} \to 1. \tag{3.16}$$

Using expressions like this, not only for the ϵ_{β_r} but also for the ϵ_{ce}, $\epsilon_{\beta\gamma}$,

ϵ_{c_r}, and ϵ_{γ_r} if these are not constant, it is then easily shown that Eqs. 3.15 reduces to the following simple coincidence formula:

$$N_\beta = N_0 f(N_c/N_\gamma) \rightarrow N_0 \quad \text{as} \quad N_c/N_\gamma \rightarrow 1. \qquad (3.17)$$

(cf. Houtermans and Miguel, 1962; Williams and Campion, 1963; Baerg, 1966). The function, $f(N_c/N_\gamma)$ is of unspecified general form, and, in practice, is defined simply as a polynomial in N_c/N_γ although in practice the functions are usually linear or of low order in this variable. The coincidence measurement then consists in measuring N_β, N_γ, and N_c for various levels of beta-detection efficiency, and extrapolating the function, that fits the data, to unit value for N_c/N_γ.

It is particularly significant that Eq. 3.17 is an expression in observables alone and does not depend directly on decay-scheme parameters. The effects of any such dependence are now absorbed in the parameters of the "efficiency function", $f(N_c/N_\gamma)$, and these are to be determined experimentally by varying the beta-detector efficiency over a range for which expressions like Eq. 3.16 are valid.

3.2.8 Experimental Techniques for Complex Beta-Gamma Decay

While for simple decay schemes beta detectors of various kinds are often used, for complex beta-gamma decay it is necessary that the detector be of 4π geometric efficiency (Baerg, 1966, 1967, 1973b). In the case of the $4\pi\beta$ proportional counter, it is also useful if it can be operated at elevated pressures (~25 atm; 2.54 MPa) (Baerg et al., 1967; Baerg, 1973b). In other respects, the counting system can be of conventional design (Section 3.2.1).

3.2.8.1 *Efficiency Variation.* Several methods can be used to vary the $4\pi\beta$-detector efficiency to derive a function that can be reliably extrapolated to unit detection efficiency to deduce N_0. It must be emphasized, however, that careful consideration must be given to ensure that conditions defined by equations like 3.16 are fulfilled. Detailed discussions on these matters appear in the references already noted. Briefly, however, acceptable methods are restricted to variation of the physical effects intrinsically responsible for counting losses, namely source self-absorption, source-mount absorption, and beta-channel discrimination level. With any of these methods, thin sources of the highest counting efficiency should be prepared but for self-absorption variation a series of sources is usually prepared, varying in the amount of added carrier. When source-mount absorption is to be varied, a single source is used and measurements made after each of successive additions of absorber foils. With beta-channel discrimination a single source is also used, but it is necessary to operate the 4π proportional counter at elevated pressure to ensure that counting losses occur, as a result of discrimination, only from the low-energy portions of the beta spectra (Baerg *et al.*, 1967; Baerg, 1973b). Measurements are then taken after each of successive changes in discrimination level. Alternatively, multi-channel analyzers can be used to acquire the efficiency-function data in a single measurement (Smith, 1975).

3.2.8.2 *Gamma-Channel Energy Response.* In acquiring the data, the gamma-channel energy response should be chosen to provide data sets that are as nearly linear as possible. Again, the criteria to be considered in this choice are discussed in detail in the references noted above. Usually a gamma-channel window, set to respond to the photopeak of the gamma ray associated with the main beta branch, would be appropriate. Data sets acquired with different gamma-channel responses are useful in searching for possible systematic error. As an example, results of measurements on a ^{134}Cs source are shown in Figure 22, using gamma-channel energy responses as noted. The several data sets should yield extrapolation values for N_0 consistent with defined uncertainties. If they do not do so, then it is likely that experimental requirements (defined by equations such as 3.16) are not properly fulfilled.

Fig. 22. Efficiency-variation plots obtained with differing gamma-channel gates for ^{134}Cs calibration, i.e., N_β, the beta-channel count rate, vs. N_c/N_γ ($=\epsilon_\beta$), the ratio of coincidence- to gamma-channel count rates.

3.2.8.3 *Multiple-Channel $4\pi\beta$-γ Counting.* Extension of $4\pi\beta$-γ counting systems for use of several different gamma channels offers the possibility of linearizing data sets for complex β-γ decay and so improving the reliability of extrapolation procedures. This concept was first introduced for coincidence counting in cases where all of n beta branches are associated with gamma radiation, each detectable in at least one of n or more different gamma channels (see Smith and Stuart, 1975, Bouchard and Vatin, 1977, and also Smith and Williams, 1971). Thus, for the n branches, a set of equations like 3.15 may be obtained, but expanded to include the n different expressions each for the gamma and coincidence rates (N_{γ_r} and N_{c_r}). These involve the n different beta-branch efficiencies, ϵ_{β_r}, in linear combination. Assuming constant values for the ϵ_{γ_r} and ϵ_{c_r} each of the ϵ_{β_r} is then defineable in terms of a linear combination of the observed $(N_c/N_\gamma)_r$ ratios. Consequently, with the further assumption that the values for ϵ_{ce} and $\epsilon_{\beta\gamma}$ remain fixed as the beta-channel efficiency is varied, N_β can also be written as a linear function in the n different $(N_c/N_\gamma)_r$ ratios,

$$N_\beta = N_0 - \sum_{r=1}^{n} C_r \left(1 - \frac{N_c}{N_\gamma}\bigg)_r\right) \to N_0, \quad \text{as} \quad \left(\frac{N_c}{N_\gamma}\right)_r \to 1,$$

where the C_r are constants. It would remain only to ensure the measurements conform to the conditions needed to satisfy the assumptions made and then to deduce the linear function empirically.

Another development (Baerg, 1981), dependent on fewer constraints, is also applicable to a broader range of decay modes including ground-state beta transitions (without gamma radiation) and pure beta emission in efficiency-tracing methods. The approach has been described for multiple gamma-channel anti-coincidence counting but is in principle equally available for coincidence methods. Use is made of a set of condition equations like 3.16 for all detection efficiencies. These are combined with equations like 3.15, expanded to include the equations for several different gamma channels (m in number) in an anti-coincidence arrangement. This leads to a new set of equations, polynomial in an arbitrary s-th-branch detection efficiency. Thus,

$$N_\beta = N_0 - \sum_{k=1}^{m} P_k(1 - \epsilon_{\beta_s})^k$$

where the P_k are constants, and for the anti-coincidence gamma channels,

$$\left.\begin{array}{c} Y_1 = \displaystyle\sum_{k=1}^{m} B_{k1}(1 - \epsilon_{\beta_\bullet})^k \\ \cdot \\ \cdot \quad \cdot \quad \cdot \quad \cdot \quad \cdot \\ \cdot \\ Y_m = \displaystyle\sum_{k=1}^{m} B_{km}(1 - \epsilon_{\beta_\bullet})^k \end{array}\right\} \rightarrow 0, \qquad \text{as} \quad \epsilon_{\beta_\bullet} \rightarrow 1,$$

for which the B_{ki} are constants. The statistically significant order, k, for all of these equations is expected to be the same and not to exceed m, the number of separate anti-coincidence gamma channels, as indicated by the summing limits. Then this set of equations for the Y_i provides a solution for each of the power series terms, $(1 - \epsilon_{\beta_\bullet})^k$, as a linear combination of the observations Y_i. Substituting in the equation for N_β, this becomes a function linear in the Y_i, namely

$$N_\beta = N_0 - \sum_{i=1}^{m} C_i Y_i \rightarrow N_0, \qquad \text{as the} \quad Y_i \rightarrow 0.$$

Analogous expressions for other forms of measurement follow, as, for example, with multiple-channel coincidence counting,

$$N_\beta = N_0 - \sum_{i=1}^{m} C_i' \left(1 - \frac{N_c}{N_\gamma}\right)_i \rightarrow N_0, \qquad \text{as the} \quad \left(\frac{N_c}{N_\gamma}\right)_i \rightarrow 1,$$

where C_i and C_i' are constants in the two preceding equations.

Experience shows that data obtained in coincidences or anticoincidence counting with but a single gamma channel, seldom leads to functions of statistically significant order greater than two or three. Therefore two or three separate gamma channels for complex decay should usually lead to linear data sets. It is only necessary, providing there is an adequate multiplicity of β-γ branching, to ensure that efficiency variation remains consistent with interdependence equations like 3.16.

3.2.8.4 *Data Analysis.* For simple β-γ decay graphical analysis of data sets, using linear extrapolation to unit value in N_c/N_γ, may sometimes be adequate but statistically valid extrapolation can only be made by way of iterative least-squares data fitting (Adams and

Baerg, 1967; Baerg, 1971). In the analysis, due account must be taken of errors on both the algebraically dependent and independent variables as well as correlations between them. In addition, the order of the fitted function should be determined solely by the statistical errors on the data through use of the χ^2 test for goodness of fit. The calculations are therefore rather elaborate and require the use of adequate computer facilities. Multiple-channel counting requires no significantly more difficult analytical procedures.

3.2.9 *Applications of Coincidence Counting*

Although the preceding discussions refer to beta-gamma decay, the coincidence methods are applicable to other modes of decay. Thus, the tracer method for pure beta emitters (Merritt *et al.*, 1960; Campion *et al.*, 1960; Baerg *et al.*, 1964; Williams, 1964) may be regarded as a variation on that for more complex beta-gamma decay, with the pure beta-emitter component of the mixed source considered as a ground-state branch. Electron-capture decay followed by prompt gamma-ray emission (*e.g.*, [54]Mn, Weiss, 1968) presents little difficulty except for very low-energy decay (*e.g.*, [7]Be, Taylor and Merritt, 1962) since the Auger-electron x-ray cascade, detected in the beta-counter, may usually be treated as single-branch beta decay in simple beta-gamma radioactivity. Alpha-gamma decay (*e.g.*, [241]Am, Rytz, 1964) may be treated in like manner.

Greater care is required for mixed decay involving branch spectra differing markedly in shape. This situation arises, for example, with electron capture and low-energy conversion electrons in [197]Hg (Baerg *et al.*, 1967; Taylor, 1967), [125]I (Taylor, 1967), [195]Au (Adams, 1969), [57]Co (Garfinkel and Hutchinson, 1966; Troughton, 1966; Williams and Birdseye, 1966; Baerg, 1973a) or in mixed electron-capture-positron decay (*e.g.*, ^{68}Ga, Smith and Williams, 1971).

Coincidence measurements on pure electron-capture decay (not accompanied by gamma-ray emission) are not yet in a well-developed state. Tracer methods and related techniques have been used (Petel and Houtermans, 1967; Steyn, 1967), as have direct coincidence measurements between x rays and Auger-photon cascades (Troughton, 1967), but rather elaborate corrections are required. Efficiency-variation methods appear to be possible (Baerg, 1973c) and coincidences with internal bremsstrahlung have also been considered (Spernol *et al.*, 1973).

More detailed procedures and reviews on coincidence measurements may be found in IAEA (1967a) and in Herceg Novi (1973).

3.3   Sum-Peak Coincidence Counting

If a radionuclide emits two photons in coincidence, with no direct transitions to the ground state, a measurement of its disintegration rate may be made in many cases by the method of sum-peak coincidence counting. So far, this method has been applied using a NaI(Tl) photon detector, with or without a well, or two NaI(Tl) crystals arranged in 4π geometry.

In the simplest application of this method (Brinkman *et al.*, 1963a; Eldridge and Crowther, 1964), one detector is used, to which the associated electronic circuitry is so connected as to record (i) the total number of photons interacting in the detector per unit time, N_t (integral counting rate), and (ii) the photopeak and sum-peak counting rates for the two photons of different energy.

Consider a radionuclide (Figure 23) that emits a photon, γ_1, in coincidence with a photon, γ_2, with photopeak efficiencies, ϵ_p' and ϵ_p'', respectively, and total detection efficiencies, respectively, ϵ_t' and ϵ_t''. Then the count rates under the γ_1 and γ_2 peaks, and the $\gamma_1 + \gamma_2$ sum peak, A_1, A_2, and A_{12}, respectively, can be expressed as follows:

$$A_1 = N_0\epsilon_p'(1 - \epsilon_t'') \tag{3.18}$$

and similarly,

$$A_2 = N_0\epsilon_p''(1 - \epsilon_t') \tag{3.19}$$

and,

$$A_{12} = N_0\epsilon_p'\epsilon_p'' \tag{3.20}$$

where N_0 is the disintegration rate.

The probability of recording neither γ_1 and γ_2 in coincidence is the

Fig. 23. Decay scheme for radionuclide that may be standardized by sum-peak coincidence counting.

product $(1 - \epsilon_t')(1 - \epsilon_t'')$. Therefore,

$$N_0 - N_t = N_0(1 - \epsilon_t')(1 - \epsilon_t''), \qquad (3.21)$$

whence, using Eqs. 3.18 to 3.20,

$$N_0 = N_t + \frac{A_1 A_2}{A_{12}}. \qquad (3.22)$$

N_t is determined by extrapolation of the integral count rate to zero discriminator bias.

The sum-peak method, using a single NaI(Tl) detector, has also been extended by Brinkman et al. (1963b, 1965) and Brinkman and Aten (1963, 1965) to the measurement of pure β^+ emitters and radionuclides with complex decay schemes. Harper et al. (1963) and Eldridge and Crowther (1964) have applied the sum-peak method to the assay of ^{125}I. This radionuclide decays by electron capture to the 35-keV state of ^{125m}Te, which then promptly decays to the ground state, either by gamma-ray emission or by electron conversion, the latter resulting in K x rays of ~29 keV. The pulse-height distribution (Figure 2, Eldridge and Crowther, 1964) from a NaI(Tl) crystal has an unresolved "singles" and a sum peak with areas A_I and A_{II}, respectively.

Eldridge and Crowther (1964) show that Eq. 3.22 can be cast in the form:

$$N_0 = \frac{P_1 P_2}{(P_1 + P_2)^2} \frac{(A_I + 2A_{II})^2}{A_{II}}, \qquad (3.23)$$

where P_1 is the probability per decay of the K x ray in the electron-capture transition, and P_2 is the sum of the probabilities per decay of the 35-keV gamma ray and the K x ray arising from internal conversion. The derivation of Eq. 3.23 from Eq. 3.22 is based on the assumption that there is negligible loss of energy by Compton scattering, so that $A_1 = A_2 = A_I/2$ and $N_t = A_I + A_{II}$. Using $P_1 = 0.68$ and $P_2 = 0.75$, $P_1 P_2/(P_1 + P_2)^2 = 0.249$ and is quite insensitive to possible uncertainties in P_1 and P_2.

Sutherland and Buchanan (1967) have pointed out that incorrect results for N_0 are obtained by the sum-peak method if the detection efficiency for photons emitted from different parts of the source is not a constant. For example, they demonstrate that a sum-peak measurement performed on a 25-ml sample containing ^{60}Co in a 25-mm-diameter vial placed on a 3-inch × 3-inch NaI(Tl) detector results in a value for the activity that is about 25 percent in error due to this effect.

Except in the case of very low-energy gamma-ray emitters (Eldridge and Crowther, 1964), the requirement for an accurate estimate of the summed Compton recoils, which must be subtracted from under the "singles" photopeaks, is the limiting factor in the attainment of high accuracies.

An alternative approach (Hutchinson *et al.*, 1973a), utilizing two large crystals in 4π geometry (Figure 87) and operated in the sum, coincidence, and anticoincidence modes, is capable of producing accuracies comparable to the coincidence method. The chief reasons for this are that (i) the total efficiency, N_t/N_0, is extremely high, and therefore, many subtle corrections, such as for angular correlation, are negligible; and (ii) the possibility of operating the two crystals in the various modes permits the accurate separation of total absorption events from otherwise indistinguishable coincident Compton events under the "singles" peaks.

The experimental arrangement for this method is shown in Figure 24. Figure 25 shows superimposed spectra of ^{94}Nb collected in the coincidence and anticoincidence modes. The three peaks correspond to 0.70- and 0.87-MeV "singles" peaks and the summed (0.70 + 0.87)-MeV peak respectively. Because each gamma ray has an equal chance of scattering from the top or bottom detector, the Compton summed continuum is the same for each mode. Thus the γ_1 and γ_2 photopeak count rates in the anticoincidence mode are obtained by means of a subtraction of the coincidence from the anticoincidence spectra. A few "singles" are recorded in the coincidence mode due to gamma rays backscattered from one detector into the other, where they lose the remainder of their energy, with the coincident gamma rays of each such "single" escaping detection. A measure of such "singles" events can be obtained from measurements of monoenergetic gamma-ray-emitting radionuclides.

To take advantage of the accurate measurement of the photopeaks $^aA_1$ and $^aA_2$ now possible in the anticoincidence mode it is necessary

Fig. 24. Block diagram of the experimental arrangement and electronic circuitry used for sum-peak coincidence counting. The gate can be turned *off* to give the summing mode configuration, or *on* to give the coincidence or anticoincidence mode depending on whether the gating pulse is used to pass or block, respectively, the input pulse.

Fig. 25. Superimposed coincidence and anticoincidence spectra of ^{94}Nb, taken for the same live times.

to recast Eq. 3.22 in terms of measured quantities. It can be shown that, when this is done (Hutchinson *et al.*, 1973a), Eq. 3.22 becomes

$$N_0 = N_t + \frac{^aA_1\,^aA_2}{2\,^aA_{12}} \tag{3.24}$$

where N_t is still measured in the sum mode with no coincidence or anticoincidence conditions and $^aA_{12}$ is the count rate under the sum peak in the anticoincidence mode.

Further examples of the application of the sum-peak coincidence method with two large NaI(Tl) crystals in 4π geometry, to the assay of ^{22}Na, ^{26}Al, ^{60}Co, ^{88}Y, and ^{207}Bi are given in Hutchinson *et al.* (1973a).

3.4 Photon-Photon Coincidence Counting

Although the technique of gamma-gamma coincidence counting is more limited in application than that of beta-gamma coincidence counting, it has the advantage that sources in the form of wires or pellets can be readily standardized without the preparation of thin samples such as are required for particle counting. Also, standardizations may be made by this method on gamma-ray-emitting radionuclides in sources containing beta-ray-emitting impurities, which would make beta-gamma coincidence counting unusable.

In the gamma-gamma coincidence method there are the following important considerations: (a) the relative efficiency of each of the counters for the two gamma rays, (b) possible spurious coincidences due to the Compton scattering of a gamma ray from one counter into the other, (c) possible angular correlation of the gamma rays, and (d) the attenuation of this correlation due to the finite sizes of the source and counters. Treatments of the problem, for low detection efficiencies, have been given by Meyer et $al.$ (1959), Peelle and Maienschein (1960), and Kowalski (1965). A description of x-ray-x-ray coincidence counting of ^{125}I and ^{197}Hg, which includes a careful accounting for summing corrections, has been given by Taylor (1967). The method is illustrated by reference to its use in the standardization of ^{60}Co (Hayward et $al.$, 1955). Summing corrections, that may be significant at higher efficiencies, were not discussed in this paper.

The disintegration rate in terms of observable quantities, with second-order terms neglected, for a short channel dead time, and resolving time τ_r, as given by Hayward et $al.$ (1955) is

$$N_0 = \frac{(N_1 - N_{B1})(N_2 - N_{B2})}{N_c - N_{Bc}} F(\phi)f(\theta) \left(1 + 2\tau_r \frac{N_1 N_2}{N_c}\right), \quad (3.25)$$

where N_1 and N_2 are the counting rates in each of two similar detectors, N_c is the coincidence counting rate, and $f(\theta)$ is a function depending on the geometry and on any angular correlation between the coincident gamma rays (see Feingold and Frankel, 1955). N_{B1}, N_{B2}, and N_{Bc} are the respective background counting rates, and $F(\phi)$ is a correction factor, discussed below, which arises from the change in detection efficiency with gamma-ray energy.

Except when two photons differ greatly in energy (e.g., x rays and higher-energy gamma rays), it is impossible to make one detector sensitive only to the first radiation and the other detector sensitive only to the second. In the case of two gamma rays of nearly equal energy, such as occurs in the decay of ^{60}Co, it is convenient, by appropriate setting of the bias in each channel, to adjust the efficiencies to be nearly the same for each radiation, as determined by integration under the pulse-height-distribution curves. This can be done, as shown in Figure 26, where subtraction of the corrected pulse-height distribution from the 1.11554-MeV gamma ray occurring in the

Fig. 26. Calculated pulse-height distributions of 1.173 and 1.332-MeV gamma rays for a NaI(T1) detector system. When added, they give the distribution produced by ^{60}Co. In γ-γ coincidence counting of ^{60}Co, the integrated areas for the two spectra above the bias point are equalized by adjusting the bias voltage.

decay of ^{65}Zn, has been used to separate the pulse-height distributions of the two gamma rays due to ^{60}Co. The bias point is chosen so that the areas under the pulse-height-distribution curves, for each counter, are approximately equal. If ϕ_1 is the ratio of the areas for counter 1, and ϕ_2 is the ratio for counter 2, it is easily shown that the coefficient modifying the apparent disintegration rate is

$$F(\phi) = \frac{\phi_1 + \phi_2}{(\phi_1 + 1)(\phi_2 + 1)} \qquad (3.26)$$

When the ϕ's are equal to unity the coefficient is 1/2. If, for example, the ϕ's are adjusted to no better than 1.2, then the coefficient is 1/2.02; i.e., if a 20-percent error in setting the bias is made, a 1-percent error in the coefficient occurs. With reasonable care, therefore, good accuracy can be achieved.

The setting of the biases by the above criterion usually results in biases so high that spurious coincidences due to scattered gamma rays will be eliminated, except perhaps in situations where the angular separation of the two detectors with respect to the source is small. A 1-MeV gamma ray when scattered through an angle of 90° to 180° is degraded in energy to about 200 keV.

As in the beta-gamma coincidence method described in Section 3.2, the resolving time of the coincidence mixer circuit must be sufficiently long to avoid any loss of true coincidences.

3.5 4π Proportional Counting

3.5.1 *General*

Two decades ago one of the most widely used methods for the activity standardization of alpha- and beta-emitting nuclides was that of 4π counting. In its heyday the 4π counter was the preferred instrument for the assay of such radionuclides as ^{32}P, ^{35}S, ^{131}I, and ^{241}Am. When a range of values of 1 or 2 percent in international comparisons was considered to be good, many national laboratories reported results obtained by the method of 4π counting (see, for example, ICRU, 1963; Roy and Rytz, 1965, for ^{35}S; and Rytz, 1964, for ^{241}Am). Now, however, its chief use is as a high-efficiency detector in the method of $4\pi\beta$-γ coincidence counting.

As mentioned in Section 1.5.2, the 4π counter is a defined solid-angle counter, with the highest possible geometric efficiency. In the past it has been operated both in the Geiger and proportional regions, but now is used practically only in the latter (see Section 2.2) because the much lower dead times permit the use of high count rates, without large dead-time corrections.

In the method of 4π counting, a radioactive source, on a thin plastic film, is located at the center of symmetry of two detectors, each subtending a solid angle of essentially 2π steradians at the source.

The intrinsic efficiency of a well-designed 4π counter is assumed to be 100 percent for charged-particle radiation (Hawkins *et al.*, 1952; Seliger and Schwebel, 1954). Departures from 100-percent efficiency can be attributed to absorption by the source and the source mount.

Several designs of 4π counters have appeared in the literature, but two very acceptable designs are illustrated in Figures 27 and 28.

The National Bureau of Standards 4π counter, comprised of two stainless-steel hemispheres, is shown in Figure 27. The active source is deposited on the center of a metallized collodion film supported on

Fig. 27. National Bureau of Standards 4π proportional counter; spherical type. (From Mann and Seliger, 1958)

(a)

(b)

Fig. 28. Sketches of a simple slide-operated 4π counter adaptable to operation at elevated pressures.

an annular aluminum foil that divides the counter, when assembled, into two halves. Two loops of 0.001-inch-diameter wire, usually of stainless steel, form the anodes in each hemispherical section. Methane, or a mixture of argon (90 percent) and methane (10 percent), is used as a flow gas at atmospheric pressure.

Sketches of a "pill-box" type of 4π gas-flow proportional counter of a kind used at the National Research Council of Canada and other laboratories are shown in Figure 28. The source mount can be moved in and out of the counter by means of a simple slide, and the design is easily adaptable to use at pressures up to 75 atmospheres (7.6 MPa) (Baerg *et al.*, 1967; Baerg, 1973b), especially for use with 4πβ-γ coincidence counting systems (Section 3.2). The counter gas is usually a mixture of argon (90 percent) and methane (10 percent), purified by passing over heated granular calcium metal. For most beta-particle sources, the counting-gas pressure is maintained, typically, at 20 atmospheres (2.03 MPa).

As the source mount forms an integral part of the counter cathode, it must be sufficiently conducting to provide adequate field strength at the source and to prevent local accumulation of electric charge. At the same time the mount must be as thin as practical to reduce absorption losses. A common technique is the use of a thin plastic film, which can be made as thin as 1 or 2 $\mu g/cm^2$ (see, for example, Pate and Yaffe, 1955a), rendered conducting by the vacuum evaporation of a suitable metal onto the surface. Gold is commonly used as it is stable to most chemicals and can be readily evaporated; but other elements and alloys have been used. For routine work, a better film-survival rate is obtained with plastic films of 5, or even 10, $\mu g/cm^2$. Film-preparation techniques are described in Section 5.2.1.

The output pulse from proportional counters is such that external amplification is necessary. The pulse height may vary over a considerable range, perhaps three orders of magnitude, and hence, good overload properties may be required for the circuitry. The possibility of overload may be reduced by the use of a low-noise charge-sensitive preamplifier which permits the use of lower gas gain. As the noise level of such a preamplifier is equivalent to a pulse of about 200 electrons from the detector, the gas gain need not exceed a factor of 500 to detect even single electrons. Commercially available double-differentiating (RC or delay-line) main amplifiers can provide both good overload recovery and adequately short dead times (about 1 μs). If there is a variable dead time present anywhere in the counting system, it is usually overcome by artificially introducing a constant instrumental dead time that is longer than any possibly occurring dead time. It should be noted that for disintegration-rate measurements with, say, a $4\pi\beta$-proportional counter, it is not necessary for the amplifier to be linear, since an indication of an event is all that is required.

The measurement of the activity of a source is normally made by plotting a curve of counting rate either against the potential applied to the counter, keeping the discriminator bias level constant, or against the discriminator bias for a fixed counter potential. (The first alternative must be used with Geiger-Müller counters.) The plateau of a 4π proportional counter may extend for some 200 volts of the applied potential with a beta-particle source, and its slope gives some indication of an increase in detection efficiency with voltage. The rise at the end of the plateau appears to indicate the onset of spurious events (Campion, 1973). With alpha particles the plateau begins at much lower applied voltages, because the ionization produced is usually much greater than for beta particles, but it extends to essentially the same upper limit. The slope may be of the order of 0.1 percent per 100 volts. Operation at the center of the plateau is the usually accepted practice. The counting rate so obtained must be treated for corrections described in Sections 3.5.2 to 3.5.5 below.

3.5.2 *Dead Time*

The dead times of proportional counters are of the order of 1 μs. The subject of dead-time corrections is treated in detail in Section 2.7.

3.5.3 *Background*

The background count rate is readily obtained by using a blank source mount in the detector. The background rate is then subtracted from the observed counting rate *after* corrections for dead-time losses (see Section 2.7.3). Part of the background is due to natural activity in the materials from which the counter is constructed, and a useful survey of the "cleanliness" of such materials is given by Grummitt *et al.* (1956), DeVoe (1961), ICRU (1963), Weller (1964), and Rodriguez-Pasqués *et al.* (1972). However, the short dead times of proportional-counter systems permit such high counting rates to be used (with small dead-time losses) that the background-rate correction can be *relatively* small, although not necessarily negligible, when low-level samples are being counted.

3.5.4   *Source-Mount Absorption*

If necessary, a correction for the absorption in the source mount should be made. The analysis of Pate and Yaffe (1955c) is recommended for the most precise work when counting beta particles.

Results obtained by the method of 4πβ-γ coincidence counting (Section 3.2) now give us insight into the efficiency of the 4π proportional counter under different conditions of source preparation with different radionuclides.

3.5.5   *Source Self-Absorption Correction*

Apart from the method of β-γ coincidence counting (Section 3.2), it is impossible to determine the source self-absorption correction with an accuracy comparable to that for other corrections in 4π counting. The technique of progressively diluting a stock solution until the observed activity expressed in terms of a given mass of the original solution reaches a constant maximum value, merely indicates that no further reduction of source self-absorption is obtainable by dilution, rather than that the correction is zero. Calculated estimates based on the size of the crystals forming the active deposits and on absorption coefficients have been made by Meyer-Schützmeister and Vincent (1952) and Seliger and Schwebel (1954), the latter by electron-shadow micrography of potassium-iodide sources. Pate and Yaffe (1956) and Yaffe and Fishman (1960) have used a vacuum-distillation method to set up self-absorption curves applicable to this source-preparation technique. Carswell and Milsted (1957) have used a spray technique to produce very thin sources, but this technique does not permit quantitative deposition; it is possible, however, in the case of a gamma-ray emitter to derive the activity per unit mass of solution deposited on the source mount by a gamma-ray comparison with a quantitatively deposited source. Electrospraying techniques have been described by Merritt *et al.* (1959). Special methods of source preparation have also been discussed by Lowenthal and Wyllie (1973a). The preparation of uniform sources with low-self absorption is discussed further in Section 5.2.1. This question is treated in further detail for alpha, beta, and photon emitters in Section 3.5.6.

3.5.6   *Applications*

Proportional counters with 4π geometry have been used as high-efficiency detectors of alpha particles, beta particles, and electrons, and, at elevated pressures, of x rays. Some instances of such uses are given in the following subsections.

3.5.6.1 *4πα Proportional Counting.* Provided that the effective thicknesses of the source and source mount are below the range of the alpha particles in those media, good results can be obtained in the assay of alpha-particle sources. In a 1963 international intercomparison of measurements of ^{241}Am samples, dispensed from the same master solution, 8 of 19 participating laboratories used the method of 4π proportional counting (Rytz, 1964). The range of these eight measurements was 2.5 percent, but if one outlier is removed, the range is 0.85 percent for the remaining seven measurements. The average of measurements made by 12 laboratories using $4\pi\alpha$-γ coincidence counting was 0.4 percent higher than the average of the $4\pi\alpha$ measurements, a result indicating $\approx$0.4-percent self-absorption in the latter measurements, and a detection efficiency of $\approx$99.6 percent.

3.5.6.2 *4πβ Proportional Counting.* In the assay of negatron and positron emitters the problems encountered arise chiefly from source absorption. Auger electrons can also be detected with low efficiency but are usually removed by the interposition of thin metal foils.

Twenty years ago when $4\pi\beta$ counting was probably the most widely used method in radioactive metrology, the treatment of source self-absorption was largely empirical. In the international intercomparison of ^{35}S measurements, the National Bureau of Standards, for example, relied on preparing large numbers of sources, some 50 in number, and reporting the maximum value on the assumption that the source for this value suffered negligible absorption (Roy and Rytz, 1965). Seliger *et al.* (1958) used 121 different sources to measure the ^{35}S activity in the determinations of the average beta-particle energy in the decay of ^{35}S. An alternative method was that used by Seliger and Schwebel (1954), mentioned earlier, in which electron-shadow micrography was used to estimate the average potassium-iodide-crystal thickness to derive an absorption correction for ^{131}I sources. The state of the art at that time is well exemplified by the papers of Pate and Yaffe (1955a, b, c, and d; 1956).

With the advent of $4\pi\beta$-γ coincidence counting, a powerful tool was made available whereby the value of ϵ_β, the total efficiency for detecting beta particles, could be simply derived from the ratio of the beta-channel count rate, N_β, to the disintegration rate, N_0. Gunnink *et al.* (1959) and Merritt *et al.* (1959) used the method to measure the source self-absorption for beta particles emitted from several radionuclides. The latter authors found that, for sources having a mean superficial density of 2 μg cm^{-2}, the source self-absorption varied from about 1 percent for ^{24}Na beta-particles (β_{max} = 1.39 MeV) to between 20 and 40 percent for the 89-keV beta transition in ^{134}Cs, depending on the method of source preparation. An appreciable decrease in self-absorption was found by using spreading agents (e.g., insulin as a wetting agent, and colloidal silica as a seeding agent in the crystallization process), particularly for the less energetic beta-particle emitters. (See also Section 5.2.1.)

By incorporating two radionuclides into the same chemical compound, one a beta emitter and the other a previously standardized beta-gamma emitter, Merritt et al. (1960) have used $4\pi\beta$-γ coincidence counting to determine the source self-absorption for ^{35}S. (See also Section 3.2 and Campion et al., 1960.) Subsequently, this variant of the efficiency-tracing method was extended by Baerg et al. (1964) to two radionuclides that were not amenable to formation of a single chemical compound. The mixed efficiency-tracing method was successfully applied by Baerg and Bowes (1971), Merritt and Taylor (1971), and by Lowenthal et al. (1973) to intercomparative measurements of ^{63}Ni that had been standardized at the National Bureau of Standards by microcalorimetry (Barnes et al., 1971).

3.5.6.3 *4πx Proportional Counting.* A 4π-proportional counter, filled with a mixture of 90-percent argon and 10-percent methane at pressures ranging from 1 to 3.5 atmospheres, was used by Allen (1957) to measure the emission rates of x rays from electron-capturing nuclides. By plotting count rate as a function of reciprocal pressure, and then extrapolating to zero reciprocal pressure, he was able to determine the x-ray-emission rate for infinite pressure, for which the counter was assumed to be 100-percent efficient. From the count-rate reciprocal-pressure curves the K-x-ray detection efficiency could then be deduced for one atmosphere. Allen measured K-x-ray-emission rates for a number of electron-capturing nuclides, absorbing out residual Auger electrons with about 200 μg cm^{-2} of aluminum foil, and he also measured, and corrected for, the absorption of x rays in these foils.

From a knowledge of the K-shell fluorescence yield (Figure 29), the K-capture rate can be calculated. Theoretical or experimental values of the K/L, K/M, etc., capture ratios can then be used to calculate the total electron-capture activity. Taylor and Merritt (1963) combined high-pressure 4πx proportional counting with 4π(e,x)-γ coincidence counting of ^{51}Cr, ^{54}Mn and ^{65}Zn to obtain accurate ω_K values for V, Cr and Cu which are included in Figure 29.

Garfinkel and Hutchinson (1962) have also used a pressurized 4πx counter to measure source self-absorption in the determination of K-x-ray-emission rates from radionuclides that decay by electron capture

Fig. 29. Fluorescence K-shell yields, ω_r, as a function of atomic number, Z: (a) according to Kostroun et al. (1971); (b) a best fit to selected experimental data; and (c) critically evaluated experimental results. (From Bambynek et al., 1972)

to the ground state. In this method, in which the count rates in the two halves of the 4π counter are measured separately, Auger electrons are used as a tracer in order to determine the K-x-ray absorption to ±0.3 percent for ^{55}Fe sources. Because the Auger-electron efficiency is a much more sensitive function of source thickness than K-x-ray efficiency, a rough measure of the fraction of absorbed Auger electrons is shown to give an accurate measure of the fraction of absorbed K x rays. Lowenthal et al. (1982) have measured the activity of sources of ^{55}Fe using a 4π proportional counter, the x-ray Auger-electron efficiency of which was derived from the corresponding efficiencies of the neighboring nuclides ^{51}Cr and ^{54}Mn.

3.5.6.4 *$4\pi e^-$ Proportional Counting.* The use of a 4π counter in the measurement of atomic-electron radiations is usually confined to that of internal-conversion electrons. In this case the problems of source self-absorption are not normally so great as there is no low-energy component of conversion electrons. It is often desirable, however, to obtain the complete electron spectrum, as, for example, in the case of ^{109}Cd, where the x rays and Auger electrons arising from electron capture are not in coincidence with the conversion electrons arising from the decay of ^{109m}Ag.

Legrand et al. (1973) have constructed a 4π proportional counter capable of operating up to a pressure of 70 atmospheres, and have used it to measure the spectra of conversion electrons arising in the decay of ^{99m}Tc, ^{109}Cd, and ^{139}Ce. The sum of events under the K and L internal-conversion peaks in a given time, together with a knowledge of the K and L internal-conversion coefficients, the K: L: M ratio, and the probability of the given gamma-ray transition per decay, enables the activity of the parent radionuclide to be determined. Alternatively the activity can be computed directly using the internal-conversion probabilities. "Intensity" is given in Appendix A.3.

Another approach to the same problem is to construct a large 4π counter operating at atmospheric pressure. This was adopted by Troughton (1977) who constructed a cylindrical 4π counter having a diameter of 30 cm. She obtained a ^{109}Cd-^{109m}Ag spectrum, shown in Figure 30, in which the "valley" between the x-ray and K-conversion-electron peaks dropped to practically zero count rate, and was thus able to determine the number of events under the K and $L + M$, etc. peaks with no significant contribution from the x-ray peak. This slight difference from the spectrum of Legrand *et al.* (1973), where the count rate in the valley is still appreciable, may be due to some subtle difference in design, or mode of operation, or size of source mount.

3.5.7 *Further Reading*

For further information on selected topics of 4π proportional counting, reference should be made to the proceedings of the Herceg Novi (1973) conference.

3.6 Internal Gas Counting

The disintegration rate of low-energy beta-emitting, or electron-capturing, radionuclides that are in, or can be compounded into, gaseous form may be measured by the direct method of internal gas counting. In this method the radioactive gas is introduced, together with a suitable counting gas, into a counter operating in either the proportional or Geiger regions. Such counters usually comprise a cylindrical cathode with an anode of wire (of the order of 0.001 inch in diameter) stretched along the axis of the cylinder. For the assay of radon see Section 6.1.2.3.

The use of such counters of known physical dimensions was first reported by Miller (1947) and they were used by W. F. Libby and his colleagues (Anderson *et al.*, 1947) for determining the activity of ^{14}C in the environment. Measurements using such counters, however, required that corrections be made for wall and end effects (see, for example, Engelkemeir and Libby, 1950), due to electrons entering the wall without the creation of an ion pair, to the decrease in electric

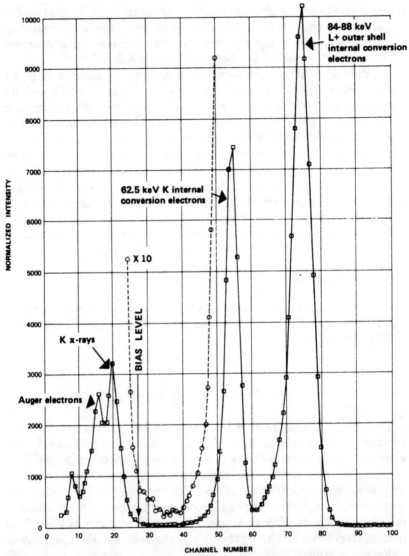

Fig. 30. ^{109}Cd-^{109m}Ag conversion-electron spectrum taken in a 30-cm-diameter cylindrical 4π gas-flow proportional counter. (From Troughton, 1977)

field intensity at the ends, and, in cases where the anode was longer than the cathode, to electrons entering the counting volume from the ends. The activities of the gaseous contents per mole were calculated from the count rate, the effective volume of the counter, and the pressure, temperature, and composition of the counting gas.

For the purpose of standardizing gases or gaseous compounds, Mann and Parkinson (1949) used a length-compensated gas-counter system. This consisted of two counters of precisely similar dimensions except for the cathode and anode lengths. To the best precision possible, the end supports for the anode wire and the cathode diameters are made the same. The difference in counting rates then gives the count rate for an ideal volume of gas contained in a cathode cylinder equal to the difference in lengths of the two counters. The counters may be operated in either the Geiger or proportional region (Mann *et al.*, 1960).

Wall effects are determined by measuring the count rates at several pressures and then extrapolating the observed count rate as a function of reciprocal pressure to zero reciprocal pressure (i.e., to infinite pressure). A correction for undetected low-energy events can be made either by a discriminator-voltage extrapolation to zero voltage or by the extrapolation of the electron, or electron plus x-ray, spectrum to zero energy. In both cases, data are taken above the level of electronic noise.

Multiwire-cathode internal gas counters with surrounding integral cylindrical ring counters in the anticoincidence mode were used by Drever *et al.* (1957) to reduce wall effects, and, recently, for the accurate measurement of electron-capture intensity ratios (Chew *et al.*, 1974a). This type of gas counter consists of a cylindrical central counter with multiwire cathode, surrounded by a ring of counters, separated from the central counter only by the multiwire cathode. The ring counters are operated in anticoincidence with the central counter, so that decay events that trigger both a ring counter and the central counter are not registered.

The pulse-height spectrum recorded from such a gas counter is a much improved representation of the energy distribution between the L and K electron-capture decay events, and accurate electron-capture intensity ratios may be determined from such spectra. Corrections must, however, be made both for x rays that escape from the central counter, and for those that enter from the ring counters or end zones (Vatai, 1970; Chew *et al.*, 1974a, 1974b).

Internal gas counting is essentially a 4π method that avoids the problem of source self-absorption. A description of the method was given at the Herceg Novi International Summer School (Garfinkel *et al.*, 1973b), and has been extensively studied by Spernol and Denecke (1964a, 1964b, and 1964c). A photograph of the National Bureau of Standards (NBS) length-compensated gas-counting system, consisting of two sets of three counters, one set of copper and one of stainless steel, is shown in Figure 31. The counting data from the six counters

Fig. 31. The NBS stainless-steel and copper length-compensated internal gas counters mounted in their base with vacuum-chamber cover and one counter removed. An automatic Töppler pump for mixing the radioactive gas sample with a suitable counting gas can be seen just behind the counters. A flask containing the radioactive gas sample can be seen in the temperature equilibrating box above the steel safe, and two mercury barometers for measuring the sample pressure and counting-gas pressure are to the left. M. P. Unterweger and J. E. Harding have recently modified the system so that cryogenic mixing and transfer can be used as an alternative to mixing and transfer by means of the Töppler pump.

are collected simultaneously and processed on a minicomputer, operating, when required, as a multichannel analyzer (Garfinkel *et al.*, 1973b).

Examples of NBS standards produced by this method are ^{3}H, ^{14}C (as CO_2), ^{37}Ar, ^{85}Kr, ^{131m}Xe, and ^{133}Xe. The first NBS ^{3}H standards were calibrated in 1954 by calorimetry, and then recalibrated by the method of internal gas counting in 1961 (Mann *et al.*, 1964). They were then recalibrated again by gas counting in 1978 (Unterweger *et al.*, 1980). These two values were then combined, for the gas counting method, with several other calibrations of these standards made by other laboratories by means of a non-linear chi-squared-minimization fit to give a new value for the half life for tritium of 12.43 ± 0.05 years.

3.7 Alpha-Particle Counting

3.7.1 *Introduction*

The use of low-geometry defined-solid-angle and 4π-proportional counters for alpha-particle counting has been treated in Section 3.5.6, pulse-ionization chambers in Section 6.1.2.3, ZnS(Ag)-coated scintillation cells in Section 6.1.2.2, silicon surface-barrier detectors and diffused-junction detectors in Section 4.5.3, and liquid-scintillation counters in Sections 3.8, 4.3.8, and 6.1.2.2. The assay of alpha-particle-emitting sources by the methods of 2π proportional and defined-solid-angle counting is the subject of this section. Methods for the calibration of alpha-particle sources by defined-solid-angle counting have been reviewed by Brouns (1968).

3.7.2 *Standardization by 2π Proportional Counting*

The measurement of alpha-emitting radionuclides in 2π geometry requires essentially the measurement with one half of a 4π system, and all considerations in 4π proportional counting apply here equally well.

The source may, however, be electrodeposited or evaporated onto a polished-metal backing that behaves as a semi-infinite medium. Alpha particles, being more massive than electrons, are less easily scattered from the backing, so that the fraction of the alpha particles scattered into the 2π counter is far less than in the case of beta particles. Also, alpha particles are emitted in monoenergetic groups with much higher energies than beta particles, which are characteristically emitted in a spectrum of energies.

Nevertheless, to obtain accuracies comparable with those obtained in 4π counting, scattering and also source self-absorption must be evaluated. In general, such an evaluation can be readily made only for sources of uniform and known thickness.

3.7.3 *Backscattering*

Backscattering, in this section, refers specifically to the effect in which alpha particles, initially moving in the direction of the source mount, are deflected into the sensitive volume of the 2π detector. Many investigators e.g., Deruytter, 1962 and Hutchinson et al., 1968, have concluded that the number of backscattered alpha particles is between zero and five percent of the number emitted into 2π steradians, depending on the thickness and uniformity of the source, the atomic number of the mount and source material, and the energy of the alpha particles.

Walker (1965) and others have shown that backscattered alpha particles move predominantly into angles of less than 5° from the plane of the source mount, as a result of *multiple* small-angle scattering primarily due to the nuclear electrostatic field. Large-angle Rutherford scattering occurs in the order of a few times in 10^5 emitted alpha particles, and is, therefore, negligible for almost all 2π counting.

Backscattering increases with increasing atomic number, Z, of the scattering material, as demonstrated in Figure 32, for which 2 μg cm^{-2} collodion films impregnated with ^{210}Po were placed on polished source mounts of a number of materials. The backscattered fraction, f_b, of the total for "weightless" sources is given by:

$$f_b = \frac{N_{2\pi}}{N_0} - 0.5 \qquad (3.27)$$

where $N_{2\pi}$ is the alpha-particle count rate in 2π geometry, and N_0 is the total alpha-particle-emission rate. Figure 33 shows the experimental values of $N_{2\pi}/N_0$ plotted as a function of Z for sources adsorbed onto polished and unpolished mounts, and the large scatter of the points demonstrates the large effect often encountered due to even minor surface irregularities on the backscattering.

It has also been shown (Crawford, 1949; Hutchinson et al., 1968)

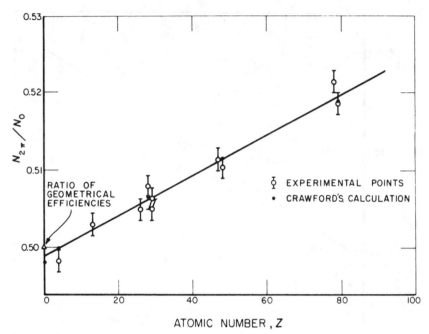

Fig. 32. Counting rate in the NBS 2π α-particle counter divided by N_0 for ^{210}Po sources impregnated in collodion on polished mounts plotted against Z, the atomic number of the backing material. The error bars represent plus and minus one standard error. The solid points are obtained from a relation developed by Crawford (1949). (From Hutchinson *et al.*, 1968)

that backscattering varies approximately as $E^{-\frac{1}{2}}$, where E is the alpha-particle energy.

3.7.4 *Source Self-Absorption*

The absorption of alpha particles by the source is primarily a function of the thickness, d_s, of the source and the range, R_s, of the alpha particles in the source material. Thus, several authors, Robinson, H., 1960; Friedlander *et al.*, 1964; Gold *et al.*, 1968; White, 1970, have shown, both theoretically and experimentally, that for a uniformly thick source, the fraction of alpha particles absorbed by the source increases proportionately to $d_s/2R_s$ for $d_s < R_s$. Figure 34 demonstrates this dependence, which can be calculated from simple geometrical considerations and the assumptions that alpha particles have a well defined range and are rarely deflected from their initial direction in the source material.

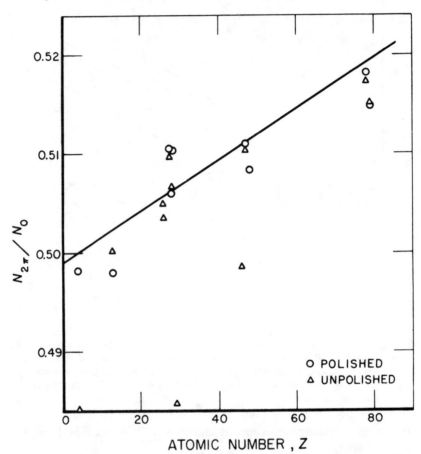

Fig. 33. Counting rate in the NBS 2π α-particle counter divided by N_0, for ^{210}Po sources adsorbed onto polished and unpolished mounts plotted against Z, the atomic number of the backing material.

3.7.5 *Calculation of Scattering and Source Self-Absorption*

These last assumptions (Section 3.7.4) represent only first order approximations, and scattering would not occur if they were strictly true.

An expression for $N_{2\pi}/N_0$, for the source configuration shown in Figure 35, is given in Eq. 3.28. A non-radioactive layer of absorber a, thickness d_a, covers an alpha-particle emitting source s, thickness d_s, deposited on a semi-infinite backing material b. Then

$$\frac{N_{2\pi}}{N_0} = \frac{1}{2}\left(1 + \frac{d_a}{R_a} - \frac{d_s}{2R_s}\right) + f_s, \qquad (3.28)$$

Fig. 34. $N_{2\pi}/N_0$ as a function of surface density in μg cm^{-2} for UF$_4$ sources. Error bars are representative of one standard deviation. (From Gold *et al.*, 1968)

THICKNESS

INERT ABSORBING MATERIAL	d_a
SOURCE MATERIAL	d_s
SOURCE BACKING	$>R_s$

Fig. 35. Schematic illustration of a covered, deposited, alpha-particle source.

where R_a and R_s are the alpha-particle ranges in absorber and source, respectively, and f$_s$ is the scattered fraction. For "weightless" sources and no absorbing material, f$_s$ reduces to f$_b$, which is the backscattering due to the backing material. These results have been discussed in Lucas and Hutchinson (1976) and Hutchinson *et al.* (1976a). Figure 36 shows their experimental values and the theoretical curve for $N_{2\pi}/N_0$ vs. d_s/R_s for sources of ^{235}UO$_2$ deposited on polished platinum.

3.7.6 *The Robinson Counter*

Robinson, H. (1960) has developed a counter, Figure 37, with an accurately defined geometrical efficiency close to 0.8π steradians, for alpha particles emitted in a direction normal to the plane of the source.

Fig. 36. Experimental values and theoretical curves for $N_{2\pi}/N_0$ vs. d_s/R_s, the source thickness expressed as a fraction of the range. The diagonal straight line corresponds to no scattering. The lower heavy curved line was calculated for $^{235}UO_2$ and the upper dashed curved line was calculated for a source of $^{238}UO_2$, both on polished platinum. Error bars represent plus and minus one standard error.

Multiply scattered alpha particles, essentially all of which are emitted at grazing angles to the source, are not detected. The difficult problem of estimating the backscattering contribution, encountered in 2π counting, is thus avoided. The detector is usually of scintillating plastic, CaF(Eu), ZnS, or CsI(Tl) (Hutchinson *et al.*, 1968), and a baffle at its center is so designed that variations in source positioning of as much as 1 cm result in negligible changes in the overall detection

CYLINDRICAL
CHAMBER

PHOTOTUBE

LUCITE LIGHT PIPE

PLASTIC
SCINTILLATOR

ALUMINUM
BAFFLE

SOURCE MOUNT

H_2 OUT

H_2 IN

Fig. 37. Schematic diagram of the NBS Robinson counter.

efficiency. Consequently, the counter can be used to measure relatively large low-activity sources with accuracies of approximately 0.5 percent. In this geometry less than 0.2 percent of the counts are lost due to source self-absorption perpendicular to the source, if its thickness is less than 0.25 R_s. Such counters, during operation, are usually evacuated or filled with flowing hydrogen at atmospheric pressure.

The detection efficiency, $\epsilon_{0.8\pi}$, of the 0.8π counter was deduced from measurements on thin sources with the 2π and 0.8π counters for different values of atomic number, Z, of the backing material (Hutchinson et al., 1968). The 2π count rate, $N_{2\pi}$, divided by the 0.8π count rate, $N_{0.8\pi}$, was plotted as a function of Z, and extrapolated to $Z = 0$ where, theoretically, $\epsilon_{0.8\pi} = N_{0.8\pi}/2N_{2\pi}$.

3.7.7 *Low-Geometry Solid-Angle Counters*

As mentioned in Section 1.5.2, the low-geometry defined-solid-angle counter, although often used just for relative measurements, has proved to be powerful as a tool for the measurement of alpha-particle-emitting radionuclides. Such a counter consists essentially of a detector, behind a diaphragm of known area, and a source mount, designed so that the source may be positioned at known distances from the diaphragm. The silicon surface-barrier-detector system, described in Section 4.5.3, is an example of such a system for *relative* measurements in which the unknown area of the detector replaces the diaphragm. Low-geometry defined-solid-angle counters have been described by Rutherford and Geiger (1908), Zumwalt (1950), Jaffey (1954), Spernol and Lerch (1965), and others.

3.8 Liquid-Scintillation Counting

3.8.1 *Detection Sensitivity*

The physical basis for the method of liquid-scintillation counting is discussed in Section 2.4.5. While liquid-scintillation counting is useful in making relative measurements between samples of the same beta-particle emitters of all energies, and the direct measurement of alpha-particle emitters, it does not compare to gas counting when it comes to the direct assay of beta-particle emitters of low energy, as the average energy that must be deposited in the scintillator to result in the ejection of one electron from the photocathode of the phototube is of the order of 2 keV, compared with the approximately 35 eV required to produce one ion pair in a gas.

This lack of sensitivity to low-energy radiations does not present a problem in the measurement of monoenergetic alpha particles, for which the detection efficiency is generally 100 percent, if losses due to all "wall effects" can be neglected. An early example of this method is the work of Basson and Steyn (1954) in standardizing ^{210}Po in solution.

A major disadvantage of alpha-particle liquid-scintillation counting has been that of poor energy resolution. With most commercial liquid-scintillation counters it is difficult to resolve the peaks of the alpha particles of interest from contributions due to impurities or beta-particle-emitting daughters. However, McDowell (1975), described a system that has much improved resolution, and he (*loc. cit.*) and McBeth *et al.* (1971) have used the pulse shape to discriminate against unwanted beta-particle pulses.

Liquid-scintillation counting is frequently used for the direct assay of high-energy beta emitters such as ^{32}P and ^{90}Y, but many low-energy beta particles escape detection. The amplitudes of output pulses from the phototube will therefore give a spectrum, from which an integral discrimination curve may be obtained that, at low energies, will fall below that expected from the true beta-particle spectrum. For high-energy beta emitters, the proportion of such low-energy beta particles escaping detection may be sufficiently small that a simple linear extrapolation of the plot of integral-count-rate against discriminator-voltage, to zero discriminator volts, will give a value of the activity to well within the accuracy required. Other techniques are required for measuring low-energy beta-particle emitters and these are discussed below.

For radionuclides of low and intermediate average beta-particle energy, it is necessary to consider the distribution of pulses arising from decay events in the scintillator below 20 keV. The theory for an ideal scintillation counter was reviewed by Gale and Gibson (1966) who provided a detailed study of the statistical processes involved in converting electron energy deposited in a scintillator to a pulse-height distribution from an electron-multiplier phototube. Subsequent papers by Gibson and Gale (1968), Gibson (1972; 1976) and Gale (1972) established models required for describing this distribution by the use of beta and gamma functions. The introduction of the high-gain phototube (RCA type 8850) has largely removed the need for such functions and a model based upon the Poisson distribution with small corrections was described by Gibson (1980) and has been further refined by Grau Malonda (1982) and by Grau Malonda and Garcia-Toraño (1983). Neither the Gibson and Gale, nor the Grau Malonda and Garcia-Toraño methods is a direct method as they measure activity in terms of a previously calibrated reference source of activity. These are therefore discussed in Section 4.3.9. However, as Houtermans (1973) has pointed out, "any extrapolation method is just as good as the physical model justifying the adopted shape of the extrapolated function." The arbitary choice of linear or logarithmic (Flynn et al., 1971) extrapolations cannot be justified just because the result obtained agrees with that obtained by other direct methods. Each method must be separately justified.

3.8.2 Zero-Detection Probability

Practically, the scintillation counter can be thought of as a first converter (electron energy into light), a second converter (light into photoelectrons at the cathode of the phototube) and an amplifier (the dynodes). Each stage of the whole process of converting energy into an electronic pulse involves some distortion or uncertainty so that an energy input E will produce, on average, a pulse of height V with an uncertainty of one estimated standard deviation, $1s$, which can be equal to 50 or 100 percent of V. In addition, an input of $2E$ will not necessarily produce an output of $2V$ because the system is nonlinear. These distortions and nonlinearities can be measured and explained by mathematical models.

Starting with a single radiation-energy-depositing event in the scintillator solution, light quanta will be emitted of which only a small fraction will interact with the photocathode of the phototube to produce photoelectrons. In the latest phototubes, the quantum efficiency of the photocathode is about 30 percent. The deposition of energy along the track of a 5-keV electron in the scintillator, if totally converted into light at 425 nm (maximum emission of a typical scintillator), would produce 1700 quanta. Typically, however, between 95 and 99 percent of the energy goes into heating the scintillator, and only the remaining 1 to 5 percent is produced as light. In practice, in many scintillators, the scintillation efficiency is approximately 1 percent, so that a 5-keV electron would produce about 17 light quanta, with about five photoelectrons (*i.e.*, about 30 percent) incident upon the first dynode.

This is not quite the complete story because lower-energy electrons with higher rates of energy deposition (LET) in the scintillator will also reduce the efficiency of conversion into light. At 5 keV this "ionization-quenching" effect (Birks, 1964) amounts to about 40 percent, so that a mean number of about three photoelectrons will strike the first dynode for 5 keV of energy deposited in the scintillator. These three electrons will be amplified into a pulse of perhaps 10^7 electrons at the anode of the phototube with further amplification by external electronics. However, three is the mean, $\bar{r}$, of an assumed Poisson distribution in which 0, 1, 2, 3, ..., r, photoelectrons may strike the first dynode. The probability of obtaining r photoelectrons incident upon the first dynode is therefore

$$P(\bar{r}, r) = e^{-\bar{r}} \frac{(\bar{r})^r}{r!},$$

where r! is factorial r. The probability that no photoelectron will strike the first dynode for $\bar{r} = 3$, is, therefore, $e^{-3} = 0.050$ which is the zero-detection probability for this energy in the system.

Wright (1954) was apparently the first to invoke the hypothesis of zero-detection probability for monoenergetic particles. It was also applied by Tissen (1958) to continuous beta-particle spectra, and was tested by Lukens (1961), using ^{3}H and ^{14}C, to confirm the validity of assuming the Poisson distribution. Vatin (1980) reports that "approximating to the allowed form of decay of radionuclides of low atomic number, and on the hypothesis of linear response of the scintillator" Tissen found that the detection efficiency depended only on the quality factor (photoelectrons per unit loss of energy per detected particle), and on the maximum energy of the continuous spectrum. Vatin also reports that "values calculated in this way for ^{3}H and ^{14}C are found to be in good agreement with current experimental results".

Horrocks and Studier (1961) were also amongst the early authors to consider the effects of both zero-detection probability and spectral shape on the accuracy with which activity could be obtained. By assuming a figure of merit of 1.5 keV per photoelectron at the photocathode they were able to obtain excellent agreement between the theoretical and experimental counting efficiencies for both single and coincident operation of two phototubes with ^{3}H and ^{241}Pu. Because of complications due to afterpulsing they failed to obtain such good agreement with ^{106}Ru.

In some cases for r equal to 1, or 2, such a loss of efficiency could occur again at the second dynode, but with modern high-gain phototubes this is highly improbable. If P $(\bar{r}, 0)$ is the probability of zero detection, then $1 - P(\bar{r}, 0)$ is the probability, or efficiency, for detection of photons incident on the photocathode. The relative error contributed by the inefficiency $P(\bar{r}, 0)$ to the assay of beta-emitting nuclides by the method of extrapolating the integral pulse-height spectrum to zero energy is clearly smaller for those nuclides with the larger values of maximum beta-particle energy, $E_{\beta\,max}$. $P(\bar{r}, 0)$ can be as large as 50 percent in the case of ^{3}H (Gibson and Gale, 1968).

Houtermans (1973) experimented with an RCA 8850 phototube and a source of incandescent light that emitted photons of energy only sufficient to produce single electrons from the photocathode, and obtained the output pulse-height spectrum for such single electrons (so-called "monos") from the phototube. By applying this spectrum to r equal to 1, 2, 3, 4, and 6 photoelectrons emitted simultaneously by the photocathode, Houtermans generated the spectra that should be produced by $\bar{r}$ equal to 1, 2, 3, 4, 6 and 8 "monos". He also measured the pulse-height spectrum for the x rays and Auger electrons emitted in the decay of ^{55}Fe and showed it to be a "mono" spectrum with $\bar{r}$ having a value between 2 and 3. These elegant results constitute a strong confirmation of the basic assumption of a Poisson distribution of photoelectrons incident upon the first dynode of the phototube. Similar results for the x rays and Auger electrons emitted in the decay of ^{54}Mn and ^{55}Fe have recently been obtained by Yamashita *et al.* (1983). In the case of ^{54}Mn, unwanted background pulses were eliminated by gating the output of the phototube by means of the 835-keV ^{54}Mn gamma rays that are emitted in coincidence with the x rays.

The consequences of the finite probability of zero detection in liquid scintillation counting have been described in detail in a comprehensive review by Gibson (1980).

One limitation of these techniques is the production of afterpulses which are caused by feedback processes, mainly optical and ionic, which limit the lowest value of discriminator setting that can be used, even for high-energy beta-particle emitters. A review of methods used to overcome the problems is given by Smith (1980) and the following techniques assume that this problem can be solved.

3.8.3 *Double- and Triple-Coincidence Methods*

Kolarov *et al.* (1970) proposed a technique using a coincidence system, and comparing the coincidence counting rate with that from each of the phototubes. The method is limited to material of high radioactivity because of the noise level in the single-channel counting. It also suffers from the limitation that it generates three equations with three unknowns, which, while soluble by iteration may not converge satisfactorily to an accurate value of activity. Thus Vatin (1980) quotes a value for an NBS tritium standard that differs by four percent from the NBS value. The addition of a third phototube by Pochwalski *et al.* (1981a) enables the ratio of triple coincidences to the sum of triple and double coincidences between all three phototubes to be measured. The efficiency for counting triples is $\epsilon_T = N_T/N_o$ and doubles is $\epsilon_D = N_D/N_o$ where N_o is the disintegration rate in the sample and N_T and N_D are the triple- and double-coincidence counting rates respectively. The Triple-to-Double Coincidence Ratio (TDCR) is then

$$K = \frac{\epsilon_T}{\epsilon_D} = \frac{N_T}{N_D}.$$

Pochwalski *et al.* (1981a) changed the optical efficiency by altering the tension on a spring around the scintillation vial and so produced variations in ϵ_T and ϵ_D, with K. As K increases and approaches unity the individual efficiencies tend to unity and N_T and N_D approach the disintegration rates. They then fitted fourth-power polynomials to the values of ϵ_T and ϵ_D and extrapolated these to $K = 1$. Pochwalski *et al.* (1981b) also calculated the values of ϵ_T and ϵ_D, using the theory of Gibson and Gale (1968), as a function of K; these agreed well with the experimentally measured values. In the latter case, due allowance must be made between color and chemical quenching, for changes in background and for the large uncertainties for samples of low radioactivity.

External optical filters (serving the same purpose as the coiled spring) have also been used by other authors (*e.g.* Flynn *et al.*, 1971; Song *et al.*, 1980) for the purpose of extrapolating integral pulse-height spectra to zero pulse height in liquid-scintillation systems. Such a method is analogous to changing the gain of an electronic amplifier system.

3.8.4 *Beta-Gamma Coincidence-Counting Techniques*

Liquid-scintillation detectors have also been used for the x-ray, alpha-particle, and electron-detection channels for 4π coincidence-counting systems (Steyn, 1973). Several participants in an international intercomparison of ^{241}Am, organized by the Bureau International des Poids et Mesures (BIPM) in 1963 (Rytz, 1964), successfully used liquid-scintillation counting for both direct alpha, and alpha-gamma coincidence measurements. Good examples of more recent work are the $4\pi(LS)\beta$-γ measurements of radioactive concentrations of ^{134}Cs solutions by Smith and Stuart (1975) and Yura (1983). International comparisons using the method of $4\pi\beta(LS)$-γ coincidence counting have also been organized by BIPM (BIPM, 1980; Rytz, 1980). A number of laboratories also made confirmatory measurements at the same time using $4\pi\beta(PC)$-γ and $4\pi\beta(PPC)$-γ coincidence counting systems.

An extensive list of references concerning the use of liquid-scintillation detectors for high-precision activity measurements is given in Vaninbroukx and Stanef (1973).

3.9 Calorimetry

The calorimetric method has been described earlier in this report as a "near-direct" method for activity measurements of an alpha- or beta-particle emitter in that it is necessary to know the alpha-particle energy, or the average beta-particle energy per disintegration. The method is usually considered to be "direct" when it is used to measure, for example, the power of a ^{239}Pu source in terms of electrical quantities. In the case of measurements of radium and its daughters, calorimetric measurements are, however, relative.

The case of radium is probably unique in that radium standards are not radioactivity standards but are *defined* mass standards, usually in milligrams of ^{226}Ra content as of a certain date, and are analogous to the *defined* standard of mass, the kilogram. All other radium standards, in varying degrees of known equilibrium with their daughters, are measured relative to the international radium standards by a comparison of some measurable quantity such as gamma-ray emission rates, or rates of energy emission.

Many calorimeters utilize two similar receptacles (twin-cup calorimeters) in which various radioactive sources can be placed, the heat generated being balanced by electric heating coils within, or wound around, the receptacles. In some cases, the rates of energy emission from the radioactive sources are balanced by passing an appropriate electric current in such a direction as to cool Peltier junctions attached to the cups of a twin-cup calorimeter. The Bunsen ice calorimeter can be used to measure the rate of energy emission, of the order of 1 watt, from more active sources. The low specific heat of materials at low temperatures (<4.2 K) permits measurement of small amounts of radioactive materials in calorimeters designed for operation in this temperature range by the Centre d'Études Nucleaires (LMRI, 1975).

Calorimetric, or microcalorimetric, methods have been used in the measurement of radioactive effects since the experiments of Curie and Laborde in 1903. It is of interest to note that the historic calorimetric measurement of the rate of energy emission from radium E by C. D. Ellis and W. A. Wooster led to the postulation of the neutrino by W. Pauli in 1931. A quite comprehensive survey of the measurement of radioactivity by calorimetric methods, with general references covering the years 1903 through 1958, was prepared for the Encyclopaedic Dictionary of Physics (Mann, 1962). Very comprehensive reviews of work subsequent to 1949 have been written by Gunn (1964, 1970), and more recent calorimetric applications were described at the Herceg Novi International Summer School (Ramthun, 1973; Mann, 1973).

Microcalorimetry has been used at the National Bureau of Standards both for the intercomparison of national radium standards and for the standardization of artificially produced radionuclides and ^{210}Po. A precision of measurement of about 0.1 to 0.2 percent is normally attainable. The twin-cup compensated microcalorimeters that were used to make these measurements (Mann, 1954a, 1954b) employ the

Peltier-cooling effect to balance the energy emission of the radioactive source.

The National Bureau of Standards radiation balance, as this twin microcalorimeter has been called, has been used to prepare standards of ^{3}H (Mann et al., 1964), ^{35}S (Seliger et al., 1958), and ^{63}Ni together with a half-life measurement using mass spectrometry (Barnes et al., 1971), and to calibrate a ^{210}Po source in connection with the measurement of the branching ratio in the decay of ^{210}Po (Hayward et al., 1955). A knowledge of this ratio, namely 1.21×10^{-5} gamma quantum per alpha particle, enables a strong ^{210}Po source to be assayed simply by measurement of the number of 0.804-MeV gamma rays from the daughter nucleus, ^{206}Pb.

NBS solution standards prepared from the calorimetrically assayed ^{63}Ni were also measured by the method of efficiency tracing (see Section 3.2.7), by the National Research Council of Canada (Baerg and Bowes, 1971), and Atomic Energy of Canada Limited (Merritt and Taylor, 1971), and by the Australian Atomic Energy Commission Research Establishment (Lowenthal et al., 1973) with complete agreement between two laboratories, and a total range of 2.7 percent in the measurements (Lowenthal et al., 1973).

The NBS radiation balance was also used to calibrate a source of radium chloride in the preparation of radium solution standards in the ranges 10^{-6} g of ^{226}Ra per 5 ml, and 10^{-11} g of ^{226}Ra per 100 ml, of carrier solution (Mann et al., 1959). In this calibration, reproducibility within the range of 0.004 mg in 6.106 mg in the mass of ^{226}Ra was obtained. The name "radiation balance" is based on the fact that a constant source of power in one cup, either radioactive or electrical, can be balanced against one or more similar sources in the other cup of the twin-cup microcalorimeter. An apt example of this aspect of the potential of the NBS radiation balance was its use in calibrating, relative to the NBS 38-mg Hönigschmid radium standard, the 100-mg radium source used by Attix and Ritz (1957) in their measurement of the gamma-ray constant of radium. The Hönigschmid standard was balanced first against a 65-mg radium source which, in turn, was balanced against the 100-mg source.

An isothermal calorimeter used for the calibration of α- and β-particle sources has been described by Lowenthal and Hicks (1966).

With the advent of nuclear power, calorimetry is also used to assay plutonium derived from spent nuclear fuel. Calorimeters of the type developed by Eichelberger et al. (1954), from the original design of Rutherford and Barnes (1904), and Rutherford and Robinson (1913), have been widely used to assay many special nuclear materials, and recommended procedures for making such measurements have been published by the American National Standards Institute (ANSI, 1975).

Ditmars (1976), using the NBS Bunsen ice calorimeter, has calibrated approximately 0.23-watt and 1.4-watt sealed sources of special nuclear material, comprised chiefly of ^{238}Pu, but also containing ^{239}Pu, ^{240}Pu, ^{241}Pu, ^{242}Pu, and ^{241}Am. To assay the individual nuclides, the abundances of americium and each of the plutonium isotopes present in such special nuclear materials must be determined by either mass spectrometry or gamma-ray spectrometry (ANSI, 1975). A number of half-life and specific-power measurements utilizing the calorimetric method are also referenced in ANSI (1975), and half-life measurements of ^{239}Pu and ^{240}Pu involving calorimetric measurements at Mound and the Lawrence Livermore Laboratory, are described in the reports by the U. S. Half-Life Evaluation Committee (Lucas and Mann, 1978; 1984). Other developments have been described by Ramthun (1973).

3.10 Mass and Resonant-Ionization Spectrometry

The measurement of radioactivity by mass spectrometry exploits the radioactive decay law, $dN/dt = -\lambda N$. If λ is known and N is measured, then the activity, dN/dt, may be calculated. Conversely, the half life of a long-lived radionuclide can be determined, if dN/dt and N are measured. Examples are the measurements of the half lives of ^{14}C and ^{63}Ni, respectively 5730 years and 96 years, which are too long to be determined by direct observation. In the case of ^{14}C, the activity was measured by internal gas counting (Section 3.6), and the abundance of ^{14}C relative to ^{12}C and ^{13}C in the quantitative carbondioxide filling in the gas counters was determined by mass spectrometry (Watt et al., 1961; Mann et al., 1961; Olsson et al., 1962; Hughes and Mann, 1964). The activity of the ^{63}Ni, in the form of a small 1.3-g cylinder of nickel containing about 1.3 Ci of ^{63}Ni, was determined by microcalorimetry (Section 3.9). The rate of energy emission was about 133 μW, and the activity was calculated using a computed average energy of 17.23 keV per disintegration. The nickel cylinder was then dissolved and the abundance of ^{63}Ni relative to the stable isotopes of nickel was determined mass-spectrometrically (Barnes et al., 1971). From these measurements, both radioactivity standards of ^{63}Ni and a value for its half life were obtained.

Mass spectrometry has been used to measure the half lives of ^{3}H by measuring the rate of growth of ^{3}He (Jones, 1955); of ^{81}Kr, in conjunction with an isotope separator and the measurement of activity (Eastwood et al., 1964); of ^{129}I together with activity measurements (Katcoff et al., 1951); of ^{134}Cs and ^{137}Cs by measuring the number of ^{134}Cs and ^{137}Cs atoms as a function of time over a period of 2 and 11 years, respectively, without activity measurements (Dietz and Pachucki, 1973); and of many of the transactinium elements between ^{230}Th and ^{252}Cf by a number of laboratories (see for example, Surin and Formishkin, 1983). Half-life measurements by mass spectrometry of ^{239}Pu and ^{240}Pu (by the methods of daughter ingrowth and isotopic dilution) carried out at the Argonne and Los Alamos National Laboratories are described in two reports of the U. S. Half-Life Evaluation Committee (Lucas and Mann, 1978; 1984). The half life of the ^{241}Pu has recently been measured by de Bievre et al. (1983), using isotopic-ratio mass spectrometry.

"Burn-up standards" of ^{137}Cs have been prepared at the National Bureau of Standards by use of mass spectrometry, by the method of mass-spectrometric isotope dilution using ^{133}Cs. These standards are calibrated for concentration, in terms of ^{137}Cs atoms per gram of solution, for the measurement of burn-up of nuclear fuels. Results are quoted with an estimated overall uncertainty of ±0.5 percent, and a 99-percent confidence interval of the mean of 0.1 percent.

In recent years greater sensitivities have been achieved in the measurement of activity by counting the numbers of atoms present, rather than the number of such atoms that decay in a given time. The larger the half life T, then the larger does N become relative to dN/dt. Using high-energy mass spectrometry (with a cyclotron or Van de Graaff tandem accelerator as ion source) the sensitivity is so enhanced that artifacts can be dated by the radiocarbon method using milligrams rather than grams of the sample. While the sensitivity is so much greater, the accuracies attainable by measuring the numbers of ^{14}C atoms rather than their decay rates are still not yet as good. Interesting historical reviews of the development of the method of high-energy mass spectrometry, with many references, have been given by Muller (1979) and Currie (1982).

Another method in which atoms are counted, rather than their rates of decay, is that of resonant-ionization spectrometry using appropriately tuned lasers to ionize the atoms of interest. These are then extracted from the ion chamber, accelerated and electromagnetically analyzed and quantified. Recent developments using this technique have been described by Hurst et al. (1980).

3.11 Standardization of Gamma-Ray Emitters by Dosimetry Measurements

3.11.1 Measurements by Ionization Chambers and Sources Calibrated in Terms of Exposure

A useful relationship exists between the activity of a radionuclide source and the exposure rate at some stated distance. In order to derive this relationship, we note that the exposure rate (due to a gamma-ray fluence), at a given point in space, is the rate at which gamma-ray energy is transformed into ionization in free air, per unit mass of air.[5]

Consider a source with activity A nuclear transitions per unit time. Let P_i be the fraction of the nuclear transitions which give rise to photons of energy E_i. Then the gamma-ray energy fluence rate at distance l from the source, due to photons of energy E_i, is $A P_i E_i / 4\pi l^2$. The rate at which the energy fluence is transformed into kinetic energy of moving electrons, per unit volume of air, is obtained by multiplying the energy fluence by the linear energy-absorption coefficient in air, $\mu_{en,i}$. Kinetic energy per unit volume is converted into kinetic energy per unit mass of air by dividing by ρ, the density of air.

---

[5] The approximate definition of exposure given here is adequate for the heuristic derivation of the desired relationship between activity and exposure rate. A rigorous and complete definition of exposure, and also of the other dosimetric quantities used in Section 3.11, is given in ICRU (1980). Because the SI unit of exposure (C/kg) is considered to be inconvenient for some purposes, consideration is being given to the use of another quantity in place of exposure. ICRU Report 33 (ICRU, 1980) uses the symbol Γ_δ to represent the air-kerma rate-constant, in place of the exposure-rate constant. Exposure X is related to air kerma K_{air} by the equation

$$K_{air} = X(W/e)/(1 - g)$$

where g is the fraction of the kinetic energy of the secondary electrons that is expended in air as bremsstrahlung. For photon energies below a few MeV, $(1 - g) \geq 0.995$. The rate constants are related in the same manner as K_{air} and X:

$$_K\Gamma_\delta = {_X}\Gamma_\delta (W/e)/(1 - g)$$

where $_K\Gamma_\delta$ and $_X\Gamma_\delta$ are respectively the air-kerma-rate and the exposure-rate constants.

With these equations it is easy to change from exposure in SI units to air kerma in SI units. Note that Eqs. (3.29) and (3.31) are simplified if the change is made to air kerma, because W/e disappears from the right side of these equations. Equation (3.30) is otherwise unchanged if $\dot{K}_{air}$ and its rate constant are substituted for $\dot{X}$ and its rate constant. In Eq. (3.32), $k_1 = 1/4\pi$ if exposure is changed to air kerma.

When exposure X is given in roentgens (R), the air kerma in grays (Gy) is given by

$$K_{air} = 2.58 \times 10^{-4} X(W/e)/(1 - g) \simeq 8.75 \times 10^{-3} X.$$

Finally we convert to the rate of creation of charge per unit mass of air by dividing by W/e, where W is the mean energy expended in air to form one ion pair and e is the electronic charge. Thus, the exposure rate at the specified point, due to photons of energy E_i, is

$$X_i = \frac{A P_i E_i \mu_{en,i}}{4\pi \rho l^2 (W/e)}. \tag{3.29}$$

It is assumed that (1) the distance l is sufficiently large so that the exposure rate is independent of source size, and (2) there is neither attenuation nor buildup of the photon fluence in the distance l.

In general, radionuclides emit photons over a range of energies, and a summation over these energies must be made to obtain the exposure rate due to the source. Some of the photons may, however, be of such a low energy that they are absorbed in the source itself or in the intervening air, so that the exposure rate of interest may involve only photons with energies above some minimum value, represented by δ. If, then, this summation is carried out in Eq. 3.29 and the result solved for A, we get

$$A = l^2 \dot{X}_\delta / \Gamma_\delta, \tag{3.30}$$

where

$$\Gamma_\delta = \frac{1}{4\pi (W/e)} \sum_i P_i E_i (\mu_{en,i}/\rho). \tag{3.31}$$

The summation is carried out over those values of i for which $E_i > \delta$. Equation 3.30 is the desired relationship between activity and exposure rate. The *exposure-rate constant* Γ_δ (ICRU, 1971) is the exposure rate at unit distance from a source of unit activity due to photons of energy greater than δ, subject to the two conditions stated following Eq. 3.29. The exposure-rate constant is characteristic of the radionuclide and independent of the material properties of a particular source. Having specified δ, Γ_δ can be calculated from the decay parameters of the radionuclide. In choosing that value of δ most useful for a particular application, the attenuation properties of the source, the intervening medium, and the detector or object irradiated should be taken into account; photons of too low an energy to make a significant contribution to the measurement or irradiation should be excluded by choosing an appropriate value of δ.

For $\delta = 0$, the exposure-rate constant Γ_0 includes all the photons emitted by the radionuclide, including gamma rays and x rays. Except for the inclusion of x rays, Γ_0 is the same as a parameter that has been variously known as the specific gamma-ray emission, the specific

TABLE 8—*Typical correction factors for ionization-chamber gamma-ray measurements*

Radionuclide	Absorption in typical ampoule	Chamber wall absorption relative to radium	Overall correction factor
^{170}Tm	1.07	1.10	1.18
^{99m}Tc	1.05	1.07	1.12
^{131}I	1.025	1.025	1.05
^{137}Cs, ^{82}Br, ^{60}Co	1.02	1.00	1.02
^{24}Na	1.015	1.00	1.015

gamma-ray constant, and the gamma-ray dose-rate constant, is generally represented by Γ, and has been tabulated by a number of authors (NCRP, 1961, Appendix A; Hine, 1956, Appendix; Johns and Cunningham, 1969, Table A-8; Shalek and Stovall, 1969).

If the decay scheme of the radionuclide is known (i.e., for known values of P_i and E_i) and once the value of δ has been specified, Γ_δ can be calculated from Eq. 3.31. The calculation can be performed using the mass energy-absorption coefficient $\mu_{en,i}/\rho$ (ICRU, 1964; Hubbell, 1982), in which case Eq. 3.31 takes the form

$$\Gamma_\delta = k_1 \sum_i P_i E_i (\mu_{en,i}/\rho), \tag{3.32}$$

where the numerical value of k_1 depends on the choice of units. Alternatively, the calculation can be performed using the linear energy-absorption coefficient, shown in Figure 38 as a function of energy, in which case Eq. 3.31 takes the form

$$\Gamma_\delta = k_2 \sum_i P_i E_i \mu_{en,i}, \tag{3.33}$$

where $k_2 = k_1/\rho_{air}$, and ρ_{air} is the density of air for ambient temperature and pressure. At one standard atmosphere and 20°C, $\rho_{air} = 1.205$ mg/cm^3. The presently recommended value of W is 33.7 eV per ion pair formed in ambient air (ICRU, 1979a), which gives for the average energy expended per unit charge $W/e = 33.7$ J/C. If SI units are used (see Section 1.3)

$$k_1 = 1/4\pi(W/e) = 2.36 \times 10^{-3} \text{ C/J}. \tag{3.34}$$

In this case, the energy per nuclear transition $P_i E_i$ has units of joules per becquerel second[6] (J/Bq s), $\mu_{en,i}/\rho$ has units of meters squared per kilogram (m^2/kg), and the exposure-rate constant Γ_δ has dimensions of exposure times length squared per nuclear transition, corresponding to units of coulombs per kilogram times meters squared

---

[6] Since the becquerel is one nuclear transition per second, one becquerel second is one nuclear transition, and 1 J/Bq s = 1 J per nuclear transition.

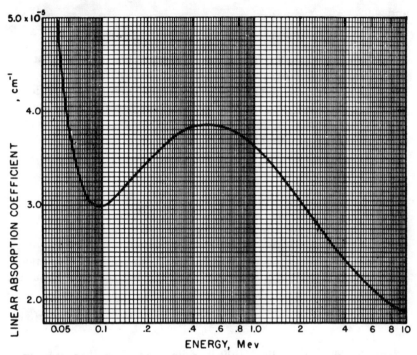

Fig. 38. Linear energy-absorption coefficient for photons in air at standard temperature and pressure. (From data in ICRU, 1961)

per becquerel second ((C/kg) m²/Bq s). When using Eq. 3.30 with SI units, length l is in meters (m), exposure rate, $\dot{X}$, is in coulombs per kilogram per second (C/kg s), and the activity, A, is obtained in becquerels (Bq). Equations 3.29 and 3.31 are correct for SI units as written; if other units are used, an additional numerical factor may be necessary, as indicated in Eq. 3.32 and Eq. 3.33.

It is frequently convenient to use SI units for all quantities except the photon energy E_i, for which the unit mega-electron volts (MeV) is used. For this case $k_1 = 3.78 \times 10^{-16}$ C/MeV.

A frequently used combination of units is activity, A, in millicuries (mCi), exposure rate, $\dot{X}$, in roentgens per hour (R/h), distance, l, in centimeters (cm), and photon energy, E_i, in mega-electron volts (MeV). Then the product P_iE_i has units of mega-electron volts per becquerel second (MeV/Bq s), $\mu_{en,i}/\rho$ has units of centimeters squared per gram (cm²/g), and $\mu_{en,i}$ has units of reciprocal centimeters (cm⁻¹). Using these units, the numerical constants in Eq. 3.32 and Eq. 3.33 become

$$k_1 = 195.2 \text{ R g Bq s/mCi h MeV}$$

and $\hfill$ (3.35)

$$k_2 = 1.620 \times 10^5 \text{ R cm}^3 \text{ Bq s/mCi h MeV}.$$

Using these units, the exposure-rate constant Γ_δ has units of roentgens centimeters squared per millicurie hour (R cm^2/mCi h).

It is seen in Figure 38 that the linear energy-absorption coefficient $\mu_{en,i}$ is approximately constant for photon energies from about 0.06 to 2 MeV. As a result, in this photon energy range, the exposure-rate constant is approximated to within about ± 15 percent by

$$\Gamma_\delta \approx 5.2 \sum_i P_i E_i, \hfill (3.36)$$

where E_i is in mega-electron volts (MeV), and Γ_δ is in roentgens centimeters squared per millicurie hour (R cm^2/mCi h).

The exposure-rate constant Γ_δ can be calculated from Eqs. 3.32, 3.33, or 3.36, using the appropriate numerical value of k_1 or k_2. Alternatively, the exposure-rate constant Γ_i per photon of energy E_i can be read directly from Figure 39, and then Γ_δ obtained from the equation

$$\Gamma_\delta = \sum_i P_i \Gamma_i, \hfill (3.37)$$

where, as usual, the summation is taken over those photon energies for which $E_i > \delta$. The units of Γ_δ are the same as the units of Γ_i.

If a calibrated ionization chamber is available to measure the exposure rate $\dot{X}_\delta$ at a known distance l from a radionuclide source for which the exposure-rate constant Γ_δ is known or can be calculated, the activity can be calculated from Eq. 3.30. Suitable ionization chambers can be calibrated in terms of exposure rate at the National Bureau of Standards (NBS), or at other calibration laboratories which possess secondary exposure standards which have been calibrated by NBS. Exposure calibration factors are supplied for certain photon energies (e.g., those of the gamma-ray beams of ^{60}Co and ^{137}Cs), or for filtered x-ray beams generated by potentials less than 300 kV. In general, ionization-chambers are to some extent energy dependent, and it may be necessary to interpolate between known calibration factors in order to obtain a calibration factor appropriate for use with the calculated value of Γ_δ. In addition, it may be necessary to correct for attenuation of the gamma-ray beam in the source, attenuation and buildup in the intervening air, and scatter from the environment. It should be noted that attenuation and scatter in the ionization chamber wall is included in the calibration factor supplied by the calibration laboratory; no correction for these effects is necessary unless the

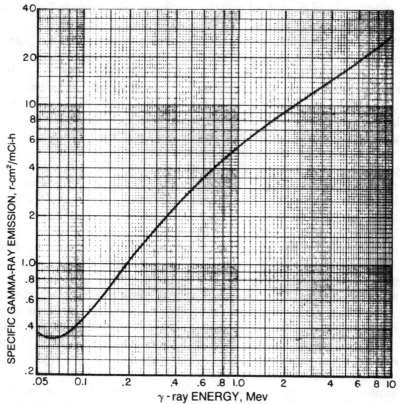

Fig. 39. Exposure-rate constant Γ for a point source emitting one photon of energy E per nuclear transition, as a function of E. The values of Γ were calculated by K. W. Geiger for NCRP Report No. 28 (NCRP, 1961) from the data given in Table 8.1 of NBS Handbook 78 (ICRU, 1961), using a value $W = 34.0$ eV per ion pair. If the presently recommended value of 33.85 eV per ion pair (ICRU, 1979a) had been used, the value of Γ shown in the graph would be 0.4 percent larger. Values of Γ_δ are obtained by summing the individual Γ values using Eq. 3.37. The ordinate must be multiplied by 1.94×10^{-19} to convert to the SI units C m^2 kg^{-1}, or by 0.194 to convert to aC m^2 kg^{-1}.

chamber is used under different circumstances from those in which it was calibrated.

An alternative to the use of a calibrated chamber is the use of gamma-ray sources or photon beams calibrated in terms of exposure rate. Encapsulated brachytherapy sources of ^{137}Cs and ^{60}Co are calibrated by NBS in terms of exposure rate at 1 meter in air; encapsulated and filtered sources of radium are calibrated by NBS in terms of mass of radium, and values of Γ_δ for radium sources are given in the tabulations of Γ in references given after Eq. 3.31. X-ray and gamma-ray beams used in medical practice are commonly calibrated in terms of exposure rate using instruments, the calibration of which is directly traceable to NBS exposure standards. Thus, a variety of photon beams is available for calibrating an ionization chamber in terms of exposure rate, the disadvantage of this method being that the calibration factor so obtained is one or more steps further removed from NBS standards than would be the case with a chamber calibrated at a calibration laboratory. If this method is used, it is highly desirable to use several sources or beams, in order to assess the energy dependence of the chamber being calibrated and thus choose a calibration factor suitable to the photon spectrum of the activity being measured. It may be necessary to correct for attenuation and scatter in the source, the intervening air, and the environment, but not for that in the walls of the ionization chamber.

Still another alternative is the use of a gamma-ray source calibrated in terms of activity. If such a source, with a known exposure-rate constant, is available, the activity of another gamma-ray source can be obtained by comparing the two using an ionization chamber. The comparison gives the unknown activity in terms of the known activity by use of Eq. 3.30, and involves the square of the ratio of the distances, the inverse of the ratio of the exposure-rate constants, and the ratio of the measured currents. The last ratio is equal to the ratio of the exposure rates, after correcting for differences in ionization chamber response due to differences between the photon spectra. It may also be necessary to correct for self absorption in one or both of the sources. Radium standards are often used, since accurately standardized sources are widely available and are of long half life. Table 8 summarizes the corrections needed in determining the activity of a few radionuclides by comparison with a radium standard. The corrections for absorption in a typical small ampoule and in the wall of the ionization chamber are seen to be small except at the lowest photon energies.

In carrying out ionization-chamber measurements, care should be taken that the wall thickness is sufficient to exclude all electrons arising outside the chamber walls. The operation of ionization chambers is also discussed in Sections 2.1 and 4.4.

3.11.2 *Measurement by Ionization Chamber of Known Volume*

The use of an ionization chamber of known volume for the determination of activity was first described in detail by Gray (1949). The rate at which energy is absorbed within the air of a cavity chamber can be calculated from the ionization produced, assuming a value for the mean energy expended in air per ion pair. If the cavity is small enough to produce no appreciable disturbance of the electron fluence generated by the gamma rays in the neighborhood of the cavity, the rate of energy absorption per unit mass in the chamber wall is related to that in the air by the ratio of the mean stopping power of the wall relative to that of air, provided the chamber walls are thick enough to exclude all electrons arising outside the walls. The rate of energy absorption in the chamber wall can be related to the source activity in a manner closely analogous to the relationship derived in Section 3.11.1 between activity and exposure rate, except that the present case involves the mass energy-absorption coefficients of the chamber wall material instead of air and does not involve the mean energy expended in air per ion pair. Equating the two independent calculations of the rate of energy absorption in the chamber wall allows the source activity to be expressed in terms of the measured ionization current. The relevant equations have been given by Gray (1949), and in considerable detail in an investigation of the relationship between source activity, source power, and exposure rate at unit distance, by Petree et al. (1975) and Roux (1974).

A determination of activity by this method involves not only a knowledge of the decay scheme of the radionuclide, but also the accurate measurement of the source-to-chamber distance, chamber volume, air density, and ionization current. Allowances must be made for absorption of the gamma rays in the source, chamber wall, and intervening air, and for the relative stopping power of wall to air. The measurement of effective distance presents no great difficulties; the inverse square law holds until the distance is only a few times the chamber dimensions (Sinclair et al., 1954; Kondo and Randolph, 1960). Dependence on the distance can be eliminated entirely by using a spherical extrapolation chamber surrounding the source (Gross et al., 1957b). Absorption within the source can be allowed for by the methods of Evans and Evans (1948). Absorption in the chamber walls can be estimated from observation of the effect of increasing the wall thickness. There is an uncertainty, however, leading to an error that is usually less than 1 percent, in deciding the wall thickness to which the observations should be extrapolated, since the secondary electrons arise at various depths within the walls (Attix and Ritz, 1957; Loftus and Weaver, 1974). The correction for relative stopping power can be made quite small by choosing a wall material whose atomic number is close to air; carbon in the form of graphite is often used for this purpose.

The use of an ionization chamber of known volume to determine the activity of a gamma-ray source is considerably more demanding than the use of a calibrated chamber, and is likely to be the method of choice only when special circumstances make traceability to a dosimetry standard undesirable.

3.12 Loss-of-Charge Method

Measurements of the charge carried away from a source of radium have been made by several investigators (Wien, 1903; Rutherford, 1905; Makower, 1909; and Moseley, 1912). Subsequently, however, this method was applied to the standardization of beta-emitters (Clarke, 1950; Gross and Failla, 1950; Keene, 1950). Using a spherical geometry, Clarke (1950) achieved agreement with National Bureau of Standards measurements of standard samples of ^{32}P to within 5 percent, but the method has not since been in general use. Gross and Failla (1950) used a parallel-plate geometry where the source was a very thin deposit of the emitter in question on one plate of a parallel-plate collecting system, the second plate being sufficiently close to balance the secondary electron emission and backscattering effects. Low-energy secondary electrons were bent back by means of a weak magnetic field, and the measurement was made in a vacuum to eliminate air ionization. Agreement with the National Bureau of Standards measurements of standard samples of ^{32}P to within ± 2 percent was obtained by this method by Gross and Failla over a number of years (Seliger and Schwebel, 1954).

3.13 Calculation of Activity of Artificially Produced Radionuclides From Irradiation Data

The disintegration rate of radionuclides produced by bombardment with neutrons or charged particles may be calculated if the cross section for interaction with the target nuclide (stable) and the flux of bombarding particles are known (Hughes, 1953; Halliday, 1955). Neutron reactions only will be considered here, since their treatment is typical of particle bombardments.

Thermal, or "slow" neutrons in a reactor are most often used for such activation. The neutron flux is usually determined by use of a monitor containing a known amount of a nuclide whose thermal-neutron-activation cross section is well established. Cobalt (37.2 barns; 37.2×10^{-24} cm^2), gold (98.8 barns), and manganese (13.3 barns) are convenient standards. A tabulation of cross sections is available (Mughabghab and Garber, 1973).

For calculation of the neutron flux density by use of the monitor, one may use the equation

$$f = \frac{A_m}{N_m \sigma_m e^{-\lambda_m t_d}(1 - e^{-\lambda_m t_b})}, \tag{3.38}$$

where

f = flux density, in cm^{-2} s^{-1},

A_m = disintegration rate of monitor, in s^{-1},

N_m = number of atoms of stable monitor nuclide,

σ_m = thermal-neutron-activation cross section of monitor nuclide, in cm^2,

λ_m = decay constant of monitor activity, in s^{-1},

t_d = time between end of bombardment and measurement, in s, and

t_b = bombardment time, in s.

The disintegration rate of the desired radionuclide produced by bombardment of an appropriate target may be calculated from the relation

$$A = N\sigma f e^{-\lambda t_d}(1 - e^{-\lambda t_b}), \tag{3.39}$$

where N and σ now refer to the target and A and λ to the activity produced. The introduction of the sample must not affect the flux distribution appreciably.

In general, the use of bombarded samples as standards is not as accurate as most of the methods previously discussed, chiefly because of uncertainty in knowledge of the energy distribution of neutrons in reactors. The usually well-known thermal-neutron cross section therefore has to be modified to include activation by epithermal or fast neutrons. Further, it is necessary to measure the disintegration rate of the monitor; such standards might therefore be considered to be derived standards.

4. Indirect Or Comparative Measurements of Activity in Radioactive Decay

4.1 General

The simplest way to prepare a calibrated relative (or derived, or indirect, or "secondary") standard of radioactivity in terms of a directly measured (or "absolute," or "primary") standard is to measure, by means of a substitution method under reproducible conditions, the relative activities using any instrument that records either individual radiations or else measures their ionizing effects. With instruments of adequate precision, such methods can give relative values having an estimated standard deviation of from 0.03 to 1.0 percent, which is often less than the uncertainty with which the activity of the direct standard is known. It is also to be understood that within the meaning of "reproducible conditions" are to be included the geometry of the equipment and the disposition and quantity of any material absorbing or scattering the radiations. In these respects, the characteristics of the source itself are no less important than those of the measuring instrument.

The recording device used in such a substitution method can be a gold-leaf or quartz-fiber electroscope; a beta-particle ionization chamber or gamma-ray ionization chamber used in conjunction with a Lindemann-Ryerson vibrating-reed, or Keithley electrometer, or any other low-current-measuring device; or any kind of alpha-, beta-, gamma-, or x-ray detector.

4.2 Radionuclide Identification and Assay by Energy Spectrometry

4.2.1 General

As discussed in Section 1, the decays of radioactive nuclei result in the emission of charged particles (alpha and beta particles, and electrons), gamma rays, and x rays due to interactions with bound atomic electrons. In negative beta decay the deexcitation energy being shared with an antineutrino, gives rise to a spectral distribution of beta-ray energy. The alpha particles and photons emitted exhibit discrete energies corresponding to the amount of energy given off as the nucleus or atom deexcites from one allowed energy state to another. Thus, if we observe the energy spectrum of alpha particles or gamma rays emitted in the decay of a given radionuclide, a pattern of energies and intensities will be seen that is unique. Just as the discrete structure of optical spectra resulting from electronic transitions in atoms forms the basis for elemental analysis, charged-particle- and photon-energy spectrometry may be used to analyze radioactive material for radio-nuclidic content.

Pulse ionization chambers and gas proportional counters (see Section 2.2) provide the means for measuring the energies of charged particles. Because the range of both alpha particles and beta particles is finite in a gas medium, these counters offer a means of measuring the energy of individual particles by pulse-amplitude spectrometry. As the particle ionizes the gas in the counting chamber, the collection of charge resulting from the total energy absorption of an incoming particle will provide a measure of the kinetic energy of that particle. The electrical pulse appearing momentarily at the anode of the counter chamber will have an amplitude proportional to the energy of the particle detected. As described in Sections 2.4 and 2.5, both scintillation and semiconductor detectors exhibit the same energy-measurement properties. All these devices are extensively used for the measurement of both charged-particle- and photon-energy spectra.

4.2.2 Interaction of Photons with Matter

A knowledge of the basic processes by which a photon interacts with matter is essential to an understanding of the response of a gamma-ray spectrometer and the techniques of pulse-amplitude (or pulse-height) analysis. Although many processes are involved in the chain of events that produces an electrical pulse at the output of the detector, the major features of a differential pulse-amplitude spectrum resulting from the detection of gamma rays may be interpreted in terms of the basic interactions that occur within the detector. As indicated in Section 1.4, the three primary processes by which photons interact with matter are (1) the photoelectric effect, (2) Compton scattering, and (3) the production of positron-electron pairs. These processes are represented diagrammatically in Figure 40. Photoelectric absorption

Fig. 40. Illustration of major gamma-ray interaction processes.

predominates in the low-energy region with its probability decreasing rapidly with increasing energy. Figure 41 is a plot of the probability for interaction of photons in NaI as a function of primary photon energy. The probability for interaction by the Compton process is seen to vary slowly up to several MeV in energy, while the probability for interaction by the pair process increases rapidly above the threshold at 1.022 MeV to become the predominant process at higher energies.

4.2.2.1 *Photoelectric Effect.* In the photoelectric process, absorption occurs by interaction between the photon and an inner-shell bound electron in the absorbing material. The energy of this ejected electron will be equal to the difference between the energy of the incident photon and the binding energy of the shell from which the electron was ejected. As a result, the atom is left with an electron vacancy, resulting in the emission of x rays and Auger electrons. This series of events occurs in less than 10^{-12} s, and if the photoelectron and other radiations expend all their energy in the detector the total energy of the primary photon is absorbed in the detector. Thus, a monoenergetic photon interacting by this process produces electrons totalling in energy that of the incident photon. If this were the only process for energy loss, the response of the detector would be quite simple.

4.2.2.2 *Compton-Scattering Process.* In Compton scattering, photons suffer partial losses of energy when scattered by electrons. In this process scattering generally occurs with outer-shell atomic electrons that are essentially free, and the energy of the incident photon is shared between the electron and the scattered quantum.

In the Compton process the energies of the scattered photon and electron, E'_γ and E_e, respectively, are given by the following relationships:

$$E'_\gamma = \frac{E_\gamma}{1 + \alpha(1 - \cos\theta)}, \tag{4.1}$$

$$\text{and} \quad E_e = E_\gamma - \frac{E_\gamma}{1 + \alpha(1 - \cos\theta)}, \tag{4.2}$$

Fig. 41. Plot of linear attenuation coefficient for NaI as a function of gamma-ray energy showing contributions from photoelectric effect, Compton scattering, and pair production.

where E_γ is the incident gamma-ray energy, α is the ratio E_γ/mc^2, and θ is the angle between the direction of the primary and scattered photons. From these relationships it may be deduced that a Compton-electron energy spectrum will extend from zero energy ($\theta = 0°$) up to a maximum energy ($\theta = 180°$) which is somewhat less than the energy of the incident photon. The energy of the scattered photon then extends from the original photon energy down to a minimum value that is always less than $mc^2/2$ (0.256 MeV). Figure 42 shows a Compton-electron energy distribution obtained by integrating the differential-scattering cross section over all angles from a primary photon

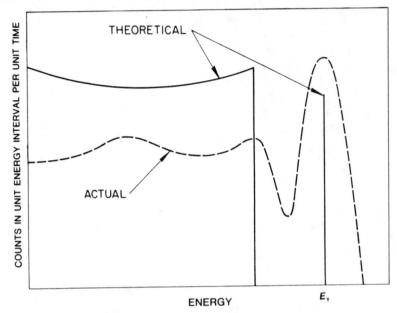

Fig. 42. Theoretical and actual electron-energy distributions for single Compton and photoelectric interactions in a 3″ × 3″ NaI(Tl) detector using a 0.511-MeV photon source. (From Heath, 1964)

energy of 0.511 MeV. A monoenergetic source of gamma radiation will produce such an energy distribution of electrons as a result of inter-action by the Compton process.

4.2.2.3 *Pair Production.* If the incident photon has an energy in excess of the rest mass of a positron-electron pair (1.022 MeV), then pair production is possible. In this process, which occurs in the presence of the Coulomb field of a nucleus, the gamma ray disappears and a positron-electron pair is created; their kinetic energy will be equal to the total energy of the gamma ray minus the rest energy of the two particles ($2mc^2$). When the positron comes to rest in the field of an electron, annihilation of the two particles occurs with the emission of two photons equal in energy to the rest mass of the particles (0.511 MeV). Interaction by the pair process in a detector will therefore deposit energy at least equal to the kinetic energy of the positron-electron pair. As will be discussed below, the possibility also exists of detecting one or both of the annihilation quanta by either the photo-electric or Compton process. These alternatives result in a complex electron-energy distribution for the pair process. A series of such events can result in the deposition in the detector of any energy between $E_\gamma - 1.022$ MeV and E_γ, the energy of the primary photon.

The total probability for detection of a gamma ray, expressed as the total linear attenuation coefficient (less the coherent contribution), will then be given by:

$$\mu = \mu_{\text{phot}} + \mu_{\text{Compt}} + \mu_{\text{pair}} \tag{4.3}$$

The total linear attenuation coefficient for NaI(Tl) and the contribution from the photoelectric, Compton, and pair processes is shown in Figure 41.

4.2.3. *The Pulse-Height Spectrum*

To provide a basis for describing a pulse-height distribution observed at the output of the scintillation or semiconductor detector in terms of the basic processes described above, let us briefly review the succession of events that produce it. Assume that a monoenergetic source of gamma radiation of approximately 0.5 MeV is incident upon a NaI(Tl)-crystal-phototube combination or semiconductor detector. Inspection of the linear attenuation coefficient for NaI(Tl) shown in Figure 41 indicates that these gamma rays will interact with the NaI(Tl) detector by both Compton scattering and the photoelectric effect in the ratio of about 6:1. If a sufficient number of gamma rays is detected to give a reasonable statistical sample, an electron-energy distribution similar to that portrayed by the solid line in Figure 42 will be produced within the volume of the detector. We see a monoenergetic line of photoelectrons and a continuous distribution of Compton electrons from zero energy up to a sharp cut-off at an energy somewhat below the "photoline." Shown for comparison is the pulse-height distribution from a $3'' \times 3''$ NaI(Tl) detector obtained with a differential pulse-height analyzer. Although the pulse-amplitude spectrum is similar in character, we observe that the energy distribution has been smeared by what appears to be a Gaussian function. This is due to fluctuations in the light output from the phosphor and to the statistical nature of the processes occurring in the electron-multiplier phototube.

The peak resulting from total energy loss (photopeak) is the distinguishing characteristic of all spectra. The location of this peak and its intensity are used to determine the energy and intensity of gamma rays producing a given pulse-height distribution. The energy response of NaI(Tl) is approximately linear, hence the peak location is about proportional to the photon energy. Non-linearity effects in NaI(Tl) are discussed in Heath (1964). The width of this peak is a measure of the energy resolution of the detector—a subject that will be treated in more detail in a later section.

In addition to the amplitude smearing of the original electron spectrum, it is apparent that a much larger fraction of the total number of events has appeared in the "photopeak" than would be predicted from the ratio of the photoelectric and Compton cross sections. In a detector of this size, the probability that a Compton-scattered gamma ray will escape the volume of the detector is somewhat reduced and many multiple events (e.g., Compton scattering followed by a photo electron) can occur that result in total energy transfer. As this chain of events occurs well within the decay time of the phosphor, the energy loss from all events will sum to produce a pulse in the full-energy peak. As will be shown later, the relative probability for total energy loss in the detector due to the occurrence of multiple events will increase as the dimensions of the detector are increased.

The complicated pulse-height spectrum produced from monoenergetic photons incident upon a detector presents the basic problem in the interpretation and analysis of data obtained with a gamma-ray spectrometer. The variation in the shape of the pulse-height spectrum as the incident gamma-ray energy varies is shown in Figure 43. The lower portion of Figure 43 shows the energy positions of four gamma rays of 0.06-, 0.32-, 0.83-, and 1.92-MeV incident energy, all of equal intensity. Above this are the pulse-height spectra for each of these gamma rays obtained with a 3″ × 3″ NaI(Tl) detector. At 0.06 MeV, gamma rays interact with the detector almost entirely by the photoelectric process and the detector response is essentially a single peak, nearly Gaussian in shape. The peak of lesser intensity on the low-energy side of the full-energy peak is attributed to the escape of iodine x rays from the surface of the detector following a photoelectric interaction. This effect will be discussed in more detail in a later section. For the gamma-ray energy equal to 0.32 MeV, the pulse-height distribution becomes more complicated. In addition to the photopeak, the continuous distribution of pulses resulting from the detection of Compton electrons is seen. For the gamma ray of energy 0.83 MeV, the fraction of pulses in the photopeak is reduced as the Compton cross section becomes still more dominant. At 1.92 MeV, even more features appear in the pulse-height distribution. In addition to the Compton-electron distribution and the full-energy peak, several satellite peaks are superimposed upon the Compton distribution as a result of interactions involving the pair-production process. In this process, all energy in excess of the 1.022-MeV threshold is converted into kinetic energy of the positron-electron pair, which will normally be deposited in the crystal. If the energies of both annihilation quanta are totally absorbed in the crystal, then the energy deposited by all

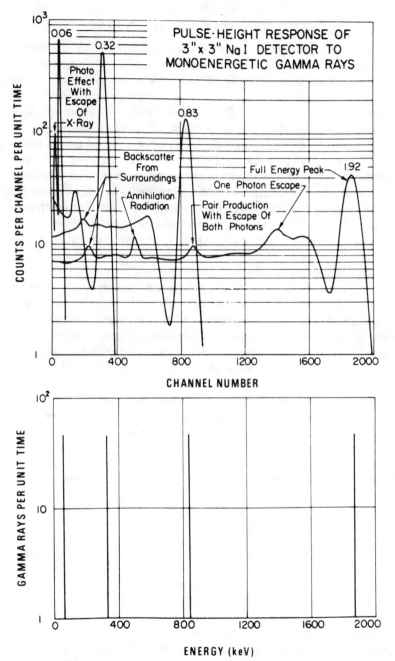

Fig. 43. Pulse-height distributions obtained with a 3″ × 3″ NaI(Tl) detector from monoenergetic gamma-ray sources of 0.06, 0.32, 0.83, and 1.92 MeV normalized to equal emission rates. Pertinent features of these spectra are noted, including the decrease in detector efficiency with increasing energy. (After Heath, 1964)

147

processes will produce a pulse in the full-energy peak. Subsequent annihilation of the positron will create two 0.511-MeV photons, one or both of which may escape from the crystal without detection. These events correspond, respectively, to the "single-escape" and "double-escape" peaks at 0.511 and 1.022 MeV below the full-energy peak (i.e., at 1.41 and 0.90 MeV). The total result of interactions involving the pair-production process will then be a distribution of pulses ranging from the full-energy peak to 1.022 MeV less than the photopeak, including the three prominent peaks just described. The backscatter peaks and the annihilation-radiation peak at 0.511 MeV arise from interactions with material surrounding the detector. For comparison, Figure 44 shows a pulse-height spectrum of gamma rays emitted in the decay of ^{88}Y, obtained with a Ge(Li) spectrometer. Apart from the improved energy resolution, the general character of the detector response is seen to be similar to that obtained with NaI(Tl) detectors. The lower atomic number of germanium and the smaller volumes used in these detectors as compared with NaI(Tl) scintillators lead to differences in the detector response to monoenergetic radiation. The lower atomic number results in a photoelectric cross section which is smaller by a factor of 10 to 20 and hence a photoelectric interaction is less probable. Total energy absorption in single or multiple interactions is also less probable because of the smaller volumes of the germanium detectors. Thus a much smaller fraction of the total energy deposited in the detector appears in the full-energy peak. The spectra in Figs. 15 and 44 show that the improved resolution of the semiconductor detector compensates for the smaller fraction of energy deposited in the peak and the peaks rise above the underlying continuum. Apart from this, the general features of the spectra are similar.

4.2.4 Pulse-Height Spectrometers

As outlined in the previous section, the proportional response of solid scintillators and semiconductor detectors to energy deposited in them by single ionizing events affords a means of measuring the energy of photons interacting with them. The deposition of the total energy of a single photon results in the production of a small amount of electrical charge that is collected at the output of the detector as a "pulse" of current whose amplitude is proportional to the energy of the detected photon. A modern pulse-height analysis system employed for particle- or photon-energy spectrometry consists of the following functional elements as shown in Figure 45: (1) scintillation or semiconductor detector; (2) linear pulse amplifier; and (3) a multichannel

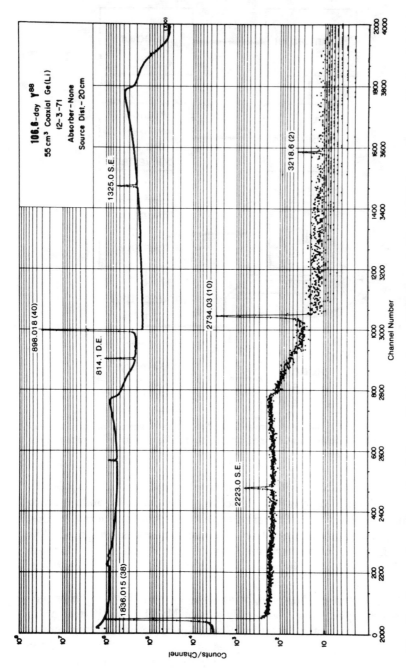

Fig. 44. Gamma-ray spectrum of radiation emitted in the decay of ^{88}Y observed with a Ge(Li) detector system. The letters S.E. and D.E. refer to single and double escape peaks, respectively, and the values in parentheses are the uncertainties in the last digits. (From Heath, 1974)

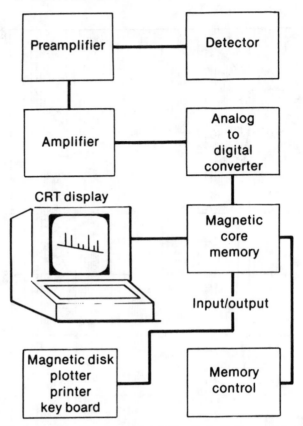

Fig. 45. Block diagram of a conventional "hard-wired" gamma-ray spectrometer system used for pulse-amplitude analysis of the output from scintillation and semiconductor radiation detectors.

pulse-height analyzer with suitable display and readout devices. Although a detailed description of nuclear electronics is beyond the scope of this discussion, a brief description of the design considerations that are important to the operation of these systems is presented. The reader is referred to references Chase, 1961; Heath, 1964; Fairstein and Hahn, 1965; and Nicholson, 1974, for a more complete treatment of this subject.

4.2.4.1 *Detector.* Pulse-height analysis systems for photon or particle-energy spectrometry generally employ NaI(Tl) scintillation or Ge and Si semiconductor detectors. Special-purpose high-voltage power supplies are required for both types of detector. In the case of the scintillation detector, a very stable high-voltage source is needed to provide accelerating voltage for the electron-multiplying structure of the phototube. Semiconductor devices require a low-noise source of several thousand volts to insure collection of charge created in the detector. The operation of semiconductor detectors at liquid-nitrogen temperatures requires the use of a vacuum-insulated (Dewar) container. The semiconductor device itself is contained in an evacuated chamber, mounted on the end of a long metal rod, or "cold finger," designed to conduct heat from the detector to the liquid nitrogen. A typical detector arrangement for a Ge detector is shown in Figure 46. As indicated, the vacuum in the detector housing is usually maintained by absorption-pumping techniques. In the measurement of radioactivity, it is often necessary to operate these detectors in a large shielded enclosure to reduce the probability of detecting background radiation. A typical shield installation for a laboratory spectrometer is shown in Figure 47.

4.2.4.2 *Pulse Amplifier.* A linear pulse amplifier is used to amplify the small electrical pulses created in the detectors to a level sufficient for measurement of pulse amplitude. In the scintillation detector, the phototube not only converts the light emitted by the scintillator to an electrical impulse, but serves as a very fast, low-noise amplifier. Phototubes that are used with NaI(Tl) scintillation detectors have current gains ranging approximately from 10^6 to 10^8. The very minute

Fig. 46. Illustration of typical method employed in the operation of semiconductor radiation detectors. To maintain the integrity of Ge(Li) detectors and to achieve optimum electronic-noise levels these detectors must be maintained and operated at cryogenic temperatures. HPGE and Si(Li) detectors are operated at 77K but may be stored at room temperature.

Fig. 47. Photograph of typical shield arrangement used in the operation of a laboratory Ge gamma-ray spectrometer.

amount of energy associated with the detection of single photons in radiation detectors requires that special care be employed to amplify these signals with no addition of electrical noise from the amplifier circuits. The principal effect of the addition of noise to the signal will be a deterioration in the energy resolution of the system. In the case of semiconductor detectors, the level of signal produced at the output of the detector is so minute that very sophisticated techniques must be employed in the design of low-noise-amplifier circuits to realize the energy-resolution capabilities of these devices.

A basic understanding of the importance of the characteristics of amplifiers used in these systems is important to an appreciation of the pulse-amplitude measurement technique. To achieve optimum signal-to-noise ratios, the first stages of the amplifier system are frequently mounted inside the detector cryostat (as shown in Figure 46) with field-effect transistors (FETs) operated at temperatures in the range of 100 to 130K to reduce the thermal-noise level in the transistors. The development of this technology, pioneered for nuclear-radiation detectors, has found other applications in the detection and measurement of very minute electrical signals.

Following the first stages of amplification in the "preamplifier," the pulses are further amplified and shaped by integration and differentiation circuitry. Figure 48 shows examples of pulses from a typical detector at various stages in the amplification process. Part (a) of this figure shows pulses at the output of the preamplifier. Typically, pulses at this point in the system are characterized by a fast-rising leading edge (<0.1 μs) and a slow decay (typically 100 μs). Parts (b) and (c) show typical output pulses from a linear pulse amplifier, after they have been shaped to produce symmetrical pulses that return to zero in a finite time. Part (b) is the pulse shape normally associated with low-noise amplifiers used with semiconductor detectors. Since it rises and returns to zero and ideally does not cross the zero level, it is termed a unipolar or "single-differentiated" pulse shape. Part (c) illustrates a bipolar or "double-differentiated" pulse shape that is

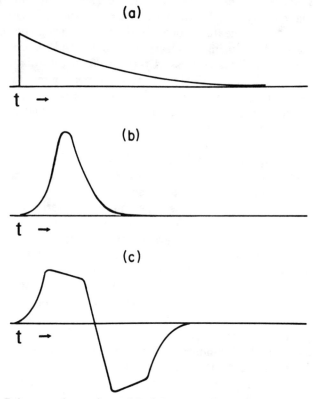

Fig. 48. Pulse wave forms observed in detector-amplifier systems: (a) preamplifier output pulse, (b) unipolar output pulse from shaping amplifier using RC circuit elements, (c) bipolar or "double-differentiated" shaping amplifier output.

frequently employed with NaI(Tl) scintillation detectors. There are many experimental variables which influence the choice of pulse-shaping networks and specialized pulse-conditioning circuitry employed in pulse-counting systems. Of these, the most important are noise performance (related to energy resolution) and pulse width (related to time resolution of counting systems). The choice of circuit parameters that determine the pulse width can be varied to satisfy the measurement requirements. Typically, NaI(Tl) spectrometers use bipolar shaping to produce pulse widths of 2 to 3 μs. In the use of semiconductor detectors, the shaping "time constants" will generally vary from 2 to 10 μs. This results in pulses with a width at the base which varies from 5 to 25 μs, with consequent increased counting losses, compared with NaI(Tl). The general subject of counting losses is discussed in Section 2.7. Specific counting-loss problems in the use of semiconductor detectors are discussed in Cohen (1974).

4.2.4.3 *Multichannel Pulse-Height Analyzer.* The multichannel pulse-height analyzer (PHA), the heart of the pulse-height-analysis system, provides a means of quantitatively measuring the maximum amplitude reached during an electrical impulse and recording this as a digital number. This process is termed "analog-to-digital conversion," and the circuitry developed for such measurements is termed an "analog-to-digital converter" (ADC). The approach that is used is to divide the voltage range to be measured into n equal intervals or bins, so that a 0- to 10-V range of possible pulse sizes can be divided into 10, 100, or 1000 equal intervals. In this manner, if the peak amplitude reached for a single pulse falls within one of these intervals, or channels, its amplitude is assigned the number of that interval within the 0- to 10-V-amplitude range.

The first instruments of this type employed a series of amplitude discriminators that would indicate whether or not the amplitude of a pulse fell between two voltage levels. For example, a 0- to 10-V range might be divided into 10 equal divisions of 1 V. A pulse having an amplitude between 5 and 6 V would then exceed the voltage levels of the first five discriminators but not the sixth discriminator. An anti-coincidence circuit was then employed to produce an output only from the discriminator at the 5-V level. Thus, the term "channel" signifies that a pulse has a measured maximum amplitude that resides somewhere between the boundaries of voltage increment associated with channel number 5. As the art of pulse-amplitude analysis advanced and detector energy resolution improved, there was a need for improved accuracy in amplitude measurement and a requirement for dividing the voltage range into finer increments or more "channels." Gamma-ray spectrometers using NaI(Tl) scintillation detectors generally divide the amplifier output into 256 or 512 channels or voltage increments. High-resolution Ge spectrometers frequently employ an amplitude scale with 4096 or even 8192 channels.

The modern pulse-height analyzer most frequently used in nuclear spectroscopy utilizes a technique for analog-to-digital conversion that is based upon an amplitude-to-time concept first developed by Wilkinson (1950). In this type of ADC, the pulse input to the circuit is first "stretched" or held at its maximum amplitude as shown in Figure 49c. A capacitor is then charged to a voltage proportional to this peak amplitude and allowed to discharge at a constant rate, and so produces a linear sloping ramp as shown in the figure. During the voltage

Fig. 49. Timing diagram of time-conversion type analog-to-digital converter (ADC) indicating sequence of events in the processing of a pulse to produce a digital number indicating relative amplitude of the maximum voltage.

discharge, pulses from a constant frequency oscillator are counted in a scaler for the time required for the capacitor to discharge to zero, as shown in Figure 49c, d, and e. In this manner a digital number is accumulated in the scaler, the value of which is proportional to the amplitude of the stretched pulse.

The remaining operations of Figure 49 indicate logic signals controlling the accumulation of data in the analyzer memory. As previously described, a pulse-amplitude spectrum is obtained by measuring the amplitude of a large number of pulses and accumulating a histogram or pulse-amplitude distribution that is a tally of all pulse amplitudes. Figure 50 is a simplified block diagram of a multichannel pulse-

Fig. 50. Simplified block diagram of a multichannel pulse-height analyzer used in photon spectrometry.

height analyzer. As shown, it consists of three functional segments: (1) an ADC to provide a means of measuring pulse amplitude; (2) a magnetic-core or semiconductor memory consisting of a number of registers (one for each amplitude channel) to tally the number of pulses having an amplitude within a given voltage increment; and (3) an input/output section that provides for oscilloscope display of data and output of digital information in printed form or on other data storage media such as paper tape or magnetic tape. A photograph of a laboratory pulse-height analyzer is shown in Figure 51.

To assist in understanding the operation of a PHA, reference is made to the timing diagram shown in Figure 49, which illustrates the sequence of events that occurs when a pulse appears at the input of the analyzer. At the end of the capacitor discharge the gated scaler contains a number that represents the amplitude of the pulse, as shown in Figure 49c, d, and e. This number corresponds to one of the "channels" in the memory. The memory "bin" of this channel is identified and incremented by "1" to complete the measurement of an individual pulse. In this manner, as events occur in the detector, the amplitude of each pulse is measured and recorded in the PHA memory.

Fig. 51. Photograph showing typical commercial laboratory pulse-height analyzer system.

The measurement of many events will then result in the collection of a differential-amplitude spectrum.

The fastest analyzers in use today use the method of *successive approximation* to measure the final output-pulse height. This method is based on building up a "comparison" amplitude by means of very fast binary increments and comparing this with the actual output-pulse amplitude, using a comparator. The great advantage of this method is its high speed and the time required to process a pulse which is independent of its amplitude.

Digital buffers enable the processed pulse-amplitude data to be stored and then transferred by an interface to commercially available microcomputers.

Once again it must be emphasized that, in this explosively developing era of microelectronics and integrated circuits, it is useless to try to give, within the limited scope of this report, a full and comprehensive description of the very complex pulse-processing equipment now available. As is pointed out in Section 6.5, at the present time the appropriate function is selected and purchased.

4.2.4.4 *Instrument Dead Time.* An examination of the timing diagram of Figure 49 indicates that the input to the analog-to-digital converter, Figure 49g, is closed during the processing of a pulse, so that any pulses arriving at the input of the ADC during this "dead time" will not be processed. If we are interested in the quantitative measurement of radiation striking the detector, this "counting loss" must be accounted for. The amount of "dead time" required in this type of ADC will be dependent upon the amplitude of the pulse which is being measured in terms of clock cycles. Typical ADCs use an oscillator frequency of 100 MHz, so that the time required to measure a pulse amplitude will vary from 0 to 40 μs for a 4000-channel amplitude scale. To this we must add the fixed time for memory addition and other "housekeeping" operations, which typically require 5 to 10 μs. In view of the difficulties involved in calculating an effective "dead time" or fraction of clock time during which incoming pulses cannot be measured, specialized circuitry is incorporated into PHAs to correct experimental data for this instrumental dead time. This correction for counting losses is accomplished with a crystal-controlled clock oscillator whose output pulses are stored in a register during a measurement. During the processing of pulses by the ADC the clock oscillator is turned off. Thus, in the absence of pulses at the input of the ADC, this register accumulates "real clock time." During a measurement, the action of the gate will only allow the clock to store pulses during the amount of time that the ADC is available for processing of data. In this manner the clock register accumulates elapsed "live time" to provide a running correction for loss of time due to ADC processing. In present pulse-height analyzers, these "live timers" provide automatic correction of ADC counting losses with an accuracy better than 1 percent over the normal operating range of input counting rate. Corrections for combined dead-time and pulse-pile-up losses have been considered in detail by Cohen (1974). Also see BIPM, 1981a and 1981b for extensive bibliographies on these effects. For other methods of determining losses, see Sections 2.7.2.3 and 2.7.2.4.8.

4.2.5 Experimental Effects in Photon Spectrometry

In the analysis of x- and gamma-ray spectra obtained with scintillation and semiconductor systems, there are a number of experimental effects that must be considered in order to obtain correct photon energies and intensities. Important considerations include: radiation scattered into the detector from surrounding material, distortion of the pulse spectrum at high counting rates due to random coincidence effects, and limitations in the electronic system. A brief description of these problems is presented to provide insight into these experimental problems.

4.2.5.1 *Effect of Environment on Detector Response.* If we could isolate the source and detector from all surrounding material, the shape and magnitude of the observed pulse-height spectrum would be dependent only upon the energy of the gamma ray, the physical properties of the source and detector, and the geometrical arrangement. In practice, the shape of the observed pulse-height distribution is influenced by many factors. As the response of the detector to its experimental environment must be understood in any attempt to analyze data, let us examine the many experimental variables that can influence the response of the detector. Figure 52 shows the experimental pulse-height spectrum resulting from the detection of a monoenergetic gamma-ray source (0.478 MeV) by a 3″ × 3″ NaI(Tl) detector. This spectrum was obtained under experimental conditions that had been optimized to reduce extraneous effects to a minimum. The spectrum has been measured with and without a 670 mg/cm^2 polystyrene absorber that is normally interposed between detector and source to prevent the entry of beta particles into the detector. The spectrum shows extraneous contributions that do not represent the response of the detector to gamma radiation directly incident on the detector.

The small peak appearing on the Compton distribution results from gamma rays Compton-scattered from the surrounding materials in the

Fig. 52. Experimental response of NaI(Tl) detector to a monoenergetic gamma-ray source showing contribution to observed pulse-height spectrum from scattering and other experimental effects. (From Heath, 1964)

Fig. 53. Pulse-height spectrum obtained with a 3″ × 3″ NaI(Tl) scintillation detector from a monoenergetic gamma-ray source, ⁵⁴Mn, measured in three different detector-shield configurations, illustrated in Fig. 54. (After Heath, 1964)

detector shield, air, source, and source mount. The calculated energies of these scattered gamma rays is equal to or less than 0.25 MeV. The magnitude and shape of the contribution from scattered radiation is dependent upon the geometrical arrangement. Figure 53 shows the pulse-height spectrum for a 0.835-MeV gamma ray measured in the three different shield configurations illustrated in Figure 54. Inspection of the scatter spectrum shows that the intensity and shape of the distribution of pulses attributable to scattering is inversely proportional to the size of the shield and to the Z of the absorbing material. The reduction in scattering from lead as opposed to iron is due to the relatively high photoelectric-absorption cross section of the high-Z material. It is of interest to inspect the shape of the scattered spectrum from the small iron shield. Note the presence of two distinct peaks in this distribution. These correspond to 180° single scattering and to processes involving two successive large-angle Compton events before striking the detector. An important deduction from these measure-

Fig. 54. Detector-shield configurations used to illustrate the effects of scattered radiation shown in Fig. 53. The large lead shield encloses a 32-inch-sided cubic volume. The dimensions shown in the smaller shield represent those of the two 6″ × 6″ lead and iron shields. The inside surfaces of the large lead shield are graded with cadmium and copper as shown in the insert. (After Heath, 1964)

ments is that appreciable scattering from surrounding material considerably degrades the quality of the spectrum. The presence of a large contribution to the pulse-height spectrum from scattering will reduce the "uniqueness" of the energy response.

4.2.5.2 *Summation Effects.* If Figure 52 is examined in the region above the photopeak, a continuous distribution of pulses is seen that extends up to approximately twice the amplitude of the full-energy peak. This contribution to the monoenergetic energy response of the detector results from random time-coincidence between pulses. Because radiations emitted from a radioactive source are randomly distributed in time, there is a finite chance that two events may occur in the detector within the resolving time of the electronic circuitry. If this occurs, then the amplitude of the resultant pulse will be the sum of the amplitudes of the two separate events. The count rate in this "random sum spectrum" will be related to the source intensity and the system resolving time by the following simplified relationship:

$$N_{\text{rss}} = 2\tau N^2, \tag{4.4}$$

where N is the input pulse rate, τ is the resolving time of the electronic system for pulse pairs, and N_{rss} is the total sum-pulse rate in the sum spectrum. Figure 55 illustrates the superposition of two pulses that occur within the sampling time of the pulse-height analyzer. Two pulses, with amplitudes V_1 and V_2 are shown partially superimposed in time. The sum of these two pulses is shown by the dashed line. Clearly, the shape of the resultant pulse will vary with the time interval between the two pulses and their respective amplitudes. The complex pulse that results can differ markedly from the normal pulse shape presented to the pulse-height analyzer. Depending upon the method used in the analog-to-digital converter to determine pulse amplitude, the recorded amplitude for such pulses can vary from V_1 to the highest amplitude reached in the region of overlap. The result is the observed "smeared" pulse-height distribution. Pulses in this sum spectrum represent the detection of photons directly incident upon the detector

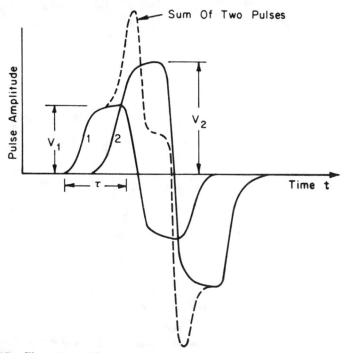

Fig. 55. Illustration of the effect of superposition of two pulses within the sampling time of the ADC. Variations in overlap will result in smearing of the random-coincidence sum spectrum. (From Heath, 1964)

and so must be considered to be part of the energy response of the detector.

This random sum spectrum should not be confused with the so-called "coincidence sum spectrum," which is illustrated in Figure 56. This figure shows the spectrum of radiations emitted by ^{94}Nb. In the decay of a nucleus of this radionuclide, two gamma rays are emitted simultaneously. In addition to the response of the detector to the two gamma rays individually, we also see a distribution of pulses that results from the simultaneous detection of both coincident gamma rays. The resultant sum spectrum is really a convolution of the two response functions for the individual gamma rays. The most prominent

Fig. 56. Pulse-height spectrum of radiation emitted in the decay of ^{94}Nb showing the "true" coincidence spectrum (calculated sum spectrum) due to cascade emission of gamma rays in the decay of a single nucleus. (c.f. Fig. 25.) The source is located 10 cm from a 3″ × 3″ NaI(Tl) detector. (After Heath, 1964)

feature of this spectrum is the so-called sum peak that results from the coincident detection of pulses from the photopeaks of the two coincident gamma rays. The probability for the detection of pulses in this sum spectrum, N_{css}, is given by the following expression:

$$N_{css} = N_0 \epsilon_1 \epsilon_2 \overline{W}(\theta), \qquad (4.5)$$

where N_0 is the disintegration rate of the source, ϵ_1 and ϵ_2 are the total detector efficiencies for the detection of gamma rays 1 and 2, respectively, and $\overline{W}(\theta)$ is an averaged factor to account for the angular distribution of the coincident gamma rays. Corrections for peak losses and gains due to coincident summing in semiconductor detectors are given by McCallum and Coote (1975).

To summarize, we have seen that the shape of the pulse-height spectrum resulting from the detection of monoenergetic gamma rays may be influenced by many parameters. These include source preparation, source-detector geometry, size and shape of the detector, input count rate to the spectrometer, and the environment in which the measurement is performed. In addition, other radiations emitted by the source and the time relationships between emitted radiations may produce extraneous effects that must be considered in any analysis of the data. For these reasons it is most important that serious thought be given to the design of the experimental arrangement.

4.2.5.3 *Problems at High Counting Rates.* In the use of high-resolution spectrometers, experimental conditions are frequently far from ideal for optimum performance of the electronic systems employed. Perhaps the most difficult problems are those that arise when high counting rates are encountered. Without the use of special precautions, operation at high counting rates will result in serious shifts of zero baseline and system gain, and degradation of resolution. Most of these observed effects are the result of fluctuations in the zero reference baseline at the input of the analog-to-digital converter, produced by random fluctuations. Some insight into these effects and the nature of the specialized circuitry employed to correct these problems is provided by reference to Figure 57. To achieve optimum signal-to-noise ratio in high-resolution systems, a monopolar pulse shape with equal RC integration and differentiation time constants is generally employed (Figure 57). The output pulse shape from such a filter network will exhibit a more-or-less Gaussian shape with equal rise and fall times. Following the pulse there will be a negative undershoot with a long recovery time constant (generally several hundred microseconds). At low count rates, the pulse will have returned to the original baseline and will have no influence on succeeding events. If, however, at high count rates, another pulse (indicated by the dotted

Fig. 57. The improved response of spectrometer systems with two circuit modifications. Parts (a) and (b) illustrate the response of a system at low count rates, with and without pole-zero cancellation, while (c) and (d) illustrate high-count rate performance of a system with and without d.c. restoration, in addition to pole-zero cancellation. (After Heath, 1969)

line) occurs, the negative tail of the preceding pulse will result in the ADC recording a reduced amplitude. At reasonable rates (a few thousand pulses per second) the net effect of this will be an asymmetry of spectral peak shapes as indicated at the far right of Figure 57a.

A circuit technique, first proposed by Nowlin and Blankenship (1965), quite effectively reduces the undershoot from a monopolar pulse. The net effect of this technique, called pole-zero cancellation, is shown in Figure 57b. Here we see that the undershoot returns more quickly to the original baseline, so that the next pulse can follow more closely without reduction in amplitude. It should be pointed out that pole-zero cancellation must be accomplished on the complete preamplifier-amplifier system to be effective. The use of such networks in low-noise-amplifier systems is now quite common and, if adjusted properly in combination with a given detector, will provide considerable reduction in spectrum degradation from undershoot at moderate counting rates.

At high counting rates (in excess of 5000 s^{-1}), we still have problems from fluctuations in the baseline as shown in Figure 57c. In this figure, we have reduced the time scale to show the long-term character of the baseline shift. As a result of residual charge on coupling capacitors within the a.c.-coupled amplifier system, the baseline will vary in a random manner to produce a net negative shift in the zero reference and a general broadening of peaks in the spectrum, toward lower energy, as indicated on the right side of the figure. This effect can be reduced by the use of d.c.-baseline-restoring circuitry. Typical circuits have been described by Chase and Poulo (1967). The action of these circuits is to remove the random fluctuations in the zero reference level. It should be stated that the input time constant on the restorer must be optimized for best performance at high rates. The most recent d.c. restorers have switch selection of this time constant to permit selection of the best operating conditions for a given experiment, although noise fluctuations will cause a small loss in resolution at low pulse rates for any setting. The result of restoring action, properly optimized, is illustrated in Figure 57d. Here we see that the zero reference baseline has been maintained at a constant level and the degradation of the peak resolution, seen without the restorer, has been largely removed.

4.2.6 Quantitative Analysis of Gamma-Ray Spectra

4.2.6.1 *Detector Efficiency.* The total detection efficiency ϵ_t, the fraction of gamma rays of energy E_γ, emitted from the source that interacts with the detector, can be calculated for known values of the linear absorption coefficient μ_{en}, for NaI(Tl) under a well-defined source-detector geometry. The equation derived for the case of a point source of radiation located on the extended axis of a right circular cylindrical detector is shown in Figure 58. In this expression, t_0 is the thickness of the detector, r_0 is the radius of the detector, and h_0 is the perpendicular distance between the source and the top surface of the detector. Extensive calculations have also been made for point, disk, and line sources located on the central axis of the detector (Heath, 1964).

In a solid material such as NaI(Tl), which has a density of 3.67 g/cm³, the sensitive volume is very clearly defined. Because the amount of material that a secondary electron, produced in the solid phosphor, must traverse to leave a measurable amount of energy is negligible, edge effects are insignificant.

$$\epsilon_t = 1/2 \left\{ \int_0^{\tan^{-1}\frac{r_0}{h_0 + t_0}} \left[1 - e^{-\mu_{en}\frac{t_0}{\cos\theta}} \right] \sin\theta \, d\theta \right.$$

$$\left. + \int_{\tan^{-1}\frac{r_0}{h_0 + t_0}}^{\tan^{-1}\frac{r_0}{h_0}} \left[1 - e^{-\mu_{en}\left(\frac{r_0}{\sin\theta} - \frac{h_0}{\cos\theta}\right)} \right] \sin\theta \, d\theta \right\}$$

Fig. 58. Expression for the calculation of the total detection efficiency, ϵ_t, of a right-circular cylindrical detector for a point source of radiation. (From Heath, 1964)

4.2.6.2 *Photopeak, or Full-Energy-Peak, Efficiency.* Consider a point source of monoenergetic radiation and a detector of specified size that is isolated from all material that might scatter gamma rays into the detector that would not originally have been emitted into the cone of solid angle subtended by the detector. The number of events in the observed pulse-height spectrum will then be related to the emission rate from the source by the equation given in Figure 58. If, however, there is a contribution of scattered radiation to the detector response, it is evident that this relationship will hold only if this contribution is negligible. In view of the complex nature of the detector response, it is more convenient and precise to work in terms of the photopeak efficiency. This efficiency, ϵ_p, is defined as the probability that a gamma ray of energy E_γ, emitted from the source, will appear in the photopeak of the observed pulse-height spectrum. The efficiency and the photopeak area provide a measure of emission rate that is independent of scattering geometry.

4.2.6.3 *Peak-to-Total Ratio.* It would be difficult to calculate ϵ_p directly because of the large number of multiple processes that occur in a scintillation detector. For this reason it is convenient to use the following expression:

$$\epsilon_p = \epsilon_t P \tag{4.6}$$

where P is defined as the fraction of the total number of events in the pulse-height spectrum that appears in the photopeak (the peak-to-total ratio). The peak-to-total ratio has been determined experimentally by careful measurement of selected sources under experimental conditions that reduce scattered radiation to negligible levels. As an example, a plot of experimental values of the peak-to-total ratio for a $3'' \times 3''$ NaI(Tl) detector as a function of energy is shown in Figure 59.

4.2.6.4 *Emission-Rate and Energy Measurements.* The emission rate of a single gamma ray will be given by the following relationship

$$N_0 = \frac{N_p}{\epsilon_t P A}, \tag{4.7}$$

where N_0 is the number of gamma rays emitted in unit time from the source, N_p (as shown in Figure 60) is the area under the photopeak, ϵ_t is the total detection efficiency for the source-detector geometry used, P is the appropriate value for the peak-to-total ratio, and A is a correction factor for absorption of the gamma radiation by any beta-particle absorber used in the measurement. This absorption correction is usually determined experimentally in order to account properly for measurement geometry. Values for ϵ_t, P, and A have been determined with considerable precision for a 3″ × 3″ cylindrical NaI(Tl) scintillation detector and are tabulated in Heath (1964).

As outlined above, a monoenergetic source of photons will produce a unique pulse-height distribution in a scintillation- or semiconductor-spectrometer system from which the energy and intensity of the photon source can be deduced. For most radionuclides the gamma-ray spectrum observed is generally more complex than that for a monoenergetic source. The observed detector response represents the summation of the response of the detector to individual gamma rays. To illustrate the methods required to interpret complex spectra, the response of a NaI(Tl) spectrometer to a source emitting several gamma

Fig. 59. Plot of experimental values of P, the peak-to-total ratio for a 3″ × 3″ cylindrical NaI(Tl) scintillation detector for source-detector distances of 3 and 10 cm. (From Heath, 1964)

Fig. 60.   Illustration of peak-to-total ratio concept for the NaI(Tl) pulse-height spectrum of ^{54}Mn. (From Heath, 1964)

rays of different energy is shown in Figure 61. The radiation source was a composite of three radioisotopes each emitting a single gamma ray. As shown, the three sources were measured individually and in combination. If reference spectra for each radionuclide are available, then the composite spectrum can be decomposed by successively subtracting each component spectrum, normalizing in turn to the highest-energy peak. This "stripping" process can be used to analyze complex spectra to obtain the energies and intensities of component gamma rays.

4.2.6.5   *Efficiencies of Semiconductor Detectors.* Although the general problem of quantitative measurement of radiation with these

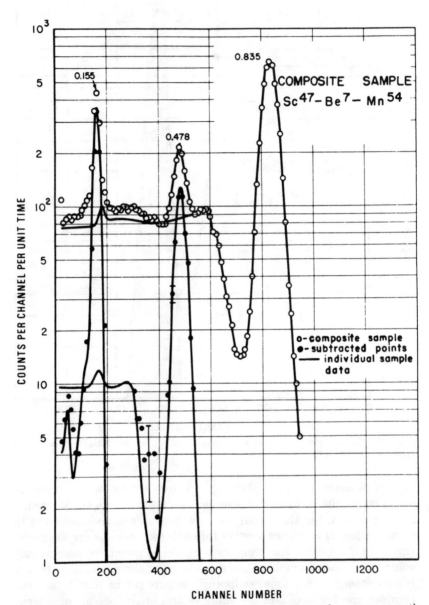

Fig. 61. Pulse-height spectrum showing detector reponse to three monoenergetic gamma-ray sources measured individually and in combination to illustrate approach to the analysis of complex gamma-ray spectra.

devices is identical to that of scintillation detectors, the detection efficiency is usually determined experimentally by using a selected set of gamma-ray sources with precisely known gamma-ray emission rates (McNelles and Campbell, 1973; Debertin et al., 1976; Hirshfeld et al., 1976). The gamma-ray spectrum of each source is measured at a specific source-to-detector distance. Unlike the large-volume scintillator, where the sensitive volume of the detector can be determined with high precision, the semiconductor device has a poorly-defined sensitive volume. Significant edge effects and scattering components in the response function make it difficult to define the complete response of such a detector to monoenergetic radiation with sufficient precision. For this reason it is customary to determine the photopeak efficiency as a function of energy and measurement geometry by using standard reference sources. A typical photopeak, or full-energy peak, efficiency curve, as a function of photon energy, for a large coaxial Ge(Li) detector is shown in Figure 62. This plot presents the photopeak efficiency for this detector for point sources of radiation measured at a distance of 10 cm from the detector face. To permit calibration of Si(Li) and Ge detectors, calibrated photon sources are available from the National Bureau of Standards and several commercial suppliers.

Fig. 62. Full-energy-peak efficiency measured, as a function of photon energy, for a 50-cm^3 coaxial Ge(Li) detector.

A set of standards that has been widely used to calibrate these detectors is shown in Table 9.

Source-detector geometry is an important consideration in the calibration of detector systems and must be reproducible with good precision. Rate-dependent factors, especially dead-time and summation effects, are also important in accurate efficiency determinations. These are described in Sections 4.2.4.4, 4.2.5.2, and 4.2.5.3. Section 2.7 treats rate-dependent corrections in general and Section 2.7.2.4.8 deals specifically with their application to gamma-ray spectrometry.

TABLE 9—*Photon point-source standard reference materials (SRM's) available from the National Bureau of Standards*[a]

Parent radionuclide	Half life	Photon energy keV	Photon probability per decay[c]
Gamma-Ray Sources[b]			
^{241}Am	432.2 ± 0.5 y	59.54	0.359 ± 0.004
^{57}Co	271.7 ± 0.2 d	122.06	0.856 ± 0.007
^{207}Bi	32.2 ± 0.9 y	569.70	0.977 ± 0.004
		1063.66	0.74 ± 0.03
^{137}Cs	30.0 ± 0.2 y	661.66	0.851 ± 0.003
^{94}Nb	(2.03 ± 0.16) × 10^4 y	702.64	1
		871.12	1
^{60}Co	5.2714 ± 0.0005 y	1173.24	0.99900 ± 0.00020
		1332.501	1
X-Ray Sources			
^{55}Fe	2.72 ± 0.02 y	5.89 ($K_{\alpha2}$), 5.90 ($K_{\alpha1}$), 6.49 (K_β)	
^{93m}Nb	16.0 ± 0.5 y	16.52 ($K_{\alpha2}$), 16.62 ($K_{\alpha1}$), 18.62 (K_β)	
^{125}I	59.6 ± 0.2 d	27.20 ($K_{\alpha2}$), 27.47 ($K_{\alpha1}$), 31 (K_β), 35.49 (γ)	
Gamma-Ray Sources of Radionuclides with Complex Decay Schemes			
^{75}Se	119.8 ± 0.1 d		
^{133}Ba	10.5 ± 0.1 y	See Appendix A.3	See Appendix A.3
^{152}Eu	13.4 ± 0.9 y		
^{228}Th[d]	1.913 ± 0.002 y	—	—
Mixture of:			
^{125}Sb	2.758 ± 0.001 y	Emission rates are specified	
^{154}Eu	8.55 ± 0.06 y	at 19 selected energies	
^{155}Eu	4.72 ± 0.02 y	between 27.4 and 1596.5 keV	

[a] Detailed decay schemes for most of these radionuclides are given in Appendix A.3.

[b] Photon probabilities per decay are not given in this table for those SRM's which are certified in terms of photon-emission rate. Such data are available in Appendix A.3 except for ^{93m}Nb.

[c] In order of increasing gamma-ray energy.

[d] ^{228}Th is primarily used as a standard because of the 2.614533-MeV gamma ray from ^{208}Tl later in the chain.

4.2.7 Methods of Spectral Analysis—NaI(Tl) Detectors

Typical pulse-height spectra obtained from gamma-ray spectrometers may result from the detection of a discrete set of monoenergetic gamma rays, or a gamma-ray continuum such as bremsstrahlung, or both in combination. The method of analysis to be applied in a given case will depend upon which type of spectrum is being measured.

4.2.7.1 *Graphical Analysis by Subtraction.* Graphical analysis has been the basic method employed for the analysis of pulse-height spectra by nuclear spectroscopists for some time. A plot of the pulse-height distribution is inspected and the component with the highest energy identified by its photopeak. The pulse-height distribution corresponding to the experimental response of the spectrometer to this component is then subtracted from the composite spectrum as indicated in Figure 61. The resulting difference is then inspected and the process repeated, in succession of decreasing energy, for each remaining component identified in the spectrum. This method is rather subjective and the accuracy of the result is highly dependent upon the number of gamma rays present in the spectrum, their relative intensities, the photopeak-to-total ratios, and the skill of the analyst. Estimation of error is highly dependent upon many experimental factors, such as gain shifts, different scattering conditions, and pulse pile-up.

4.2.7.2 *Linear Least-Squares-Fitting Methods.* For the analysis of spectra that exhibit well-defined photopeaks the so-called least-squares method of analysis offers a useful alternative to the graphical technique described above. This method can be programmed for computer use and has proved to be very satisfactory for the analysis of many classes of data (Heath, 1966). It assumes that the pulse-height spectrum to be analyzed consists of a sum, from 1 through c, of the gamma rays of known energy from the radionuclides j, each represented as a pulse-height spectrum of i channels, a through n. It is further assumed that *standard spectra*, representing the response of the detector to these radionuclides, are available for comparison. The total count rate in channel i, C_i, can be represented by the composite of these standard spectra, where the count rate in channel i from the jth component will be S_{ij}. If the standard spectra are all normalized to the same disintegration rate, then it is necessary to include a normalizing factor, α_j, which is the ratio of the intensity of component j in the composite to the normalized standards. We may then write

$$C_i = \sum_{j=1}^{c} \alpha_j S_{ij} + \delta_i, \qquad (4.8)$$

where δ_i is the residual, or departure of the synthesized value from the observed rate in the ith channel.

If it is assumed that the only error is the random fluctuation in the channel count, the principle of least squares may be applied. Then the sum of the weighted squares of the residuals, δ_i, can be minimized over all channels, *i.e.*,

$$\text{minimize} \sum_{i=a}^{n} \left(C_i - \sum_{j=1}^{c} \alpha_j S_{ij} \right)^2 w_i, \qquad (4.9)$$

where w_i is a weighting factor.

This yields a set of linear simultaneous equations that may be solved to obtain values for the α_j's. A matrix technique that can be readily programmed for machine solution is a most convenient method for solving these equations.

A number of computer programs using the simple linear least-squares technique have been applied to this problem (Salmon, 1961; Schonfeld *et al.*, 1966; Trombka and Schmadebeck, 1968). More sophisticated non-linear regression techniques have been developed that yield higher precision. An example is a program called SHIFTY (Helmer *et al.*, 1967b; see also Heath *et al.*, 1967) that allows the gain, as well as the intensity, of a whole spectrum, or of the portions of the spectrum for each radionuclidic component, to be varied during the fitting process.

This program computes intensities by a linear least-squares program. In addition, the best gain scale for each component is determined by a "direct-search" technique that operates as follows: If a composite is being fitted with two components and both intensities and both gains are variable, the intensities of the two spectral components are computed using an initial value for the energy scale. Then, the sum of the weighted squares of the residuals is computed. With initial computed intensity, the first component is gain-shifted by a small increment, and a new value of residuals is computed. If the new value for the residual is smaller than the previous value, the new gain is retained; if not, a small decrement is tried. Next, the second component is increased or decreased in gain by small amounts. After the best gains among this set have been determined in this manner, a new set of intensities is calculated. This procedure continues until the shifts cease to have significant effect on the comparison between the original data and the sum of assumed components.

Figure 63 shows an example of an application of this program. A five-component spectrum containing ten gamma rays was analyzed with this program by using a library of experimental spectra as

Fig. 63. Example of least-squares analysis of a five-component spectrum by using computer-data-analysis techniques.

response functions for the individual radionuclidic components. For this purpose, the reference spectra must be first normalized to a standard gain or energy scale. To demonstrate the method, the five components were first measured individually, and were then combined to produce a composite spectrum. As indicated by the residuals in the figure, the least-squares analysis shows little systematic error.

4.2.8   *Methods of Spectral Analysis—Ge Detectors*

As previously indicated, spectra obtained with Ge detectors, Ge(Li) or HPGe, are different from those obtained with scintillation detectors in several ways. First, the energy resolution is greatly enhanced so that individual photopeaks are usually resolved. For this reason an important simplification can be made in the interpretation of pulse-height spectra obtained with these detectors. The essential information, gamma-ray energy and intensity, is represented by the position and magnitude of individual photopeaks.

In Figure 64a two cases are shown. On the right-hand side of the figure a single peak is shown, with all experimental quantities defined. A Gaussian function and a straight-line continuum (background) have been fitted to the experimental points. On the left-hand side the peak is more complex, consisting of two unresolved components. The best representation of this region of the spectrum would be two peak-response functions plus an underlying continuum. The fitting procedure involves the use of a "non-linear least-squares" technique. The simple approach to the analysis of well-resolved single peaks in a pulse-height spectrum is to represent the peak response of the detector in a spectrum by a Gaussian function, $Y(x)$, of the form:

$$Y(x) = y_0 e^{-(x-x_0)^2/b_0} \tag{4.10}$$

where x is the channel number, x_0 is the midpoint of the peak, b_0 is related to the width of the peak, and y_0 is the peak maximum. The underlying background is then represented by a straight line function or a polynomial. By using a linear least-squares procedure, data points in the region of the peak may be fitted to a combined Gaussian distribution and a straight line or other functional form (for the background continuum) in order to determine the most probable location of the peak on the horizontal axis to give the gamma-ray energy, and the area under the peak to give the intensity of the gamma ray.

In the case of unresolved multiplets such as that illustrated by the lower-energy peak in Figure 64a, a more detailed representation of detector peak response is required if the area under the peak is to be accurately computed. Figure 64b illustrates the peak response of a large coaxial Ge detector to 1.332-MeV gamma rays emitted in the decay of ^{60}Co. Observing the experimental points, the peak appears to be essentially Gaussian in shape on the high-energy side, becoming asymmetric on the low-energy side, slowly leveling off at a value greater than the baseline on the high-energy side of the peak. This characteristic shape is typical at moderate counting rates. As the counting rate is increased, asymmetry will also appear on the high-energy side of the peak as a result of "pile-up" (BIPM, 1981a; Andai and Jedlovsky, 1983) or coincidence summing effects in the electronic system.

Fig. 64a. Segment of a Ge gamma-ray spectrum. The first peak illustrates an unresolved pair. The second peak indicates the essential parameters used in the analysis of a spectrum to obtain peak location, width, and area.

Fig. 64b. Response of a large coaxial Ge detector to the 1.332-MeV gamma rays from the decay of ^{60}Co, showing low-energy tailing. The result of a least-squares fit of the data using several functions to represent the detector peak response is indicated.

The shape on the low-energy side of the peak is due to one or more of the following conditions: (1) incomplete charge collection due to trapping by defects in the detector material, (2) escape of secondary electrons or photons from the detector sensitive volume, (3) Compton scattering of the primary photons in surrounding material, and (4) statistical noise effects in the detector and electronics. In addition, an underlying continuum may be present due to Compton electrons from the scattering of higher-energy gamma rays present in the spectrum.

In the analysis of complex spectra with unresolved peak multiplets, an attempt is made to account for the tailing observed on the low-energy side of the detector peak response. A detector peak response such as that shown in Figure 64b may be represented by several components. The main component is a Gaussian function, that is not shown in Figure 64b, but which will fit inside the experimentally measured peak, much as a finger fitting inside a glove, with a peak height of about 90 percent of that of the measured peak. The low-energy asymmetry is then represented by a short-term function (1) corresponding to the major part of the asymmetry, and a long-term function (2) corresponding to the slowly decaying tail. The last component shown, the step function (3), represents Compton-scattered photons and coincidence summing of multiple photon events. The background continuum, which is not shown in Figure 64b may be represented by a polynomial extending to low energies from the high-energy side of the step function (3). In Figure 64b the magnitudes of these components have been computed by non-linear least-squares fits to best represent the system.

The fitting of a photon peak in this manner enables one to locate the peak channel of the multichannel spectrum for the photon of interest. It is, however, necessary to extract from the raw data the emission rate and energy of each photon, and then to proceed, in many cases, to an identification of the radionuclide in the source giving rise to the observed spectrum. To obtain this information from the raw data it is necessary to perform the following operations, which are described in more detail in the next sections:

(1) locate the median positions of the peaks in the pulse-height distribution by fitting a number of component distributions as described above;

(2) establish a precise energy vs. channel-number scale by suitable calibration techniques;

(3) assign energies to the photons observed in the pulse-height spectrum from the information obtained in the first two steps;

(4) compute the areas of the peaks in question after having subtracted the underlying continuum;

(5) compare the calculated energies of the peaks with a list of energies of all the characteristic x rays and gamma rays to identify the radionuclide sources of the photons detected in the spectrum;

(6) determine the concentration of each radionuclide present in the sample from the peak areas by suitable calibration techniques.

4.2.8.1   *Energy vs. Pulse-Height Scale.* The essential problem involved in the measurement of gamma-ray energy with Ge spectrometers is to establish a precise energy vs. pulse-height scale. This requires that the departures from linearity of the pulse-height analysis system be measured with a high degree of precision, to typically 0.01 channel, to permit interpolation. For this purpose it is customary to use radioactive sources emitting photons of precisely known energies and to identify the peak channel number corresponding to each photon energy. A typical calibration curve is shown in Figure 65, in which the departure from linearity over a range from about channel 200 to channel 4000 is not greater than ±0.3 channel.

Having calibrated the entire spectrometer system for non-linearity of response to energy deposited in the detector, the actual energy calibration of the system for the particular gain chosen is carried out by exposing the detector to photon-emitting reference sources with accurately known photon energies. The spectrometer system can now be used to measure photon energies and intensities over a wide energy range.

If it is required to make a more precise energy measurement of an unknown spectral line, an interpolation technique, illustrated in Figure 66, is employed. In this technique, the unknown line is recorded *simultaneously* with several lines of slightly greater and slightly less, and accurately known, energy. The energy of the unknown can then be determined as indicated in Figure 66, and the calibration curve of Figure 65, for the appropriate region of interest, can then be used to obtain a more precise interpolation for the energy of the unknown.

A list of reference calibration sources for which the energies have been measured with high precision is given in Table 10.

A block diagram of typical programs used for analysis of Ge pulse-height spectra is presented in Figure 67. After collection of a spectrum, the following procedure is used: (1) an algorithm is used to search the spectrum with a Gaussian function to determine the location of all significant peaks; (2) the contribution to each peak from the underlying background continuum is determined; (3) a least-squares fit of the data points, after background subtraction, in the vicinity of each peak is performed to determine the centroid and area of each peak;

TABLE 10—Gamma Rays Used as Reference Standards for Energy Calibration[a,b,c,d]

Parent Nuclide	Gamma-ray Energy	Parent Nuclide	Gamma-ray Energy	Parent Nuclide	Gamma-ray Energy	Parent Nuclide	Gamma-ray Energy
^{241}Am	59.537 ± 0.001	^{192}Ir	295.958 ± 0.001	^{110m}Ag	744.277 ± 0.003	^{22}Na	1274.542 ± 0.008
^{75}Se	66.060 ± 0.007	^{160}Tb	298.580 ± 0.002	^{95}Zr-^{95}Nb	756.729 ± 0.012	^{182}Ta	1289.156 ± 0.005
^{182}Ta	67.751 ± 0.001	^{133}Ba	302.853 ± 0.001	^{110m}Ag	763.944 ± 0.003	^{59}Fe	1291.596 ± 0.008
^{153}Gd	69.673 ± 0.001	^{75}Se	303.924 ± 0.003	^{95}Zr-^{95}Nb	765.807 ± 0.006	^{82}Br	1317.476 ± 0.007
^{133}Ba	80.998 ± 0.003	^{192}Ir	308.457 ± 0.001	^{82}Br	776.517 ± 0.003	^{132}Cs	1317.927 ± 0.008
^{170}Tm	84.255 ± 0.001	^{192}Ir	316.508 ± 0.001	^{95m}Tc	786.198 ± 0.004	^{124}Sb	1325.512 ± 0.006
^{182}Ta	84.680 ± 0.001	^{51}Cr	320.084 ± 0.001	^{58}Co	810.775 ± 0.009	^{60}Co	1332.502 ± 0.005
^{160}Tb	86.788 ± 0.001	^{133}Ba	356.017 ± 0.001	^{110m}Ag	818.031 ± 0.004	^{56}Co	1360.206 ± 0.009
^{109}Cd	88.034 ± 0.001	^{133}Ba	383.851 ± 0.002	^{95m}Tc	820.624 ± 0.005	^{124}Sb	1368.164 ± 0.009
^{160}Tb	93.919 ± 0.006	^{113}Sn	391.702 ± 0.004	^{82}Br	827.827 ± 0.005	^{24}Na	1368.630 ± 0.009
^{75}Se	96.734 ± 0.002	^{75}Se	400.657 ± 0.002	^{54}Mn	834.843 ± 0.006	^{182}Ta	1373.836 ± 0.005
^{153}Ge	97.432 ± 0.001	^{203}Pb	401.325 ± 0.002	^{95m}Tc	835.149 ± 0.005	^{110m}Ag	1384.300 ± 0.004
^{182}Ta	100.106 ± 0.001	^{198}Au	411.804 ± 0.001	^{56}Co	846.764 ± 0.006	^{182}Ta	1387.402 ± 0.005
^{153}Ge	103.181 ± 0.001	^{108m}Ag	433.936 ± 0.004	^{58}Co	863.959 ± 0.010	^{82}Br	1474.884 ± 0.006
^{177}Lu	112.954 ± 0.003	^{110m}Ag	446.811 ± 0.003	^{94}Nb	871.119 ± 0.004	^{110m}Ag	1475.788 ± 0.006
^{182}Ta	113.672 ± 0.001	^{192}Ir	468.071 ± 0.001	^{185}Os	874.831 ± 0.004	^{144}Ce	1489.160 ± 0.005
^{182}Ta	116.418 ± 0.001	^{7}Be	477.605 ± 0.003	^{160}Tb	879.383 ± 0.003	^{110m}Ag	1505.040 ± 0.005
^{75}Se	121.119 ± 0.003	^{192}Ir	484.578 ± 0.001	^{185}Os	880.286 ± 0.003	^{110m}Ag	1562.302 ± 0.005
^{57}Co	122.061 ± 0.001	^{85}Sr	514.009 ± 0.012	^{84}Rb	881.610 ± 0.003	^{82}Br	1650.389 ± 0.008
^{185}Os	125.360 ± 0.004	^{82}Br	554.348 ± 0.003	^{192}Ir	884.542 ± 0.003	^{58}Co	1674.730 ± 0.010
^{75}Se	136.002 ± 0.003	^{207}Bi	569.702 ± 0.002	^{110m}Ag	884.685 ± 0.003	^{124}Sb	1690.980 ± 0.006
^{57}Co	136.474 ± 0.001	^{95m}Tc	582.082 ± 0.003	^{46}Sc	889.277 ± 0.003	^{207}Bi	1770.237 ± 0.010
^{59}Fe	142.652 ± 0.002	^{228}Th	583.191 ± 0.002	^{88}Y	898.042 ± 0.004	^{56}Co	1771.350 ± 0.015
^{141}Ce	145.444 ± 0.002	^{192}Ir	588.585 ± 0.002	^{110m}Ag	937.493 ± 0.004	^{125}Sn	1806.701 ± 0.016
^{185}Os	152.431 ± 0.001	^{185}Os	592.076 ± 0.004	^{160}Tb	962.317 ± 0.004	^{56}Co	1810.722 ± 0.017
^{182}Ta	156.387 ± 0.001	^{124}Sb	602.730 ± 0.003	^{160}Tb	966.171 ± 0.004	^{88}Y	1836.063 ± 0.013
^{199}Au	158.379 ± 0.001	^{192}Ir	604.414 ± 0.012	^{124}Sb	968.201 ± 0.004	^{125}Sn	1889.895 ± 0.016
^{185}Os	162.851 ± 0.004	^{226}Ra	609.320 ± 0.002	^{84}Rb	1016.162 ± 0.013	^{84}Rb	1897.761 ± 0.014

Nuclide	Energy	Nuclide	Energy	Nuclide	Energy	Nuclide	Energy
^{139}Ce	165.857 ± 0.006	^{192}Ir	612.466 ± 0.002	^{56}Co	1037.844 ± 0.004	^{56}Co	1963.714 ± 0.012
^{182}Ta	179.395 ± 0.001	^{108m}Ag	614.281 ± 0.004	^{95m}Tc	1039.264 ± 0.006	^{132}Cs	1985.638 ± 0.008
^{59}Fe	192.349 ± 0.005	^{82}Br	619.107 ± 0.004	^{82}Br	1044.002 ± 0.005	^{125}Sn	2002.147 ± 0.013
^{160}Tb	197.035 ± 0.001	^{110m}Ag	620.360 ± 0.003	^{124}Sb	1045.131 ± 0.004	^{56}Co	2015.179 ± 0.011
^{182}Ta	198.353 ± 0.001	^{124}Sb	645.855 ± 0.002	^{207}Bi	1063.662 ± 0.004	^{56}Co	2034.759 ± 0.011
^{75}Se	198.596 ± 0.007	^{185}Os	646.129 ± 0.003	^{198}Au	1087.690 ± 0.003	^{124}Sb	2090.942 ± 0.008
^{95m}Tc	204.117 ± 0.002	^{110m}Ag	657.762 ± 0.002	^{59}Fe	1099.251 ± 0.005	^{56}Co	2113.107 ± 0.012
^{199}Au	208.206 ± 0.001	^{137}Cs	661.660 ± 0.003	^{65}Zn	1115.546 ± 0.004	^{226}Ra	2118.551 ± 0.007
^{177}Lu	208.362 ± 0.010	^{132}Cs	667.718 ± 0.003	^{46}Sc	1120.545 ± 0.005	^{144}Ce	2185.662 ± 0.007
^{160}Tb	215.646 ± 0.002	^{198}Au	675.887 ± 0.002	^{182}Ta	1121.301 ± 0.005	^{125}Sn	2201.019 ± 0.013
^{82}Br	221.480 ± 0.002	^{110m}Ag	677.623 ± 0.002	^{60}Co	1173.238 ± 0.005	^{56}Co	2212.921 ± 0.010
^{182}Ta	222.109 ± 0.001	^{203}Pb	680.514 ± 0.010	^{56}Co	1175.099 ± 0.008	^{125}Sn	2275.766 ± 0.011
^{182}Ta	229.322 ± 0.001	^{110m}Ag	687.015 ± 0.003	^{160}Tb	1177.962 ± 0.004	^{56}Co	2598.460 ± 0.010
^{185}Os	234.158 ± 0.004	^{144}Ce	696.510 ± 0.003	^{182}Ta	1189.050 ± 0.005	^{228}Th	2614.533 ± 0.015
^{228}Th	238.632 ± 0.002	^{82}Br	698.374 ± 0.004	^{182}Ta	1221.408 ± 0.005	^{24}Na	2754.030 ± 0.014
^{95m}Tc	253.068 ± 0.004	^{94}Nb	702.645 ± 0.006	^{182}Ta	1231.016 ± 0.005	^{56}Co	3201.954 ± 0.014
^{182}Ta	264.075 ± 0.001	^{110m}Ag	706.682 ± 0.003	^{56}Co	1238.286 ± 0.007	^{56}Co	3253.417 ± 0.014
^{75}Se	264.656 ± 0.004	^{124}Sb	713.781 ± 0.005	^{182}Ta	1257.418 ± 0.005	^{56}Co	3272.998 ± 0.014
^{133}Ba	276.398 ± 0.002	^{185}Os	717.436 ± 0.004	^{160}Tb	1271.880 ± 0.008	^{56}Co	3451.154 ± 0.014
^{203}Hg	279.197 ± 0.001	^{124}Sb	722.786 ± 0.004	^{182}Ta	1273.730 ± 0.005		
^{203}Pb	279.197 ± 0.001	^{108m}Ag	722.929 ± 0.004				
^{75}Se	279.538 ± 0.003	^{95}Zr-^{95}Nb	724.199 ± 0.005				

[a] From Heath (1974).
[b] From Helmer et al. (1978).
[c] From Helmer et al. (1979).
[d] From Greenwood et al. (1979).

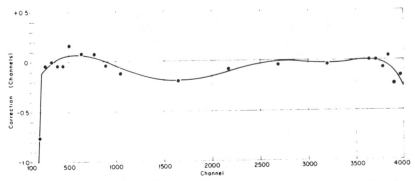

Fig. 65. Measured integral linearity of a 4096-channel gamma-ray-spectrometer system. (From Heath, 1969)

Fig. 66. Illustration of the measurement of a gamma-ray energy by simultaneous comparison with reference energy standards. ΔE is the deviation of the reference energy from the fitted function. (From Heath, 1969)

(4) instrumental linearity and energy calibration information is utilized to establish an energy *vs.* channel-number scale and the energy of each peak is determined from the peak position; and finally (5) the energies determined for each peak are compared with a reference list of known energies for all x rays and gamma rays emitted by each radionuclide to provide identification. A reference list of gamma-ray energies emitted by each radionuclide is provided in order of ascending energy for use with the "isotopic identification search routine." Some

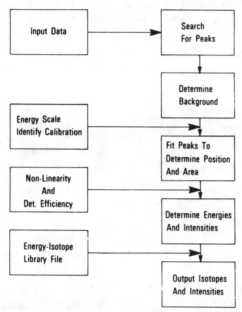

Fig. 67. Block diagram of data-analysis computer programs used in the analysis of Ge detector gamma-ray spectra for quantitative radionuclidic analysis of radioactive samples.

of the complex computer programs developed to analyze complex pulse-height spectra obtained with germanium detectors include GAUSS (Cline *et al.*, 1973), SAMPO (Routti and Prussin, 1969), GAMANAL (Gunnink and Niday, 1972), and HYPERMET (Phillips and Marlowe, 1976).

4.2.8.2 *Computer-Based Data Systems.* The problems encountered in the analysis of pulse-height data from x-ray and gamma-ray spectrometers require the application of rather sophisticated techniques to obtain the desired results. The advent of small, inexpensive digital processors has resulted in the development of automated techniques for the analysis and acquisition of pulse-height spectra. On-line computer data systems have been successfully employed for the analysis of pulse-height spectra in nuclear-physics applications for some time (Heath, 1969; Gunnink and Niday, 1976). Of particular importance has been the development of graphic techniques employing the use of large-display oscilloscopes, function keyboards, and other peripheral devices to permit operator interaction with the data to assist in the analysis procedure. As an example, Figure 68 is a photograph of a small-computer data system that utilizes a small digital processor. This system incorporates magnetic tape, a large-area CRT display oscilloscope, function keyboard, and control panel. The major difference between such a system and a conventional "hard-wired" pulse-height analyzer lies in the ability of the processor to modify information provided by the digital-to-analog converter under control of a programmed sequence of operations that may assume any form the

Fig. 68. Photograph showing a modern gamma-ray-spectrometer system employing a small digital computer for real-time data acquisition and analysis.

user desires. It is this versatility that makes a digital processor desirable for data acquisition. Figure 69 shows a block diagram of a typical processor data acquisition and analysis system suitable for gamma-ray spectral analysis. In a conventional "hard-wired" pulse-height analyzer, analysis of a pulse from the detector produces a number corresponding to a particular address in the analyzer memory. This address is uniquely defined by the memory control circuitry and the procedure is then to initiate a process that adds the number "1" to the contents of that memory location. If this same number representing the amplitude of a given pulse is entered into the accumulator of a digital processor, this information can be modified at will. For example, it might be stored in a number of different locations for different

Fig. 69. Typical computer-coupled pulse-height spectrometer.

purposes, modified to change the energy scale or to correct for non-linearities or instabilities in the electronic system, and so on.

Recent techniques in the fields of x-ray and gamma-ray spectrometry are presented in the proceedings of a seminar of the International Committee for Radionuclide Metrology (ICRM) organized by K. Debertin and held in Warsaw in May, 1981 (Debertin and Mann, 1983), and also in the proceedings of an ICRM seminar on applied radionuclide metrology organized by W. Bambynek and K. Debertin and held in Geel in May, 1983 (Mann *et al.*, 1983b).

Typical commercial pulse-height analyzers available today incorporate one or more microcomputer CPU chips which provide considerable computing capability. The use of this technology makes it possible to incorporate a number of functions into the instrument which enhance its functional capabilities. These include experimental control, display of data and data reduction with user interaction through the CRT display and control panel. For many routine applications, the measurement and interpretation of gamma-ray spectra can be accomplished entirely within a small laboratory instrument.

4.3 Liquid-Scintillation Counting

Indirect or comparative measurement of radionuclides can be advantageously carried out by liquid-scintillation counting. Although the activity of essentially all radionuclides can be measured by liquid scintillation, this technique is particularly useful for the assay of low-energy beta-emitting radionuclides such as ^{3}H, ^{14}C, and electron-capture radionuclides such as ^{55}Fe. Calibrated reference sources and activity standards specifically produced in a form that is suitable for liquid-scintillation counting are available from national and commercial laboratories: thus the National Bureau of Standards supplies several suitable pure beta emitters such as ^{3}H-H_2O, ^{3}H-toluene, ^{14}C-n-hexadecane, ^{36}Cl-NaCl, ^{32}P-H_3PO_4, ^{63}Ni-$NiCl_2$, ^{89}Sr-$SrCl_2$ and ^{90}Sr-$SrCl_2$. For most efficiency-tracing applications ^{3}H-H_2O is the standard of choice.

4.3.1 *Composition of Liquid Scintillator*

As discussed in Section 2.4.5, liquid scintillation is based on the interaction of ionizing radiation with a multicomponent mixture consisting of a solvent and one or two scintillators, the second normally the wavelength shifter. The solubilization of the sample containing the radionuclide requires the addition of solubilizing, chelating, and other agents. Although p-xylene is the most efficient solvent, toluene is more frequently used because it is more economical. In addition, if the instrument is refrigerated, the melting point of 12°C of p-xylene compares unfavorably with the melting point of toluene of −95°C.

The most commonly used primary scintillator is 2,5-diphenyloxazole (PPO). This scintillator is efficient, sufficiently soluble in common solvents, available in sufficient purity, and reasonably economical. Terphenyl is somewhat less efficient than PPO, but it is substantially more economical. Terphenyl cannot be used with dioxane-naphthalene systems because the terphenyl fluorescence band overlaps the absorption spectrum of naphthalene. There are many other scintillators with properties suitable for specific applications (Birks, 1964; Kowalski *et al.*, 1967; Peng, 1977).

One of the earliest secondary scintillators was 1,4-*bis*-(2-(5-phenyloxazolyl))-benzene (POPOP) (see Section 2.4). This scintillator is still extensively used. The dimethyl derivative of POPOP is somewhat more soluble than POPOP; however, its spectral-matching response is poor for bialkali phototubes. The most efficient secondary scintillator is presently p-*bis*-(o-methylstyryl)-benzene (*bis*-MSB), which has acceptable solubilities in common solvents.

Primary scintillators are used in concentrations ranging from 4 to 10 g/l of solvent depending upon the degree of quenching. The concentration of the secondary scintillator is 10 to 20 percent, by mass, of that of the primary scintillator.

4.3.2 *Measurement of Aqueous Samples*

As toluene and other aromatic solvents are not miscible with water, numerous attempts have been made to develop alternative mixtures in order to incorporate aqueous samples, including tritiated water, into the scintillation liquid. The oldest generally applicable technique used dioxane as the primary solvent. This solvent is miscible with water in any ratio and, if 100 to 120 g naphthalene/l is added to it, it is almost as efficient as aromatic solvents. Because of the addition of naphthalene there is, however, a practical limitation, of approximately 25 percent by volume, to the amount of water than can be added to the dioxane-naphthalene mixture. In addition, the solubility of inorganic compounds in this mixture is limited. The addition of dioxane, various alcohols, and certain other solvents to toluene makes it possible to dissolve water in toluene up to a certain limit.

These ternary mixtures are no longer used, because more convenient and more stable emulsion systems have been developed. The emulsion system consists of an aromatic solvent containing appropriate scintillators mixed with a detergent (Patterson and Greene, 1965). A large number of detergents have been tested (Lieberman and Moghissi, 1970) and the results indicate that alkyl phenyl polyethylene glycol ethers are the most desirable. Usually the alkyl group contains eight or nine carbon atoms, although mixtures of higher and lower homologs have been used. These materials are available under various brand names from a number of commercial suppliers as are, also, premixed scintillators. The great advantage of these premixed scintillators is their ability to incorporate up to 50 percent of water, by volume, while maintaining reasonable efficiencies, e.g., 20 percent for tritium in counting samples containing 40 percent of water. Also, such water-detergent-solvent emulsions have the ability to incorporate considerable quantities of various salts (Moghissi, 1970).

Solubilization or combustion of tissue samples is described in Sher et al. (1971) and in NCRP Report No. 47 (NCRP, 1976a), and will not be further described here.

4.3.3 *Measurement of* ^{14}C

Carbon-14, which is widely used in the life sciences, can be conveniently assayed by liquid-scintillation counting. Essentially all compounds containing carbon can be labeled with ^{14}C and a large number of labeled compounds are commercially available. In many cases the compound to be measured is converted to CO_2 which is subsequently introduced into a scintillation liquid. The solubility of CO_2 gas in toluene is relatively small. However, up to 50 cm^3 of CO_2 (at STP) corresponding to 27 mg of carbon can be introduced into 20 ml of scintillation liquid provided the mixture is deaerated (Horrocks, 1968). A more common technique for introduction of CO_2 into a scintillation liquid is the absorption of CO_2 by a base that is subsequently dissolved in the proper mixture. The choice of bases is often arbitrary. Accordingly, a large number of bases have been used. An efficient method developed by Woeller (1961) uses phenethylamine as the absorbent for CO_2. Woeller's mixture consisted of 46 percent toluene, 27 percent ethanol, and 27 percent phenethylamine containing appropriate scintillators. This mixture is capable of absorbing over 500 mg of carbon corresponding to approximately 930 cm^3 of CO_2 in about 22 ml of the liquid scintillator. The absorption of CO_2 in this mixture is, however, slow. Ethanolamine is also used as an absorbent for CO_2 (Jeffay and Alvarez, 1961), as is also methoxyethylamine. The latter is particularly effective because both its carbamate and carbonate are soluble in toluene.

A more convenient method was developed by Moghissi et al. (1971). This method uses NaOH and is capable of absorbing up to 100 cm^3 of CO_2. A detergent mixture consisting of 380 g of Triton QS44 per liter of Triton N101 is used. The ratio of detergent to solvent, containing appropriate scintillators, is approximately 1:2 by volume. Sodium carbonate is introduced as an aqueous solution into the mixture. As the quenching properties of NaOH and Na_2CO_3 in this mixture are similar, the quantity of added CO_2 does not influence the quenching of the mixture. Numerous other bases have been proposed and used for absorption of CO_2.

4.3.4 *Measurement of Other Radionuclides*

The activities of many other radionuclides such as ^{32}P, ^{35}S, ^{55}Fe, ^{63}Ni, and ^{125}I can be advantageously measured by liquid-scintillation counting. The techniques employed are essentially the same as those described in Sections 4.3.2 and 4.3.3. In the assay of ^{35}S, SO_2 can be used in much the same way as CO_2. There are, however, certain radionuclides that are of particular interest and, or, require specific preparation. In the measurement of the Auger electrons resulting from the decay of, say, ^{55}Fe, care must be taken to minimize quenching effects.

Alpha-particle emitters are often assayed by liquid-scintillation counting. Lindenbaum and Smyth (1971), for example, measured ^{239}Pu and ^{241}Am in animal tissues using an emulsion scintillator. Other measurements of actinides in environmental samples are described by McDowell (1971) and in Section 6.1.2. In the case of plutonium, the composition of the liquid scintillator must be carefully controlled to prevent polymerization and deposition, or precipitation, of the plutonium. The preparation of samples for liquid-scintillation counting is described in Section 5.4.1.

4.3.5 *Background*

Consideration of background in liquid scintillation is more significant than in many other systems, because certain reactions may occur in the scintillation liquid that produce light at levels exceeding that resulting from radioactive decay of the sample.

As most liquid-scintillation counters operate in the coincidence mode, background originating from thermionic emission from the photocathode is low and generally of the order of one count per minute, or less. Somewhat more significant is the contribution of electronic noise based on interactions between two phototubes that are facing each other. More details on this phenomenon, often referred to as cross-talk, as well as a discussion of other contributions to the background due to the phototubes, are given in Birks (1964).

The contribution of cosmic radiation to background is largely eliminated by shielding and in a modern liquid-scintillation counter does not exceed a few counts per minute.

There are several other phenomena that may contribute to the background. Excitation of the glass vial, vial cap, and scintillation liquid by sunlight or fluorescent light may lead to delayed light emission. The decay of this phosphorescence can take several hours and thus a high initial, irreproducible, background may be observed.

Chemiluminescence results from interactions of various compounds present in the sample. Bases are particularly known to produce chemiluminescence. This process is, however, not limited to bases; many other compounds produce light when added to the scintillation liquid (Hercules, 1970; Kalbhen, 1970).

As chemiluminescence decay is temperature dependent, preheating of the sample results in a deexcitation, whereas reduction of the temperature prolongs the decay half life (Moghissi, 1970). In either case, a lower count rate is observed, obviously for different reasons.

Both phosphorescence and chemiluminescence are monophotonic phenomena and thus are greatly suppressed in the coincidence-counting mode. However, because the light production in both processes can be large, random coincidences do occur if the sample has been exposed to prolonged or intense sunlight, or whenever chemiluminescencing chemicals are present.

4.3.6 Selection of Counting Vials

Proper selection of counting vials contributes significantly to optimization of the counting system. Methods for the determination of counting efficiency are usually based on the presumption of vials having identical properties. Garfinkel et al. (1965) have shown that differences in properties of commercially available vials can contribute to inaccuracy of measurement. These inaccuracies can be magnified if an external standardization technique is used, as such techniques are affected by differences in the physical properties of the vial.

Suggestions for improvements in vials and their properties are numerous and the reader is referred to Horrocks (1974) and Peng (1975) for reviews of the subject. In practice, only vials made from ordinary glass, polyethylene, nylon, and sometimes quartz, are used. Because of their ^{40}K content, glass vials show a higher background than plastic vials. However, permeation of various solvents through plastics is significant and can cause swelling or distortion of the vials. Table 11 contains data on the loss of various solvents through scintillation vials. Counting efficiencies are similar for all vials; however, thin-wall plastic vials tend to give slightly higher counting efficiencies than thick-wall plastic vials, which, in turn, give slightly higher efficiencies than glass vials. (See also Horrocks, 1974 and Painter, 1974.)

4.3.7 Counting Efficiency and Quenching

Due to the unpredictable nature of many materials, the counting efficiency, ϵ, must be determined for essentially every sample. If ϵ has been determined previously for an identical sample, this requirement is reduced to the routine maintenance of the instrument, i.e., corrections for instrument drift and changes in the background. However, by far the greatest number of samples require a determination of ϵ due to the differences in their composition and degree of quenching. There are four basically different methods for achieving this. The first is based on observing the effect of adding an internal spike of the same radionuclide to the sample. The second uses the shift of pulse-height spectrum to lower energies as a measure of the degree of quenching. The third involves the use of an external source of energetic gamma

TABLE 11—Loss of compounds through liquid-scintillation vials[a]

	Volume	p-xylene	Toluene	Dioxane	H₂O
	ml	mg/d	mg/d	mg/d	mg/d
Polyethylene ("Teflon" coated)	25	130	190	4	~0.5
Polyethylene	25	120	190	3	~0.1
Polyethylene (thick wall)	20	80	100	1	~0.3
Nylon	25	5	10	15	30

[a] Adapted from Lieberman and Moghissi (1970).

rays to produce Compton-electron recoils in the sample. And the last involves measuring the displacement of the Compton edge, arising from an external gamma-ray source, as a function of degree of quenching in the sample. These four basically different approaches are described in the following sections.

4.3.7.1 *The Internal-Standard Method.* This method is perhaps the most accurate for all but the most radioactive samples. It consists of a comparison between the sample count rate, C, and that of the sample plus an added activity, S, of the same radionuclide. The efficiency, ϵ, is then calculated by the following equation:

$$\epsilon = \frac{(C + s) - C}{S} = \frac{s}{S},\tag{4.11}$$

where s is the observed increase in count rate due to the activity, S, of the added standard. For the accurate calculation of ϵ, s should be equal to, or greater than C. The original technique consisted of counting a sample, adding an appropriate, but small, quantity, or "spike," of the standard, and recounting it. The added standard should therefore have a high radioactivity concentration to avoid any drastic change in the characteristics of the original counting sample. This technique has been improved by Garfinkel *et al.* (1965) for preparation of a tritiated toluene standard with an accuracy comparable to the tritiated water standard of the National Bureau of Standards. The generalized method (Moghissi and Carter, 1968) consists of the preparation of two samples that are identical in their chemical composition, except for the replacement of some of the stable isotope by an appropriate radioisotope. This technique is used only rarely because it is time-consuming, particularly for the large numbers of samples that are processed by automatic changing and counting systems. The external-standards methods are therefore widely used. These will be described in Section 4.3.7.3.

4.3.7.2 *Quenched Working Standards.* All of the following methods for obtaining the correction for quenching, described in Sections 4.3.7.3 through 4.3.7.5, require the preparation beforehand of a set of vials of "quenched" working standards of known activity, in which the volume fraction of the radioactive component is varied from, for example, 1 to 20 percent, while the total volume of sample-plus-scintillator solution is kept constant.

4.3.7.3 *The Channels-Ratio Method.* This method is based on splitting the pulse-height spectrum observed for a given sample into upper (u) and lower (l) parts by setting the discriminators at levels which will remain fixed (Baillie, 1960). The effect of quenching on this sample would be to shift the spectral energy response toward lower energy so that the observed count rate u, in window u, will decrease. The l/u count-rate ratios and counting efficiency for the whole spectrum are then calculated for each of the differently quenched working standards (See Section 4.3.7.2). These efficiencies are then plotted against the count-rate ratios to obtain a quench calibration curve, from which the efficiency of an unknown sample can be obtained by interpolation. The disadvantage of this method is that high-activity samples are required to obtain accurate values for the channels ratio in a reasonable counting time.

4.3.7.4 *The External-Standard Methods.* These methods comprise the external-standard method and the external-standard channels-ratio method described below.

4.3.7.4.1 *The External-Standard Method.* This method is based on the measurement of that component of the count rate arising from Compton electrons generated in the counting sample by an external gamma-ray source (Higashimura *et al.*, 1962). The count rates generated by the external source are measured for the working standards, prepared as described above for which counting efficiencies have been previously obtained (see Section 4.3.7.1). As in the last method (Section 4.3.7.2), a quench curve is prepared in which the "automatic-external-standard (AES)" count rate is plotted against ϵ for the working standards. The counting efficiency for an unknown sample is then obtained by reference to the quench calibration curve after a measurement of its AES count rate.

4.3.7.4.2 *The External-Standard Channels-Ratio Method.* Although the channels-ratio and external-standard techniques lack general applicability, their combination to give the external-standard channels-ratio (ESCR) method has provided a useful and convenient technique. In practice, three wide-energy channels are used; one covers the spectrum of the beta-particle-emitting nuclide and the other two are chosen above the sample spectrum for the purpose of using the external source both to prepare the quench calibration curve from the set of working standards (Section 4.3.7.2), and to measure the channels ratio for the sample in order to make the appropriate quench correction. (After each sample is counted, the external gamma-ray source is, in most commercial systems, automatically positioned near the sample for a one-minute count.)

There are several problems associated with the ESCR method. A change in the chemical composition of the samples requires a new calibration curve derived from new working standards. Vial thickness, total volume, and several other factors affect the accuracy of the results (Horrocks, 1975). Although the ESCR method has many inherent sources of error, its convenience of operation has made it a popular technique.

4.3.7.5 *Measurement of the Pulse-Height Shift.* Many of the problems of ESCR have been overcome by using the pulse-height shift-extrapolation method of Horrocks (1978). In this method, the movement of the point of inflection of the Compton distribution from a ^{137}Cs external source is measured. The location of this point of inflection on the abscissa is obtained by double differentiation of the pulse-height distribution curve. Its displacement from the previously determined location of the point of inflection for an unquenched solution gives the change in the Horrocks H number (Horrocks 1977; 1978) which is used to correct the measured count rate to that for an unquenched sample. The technique is probably accurate to within ±3 percent which should be good enough for most applications.

4.3.8 Simultaneous Measurement of the Activities of Radionuclide Mixtures by Liquid Scintillation

4.3.8.1 *General.* Although energy resolution is poor for liquid scintillators (Seliger, 1960), simultaneous measurement of the activities of several radionuclides by this technique is still possible. Measurements based on differences of half lives of two radionuclides such as ^{125}I and ^{131}I will not be elaborated upon here, as they follow general rules of radioactivity and thus are common to all radiation-detection systems. This method has been discussed by Horrocks (1974).

4.3.8.2 *Mixtures of Alpha-Particle-Emitting Radionuclides.* Horrocks (1966) has designed an alpha-particle spectroscopy system using a single phototube system. Approximately 200 μl of liquid scintillator are used to dissolve a mixture of ^{233}U and ^{238}Pu. Due to the reasonably high energy of the alpha particles emitted by these radionuclides and because of the small volume of the liquid scintillator, tolerable background count rates are obtained under the peaks. If the activities of alpha- and beta-emitting radionuclides are to be measured, it should be kept in mind that the light output of an alpha particle is only approximately one-tenth of that of a beta particle with identical energy (see Section 2.4.5.1).

4.3.8.3 *Dual-Label Assays of Beta-Particle-Emitting Radionuclides.* A widely used application of liquid-scintillation counting in the life sciences is the assay of samples containing two different beta-emitting radionuclides. In the majority of cases the samples are labeled with ^{3}H and ^{14}C, although several other pairs have been assayed, e.g., ^{3}H and ^{35}S, ^{3}H and ^{32}P, ^{14}C and ^{36}Cl. For any given pair, the extent of the overlap of the beta-particle spectra determines the accuracy of the measurement. If the $E_{\beta \max}$ for the two differ by less than a factor of two, accurate measurement is extremely difficult.

Multichannel analyzers have been used with the liquid-scintillation counter to compare the total beta spectrum for the sample with the spectra for the individual components. It is then a relatively easy matter to assign counting windows to the radionuclides with the higher and lower $E_{\beta \max}$ (Randolph, 1975). Computers have also been used to fit the composite spectrum with library spectra of the individual constituents (Oller and Plato, 1972).

In most dual-label assays made with commercial counters, however, the two counting windows must be selected by the user. Because of the importance of this type of measurement, instrument manufacturers have devoted much effort to developing automated methods for selecting the proper channels so as to reduce spectral overlap while still maintaining reasonable detection efficiencies for each radionuclide. In the simplest instance, the total beta-ray spectrum is divided into two channels by the use of only three discriminators. For ^{3}H and ^{14}C, the lower discriminator is set just above the instrument "noise" and the upper discriminator is set just beyond the ^{14}C endpoint. The spectrum is then split into channels 1 and 2 by setting the third discriminator near the ^{3}H endpoint. This last discriminator is then adjusted to allow a preselected overlap of the ^{3}H and ^{14}C spectra.

The counting rates in the two channels will be N_1 and N_2 after correction for the background in each channel. These counting rates are,

$$N_1 = H \cdot \epsilon_1(^3\text{H}) + C \cdot \epsilon_1(^{14}\text{C}) \tag{4.12}$$

and

$$N_2 = H \cdot \epsilon_2(^3\text{H}) + C \cdot \epsilon_2(^{14}\text{C}) \tag{4.13}$$

where H and C are the activities of ^{3}H and ^{14}C, respectively, and ϵ_1 and ϵ_2 are the counting efficiencies for each of the radionuclides in each of the two channels. Solving these two equations (4.12 and 4.13) gives the activities as,

$$H = \frac{N_1 \cdot \epsilon_2(^{14}\text{C}) - N_2 \cdot \epsilon_1(^{14}\text{C})}{\epsilon_1(^3\text{H}) \cdot \epsilon_2(^{14}\text{C}) - \epsilon_2(^3\text{H}) \cdot \epsilon_1(^{14}\text{C})} \tag{4.14}$$

$$C = \frac{N_1 \cdot \epsilon_2(^3\text{H}) - N_2 \cdot \epsilon_1(^3\text{H})}{\epsilon_1(^{14}\text{C}) \cdot \epsilon_2(^3\text{H}) - \epsilon_2(^{14}\text{C}) \cdot \epsilon_1(^3\text{H})}. \tag{4.15}$$

The selection of the optimum channels will depend upon the relative activities of ^{3}H and ^{14}C or any other pair of radionuclides. Kobayashi and Maudsley (1970) reviewed the methods available, one of which involves the use of separate unquenched samples of ^{3}H and ^{14}C.

The lower and upper discriminators are set as described above, and ^{3}H and ^{14}C counting efficiencies are determined in both channels at different settings of the intermediate-level discriminator. The ^{3}H efficiency in the ^{3}H channel is then plotted against the ^{14}C efficiency in the ^{3}H channel as shown in Figure 70. The point at which the curve begins to deviate significantly from a straight line through the origin, corresponds to the point beyond which the ^{14}C counting efficiency in the lower channel increases more rapidly than the ^{3}H efficiency.

Quench-correction curves must be prepared in which the ^{14}C efficiency in channel 2 ($\epsilon_2(^{14}$C)) and the ^{3}H efficiency in channel 1 ($\epsilon_1(^3$H)) are each plotted against the external-standard channels-ratio (ESCR). These curves are prepared in essentially the same manner as described for single-label in Section 4.3.7, except that two separate sets of vials, one containing ^{3}H and the other ^{14}C, are required. Alternative methods, e.g., the H number method (Horrocks, 1978), can be used to obtain

Fig. 70. Plot of ^{3}H efficiency *vs* ^{14}C efficiency in the ^{3}H channel at different settings of the intermediate-level discriminator. (After Kobayashi and Maudsley, 1970)

the efficiency for the two radionuclides in the two channels. Corrections to the background counting rates in the two channels should be included.

It was difficult to estimate the statistical errors for dual-sample counting but the inclusion of a microprocessor in many automatic liquid-scintillation counters has reduced the labor involved and it is now possible to measure samples with three labels such as ^{3}H, ^{14}C, and ^{32}P which have widely different values of $E_{\beta\,max}$.

4.3.9 *Efficiency Tracing*

Two techniques have been developed to use the method of liquid-scintillation counting to assay a pure beta-particle emitter in terms of a previously calibrated "tracer" or a beta-gamma-emitting radionuclide, or of another beta-particle emitter. In some ways this is similar to the efficiency-tracing methods referred to in 3.2.9, except that in the case of $4\pi\beta(PC)$-γ efficiency tracing the beta-gamma emitter and pure beta-emitting nuclide are both assayed in the same experiment. In liquid-scintillation the efficiency is measured in terms of previously calibrated reference sources, and in particular in terms of a tritium standard.

The two techniques used are; (i) the analogue system described by Bryant *et al.* (1967) and Jones and McNair (1970), and (ii) the methods developed and described by Gibson and Gale (1968; 1972), Gibson (1980), Grau Malonda (1980), Grau Malonda and Garcia-Toraño (1982), and Grau Malonda *et al.* (1984), and applied in Coursey *et al.* (1984). All these methods involve taking into account the beta-particle spectra of the traced and tracer radionuclides, and the scintillation efficiency.

4.3.9.1 *Efficiency Tracing by Simulated Spectra.* Alpha-particle and energetic beta-particle emitters can be satisfactorily assayed by liquid-scintillation counting using discrimination and extrapolation of the integral pulse-height curve to zero pulse height. For lower-energy beta-particle emitters, such as ^{3}H, ^{14}C and ^{241}Pu, not only does the zero-detection probability become significant, but the shape of the spectrum at low energies may not be conducive to valid linear or exponential extrapolation to zero energy.

Analogously with the β-γ tracer method used in $4\pi\beta$-γ proportional counter (see Sections 3.2.9 and 3.5.6.2), attempts have been made to carry out efficiency tracing using a liquid scintillator as the beta detector. As pointed out by Bryant *et al.* (1967), the source preparation is simple and the mixing of the tracer and the traced is assuredly more uniform than in the case of deposited solid sources. The method had indeed been tried in 1963 by Steyn and Hahne, but has not found general acceptance for accurate measurements because of the uncertainties in extrapolation to 100-percent efficiency arising from different spectral shapes and from the zero-detection probability.

Vatin (1980) quotes a private communication to the effect that the analogue device has since been replaced "by a computer simulation (that) is routinely used for the standardization of the pure β-particle emitters."

This method is analogous to the $4\pi(PC)$-γ tracer method in that the tracer does not have to have been previously calibrated, so it might be considered to be a direct method. On the other hand it is convenient to include it here in Section 4.3.9 with the somewhat similar technique to be described in the next section.

To avoid the problem of spectral shape and zero-detection probability referred to above and to provide a "transform function" between the beta-particle spectrum and the pulse-height spectrum, Bryant *et al.* (1967) simulated the spectral energy distribution by means of a mechanical device consisting of a cam-operated shutter through which variable-intensity light flashes from a cathode-ray tube were transmitted. The grid of the cathode-ray tube was biased beyond cut-off and flashes of light were generated on its screen by applying 200-ns voltage pulses to its grid. A series of diffusers, filters and then a 45° diffuse reflector replaced the sample vial. The cam was so shaped that during each rotation the aperture shutter was varied in such a way that the intensity of the flashes reaching the cathode of the phototube reproduced the energy spectrum of the particular beta-particle emitter in the scintillant. In use, the light intensity of the simulator is adjusted to match the energy range of the beta particles emitted by the radionuclide in the scintillant.

The device was tested using a mixture of ^{60}Co and ^{95}Nb in the scintillant, and two cams were constructed, one to simulate ^{60}Co and the other ^{95}Nb. Efficiency measurements were then made for the two simulated spectra, with the flash intensity appropriately matched to the maximum energy of each nuclide. The efficiency measured for simulated ^{60}Co was then plotted against that for ^{95}Nb. The next step was to replace the simulator by the vial containing the mixture of radionuclides, and the detection efficiency for each was measured by coincidence counting using the relationship N_c/N_γ. The results obtained at different efficiencies, for different discrimination or amplifier gain, were then plotted on the curve previously obtained using the simulated spectra and showed excellent agreement in a range between a lower efficiency of about 0.25 for ^{60}Co and 0.05 for ^{95}Nb and an upper efficiency of 0.90 for both nuclides.

Jones and McNair (1970) later tested the device by measuring the activities of two samples of ^{14}C-n-hexadecane, one calibrated by the National Bureau of Standards and the other calibrated by the Radiochemical Centre against an NBS-standardized solution of ^{14}C-sodium carbonate, and of two solutions of ^{35}S calibrated at the National Physical Laboratory. For this purpose cams were constructed to simulate the beta-ray spectra of ^{14}C and ^{35}S.

The ^{14}C-n-hexadecane samples were assayed using ^{95}Nb as tracer and the ^{35}S solutions were assayed using ^{60}Co and ^{95}Nb as tracers. Taking the NBS ^{14}C-n-hexadecane as an example, the procedure adopted was first to plot the curve of ϵ_β for the traced pure beta emitter, vs. $\epsilon_{\beta-\gamma}$, for the tracer nuclide, ^{95}Nb, using the analogue system. Then with a mixture of the two nuclides in the scintillant and for different efficiencies, obtained by altering the gain or discrimination, N_β for the traced nuclide and $\epsilon_{\beta-\gamma}$ for the tracer nuclide were measured, the latter being N_c/N_γ obtained by coincidence counting. The value of ϵ_β can now be derived from the ϵ_β vs. $\epsilon_{\beta-\gamma}$ curve given by the analogue system. The activity of the pure beta emitter is then equal to N_β/ϵ_β. The measurements were repeated for different efficiencies between 0.5 and 0.9 and for different samples to give an average value of the activity concentration equal to 1.062 ± 0.001 μCi/g compared with the NBS value of 1.063 ± 0.010 μCi/g.

The spread of the many results obtained in each of the three sets of measurements was of the order of 1 to 3 percent, but, apart from those for one of the NPL solutions, the average values agreed to within 0.3 percent with those of the standardizing laboratories.

4.3.9.2 *Zero Detection, Quenching, Wall Effects and Efficiency Tracing.* When radiation, alpha or beta, above a certain threshold energy, deposits energy in a liquid-scintillation-detector system photoelectrons will be emitted from the cathode of the electron-multiplier phototube and most, if not all, of them will be accelerated to the first dynode. To quantify this process Birks (1975) used a "figure of merit" that related the pulse amplitude at the phototube anode to the energy deposited in the solvent of the scintillator (see §2.4.5.1). This was useful in making comparisons between different configurations of liquid-scintillation-detector systems.

Other figures of merit have been utilized by different authors, such as:

(i) P, the average number of electrons $\bar{N}(E)$ at the first stage of the phototube divided by the energy deposited in the scintillator by an electron of energy E, (Gibson and Gale, 1968);

(ii) η, the energy dissipated in the scintillator divided by the average number of monos arriving at the first dynode (Houtermans, 1973); and

(iii) M, the energy deposited by an electron of energy E in the scintillator, corrected for ionization quenching and wall effects, divided by the average number of photoelectrons $\bar{N}(E)$ incident upon the first dynode of the phototube (Grau Malonda and Garcia-Toraño, 1982).

Qualitatively, the figures of merit (ii) and (iii) are similar to each other and are the reciprocals of (i). Houtermans (1973) remarked that in the case of (ii) the choice of the term "figure of merit" was unfortunate because the smaller it is the better is the system.

Both the figures of merit P (i) and M (iii) are relatively independent of energy, *provided that in each case* the energy deposited in the liquid scintillator is corrected by a factor $Q(E)$ for ionization quenching (Birks, 1958), and a factor $W(E)$ for the escape of incident radiation from the liquid scintillator through the interfaces between the scintillator on the one hand and the surrounding walls and upper liquid-gas surface on the other (Gibson and Gale, 1968; Grau Malonda and Garcia-Toraño, 1982).

These corrected energies then become $EQ(E)\ W(E) = E\ F(E)$ (Gibson and Gale, 1968), and $E\ Q(E)\ W(E) = E\ X(E)$ (Grau Malonda and Garcia-Toraño, 1982). These correction factors have been computed, for $Q(E)$ using an empirical equation due to Birks (1958) and for $W(E)$ using a formula derived by Benjamin *et al.* (1962). The latter have also recently been computed by a Monte-Carlo method (Grau Malonda and Los-Arcos, 1983).

These correction factors are in the nature of efficiencies, and their product, if they represented the only quenching effects, would be the efficiency for converting the deposited energy in the scintillator to energy available for the production of light. But there are other possible sources of quenching, such as color or chemical, that may or may not be present. With care neither needs to be important.

It was stated above that the ionization-quenching correction $Q(E)$ was usually computed from the empirical equation of Birks (1958). This equation was based on Birks' observation that the light output per unit distance along the path of a beta particle emitted in the scintillator solution, dL/dx at energy E was given by

$$dL/dx = S\ dE/dx/(1 + kB\ dE/dx)$$

where S is the scintillation efficiency, k is the rate constant for ionization quenching, and $B\ dE/dx$ is the linear ionization density. The total light output, if the beta particle stops in the solution, is

$$L = S \int_0^E dE/(1 + kBdE/dx),$$

and the unquenched fraction $Q(E)$ at E is given by

$$Q(E) = (1/E) \int_0^E dE/(1 + kBdE/dx).$$

The wall-effect correction first used by Gibson and his colleagues was that due to Benjamin et al. (1962). A more sophisticated computation of this correction, for use with electron-capturing nuclides, was made by Gibson and Marshall (1972). Both of these are, however, described by Gibson (1980) in the BIPM Monograph-3, in which Figure IV-6 shows the ionization-quenching and wall-effect corrections used by Gibson and Gale in 1968. But whichever wall-effect correction is used, as will be shown below, seems to be immaterial. Gibson and Gale (1968) also allowed for the theoretical spectral shape and zero-detection probability in extrapolating the integral-discrimination curve to zero pulse height and obtained values of activity for standard reference samples of ^{3}H, ^{14}C and ^{36}Cl from the National Bureau of Standards that were in agreement to within 3 percent, and obtained agreement for the same three radionuclides that were in agreement with values obtained by the Radiochemical Centre that agreed to within 9 percent.

Grau Malonda and Garcia-Toraño (1982) approach the problem by changing the overall efficiency and figure of merit by increasing the chemical quenching by the use of carbon tetrachloride. If the configuration of the liquid-scintillation-counting system is left completely unchanged from one measurement to the next and the liquid-scintillator solutions are exactly identical, then there will be no change in the figure of merit of the system between measurements of different aliquants of the master solution of the same radionuclide.

In Section 3.8.2 the response efficiency at the first dynode of the phototube to single pulses, with an average of $\bar{r}$ photoelectrons incident upon it, was

$$1 - e^{-\bar{r}},$$

the exponential term being the zero-detection probability which, as was there stated, became negligibly small for values of $\bar{r}$ greater than 7. Using the figures of merit P or M gives the response of the first dynode of the phototube to the deposition of energy E in the liquid scintillator as either

$$1 - e^{-PEF(E)} \text{ or } 1 - e^{EX(E)/M}$$

for a single phototube. The squares of these responses apply for two phototubes in coincidence. As mentioned above it is interesting to note the small effect that the wall correction has on these response functions. The form of each has been plotted in Figure IV-6 of Gibson (1980) and values of the corrected figure of merit for ^{3}H, ^{14}C and ^{36}Cl have been given in Table 4 of Gibson and Gale (1968); they are between 0.44 and 0.45 photoelectrons per keV. Typical values for $1 - \exp[-P\,E\,F(E)]$ are

0.76	for	$E = 5$ keV
0.96	for	$E = 10$ keV
0.9993	for	$E = 20$ keV
1.0000	for	$E = 200$ to 3,000 keV
0.9996	for	$E =$ about 3,500 keV

Reference to Figure IV-6 of Gibson (1980) shows that $W(E)$ starts dropping from unity at about 20 keV to 0.55 at 2,000 keV, but this drop is completely overriden by the increase of E in the exponential term.

In the paper of Grau Malonda and Garcia-Toraño (1982) use is made of the theoretical spectral distribution to avoid the difficulties that arise when an extrapolation of the integral spectrum to zero pulse height is attempted. This direct use of the theoretical spectral distribution is, in a sense, another form of the "efficiency transform" used by Bryant et al. (1967) in the shape of their cams.

In order to involve the theoretical beta-ray spectrum in the whole liquid-scintillation-counter process, from deposition of energy to the response of the phototube, it is necessary to consider an amount of energy between E and $E + dE$, deposited by an emitted radiation. Following Grau Malonda and Garcia-Toraño (1982), the average number of electrons, $\overline{N}(E)$, incident upon the first dynode of a single phototube system, for energy deposition E in the scintillator is

$$\overline{N}(E) = \frac{E\,X(E)}{M},$$

and the efficiency for detecting a single beta particle, above a given threshold energy, in the scintillator is given by

$$1 - e^{-E\,X(E)/M}.$$

The overall efficiency of the scintillator system is then

$$\epsilon = \frac{\displaystyle\int_0^{E_{\beta max}} f(E)(1 - e^{-EX(E)/M})\, dE}{\displaystyle\int_0^{E_{\beta max}} f(E)\, dE}$$

where $f(E)$ represents the beta-ray distribution as a function of energy based on the Fermi function.

As was noted above, the value of figure of merit, M, is determined only by the configuration and characteristics of the scintillator system. In the method of measurement used by Grau Malonda and his colleagues a previously calibrated standard, usually ^3H, serves as a "tracer" to calibrate the system and to obtain a value for M which will allow for any other than ionization quenching that may be residual in the scintillator. By using as nearly as possible identical vials, solvent and scintillant M will remain constant for the system. Color quenching is, however, introduced into both the tracer and traced solutions in order to measure the variation of both efficiency and figure of merit as a function of quenching.

The procedures are rather complicated but are explained in the block diagram of Figure 2 of their paper (Grau Malonda and Garcia-Toraño, 1982), to which reference should be made. But briefly the efficiency, ϵ, is measured as a function of chemical quenching, achieved by adding small amounts to a set of otherwise identical vials of scintillant with the tracer, usually ^3H. The quenching is measured using the external-standard channels-ratio method (Section 4.3.7.4.2).

The next step is to use the last equation above to calculate M as a function of efficiency. A graph of M vs. quenching can then be plotted.

The fourth step is to compute ϵ vs. M for the traced radionuclide; if, for example, the traced radionuclide is ^{14}C then the computation is made using the above equation for ϵ and the theoretical beta-ray spectrum for ^{14}C.

The fifth and last step is to calculate and plot the values of ϵ vs. ESCR for ^{14}C, using the last computed values of ϵ vs. M and the previous plot of M vs. ESCR. It is to be remembered that M remains constant for the system provided that the scintillator solutions, etc., are closely the same.

Having obtained ϵ vs. ESCR for ^{14}C, this efficiency may be used to compute the activity of ^{14}C from experimental count rates at different levels of quenching. Total estimated uncertainties for such curves for ^{14}C, ^{32}P, ^{36}Cl, ^{89}Sr and ^{185}W ranged between 2.0 and 3.5 percent (Grau Malonda and Garcia-Toraño, 1982).

Coursey *et al.* (1984) have recently applied the beta-efficiency-tracing method to standardize ^{99}Tc at the National Bureau of Standards.

4.3.9.3 *Further references.* For earlier work describing tracer measurements of pure beta emitters, including the use of liquid-scintillation counting, reference can be made to a short historical review by Steyn (1973b), who was also, with Hahne, the first to use the liquid-scintillation efficiency-tracers (Steyn and Hahne, 1963). Nearly 80 references to liquid-scintillation-counting investigations carried out from 1965 through 1972 are also given in a short review, included in the same proceedings of the Herceg Novi Summer School on Radionuclide Metrology, by Vaninbroukx and Stanef (1973). The comprehensive review in BIPM (1980) covers the period through 1980 with some 300 references.

4.3.10 *Activity measurements of electron capturers*

For lower atomic-number radionuclides decaying by electron capture, practically all the energy emitted by the x rays and Auger electrons arising from the atomic-shell vacancies created in the capture process should be absorbed by the liquid-scintillator solution. Thus the detection efficiency should be high, and is 80 percent in the case of ^{54}Mn (Vatin, 1980). In 1967 Steyn used the method of $4\pi(\text{LS})$-x($+$ Auger) coincidence counting and linear efficiency extrapolation to standardize samples of ^{51}Cr, ^{54}Mn, ^{57}Co and ^{65}Zn. He used discrimination and, or, changes in amplifier gain to change the counting efficiency (Steyn, 1967). Then at eight different discriminator or gain settings, curves of x-ray plus Auger-electron counting efficiencies were plotted as functions of x-ray energy for the four radionuclides. By interpolation, the corresponding counting efficiencies were obtained for ^{55}Fe. These values could then be used, under as identical as possible counting conditions to calibrate samples of ^{55}Fe.

As Vatin (1980) points out, measurements become more complex if electron capture is in competition with positron emission and he refers to a report on the subject by Steyn and de Jesus (1966). Special problems involved in the standardization of ^{67}Ga and ^{87}Y caused by appreciable internal conversion are discussed by Steyn (1973a).

Grau Malonda (1982) has extended the method of obtaining the counting-efficiency curve in terms of the quenching parameter, using ^{3}H, to the measurement of the activity of ^{55}Fe.

4.4 Ionization Chambers

4.4.1 *General*

The physics of gas ionization chambers has been discussed in Section 2.1. The phenomenon of the production of ions in gases under the influence of radiation has been used since the earliest experiments in radioactivity to quantify radiation. The gold-leaf electroscope led to the Lauritsen electroscope and Carmichael "Chalk River" electroscopes, the pocket dosimeter, the free-air chamber, the cavity chamber, the re-entrant (or 4π) ionization chambers, and the internal ionization chambers for the comparative assay of radioactive gases. Solid or liquid sources may be external to the chamber or placed within it. In principle, all gas ionization chambers are instruments in which an electric field is applied, by means of two insulated electrodes, across a confined volume of a gas. The electrons and positive and negative ions collected by these two electrodes constitute an ionization current that is a measure of the radiation incident upon the gas.

These instruments are widely used for the detection and assay of radionuclides and in applications to health physics, as, for example, monitors of neutrons (Manley *et al.*, 1946) and other radiations, their design and gas filling being dependent upon the type and intensity of the radiation to be measured.

Ionization chambers, once calibrated, can be used to measure or calibrate further samples of the same radionuclide (and some for which direct standards are not available, see Section 4.4.3) to give derived or secondary standards at any later date, or to check the consistency of subsequent direct measurements. All relative standardization procedures inherently require calibration of an instrument in terms of a calibration standard, and succeeding measurements of samples need to be made under conditions identical to those of calibration. Certain precautions should be observed if accuracy is to be achieved.

The geometrical factors involved (for example, the distance from source to detector and the dimensions of the sample) should be maintained constant. The arrangement of scattering material in the vicinity of the source and detector must be reproducible (for example, shielding used to protect the operator or equipment from stray radiation should not be moved since a resulting change in the scattered radiation from the shielding will change both the calibration and the background). Where completely fixed conditions cannot be achieved (for example, with unavoidable variation in sample size) corrections should be made from data obtained by investigation of changes in the pertinent parameters. The contribution of the background should always be carefully considered, and if large or variable or both, frequent measurements may be necessary. The material to be assayed should be examined for impurities; the effect of a radionuclidic impurity may be inconsequential or large, depending on the radiations of the principal and impurity radionuclides, their half lives, the amount of impurity present, and the instrument. Energy-selective systems will have to be employed to identify and quantify impurities present, and the overall ionization-chamber response corrected for the contributions, if any, due to the contaminants.

The stability of an ionization-chamber system is usually verified by measuring its response to a reference source of some long-lived radioactive material, such as ^{226}Ra in equilibrium with its daughters, whenever calibrations, or measurements, are made. In practice, one may take the activity of the sample to be proportional either to the direct response of the ionization-chamber system, or to the ratio of its response to the sample to that produced by a reference source. In the latter case, small variations in the sensitivity of the system may be of less consequence insofar as they affect the response to both the sample and reference source nearly equally. In the first method, the appropriate calibration factor, K, for a particular radionuclide is the ratio of its activity to the ionization current, or observed response R, for a source of that radionuclide in a reproducible configuration. In the second method, the relative calibration factor, K_R, is equal to AR_R/R, where R_R/R is the ratio of the measured responses, i.e., the ratio of the ionization current due to the reference source in a given geometry to the ionization current due to the particular radionuclide in a similar and reproducible geometry, and A is the activity of the radionuclide.

It should be noted that the K factor (A/R) is inversely proportional to "efficiency" that has been used for other detectors in previous sections of this report. In the case of a system that counts individual events, the count-rate response is divided by the efficiency, ϵ, to give total emission rate. The K factor has, however, classical origins and multiplies the response to give activity. Therefore, to obtain the equivalent of an efficiency curve as a function of energy, the reciprocal of the K factor should be plotted against energy. The relative factor, K_R, has the dimensions of activity and is, in fact, the "equivalent activity" referred to in Section 4.4.7.

The number of ion pairs formed per unit of path length is a function of the density of the gas in an ionization chamber. In unsealed chambers, readings may have to be corrected to standard temperature and pressure if the range of the primary radiation exceeds the dimensions of the chamber. For beta-ray and low-energy-photon emitters, a simple correction to STP may result in over-correction. To effect a correction in these two cases, one can make measurements on specific radionuclides (relative to a reference source) on days with different air densities and modify the STP correction by the empirical factor calculated from those measurements. For the most accurate work, it is desirable to have a chamber in which the gas pressure can be varied to permit investigation of the dependence of response on gas density. This correction is not necessary with sealed ionization chambers containing, for example, argon at 20 atmospheres.

Ionization chambers, employing any of a number of suitable current-measuring systems and using the precautions mentioned above, can be used to maintain standards or to make relative measurements of radioactivity with precisions as good as 0.02 percent, which is usually less than the uncertainty associated with the activity of the standard used to calibrate the instrument.

4.4.2 *Beta-Particle Ionization Chambers*

Beta-particle ionization chambers are simple in construction and in function, but in their use one might encounter difficulties, especially with low-energy beta-particle emitters, due to variations in source preparation or configuration of source containers. However, ionization-chamber measurements of evaporated sources are less sensitive to source self absorption than are Geiger-Müller or proportional-counter measurements. Each high-energy beta particle may contribute thousands of ion pairs to the ionization current, but only one count in the counter. On the other hand, the absorption of one beta particle in the source represents the negligible loss of perhaps only one or two ions from the ionization current but still represents a whole count that is lost from the counter. The relative effect on the current is, therefore, much less than it is on the count rate. One should prepare sources and perform the measurements in exactly the same manner as for the calibration of the instrument. Beta-particle ionization chambers can be used for the calibration of solid sources or of solutions.

In preparing solid sources, the following precautions should be observed:

 (1) Use highly polished, flat, pitless disks of a metal that will not react chemically with the solutions to be deposited. Solutions may be deposited on thin polyvinylchloride-polyvinylacetate

copolymer (VYNS) or collodion films, allowed to dry, then affixed to disks of any metal.

(2) Maintain the same solids content and pH for the solution of a particular radionuclide to be deposited as used in the calibration.

(3) Add a seeding or wetting agent to the liquid deposit (see also Section 5.2.1).

(4) When sources are prepared from some solutions by deposition and evaporation, there may be loss of activity due to volatilization. If this is likely, source preparation by precipitation followed by evaporation may be necessary (see Section 5.2.1).

(5) Dry sources slowly (e.g., at ambient temperature, in a gentle flow of warm air, or with a heat lamp, and take care to avoid spattering).

In preparing liquid sources:

(1) Use containers of uniform and reproducible dimensions and composition.

(2) Use the same volume.

(3) Work carefully and rapidly to minimize evaporation.

(4) Maintain the same solution composition for the sample as for the calibration sources.

In making measurements (after checking the purity of the material):

(1) Measure sources by comparison with some long-lived radionuclide. Radium-226 (in equilibrium with its daughters), contained in platinum-iridium cells, encapsulated in stainless-steel disks of the same diameter as source disks or dishes, can be used as reference sources. Inasmuch as a wide range of activities can be measured in the chamber, a set of reference sources should be intercompared, maintained, and corrected for radioactive decay.

(2) Determine corrections for source self absorption from measurements of solid sources prepared with different masses of the radioactive solution, maintaining the same source diameter by adding drops of conductivity water to the radioactive solution on the source mount, before drying. Extrapolate to zero mass. For low-solids sources this may, however, result in non-linearity due to "clumping" (See Seliger, 1960).

(3) Work in a room with controlled temperature and humidity.

(4) Use care in positioning of the source.

An early design, but a chamber still functioning reliably, is a $2\pi\beta$-ionization chamber shown diagrammatically in Figure 71. At the base of the spherical body is a cylindrical shoulder that slides over the source-mounting ring. The source-mounting ring is recessed to hold, reproducibly, disks on which aliquots of an active solution can be evaporated. The length of the 0.3-cm diameter brass collecting elec-

Fig. 71. Schematic drawing showing the construction details of the NBS $2\pi\beta$-ionization chamber. (Seliger and Schwebel, 1954)

trode has been adjusted empirically so that a point source can be displaced slightly from the center of the source mount without affecting the observed ionization current.

Widely used as a secondary instrument for both beta-particle- and gamma-ray-emitting radionuclides is the type 1383A ionization chamber[7] designed at the National Physical Laboratory (NPL), England (see Figure 72). It is a combination re-entrant cylindrical chamber for gamma-ray measurements and parallel-plate chamber for beta-

---

[7] Available from Elliot Bros. (London) Ltd., Century Works, Covington Road, London SE13 7LN.

Fig. 72. NPL combined beta-gamma reference ionization chamber with plastic jig for support of small ampoules. (From Robinson, B., 1960)

particle measurements. Designed to be manufactured to give constant characteristics, each chamber is provided with the calibration factors for a number of radionuclides (Dale, 1961; Dale *et al.*, 1961, Woods, 1970; Woods and Lucas, 1975; Woods *et al.*, 1975). Results obtainable with one such chamber are applicable to others of the same type if uniform procedures for source preparation and measurements are practiced. Both solid and liquid sources that fit into the circular depression on the slide may be used for beta-particle measurements. (See Lowenthal, 1969, for a summary review of beta ionization chambers.)

4.4.3 *Gamma-Ray Ionization Chambers*

Important relevant characteristics of these instruments are sensitivity, long-term stability, and simplicity of operation. Also, source preparation is not elaborate. These chambers can be used to maintain results of direct standardizations and to check the consistency between the results of direct measurements made at different times. They may also be used in tests of dilution and sampling methods, and to accumulate half-life data. With care, precisions of ±0.02 percent (at one estimated standard deviation) can be achieved.

Best results have been obtained with pressurized ionization chambers with reentrant source receptacles so that the geometry is almost 4π. Ionization chambers should be shielded (using Pb, Hg, or both) to reduce background and to prevent change in efficiency caused by scattering material in their vicinity. Batteries of 200 to 600 V, or stable electronic voltage supplies can supply collecting voltages sufficient to obtain saturation current. To insure reproducibility of positioning, rigid tight-fitting holders made of low-Z material are used for ampoules, samples, and reference sources.

Standards of particular radionuclides are used to effect the calibration of a chamber, by using a reference source, preferably ^{226}Ra in equilibrium, under the same defined conditions to maintain the calibration. The reference sources could be platinum-iridium cells containing the radium salt, each imbedded in the end of a polystyrene cylinder, the same diameter and height as the sources to be measured. Radium-226 in solution is unsatisfactory because of the variations in distribution of radon and its decay products between the solution and the space above the solution.

If ampoules are to be used, wall thickness and glass composition should be reproducible from ampoule to ampoule, and the height of solution should be constant in order to avoid variation in gamma-ray attenuation. Alternatively, corrections can be empirically determined for different solution heights, the function being only slightly dependent on the energy of the radiation. Again, the solutions to be assayed should have as nearly as possible the same composition as the standards used to calibrate the chamber.

It may be necessary to apply corrections for chamber saturation-current losses to measurements of high activity sources; the loss, if any, can be determined by measuring the sources at increasing collecting voltage (Weiss, 1973). Another method for detecting saturation-current losses (usually caused by ion recombination) is to prepare, by dilution, and to measure a series of sources, of equal volumes of solution in identical ampoules, of the same long-lived gamma-ray-emitting nuclide, with different radioactivity concentrations giving responses in the range of ionization current to be used. On plotting response against radioactivity concentration any departure from linearity at the higher levels of activity will indicate the occurrence of saturation-current losses through general recombination. (For the definitions of *initial* and *general* recombination, see Boag, 1966.)

It is possible to estimate the calibration factors for solution samples of some photon-emitting radionuclides for which standards are not available (because a direct measurement is difficult), provided that the photon probability per decay for each photon is known and the number of photons with energies below about 200 keV is small. First, it is necessary to determine the efficiency of the chamber as a function of gamma-ray energy, E, by measuring the response of the chamber to equal volumes of calibrated solution standards of monoenergetic photon-emitting radionuclides over the appropriate range of energy. The K factor, $K_{R,\gamma}$, relative to that of a long-lived reference source, now refers only to photon-emission rates and is given by

$$K_{R,\gamma} = P_\gamma A \frac{R_R}{R}, \tag{4.16}$$

whence

$$A = \frac{K_{R,\gamma} R}{P_\gamma R_R}, \tag{4.17}$$

where R is the reading for the calibrated source emitting a photon, of energy E, with a probability of emission per decay P_γ, A is the activity of the calibrated source, and R_R is the reading for the reference source, normally of ^{226}Ra. The readings R and R_R could be, typically, the time rate of change of voltage on a standard condenser.

If the best curve is plotted and fitted to $1/K_R$ as a function of E, it is possible to calculate the effective K factor for radionuclides emitting n photons of different energies and of known probabilities per decay, by simply taking $1/K_{R,\gamma}$ for each of these photon energies from the curve and multiplying it by the appropriate probability per decay. The activity of the radionuclide is then given by

$$A = \frac{R}{R_R} \sum_i^n \left(\frac{K_{R,\gamma}}{P_\gamma}\right)_i, \tag{4.18}$$

where the sum is carried out for all i photons. The efficiency-energy curve is often linear but a linear extrapolation below 200 keV may be incorrect. Equation 4.18 can be compared with Eq. 6.5 in Section 6.4.3.2.3 on dose calibrators. Further information has also been given by Weiss (1973).

Figure 73 is a schematic drawing of a 15-liter unsealed "4π" or "reentrant-well," atmospheric-pressure, chamber based on that de-

Fig. 73. A "4π"γ ionization chamber. (From Mann and Seliger, 1958)

scribed by Shonka and Stephenson (1949), but with slight modification in chamber design. The cost in construction is small compared with that of a pressure chamber. A thin brass cylinder is used as the collection electrode in place of the usual wire cage. The response to 5 ml of a solution (containing a gamma-ray-emitting nuclide) in a standard 5-ml ampoule is approximately 99½ percent of that of a 1-ml sample with the same total activity in a standard 5-ml ampoule. The chamber, in use, is surrounded by a cylindrical lead shield 1.5-inches thick.

Figure 74 shows the important design considerations of the Stephenson "high-precision," high-pressure ionization chamber such as the locations and construction of the insulators, guard ring, and guard-ring connector (Shonka and Stephenson, 1949). The chamber is filled with 40 atmospheres of dry argon. The high sensitivity of such a chamber necessitates enclosure in a 4-inch-thick lead housing to reduce environmental background effects.

Figure 75a illustrates the design features and internal construction of the T.P.A. Mk.II ionization chamber[8] which can be filled with gas pressures up to 20 atmospheres (Sharpe and Wade, 1951). With this chamber, the variation in sensitivity for gamma-ray-emitting nuclides is within ±1 percent for 9-cm and within 0.5 percent for 6.5-cm axial

[8] Available from 20th-Century Electronics, Ltd. (Centronic), King Henry's Drive, New Addington, Croydon CR9 0BG, England.

Fig. 74. High-precision, high-pressure ionization chamber (Shonka and Stevenson, 1949). (From O'Kelley, 1962)

displacements. The sensitivity to ^{32}P bremsstrahlung is such that one can easily measure 1 mCi of that radionuclide.

Figure 75b shows the new automated $4\pi\gamma$ ionization chamber system. The system incorporates 1-inch- and 2-inch-reentrant T.P.A. Mk.II ionization chambers. Under microcomputer control, samples

Fig. 75a. Sectional view of a one-inch-diameter reentrant-well IG = 11 pressure ionization chamber. (From Centronic, 20th Century Electronics Ltd., King Henry's Drive, New Addington, Croydon CR9 0BG, England)

217

Fig. 75b. An automated 4π-gamma ionization-chamber system.

may be placed into either chamber for measurement, according to a preselected sample sequence.

4.4.4 *Internal Ionization Chambers*

Ionization chambers into which radioactive gases can be introduced are useful for assaying such gas samples, and are often used for comparing the activities of ^{3}H and ^{14}C in the gas phase (Tolbert, 1956; 1963). The gaseous sample is admitted quantitatively into an evacuated ionization chamber of known volume, and the resulting ionization current measured. This technique is used chiefly for comparative measurements, i.e., the ionization current of the unknown sample of gas is compared with that of a sample of gas prepared in the same way and containing a known amount of activity (Marlow and Medlock, 1960). However, some direct calibrations have been conducted by this method (Wilzbach *et al.*, 1954; Rieck *et al.*, 1956; Tolbert, 1956, 1963). (For the assay of radon, see Section 6.1.2.3.)

4.4.5 *Dose Calibrators*

Dose calibrators are commercially-available ionization chambers with reentrant wells of diameter large enough to accommodate radiopharmaceuticals in vials shipped to hospital or medical laboratories. These are described in Section 6.4.3.2. Some dose calibrators have also been designed using a large plastic scintillator with reentrant chamber and digital display of phototube output current.

4.4.6 Techniques for Measuring Small Currents

Depending on the radionuclides and their activities, the currents to be measured may range from 10^{-7} to 10^{-13}A. A sensitivity of about 10^{-15}A is necessary because the currents produced by background radiation and leakage may be smaller by one or two orders of magnitude than the source ionization current (Zsdánszky, 1973). Accordingly, the current-measuring instrument must have a very high d.c. input impedance.

Many precise ionization-current measuring systems employ vibrating-reed or Lindemann-Ryerson electrometers in feedback-type current integrators that are just modern versions of the classical Townsend induction balance. Typical of these and other techniques are:

(i) Townsend induction balance with continuous compensation (Townsend, 1903; Goodier et al., 1965).

(ii) Automatic Townsend balance with stepwise compensation (Garfinkel, 1959).

(iii) Vibrating-reed electrometer as a feedback current integrator with preset time (Merritt and Taylor, 1967a).

(iv) Time-averaged voltage drop across a high resistance and a voltage to frequency converter (Walz and Weiss, 1970).

(v) Solid-state electrometers (Keithley, 1977).

For further details see, for example, Mann et al. (1980).

4.4.7 International Reference System for Measuring Activity of Gamma-Ray-Emitting Nuclides

The International Bureau of Weights and Measures (BIPM) has maintained, since about 1970, an international reference system for relative activity measurements of gamma-ray-emitting nuclides. This service is named the "Système International de Référence" (SIR). The system comprises two ionization chambers of the same type as those illustrated in Figure 75a and b (Rytz, 1977, 1978, 1983). National laboratories (as a rule, those of the member states of the Convention de Mètre) may submit two or more ampoules of their gamma-ray-emitting solution standards to BIPM for measurement. The ionization current produced by the submitted standard is compared to that produced by a radium reference source and the "equivalent activity" is calculated for the submitted solution standard. This is the stated activity of the radionuclide in the submitted standard that would have produced the same ionization current as the radium reference source at a specified time. BIPM then issues a "registration table" for each radionuclide in which is recorded the "equivalent activities" for the samples submitted by the different laboratories. Not only does this maintain international consistency, but it also indicates if the standards of a given radionuclide issued by any laboratory are consistent with time. The "equivalent activity" is the quantity that is called the relative factor, K_R, in Section 4.4.1.

The National Bureau of Standards has arranged for the production of 127,000 glass ampoules of carefully controlled dimensions, from the same melt of glass. Of these, 69,000 were reserved for the use of BIPM and other national laboratories in order to eliminate systematic differences in intercomparisons due to ampoule geometry and composition.

The International Atomic Energy Agency (IAEA) of the United Nations maintained, until recently, a similar service for its member states.

4.5 Other Counting Methods

4.5.1 *Čerenkov Counting*

Various aspects of Čerenkov counting are described in reviews by Ross, 1970 and Francois, 1973. This method is applicable, in aqueous solutions, only for beta particles having energies greater than 0.26 MeV (see Section 2.6). In practice, the sample is placed in a liquid-scintillation-counting vial, plastic or glass, 8 to 15 ml of water is added, and the sample is counted in a manner similar to liquid-scintillation counting. An advantage of the method lies in the relative ease of source preparation.

Due to the mechanism of light production in Čerenkov counting, chemical quenching, which is an important factor in liquid-scintillation counting, is non-existent in Čerenkov counting. Color quenching is a problem common to both techniques, and similar methods are used to correct for quenching.

Table 12 shows optimum counting efficiencies for various radionu-

TABLE 12—*Optimum Čerenkov-counting efficiencies for various beta-particle-emitting nuclides*[a]

Radionuclide	E	E_max	Detection efficiency
	MeV	MeV	%
^{204}Tl	0.240	0.764 (97.7%)	3
^{143}Pr	0.314	0.932	6.2
^{198}Au	0.312	0.961 (98.7%)	6.7
		0.285 (1.3%)	
^{24}Na	0.554	1.390	32.5
^{40}K	0.561	1.312	38.4
^{89}Sr	0.583	1.488	28.3
^{91}Y	0.607	1.545	35.5
^{32}P	0.695	1.710	35.1
^{90}Y	0.935	2.284	50.6
^{76}As	1.064	1.75 (16%)	
		2.41 (31%)	59
		2.97 (50%)	
^{42}K	1.430	1.996 (17.5%)	70.3
		3.521 (82.1%)	

[a] After Francois (1973). Measurements were performed at 4°C on solutions having various volumes (8 to 14 ml, 4 samples at each volume). The solutions consisted of the radionuclide and carrier in water, or, if a carrier was unavailable, in a 1-percent NaCl solution. The beta-particle "branching" is 100 percent unless stated parenthetically in the third column.

clides measured in a plastic vial. Glass vials exhibit a somewhat lower efficiency, probably because of problems associated with the light transmission of glass in the ultraviolet region.

The fact that Čerenkov radiation is emitted anisotropically means, in practice, that the usual liquid-scintillation-counting system, with its two diametrically-opposed phototubes in coincidence, will have a *lower* counting efficiency than when operated in the non-coincidence mode, or with a smaller angle between the phototubes. Some wavelength shifters, such as 4-methyl-umbelliferone, increase the coincidence-counting efficiency to greater than that for non-coincidence through a reduction of the anisotropy of the re-emitted light (Ross, 1971). However, improper use of these compounds can result in color quenching that increases with time.

Although the detection efficiency of Čerenkov-radiation counting is less than that of liquid-scintillation counting, the ease of operation and, particularly, the ability to recover liquid samples unaltered make Čerenkov-radiation counting a useful technique.

The preparation of samples for Čerenkov counting is described in Section 5.4.2, and low-level-counting applications are described in Section 6.1.3.3.

4.5.2 *Solution Counting*

Several radionuclides have been successfully counted in solution by using an inert, low-vapor-pressure solvent such as formamide, dimethyl formamide, or ethylene glycol (Schwebel *et al.*, 1951). The sample is dissolved in the inert solvent and an aliquot is pipetted into a cupped planchette. The activity is then measured in a $2\pi\beta$ windowless gas-flow counter, and compared with the activity of a solution prepared in the same manner, but containing a known amount of the radionuclide (for sample preparation, see Section 5.4.3). Formamide containing up to 10 percent by volume of water can be introduced into a proportional counter without impairment of its electrical or counting characteristics.

This "formamide" counting procedure has been found useful for the relative standardization of ^{14}C as glucose and benzoic acid, ^{32}P as sodium phosphate, ^{60}Co as cobalt chloride, ^{90}Sr as strontium chloride, ^{131}I as potassium iodide, ^{204}Tl as thallic nitrate, ^{22}Na and ^{24}Na as sodium chloride, and ^{35}S as sulfuric acid.

For any given counter and planchette it is possible also to obtain relative efficiencies for the above radionuclides. These efficiencies may then be used to assay any of these radionuclides in terms of, say, a ^{14}C source.

Proportional or Geiger-Müller counters of the dipping or jacketed type may also be used to assay aqueous solutions.

4.5.3 *Surface-Barrier Detectors for Alpha-Particle Counting*

Silicon surface-barrier detectors (see also Section 6.1.2.4) are "solid-state" devices that, in radioactivity applications, are used for the detection and analysis of alpha-particle spectra. As with ionization chambers and Ge(Li) detectors, electrons released in the "sensitive region" as the result of incident radiation are swept to the anode by an applied electric field. This charge is then deposited on a capacitor, that is shunted by a high resistance, and the resulting voltage pulse is amplified. The alpha-particle spectrum, which is usually recorded with a pulse-height analyzer, gives, typically, an energy resolution, indicated by full width of the peak at half maximum (FWHM), from a "weightless" source, of 12 to 20 keV. The counts per channel in the low-energy tail of the alpha-particle spectrum for such a system is typically of the order of 10^{-4} to 10^{-3} of the counts per channel in the peak. The low response to beta particles and gamma rays, combined with the good energy resolution and high peak-to-tail ratio, makes these detectors powerful instruments for impurity analysis and measurement.

This type of detector is also extremely useful for the relative measurement of the activities of alpha-particle-emitting sources. Because alpha particles emitted by radioactive sources are totally absorbed in the detector, the system can be calibrated by using radioactivity standards that have alpha-particle energies that are not necessarily the same as those of the unknown. For example, in a recent experiment at the National Bureau of Standards, a system calibrated with an [241]Am standard was used to calibrate [148]Gd alpha-particle point-source standards. The alpha-particle energies for the two radionuclides are approximately 5.4 and 3.2 MeV, respectively.

Gibson *et al.* (1965) have listed the advantages and disadvantages of silicon surface-barrier detectors in nuclear physics applications. An updated list drawn in part from their article, is given below:

Advantages
  a) Excellent energy resolution;
  b) Linear response over a wide range of particle energies, even though the response to β particles and γ rays is low;
  c) Insensitivity of pulse height to counting rate;
  d) Fast pulse-rise time;
  e) Essentially windowless operation;
  f) Selectivity for alpha particles with respect to other radiations;
  g) Small size, making them easy to use;
  h) Insensitivity to magnetic fields;
  i) Ability to operate down to 0°C.

Disadvantages
  a) Light sensitivity;
  b) Relative fragility—care must be taken not to touch the gold surface layer of the detector, although very recently a rugged Al coated surface-barrier detector has been developed that allows cleaning of the surface;
  c) Detector sizes are limited to a maximum of a few square inches;
  d) Must be operated in a vacuum of 50×10^{-3} mmHg (6.7 Pa) or less because surface-current leakage may impair the detector response;
  e) Variable thickness of the sensitive region, depending on bias voltage.

A word should be said in this section about diffused-junction-diode detectors whose principle of operation is the same as for surface-barrier detectors. The method of obtaining the depletion region is different, however, and has been discussed in Chapter 4 of a monograph by J. M. Taylor (1963). In spite of the fact that the energy resolution is worse than that for surface-barrier detectors, diffused-junction detectors are produced commercially because of two factors: namely, they are more rugged—the surface can be wiped—and, generally, the cost is less.

4.5.4 Assay of Radioactive X-Ray Sources

Measurements of the x-ray-emission rates of radioactive sources are sometimes needed in nuclear medicine as well as in investigations pertaining to the fundamental physics of x-ray emissions. Fink (1973) has listed 45 radionuclides that are suitable as radioactive x-ray standards emitting x rays (and gamma rays) in the energy range from 4.8 keV to approximately 150 keV. Chew et al. (1973) have described a method using a multiwire proportional counter to calibrate a number of such radionuclides. Hansen et al. (1973) have described the calibration procedure for obtaining the photopeak efficiency of a Si(Li) detector as a function of energy in this region by fitting points, obtained with radioactivity standards, to a theoretical curve. Bambynek (1967) has discussed the calibration of a defined-solid-angle, low-geometry counter for the measurement of low-energy-photon, as well as alpha-particle, sources. Such systems are being used in the production of radioactivity standards in the Laboratoire de Metrologie des Rayonnements Ionisants (LMRI), Saclay, France, and the NBS (Hutchinson et al., 1976b). Emission rates of low-energy (<10-keV) x rays have been measured using variable-pressure 4π proportional counters (Section 3.5.6.3).

The assay of low-energy x-ray sources requires that particular attention be paid to absorption by even small amounts of material in the path of the radiation, by scattering, and by x-ray fluorescence from materials in the region of the source and detector. Absorption, in 2.5 cm of air at atmospheric pressure, of 5.9-keV x rays emitted in the decay of ^{55}Fe is of the order of 6.8 percent. Backscattering of 22-keV x rays emitted in the decay of ^{109}Cd as a function of the atomic number of the source backing material is given in Hutchinson et al. (1976b). With decreasing photon energy, attenuation is caused predominantly by photoelectric interactions so that, for example, with the 5.9-keV x rays in the decay of ^{55}Fe, there is less than 0.1-percent backscattering from steel backings. Examples of fluorescence encountered in recent experiments are gold L-x-ray lines observed in a spectrum from a Si(Li) detector which has a gold covering layer (Keith and Loomis, 1976), zinc x rays emitted by sources deposited on galvanized iron, platinum L x rays produced in the edges of a platinum aperture, and so on.

Vanadium-49 K x-ray sources have been standardized by Hutchinson and Mullen (1983a), both by 2π proportional counting and by comparison with a calibrated ^{55}Fe source. The half life of ^{49}V was found to be 338 ± 5 days. The same authors (1983b) have also used a thin-walled NaI(Tl) "pin-well" detector to measure the 59.5 keV γ-ray emission rate of a quantitatively deposited source of ^{241}Am, and hence found the probability per decay for that γ ray to be 0.3582 ± 0.0012.

5. Techniques for the Preparation of Standard Sources for Radioactivity Measurements

5.1 Liquid Sources

5.1.1 *Chemically Stable Radioactive Solutions*

Most radioactivity solution standards consist of trace quantities of radionuclides of known radioactivity concentrations, but only approximately known concentrations of carriers, in flame-sealed glass ampoules. A carrier is "a substance in an appreciable amount which, when associated with a trace of another substance, will carry the trace with it through a chemical or physical process", especially a precipitation process. If the added substance is a different element from the trace, the carrier is called a non-isotopic carrier (ANSI, 1976b). The carrier concentrations and solution media for standards are chosen to satisfy two criteria: (1) that the solution standards remain stable throughout their useful lives and (2) that low-absorption sources can be prepared from them—therefore, in diluting, the diluent should, usually, have the same carrier concentration and composition as that of the standard. The minimum amount of carrier that can safely be used is found from experience. As a general guide, about 100 μg of the element per gram of about 0.1-M acid solution is adequate, but concentrations above and below this value may be used successfully for specific radionuclides. Among solutions of metals, a few exceptions to this general guide are noteworthy: niobium, tantalum, zirconium, and hafnium solutions are prepared as fluoride or oxalate complexes, and tin solutions must be in $\geq$1M HCl to avoid hydrolysis. Table 13 gives chemical data for various radionuclide solution standards and may serve as a guide for the preparation and dilution of chemically stable radioactive solutions. A few very high carrier concentrations listed in the table result from the low specific activities that were

226

commercially available for some radionuclides (e.g., ^{26}Al), and not from added carrier.

The chemistry of stable and radioactive nuclides is the same except at high levels of activity where the composition of the solvent can be substantially changed by the radiation. (As a rule of thumb, it takes an absorbed dose of about 10^8 rad to effect such changes.) If possible, inactive material added to serve as carrier should be in the same chemical form as the radionuclide, to permit them to exchange and thus follow the same chemistry. Exchange readily takes place in solutions for those elements that are normally found in only one valence state. However, it can be very slow among different valence states of multivalent elements and is best achieved through an oxidation-reduction cycle. For elements that do not have stable isotopes, those with similar chemical behavior can be used as carriers, e.g., barium for radium or lanthanum for promethium.

Elements such as krypton are usually found to be radioisotopically contaminated with man-made radionuclides. Salts of lanthanum, samarium, and lutetium contain naturally occurring ^{138}La, ^{147}Sm, and ^{176}Lu, respectively. Salts of lanthanum and cerium have been found to be contaminated with ^{227}Ac as well as ^{232}Th. Distilled water, nitric, perchloric, and hydrofluoric acids have been found to contain "smidgen[9] quantities of uranium" (DeVoe, 1961). Consequently, analytical chemical reagents should be checked for radioactive contamination, and uncontaminated salts of elements with similar chemical behavior used, if necessary.

In preparing solutions and dilutions, the chemical and physical cleanliness of all vessels used with both the radioactive and inactive solutions is paramount. If chromate or sulfate is left on glassware from cleaning solutions or is present in the water, precipitation or radiocolloidal behavior may occur with elements such as strontium and radium. Water purified by distillation combined with ion exchange should be used. This two-step procedure removes particulate matter and most dissolved substances, both ionized and non-ionized.

Some processes occur in stored solutions that result in changes in concentrations. Several of these changes may not be noted if the concentration of the solution is 10^{-3} mole per liter or more, but others produce significant alteration (Korenman, 1966). The following will be more significant at low concentrations: hydrolysis resulting in gradual precipitation, the sorption of hydrolysis products, or polymerization; oxidation of reducing agents by dissolved oxygen; change in pH if carbon dioxide is or was present; decomposition of some com-

[9] "Smidgen" in this reference equals 10^{-9}g (1 ng).

pounds when exposed to light; absorption of activity by microorganisms; solubility of glass. These effects may be diminished or eliminated by the following techniques: 1) *acidification of the solution,* which curtails colloid formation and polymerization and thereby minimizes adsorption losses; 2) *increased carrier concentration,* which in many cases reduces, or may eliminate, detectable adsorption; 3) *complex-ion formation,* which insures stable solutions of some elements such as gold, tantalum, thorium; 4) *sterilization of solutions* (by heating, if solutions are not adversely affected by heating), after flame-sealing in glass, which inhibits the growth of microorganisms; 5) *steam treatment of clean pyrex or quartz vessels,* which lowers the solubility of the glass (Korenman, 1966). A further useful reference is the "Users' Guides for Radioactivity Standards," a compilation of brief guides to chemical and counting problems for standards of many elements (NAS-NRC, 1974).

5.1.2 *Quantitative Sampling of a Solution*

To make use of solution radioactivity standards, it is necessary to take quantitative samples. This involves leaving the standard in the laboratory for a sufficient length of time to attain thermal equilibrium. Then the solution is mixed by gently shaking and inverting the ampoule to avoid error caused by condensation on the upper walls of the vessel. It is important to allow the liquid to drain from the neck after the last inversion, then to open the ampoule and transfer quickly most of the solution to another vessel, which is closed immediately. Such simple precautions will avoid the most common sources of error in sampling a solution, i.e., inadequate mixing and losses due to evaporation. If the sample is to be taken volumetrically, the vessel to which the solution is transferred is usually a small volumetric flask whereas, for gravimetric sampling, the solution is transferred to a pycnometer if the dispensing is to be done forthwith.

Whether volumetric- or gravimetric-sampling methods should be employed depends chiefly upon the certified data for the standard. If the standard to be used is certified in terms of activity, emission rate or number of atoms per unit mass of solution, samples should be taken gravimetrically unless the density of the solution has been stated or measured.

Volumetric sampling has the advantages of simplicity and low-cost equipment, but is less accurate than gravimetric sampling. Volumetric techniques have been reviewed in the literature (Moran, 1967; Merritt, 1973). For dispensing small samples for source preparation, micropipets and microburets are used and several types are shown in Figure 76. The least expensive, costing only a few cents, are capillary tubes from which the contained solution is ejected using a rubber bulb or syringe. These give a precision of from one to a few percent (Emanuel, 1973), are used only once, and discarded. Other types, such as the Grunbaum and Kirk micropipets, are normally cleaned and reused. For more accurate work they may be calibrated (Marinenko and Taylor, 1966; Moran, 1967). The same is true for micropipets or microburets in which a precision-machined plunger displaces a volume that is read on a micrometer dial. These devices are rather expensive and give good precision for larger volumes ($\approx$100 μl), but for the delivery of very small volumes ($\approx$1 μl) they give no better precision than the inexpensive pipets. In volumetric sampling, care should be taken to have all apparatus and solutions at thermal equilibrium and attention should be given to the directions for use of the particular equipment selected, as some pipets are calibrated "to deliver" a specified volume and others "to contain" it. In the latter case they must be rinsed, usually with the dilution medium of the solution being sampled.

Most radioactivity standards are certified as to activity per unit mass of solution and therefore gravimetric-sampling methods have assumed increasing importance. Furthermore, gravimetric methods can give higher accuracy, e.g., <0.1 percent (Campion et al., 1964; Merritt, 1973). The essential equipment is a high-precision balance of 20-g capacity. Merritt (1973) has emphasized that care should be taken to place the balance on a sturdy heavy table (to minimize vibration) in a location free from drafts, preferably in a temperature-regulated and humidity-controlled room; and that routine tests of balance performance and calibration procedures are necessary as an assurance against systematic error.

Mass dispensing involves difference weighings from small bottles or pycnometers. Different designs of pycnometers have been discussed by Mann and Seliger (1958), Campion et al. (1964), Merritt and Taylor (1967a), Bowes and Baerg (1970), Lowenthal and Page (1970), and Merritt (1973). The one most often used is an inexpensive small polyethylene ampoule[10] with the neck drawn out to form a capillary. Such a pycnometer is shown in Figure 76; at the lower right is the ampoule as received, and just above it is shown in the following three stages: with capillary (i) drawn out and sealed until required; (ii) cut to a suitable length for filling; and (iii) cut to a shorter length for weighing. Error caused by evaporation of liquid from such a pycnom-

---

[10] Supplied by Canus Equipment Limited, 340 Gladstone Avenue, Ottawa, Canada.

Fig. 76. Source preparation paraphernalia. In the picture above, starting at bottom left and proceeding clockwise, are a disposable capillary-tube pipet, a Grunbaum pipet, 2 Kirk pipets, a micropipet with precision-machined plunger, 2 ampoules, 2 volumetric flasks, a microburet with precision-machined plunger, 2 disposable polyethylene pipets, polyethylene pycnometers in various stages of preparation, and at bottom right, 2 source mounts of metallized thin plastic film.

eter has been shown to be negligible (Merritt and Taylor, 1967a; Bowes and Baerg, 1970). The radioactive solution (>1 g) is introduced into the pycnometer, the capillary tube is cut to a convenient length, and it is placed in the balance room for about half an hour to attain thermal equilibrium. The balance operator should be seated in front of the balance for the last 10 to 15 minutes. A simple stepwise procedure with advisable precautions has been outlined by Merritt (1973), in which the pycnometer is weighed before and after a sample of solution is dispensed; the difference between the two weighings gives the sample mass. By dispensing the solution sample to a weighed dilution flask, adding carrier solution, and finally weighing the flask plus contents, a dilution can be prepared (Merritt, 1973).

5.2 Solid Sources

5.2.1 *Quantitative Source Preparation*

Many counting methods, such as alpha-, beta-, x-, and gamma-ray or coincidence counting, require that sources be in the solid phase, normally as thin deposits covering an area of approximately 1 cm^2. The simplest way to prepare them is to deliver the solution sample to a source mount and then evaporate to dryness by placing either under an infrared lamp, or in a covered box through which warm air flows (Bowes and Baerg, 1970).

Thin electrically-conducting source mounts are necessary for $4\pi\alpha$ and $4\pi\beta$, as well as 4π coincidence-counting methods. The preparation and effectiveness of metal-coated plastic films stretched on metal support rings have been reported by Roy and Turcotte (1967). Those described are both collodion and VYNS (a polyvinylchloride-polyvinylacetate copolymer) of surface density ≈ 10 μg cm^{-2}, coated with gold or gold (82 percent)-palladium (18 percent) alloy of surface density ≈ 20 μg cm^{-2}. The preparation of other thin films has been reviewed by Yaffe (1962). As a precautionary measure, many laboratories cover the dried active deposit with a similar metal-coated plastic film to form a stronger "sandwich" source and to inhibit the loss of small particles during manipulations.

Simple evaporation of the radioactive solution is not always an adequate source preparation technique, because the deposit tends to be in the form of large crystals, i.e., it tends to be non-uniform and self-absorption can be a serious problem (Mann and Seliger, 1958; Merritt *et al.*, 1959). Furthermore, for some elements the deposit produced by simple evaporation is unstable because the resulting chemical compound is volatile or hygroscopic. In such cases, the precipitation method is effective in producing a stable source of improved uniformity. Normally a fine deposit of the radionuclide is precipitated *in situ* on the source mount by delivering the active-solution sample into a slight excess of precipitating agent (van der Eijk *et al.*, 1973). However, chemical problems encountered in specific applications may require additional steps. For example, to precipitate the radioelement in question as an insoluble hydroxide from an acid solution, it is necessary first to evaporate the radioactive solution on the source mount, then to dissolve the residue by adding a drop or two of water to the source mount, and finally to expose it to ammonia vapor during its final drying period. Some examples can be seen in

Table 13. Helpful guides in judging the volatility of a precipitate are thermogravimetric data (Duval, 1963).

A wetting or seeding agent can improve source uniformity. Among the wetting agents employed are insulin, Catanac SN,[11] Tween,[12] and Teepol[13]. An area $\approx$1 cm^2 of the source mount is covered with a dilute solution of the wetting agent and then as much as possible of it is removed. The radioactive solution sample is delivered to the hydrophilic area, and then evaporated to dryness. Seeding agents used are Ludox-SM[14] and Teflon dispersion No. 30.[15] The active-solution sample is delivered into a drop of a 1:10^4 dilution of either of these sols containing $\approx$0.2 μg solids that serve as nucleation sites where many small crystals are produced upon drying.

Still more uniform sources can be prepared by other techniques that tend to be time-consuming. Freeze-drying combined with the use of a wetting agent has given good results (Keith and Watt, 1966). New approaches employ special source mounts onto which thin layers of a seeding agent (SiO) (Bellemare et al., 1972; Lachance and Roy, 1972), or a mixture of ion-exchange resins (Lowenthal et al., 1973; Lowenthal and Wyllie, 1973a, 1973b) have been added.

Although they are not ordinarily relevant to radioactivity standards, there are other source-preparation methods that give good source uniformity but less than 100-percent yields. They have been used, however, in the assay of radioactive solutions if a yield determination can be made by relative gamma-ray counting against a quantitatively prepared source. Among these methods are the formation of monomolecular layers (Dobrilović and Simović, 1973), vacuum evaporation, electrodeposition with and without an external source of e.m.f., electrospraying, chemical transport reactions, and ion implantation, all of which are discussed in a review paper by van der Eijk et al. (1973).

An alpha-particle source should be thin, uniform, and adherent to its substrate in order to minimize or eliminate scattering in the source, degradation of the spectrum, and change in the emission rate, respectively.

Alpha-particle emission gives rise to recoil atoms or even small conglomerations of atoms, and recoil contamination of detectors can occur, especially with those radionuclides with short-lived alpha-particle-emitting daughters. (Beta-particle emission yields too little recoil energy to cause noticeable contamination under usual measure-

[11] Supplied by American Cyanamid Co., Bound Brook, New Jersey.
[12] Supplied by Atlas Powder Co., Wilmington, Delaware.
[13] Supplied by Shell Chemical Co., Houston, Texas.
[14] Supplied by E. I. DuPont de Nemours and Co., (Inc.), Wilmington, Delaware.
[15] Supplied by Interchemical Corp., Lodi, New Jersey.

TABLE 13—*Source preparation techniques and chemical data on some standards solutions*

Radionuclide	Decay modes	β_{max} for principal β (MeV)	Ionic form	Carrier solution	Approximate inactive to active ion ratio	Some methods of preparation of minimum self-absorption sources
^{3}H	β^-	0.0186	H^+	H_2O	10^{10}	
^{14}C	β^-	0.1561	CO_3^{2-}	212 g/l Na_2CO_3 in 0.001 M NaOH	10^2	
^{22}Na	β^+ ϵ γ	0.5455	Na^+	0.10 g/l NaCl in 1 M HCl	10^6	Ludox[b] added, dried in air, placed under glass exposed to infrared lamp (~50 °C)
^{24}Na	β^- $e_2^{\pm}$ [a] γ	1.3902	Na^+	0.10 g/l NaCl in 1 M HCl	10^6	Ludox added, dried in air, placed under glass exposed to infrared lamp (~50 °C)
^{26}Al	β^+ ϵ γ	1.1740	Al^{3+}	60 g/l $AlCl_3$ in 1 M HCl (see section 5.1.1)	10^5	Ludox added, dried in air, placed under glass exposed to infrared lamp (~50 °C)
^{32}P	β^-	1.7110	PO_4^{3-}	0.098 g/l H_3PO_4 in H_2O	10^6	Ludox added, dried in air, placed under glass exposed to infrared lamp (~50 °C)
^{35}S	β^-	0.1674	SO_4^{2-}	0.1 M HCl	10	Ludox added, dried in air, then placed in oven at 70 °C for several days.
^{36}Cl	β^- ϵ	0.7093	Cl^-	0.2 g/l NaCl in H_2O	20	Ludox added, dried in air, placed under glass exposed to infrared lamp (~50 °C)
^{37}Ar	ϵ			Stable argon	10^9	
^{42}K	β^- γ	3.521	K^+	0.10 g/l KCl in 1 M HCl	10^7	Ludox added, dried in air.

[a] $e_2^{\pm}$ symbolizes an internal electron-positron pair associated with the 2.754-MeV transition in ^{24}Mg.

[b] Ludox: A 1:10^4 dilution of Ludox-SM 15 (15-percent suspension of silica) is most often used. Ludox is available from E. I. Dupont, Industrial Engineering Dept., Wilmington, Delaware.

TABLE 13—Continued

Radionuclide	Decay modes	β_{max} for principal β	Ionic form	Carrier solution	Approximate inactive to active ion ratio	Some methods of preparation of minimum self-absorption sources
^{45}Ca	β^-	0.259	Ca^{2+}	0.04 g/l $CaCl_2$ in 1 M HCl	10^5	One drop ludox + 1 drop 0.001M H_3PO_4 added, dried in air, redissolved with H_2O and dried in NH_3 atmosphere.
^{46}Sc	β^- ce γ	0.357	Sc^{3+}	0.10 g/l $ScCl_3 \cdot H_2O$ or $YCl_3 \cdot H_2O$ in 0.1 M HCl	10^4	Ludox added, dried in air, redissolved with H_2O and dried in NH_3 atmosphere.
^{51}Cr	ϵ ce γ		Cr^{3+}	0.10 g/l $CrCl_3 \cdot 6H_2O$ in 0.1 M HCl	10^4	Ludox added, dried in air, redissolved with H_2O and dried in NH_3 atmosphere.
^{54}Mn	ϵ ce γ		Mn^{2+}	0.01 g/l $MnCl_2 \cdot 4H_2O$ in 1 M HCl	10^5	Ludox added, dried in air, placed under glass exposed to infrared lamp (~50 °C). Also, autoelectroplated (non-quantitative).
^{55}Fe	ϵ		Fe^{3+}	0.01 g/l $FeCl_3 \cdot 6H_2O$ in 1 M HCl	10^2	Ludox added, dried in air, placed under glass exposed to infrared lamp (~50 °C). Also, autoelectroplated (non-quantitative).
^{56}Mn	β^- ce γ	2.850	Mn^{2+}	0.01 g/l $MnSO_4$ in H_2O	10^5	Ludox added, dried under infrared lamp (~50 °C).
^{57}Co	ϵ ce γ		Co^{2+}	0.05 g/l $CoCl_2 \cdot 6H_2O$ in 1 M HCl	10^4	Ludox added, dried in air, redissolved with water, dried in air, placed under glass exposed to infrared lamp (~50 °C) for several hours.
^{59}Fe	β^- ce γ	0.466	Fe^{3+}	0.10 g/l $FeCl_3 \cdot 6H_2O$ in 1 M HCl	10^5	Ludox added, dried in air, placed under glass exposed to infrared lamp (~50 °C) for several hours.

TABLE 13—*Continued*

Radionuclide	Decay modes	β_{max} for principal β	Ionic form	Carrier solution	Approximate inactive to active ion ratio	Some methods of preparation of minimum self-absorption sources
^{60}Co	β^- ce γ	0.3179	Co^{2+}	0.05 g/l CoCl$_2\cdot$6H$_2$O in 1 M HCl	10^4	Ludox added, dried in air, redissolved with water, dried in air, placed under glass exposed to infrared lamp (~50 °C) for several hours.
^{63}Ni	β^-	0.06587	Ni^{2+}	0.08 g/l NiCl$_2$ in 1 M HCl	10^2	Ludox added, dried in stream of warm air, redissolved with water, placed under glass exposed to infrared lamp (~50 °C) for several days.
^{65}Zn	ϵ β^+ ce γ	0.330	Zn^{2+}	0.01 g/l ZnCl$_2$ in 1 M HCl	10^3	Ludox added, dried in air, redissolved with water, dried in air, placed under glass exposed to infrared lamp (~50 °C) for several days.
^{75}Se	ϵ ce γ		SeO$_3^{2-}$	0.33 g/l H$_2$SeO$_3$ in 1 M HCl	10^5	Ludox added, dried in air, placed under glass exposed to infrared lamp (~50 °C) for several hours.
^{85}Kr	β^- γ	0.687		Stable krypton	10^3	
^{85}Sr	ϵ ce γ		Sr^{2+}	0.09 g/l SrCl$_2\cdot$H$_2$O in 1 M HCl	10^6	Ludox added, dried in air, placed under glass exposed to infrared lamp (~50 °C) for several hours.
^{88}Y	ϵ ce β^+ γ	0.755	Y^{3+}	0.10 g/l YCl$_3\cdot$H$_2$O in 1 M HCl	10^3	Ludox added, dried in air, placed under glass exposed to infrared lamp (~50 °C) for several hours.
^{89}Sr	β^- γ	1.492	Sr^{2+}	0.09 g/l SrCl$_2\cdot$H$_2$O in 1 M HCl	10^6	Ludox added, dried in air, placed under glass exposed to infrared lamp (~50 °C) for several hours.

TABLE 13—Continued

Radionuclide	Decay modes	β_{max} for principal β	Ionic form	Carrier solution	Approximate inactive to active ion ratio	Some methods of preparation of minimum self-absorption sources
^{90}Sr-^{90}Y	β^- β^- ce	0.546 2.284	Sr^{2+} Y^{3+}	0.027 g/l SrCl$_2$·6H$_2$O and 0.022 g/l YCl$_3$·6H$_2$O in 1 M HCl	10^4 10^7	Ludox added, dried in air, placed under glass exposed to infrared lamp (~50 °C) for several days.
^{95}Nb	β^- ce γ ce	0.1597	Nb^{5+}	0.2 g/l Nb (HC$_2$O$_4$)$_5$ in 3 percent H$_2$C$_2$O$_4$	10^5	Ludox added, dried in air, placed under glass exposed to infrared lamp (~50 °C) for several hours.
^{99m}Tc	γ ce		TcO$_4$$^-$	3 M HCl		Ludox added and dried in atmosphere of H$_2$S.
^{109}Cd- ^{109m}Ag	γ ϵ ce γ		Cd^{2+} Ag$^+$	1 g/l CdCl$_2$·2½H$_2$O in 2 M HCl	10^4	Ludox added, dried in air, placed under glass exposed to infrared lamp (~50 °C) for several days.
^{110}Ag- ^{110m}Ag	β^- ϵ ce	2.893	Ag$^+$	1 g/l AgNO$_3$ in ~0.5 M HNO$_3$	10^4	1. Ludox added, dried in air. 2. Dried in NH$_3$ atmosphere.
^{111}In	γ ϵ ce		InCl$_2$$^+$	3 M HCl	—	Dilute solution with 0.3 M HCl, deposit solution, add a drop of H$_2$O and one of Ludox, dry in H$_2$S atmosphere
^{113}Sn ^{113m}In	γ ϵ ce		Sn^{4+} In^{3+}	0.74 g/l SnCl$_4$ in 2.5 M HCl	10^5	1. H$_2$O$_2$ and water added, dried in air. 2. Ludox added, dried in air.
^{123}I	γ ϵ ce		I$^-$	Same as for ^{125}I and ^{131}I	10^4	Same as for ^{125}I and ^{131}I
^{125}I	γ ϵ ce γ		I$^-$	0.02 g/l Na$_2$SO$_3$ + 0.02 g/l LiOH + 0.05 g/l KI in H$_2$O	10^4	^{125}I solution is deposited into a drop of AgNO$_3$ solution and dried in air.

TABLE 13—*Continued*

Radionuclide	Decay modes	β_{max} for principal β	Ionic form	Carrier solution	Approximate inactive to active ion ratio	Some methods of preparation of minimum self-absorption sources
^{129}I	β^- ce γ	0.152	I^-	NaI, NaHSO$_3$, NaOH; pH 11.5 ($\sim$2 mg Na per gram of ^{129}I solution)	1	Solution absorbed into anionic exchange paper (in the chloride form).
^{131}I	β^- ce γ	0.6063	I^-	0.02 g/l Na$_2$SO$_3$ + 0.02 g/l LiOH + 0.05 g/l KI in H$_2$O	10^6	AgNO$_3$ solution added immediately to deposit of ^{131}I solution, and dried in air. Alternatively AgNo$_3$ can be deposited first.
^{131m}Xe	ce γ			Stable xenon	10^7	
^{133}Xe	β^- ce γ	0.346		Stable xenon	10^3	Freeze to point source.
^{137}Cs-^{137m}Ba	β^- ce γ	0.514	Cs^+ Ba^{2+}	0.028 g/l CsCl in 2.7 M HCl	10^2	Ludox added, dried in air, then placed under glass exposed to infrared lamp ($\sim$50 °C) for several days. Vacuum evaporated sources (non-quantitative).
^{139}Ce	ϵ ce γ		Ce^{3+}	0.01 g/l CeCl$_3$ in 1 M HCl	10^3	
^{141}Ce	β^- ce γ	0.435	Ce^{3+}	0.1 g/l Ce(NO$_3$)$_3$ in 1 M HNO$_3$	10^6	Ludox added, dried in air, then place under glass exposed to infrared lamp ($\sim$50 °C) for several hours.
^{144}Ce-^{144}Pr	β^- ce γ	0.3155 2.996	Ce^{3+} Pr^{3+}	0.01 g/l CeCl$_3$ in 1 M HCl	10^3	

TABLE 13—Continued

Radionuclide	Decay modes	β_{max} for principal β	Ionic form	Carrier solution	Approximate inactive to active ion ratio	Some methods of preparation of minimum self-absorption sources
^{147}Pm	β^-	0.2246	Pm^{3+}	1 M HCl	—	Ludox added, dried in air, placed in desiccator (NaOH), then desiccator placed under infrared lamp for several days. Sources redissolved with water, dried in air; replaced in desiccator under infrared lamp for at least one week.
^{182}Ta	β^- ce γ ϵ ce	0.522	TaF_7^{2-}	Very dilute HF	10^5	Dried in air, redissolved with water, dried, redissolved again with water and dried in NH_3 atmosphere.
^{197}Hg	γ ϵ ce		Hg^{2+}	0.4 g/l $Hg(NO_3)_2 \cdot H_2O$ in 0.1 M HNO_3	10^4	Ludox added, dried in atmosphere of H_2S.
^{198}Au	β^- ce γ	0.9612	$Au(CN_4)^-$	0.1 g/l $KAu(CN)_4$ + 0.001 g/l KCN	10^7	Ludox added, dried in air.
^{203}Hg	β^- ce γ ϵ ce	0.212	Hg^{2+}	0.4 g/l $Hg(NO_3)_2 \cdot H_2O$ in 0.1 M HNO_3	10^4	Ludox added, dried in atmosphere of H_2S.
^{201}Tl	ϵ ce γ		Tl^{3+}	0.01 g/l $TlNO_3$ in 1 M HNO_3	10^4	Ludox added, dried in air.
^{204}Tl	ϵ γ β^-	0.7634	Tl^{3+}	0.017 g/l $Tl(NO_3)_3$ in 1 M HNO_3	10^4	Ludox added, dried in air, then placed in oven at 50°C for one week.
^{207}Bi	ϵ ce		Bi^{3+}	4 M HNO_3		
^{241}Am	α ce γ		Am^{3+}	0.001 g/l $EuCl_3$ in 0.1 M HCl		Ludox added, dried in air.

ment conditions.) Sources of high activity should be covered with metallized mylar, collodion, or VYNS films to reduce the chances of loss of material, to protect the surface from abrasion, and to avoid corrosion by the atmosphere (but see Section 5.2.2 concerning radiation damage to plastics). If sources must be used uncovered, such as for high-resolution alpha-particle spectrometry where spectrum degradation is a concern, the use of weak sources is advised to mitigate against possible problems due to radioactive contamination of the detector.

There are a number of source-preparation techniques widely employed, among which are electrodeposition, molecular plating, electrophoresis, electrospraying, vacuum evaporation, sputtering, and simple evaporation of a stable solution (see Greene et al., 1972, for lists of references to the various techniques). Each method has at least one advantage: with electrodeposition and molecular plating, recovery is almost quantitative; with electrophoresis, mg cm^{-2} amounts can be adherently deposited; by electrospraying, thin layers with large areas can be made; vacuum evaporation and sputtering give good adherence; and simple evaporation of a stable solution is quantitative. The evaporation-of-solution technique should be used with wetting or seeding agents such as those mentioned in Section 5.2.1, and in the prescribed manner. A variety of backing materials can be used such as VYNS and collodion of surface density 10 μg cm^{-2}, coated with gold or gold (82 percent)-palladium (18 percent) alloy of surface density 10 μg cm^{-2}, metallized mylar, gold, silver, platinum, and stainless steel. If the backing material is clean and smooth, and, in the case of metals, polished, the source material will adhere to it better (Greene et al., 1972).

5.2.2 Handling and Storage of Thin Sources

Thin sources are fragile and must be manipulated carefully and stored properly. In humid climates source deterioration caused by absorption of water vapor occurs and is particularly noticeable if the active deposit is a deliquescent salt. As indicated in Table 13, it can be helpful to store the sources at elevated temperatures until they have been counted.

On exposure to warm humid air for a few days, the thin metallic coating on VYNS or collodion films deteriorates, and adversely affects the electrical conductivity of the source mount. This occurs less in the case of Au-Pd-alloy coatings than for those of pure Au (Lowenthal and Smith, 1964). The problem is insidious in that the degradation usually is not visible, although in extreme cases it can be seen that the metallic coating has coalesced. The electrical conductivity of a source can be restored by the addition of another layer of metallized plastic film. The problem can be avoided by storing the sources in a dry atmosphere such as that in a desiccator.

Radiation damage to thin plastic source mounts is especially severe for alpha emitters and essentially prevents the long-term storage of such high-activity sources (>1 μCi). Although weak alpha-active sources on thin films have survived for several years, such source mounts become weaker with time. For this reason and because the health hazard associated with inhaled alpha activity is generally high, extra precaution should be taken in handling such sources. Garfinkel and Hutchinson (1970) have described sources of alpha-particle emitters for use as gamma-ray standards in which the active material is enclosed between thin gold foils that in turn are sandwiched between layers of plastic. This technique prevents radiation damage to the plastic, but, unfortunately, can be used only for cases where thicker source mounts are tolerable.

5.3 Gaseous Sources

5.3.1 *Preparation of Gases for Liquid-Scintillation Counting*

Introduction of radioactive gases into the scintillation liquid for measuring their activities is often convenient and simple. The solubility of gases is usually sufficient to permit their radioassay by liquid scintillation.

There are two distinct techniques for the radioassay of gases by liquid-scintillation counting. If accurate results are desired, the entire gas sample must be dissolved in the liquid. However, it is sometimes convenient to fill the vial only partially with the liquid and to correct for losses resulting from the radioactive decays in the gas phase above the liquid. In both cases, temperature control is important, particularly for the second method, because the solubility of gases is temperature dependent.

An important requirement for the liquid-scintillation counting of gases is the deaeration of the liquid prior to introduction of the gas. According to Shuping *et al.* (1969), the scintillation liquid should be boiled with a reflux condenser for about 30 minutes and drawn up into a syringe that has a Luer fitting and appropriate stopcocks. The counting vial, provided with a Luer fitting and removable Luer stop-cock, is evacuated and filled with the gas at a pressure that corresponds to slightly less than the limit of solubility of the gas in the solvent. Subsequently, the vial is connected to the syringe and the liquid is slowly forced into the vial. The vial is filled to the top with the scintillation liquid, the stopcock is removed, and a Luer cap is placed on the vial. Care should be exercised to avoid changes in temperature. Environmental monitoring for ^{85}Kr is routinely carried out by using this technique.

The gaseous products of oxidation and combustion of many types of samples can be trapped in suitable media such as water for ^3H, ethanolamine for ^{14}C, peroxide for ^{35}S, and then incorporated in scintillator solutions. Such methods are described in Jeffay and Alvarez (1961), Eastham et al. (1962), and Bransome (1970).

5.3.2 Preparation of Gaseous Radioactivity Standards and Samples for Internal-Gas-Counter and Internal-Ionization-Chamber Systems

Gaseous radioactivity standards are normally contained in glass ampoules, the sample being accessible by means of a glass break-seal or stopcock. The calibration is in terms of activity per mole of active gas and isotopic carrier, or total activity per sample. The latter is derived from the former simply by transferring a sample of the former gas, at a known pressure and temperature, into a vessel of known volume.

Radioactivity standards calibrated by gas counting or by internal ionization chamber (Section 4.4.4) may actually be distributed as solutions. Examples are ^3H as water, ^{14}C as Na_2CO_3 in aqueous solution (calibrated as ^3H and $^{14}CO_2$ by gas counting), and ^{14}C as benzoic-acid-7-^{14}C in toluene (calibrated as $^{14}CO_2$ in an internal ionization chamber). In the preparation of these standards, the water is reduced to hydrogen by passing the water vapor over a bed of hot zinc (Mann et al., 1964, Unterweger et al., 1980), and the sodium carbonate is converted to $^{14}CO_2$ by the action of perchloric acid (Mann et al., 1961). The benzoic acid in toluene is converted to $^{14}CO_2$ and water by combustion with a slight excess of oxygen in a Paar bomb (Marlow and Medlock, 1960). Samples of the various gases are then trapped cryogenically in glass bulbs of known volume, at known pressures and temperatures, for assay in internal gas counters (Section 3.6).

In the case of the NBS noble-gas radioactivity standards, namely ^{37}Ar, ^{85}Kr, ^{131m}Xe, and ^{133}Xe, samples of these gases are prepared for gas counting by quantitative mixing with appropriate quantities of natural argon, krypton, and xenon, respectively. ^{131m}Xe and ^{133}Xe, which are obtained from an irradiation of uranium in a nuclear reactor, are separated in the NBS isotope separator. The separated 60-keV ion beams are allowed to impinge on aluminum foil in which they are trapped (Figure 77). The aluminum strips containing the two isotopes are cut out and each aluminum strip is melted in an evacuated vycor glass tube, thereby releasing very pure samples of each radioactive xenon isotope.

In the preparation of noble-gas radioactivity standards for distribution, the radioactive gas sample is mixed with inactive isotopic carrier in a gas-handling system such as that shown in Figure 78. The mixing is performed by repeated freezing and evaporation by using either liquid nitrogen or liquid helium, the former when the vapor pressure of the radioactive gas sample at 77 K is sufficiently low for adequate freezing. Up to two hundred ampoules may be filled at one time, together with a special volume-calibrated flask (Figure 79), which is then attached to the gas-handling system of the NBS internal gas-counting system (Figure 31). If, however, the activity in the individual standard ampoules of approximately 10 ml is sufficient, the gas in them may be totally transferred, cryogenically, to the gas-counting system for measurement of the activity.

After transfer to the internal gas-counting system, the quantity of the gas sample is calculated from pressure and temperature measure-

Fig. 77. A photograph of a collector foil from the NBS isotope separator showing the implanted stable isotopes of Xe. The mass regions are visually enhanced by means of a thin coating of vacuum grease.

Fig. 78. The gas-handling rack used at NBS for mixing gaseous radionuclides with inactive carrier gas, which is usually isotopic, and for filling the ampoules shown.

ments, by using the known volumes of the gas-handling system and van der Waals' equation. It is then mixed with a known quantity of a suitable counting gas, usually methane for proportional counting, together with some 2 to 5 percent by volume of the inactive noble gas in question (or natural hydrogen for ^3H, or natural CO_2 for ^{14}C). In the process of mixing it is, however, important first to flush the previously evacuated counters with some of the counting gas mixed

Fig. 79. Partial view of ampoule rack containing approximately 200 ampoules. The calibrated volume flask is seen on the right above the vacuum stopcock.

with inactive isotopic carrier in order to mitigate against the possible subsequent adsorption of active gas onto the walls of the counters.

5.4 Preparation of Liquid Sources for Specific Methods of Measurement

5.4.1 *Liquid-Scintillation Counting*

5.4.1.1 *Solution and Emulsion Scintillators.* The principles of liquid-scintillation counting were dealt with in Section 2.4.5 and some applications of solution and emulsion scintillators were reviewed in Section 4.3. In most instances, counting samples are prepared by pipetting the scintillator formulation into the vial and then adding (either gravimetrically or volumetrically) a known amount of an aqueous solution of the sample. Procedures for accurately dispensing the aqueous sample have been reviewed recently by Merritt (1973). Techniques for preparing stable solution scintillators are given in Vaninbroukx and Spernol (1965), while precautions to be observed when preparing emulsions are given in Lieberman and Moghissi

(1970), Handler and Romberger (1973), and Benson (1976). A useful review of liquid-scintillation sample-preparation techniques, with some 250 references, has been prepared by Peng (1977).

5.4.1.2 *Suspended Scintillators.* Scintillators such as anthracene, calcium fluoride, and plastics in different shapes may be added to aqueous radioactive solutions to form two-phase counting systems. The counting efficiency with suspended-scintillator systems is more dependent on the scintillator-volume to sample-volume relationship than on the salts in, or the pH of, the solution (Bransome, 1970). One distinct advantage of this method (excluding the use of conventional ion-exchange resins) is the easy recoverability of the sample (Heimbuch, 1961).

5.4.1.3 *Gel Scintillators.* This method is applicable to insoluble or not conveniently solubilized material, to solutions with high solids content, and to materials that are strong quenchers in solution (Helf *et al.*, 1960). With the addition of a thixotropic agent to the scintillator liquid, finely divided solid samples can be suspended, a procedure that eliminates particle or sample settling. The most frequently used gelling agent is Cab-O-Sil[16] (Ott *et al.*, 1959). Benakis (1971) describes his experiences with another thixotropic agent, Poly-Gel-B[17], which offers the advantages of long-term stability of suspensions even at room temperature, ease in handling, and reduction of adsorption on surfaces of glass counting vials.

5.4.1.4 *Counting with Solid-Source Supports.* Substances that are precipitated, filtered, or absorbed on supports such as cellulose and nitrocellulose membranes, or lens, filter, and anion-exchange papers can be assayed by liquid-scintillation counting. The counting efficiency will depend on a number of variables, such as the position of the solid support in the counting vials, and the chemical and physical form of the sample material. Internal standards should be used for quench compensation; for the counting of low-energy beta-ray emitters, the procedure should be "rigorously standardized" for the specific kind of sample and support used (Bransome, 1970).

5.4.2 *Čerenkov Counting*

For materials containing beta-ray emitters with energies greater than 0.26 MeV, sample preparation for Čerenkov counting is simple and the source can be recovered. Solid samples may be prepared by dissolving the material in a suitable solvent. Detection efficiency will

---

[16] Supplied by G. L. Cabot, Inc., Boston, Massachusetts.
[17] French patent No. 1,590,762.

be dependent on the beta-ray energy of the radionuclide, the characteristics of the phototube, the geometrical and optical arrangement of the counting vials, the refractive index, and the density of the solvent (Parker and Elrick, 1970). With the addition of wavelength shifters, coincidence-counting efficiency can be improved, and the method used for lower-energy beta particles. However, sample preparation becomes more complicated. Stability of the solution is dependent on the particular wavelength shifter used, its concentration, and pH. Ross (1971) recommends 4-methylumbelliferone as the "overall best" wavelength shifter for Čerenkov counting because solutions containing it appear to be stable with changes in pH. Work by Ross (1970) gives evidence that Čerenkov-counting techniques can be extended to ^{14}C with the use of solvents of high refractive index (see also Section 4.5.1). See Peng (1977) for further information on source preparation.

5.4.3 *Formamide Counting*

Solution samples for counting in a $2\pi\beta$ windowless gas-flow counter are prepared by dissolving the radioactive material in an inert, low-vapor-pressure organic solvent such as formamide (Schwebel et al., 1951, 1954). If the radioactive material is not readily soluble in the organic solvent, it may be first dissolved in a liquid that is miscible with the organic solvent. Suitable dilutions are from 10 to 100 times, depending on the activity of the radioactive solution. Up to 10-percent water content in the formamide can be tolerated without change in the counting characteristics or efficiency of the internal gas counter. A sample of the counting solution, generally 1 ml, is pipetted into a machined counting cell of specific dimensions and counted in a $2\pi\beta$ windowless gas-flow counter. The counting solution should contain 1 percent of an appropriate carrier to prevent adsorption losses and contamination of the counting cell. If the dimensions of the counting cells used are always the same and the same volume of solution is always counted, the depth of solution is always the same and any effects due to gamma rays, for a given radionuclide, will remain constant.

6. The Assay of Radioactivity and the Identification of Radionuclides in Environmental, Medical, and Industrial Laboratories

6.1 Environmental and Low-Level Radioactivity Measurements

6.1.1 *General Considerations*

6.1.1.1 *Introduction.* Radioactivity measurements on environmental samples are usually made to assure that levels are kept below current maximum permissible concentrations. The large number of radionuclides found in the environment can be divided into four categories: namely, (i) "natural" radioactivity from uranium, thorium, and other primordial radionuclides; (ii) radionuclides produced by cosmic-ray interactions in the atmosphere; (iii) the man-made contaminants from nuclear-weapons fallout and effluent from nuclear facilities; and (iv) radioactive tracers used to measure transport phenomena. Some typical concentrations of long-lived radionuclides in ground-level air and surface soil are given in Tables 14 and 15, respectively. Table 16 lists some of the radionuclides found in effluents from nuclear-power facilities.

The instruments used in low-level radioactivity measurements are basically the same as those described in previous chapters. The special problem in low-level counting is that the ratios of counting rates of sample to background are often so low that significant modifications to the measuring systems are needed to improve sensitivity. Some extensive general references in this field are Watt and Ramsden (1964), ICRU (1972), Harley (1972), and NCRP (1976b; 1983; 1984).

Normally there are four steps in the measurement of environmental radioactivity:

(i) Selecting the measurement system;
(ii) Collecting a suitable sample;
(iii) Separating and, or, concentrating the desired species by radiochemical and, or, physical means;
(iv) Assaying the radioactivity in the source with a detector system that has been calibrated under similar counting conditions with appropriate radioactivity standards.

A general reference regarding the routine operation of an environmental monitoring program is Energy Research and Development Agency (now known as Department of Energy) ERDA, 1977. The treatment and reporting of the data obtained from such a program is described in Watson, 1980.

Although this section is primarily concerned with specific measurements systems for alpha-particle, beta-particle, and gamma-ray emitters, some general remarks concerning selection of a measurement system are in order.

6.1.1.2 *Selection of a Low-Level-Measurement System.* The choice of a measuring system for a particular application is based on a consideration of such factors as: cost, amount of sample available, type and intensity of radiation to be measured by the detector, time available for counting, magnitude of background, and system counting efficiency. Many expressions have been developed that are useful in comparing different detectors or different measurements systems. For example, Moghissi (1970) derived an expression often used for comparing liquid-scintillation-counting systems. He defined a Y value as the minimum radioactivity concentration detectable for a one-minute counting time as

$$Y(\text{pCi/g}) = \frac{\sqrt{B}}{2.22 M \epsilon}, \tag{6.1}$$

where
  B = background count rate, in counts per minute
  ϵ = counting efficiency, in counts per minute per disintegration per minute,
  M = sample mass (g),
  and the constant 2.22 is the number of disintegrations per minute per picocurie.

The three variables on the right hand side of Equation 6.1 are usually interrelated (background and efficiency depend on the sample size), and this must be considered in optimizing the sensitivity of the measurement process. It can be seen that Equation 6.1 is of the general form

$$Y(\text{pCi/g}) = \frac{\text{Constant} \times \text{Detection Limit}}{\text{Sample Size} \times \text{Efficiency}}, \qquad (6.2)$$

where Moghissi used $\sqrt{B}$ as the detection limit. In the more general case, detection limits and efficiency terms will depend on the system under consideration. Some comments concerning detection limits, sample size, and background that may aid in evaluating a measurement procedure are given in the following paragraphs.

6.1.1.3 *Detection Limits.* The problem of assigning an *a priori* detection limit to a measurement involves the theory of hypothesis testing (Section 7.1.3). Basically one must decide whether a given response is statistically different (at prescribed confidence limits) from the background response. This problem has been examined by Altshuler and Pasternack (1963), Currie (1968), Pasternack and Harley (1971), and Donn and Wolke (1977). A derivation of detection limits along the lines developed by these papers is given in Section 7.1.3, Eq. 7.9. It is sufficient to say here that the detection limit is directly proportional to the standard deviation of the background counting rate. Thus, it is not only necessary that the background be reduced as much as possible, but also that it be free from fluctuations that would increase the standard deviation of the background count. Sources of background counts are discussed in Section 6.1.1.5.

6.1.1.4 *Sample Size.* The sample size will usually be limited by such practical aspects as cost, shipping, and storage space. In the laboratory, other limitations on sample size may be imposed by the analytical procedure or physical dimensions of the detector and shielding. For each detector, there will also be an optimum sample size and shape that will depend on considerations such as the average distance of penetration of the radiation within the source, and on geometry.

6.1.1.5 *Background.* The background may be defined as the count rate observed when measuring a "blank" that simulates as closely as possible in chemical composition and physical form the source being measured. Care should be taken that the structural materials and reagents used in the preparation of the blank and the counting sample contain minimal activity. Statistical considerations involving background and minimum detectable activity are treated in Sections 7.1.2 to 7.1.4.

For measurements of low-level activities using an energy-discriminating detector in which the spectrum of the incident radiation occupies only a part of the wider spectrum of the background radiation, Povinec (1981) has proposed a method for simultaneous activity and background measurements, using just the one detector. His method is based on the assumed continuation of a previously measured correlation between background counts accrued in two spectrometer "windows," one of which also contains all the counts due to the activity. He points out the advantage of such a method in cases where counting times may be as long as a day. In such cases, short-term variations in the cosmic-ray intensity may occur during the total period of several days that may be required to measure the activity and background counts separately.

Examples of sources of background count rate are natural and fallout radioactivity, nearby nuclear facilities, cosmic radiation, instrument noise, and radioactivity in the detector materials. Some general comments concerning each of these are given in the following paragraphs.

a. Natural and Fallout Radioactivity

The common radionuclides found in the environment are listed in Tables 14 to 16. These species can contribute to the background in several ways depending on the nature of the detector. Their presence, for example, in soil, air, concrete, and other environmental matrices, gives rise to a continuous (and variable) background of radiation.

TABLE 14—Typical radioactivity concentrations of long-lived radionuclides in ground-level air

Radionuclide	Gas[a]	Particles[b]	Monthly deposition[c]
	pCi/m^3	pCi/m^3	pCi/m^3
Naturally occurring			
^{222}Rn	100	—	—
^{220}Rn	1	—	—
^{210}Po	—	0.003	—
^{210}Pb	—	0.01	—
^{7}Be	—	0.1	3,000
Man-Made			
^{239}Pu	—	3×10^{-5}	—
^{238}Pu	—	1×10^{-5}	—
^{144}Ce	—	0.01	200
^{137}Cs	—	0.002	50
^{90}Sr	—	0.001	30
^{85}Kr	17	—	—
^{3}H	1	—	20,000

[a] Mean ^{222}Rn, ^{220}Rn and ^{85}Kr values from NCRP (1976b).

[b] Data for New York City in 1972 (HASL, 1973) except ^{210}Pb values for Cincinnati in April—May 1961 (Gold et al., 1964). Units are pCi/m^3 of filtered air.

[c] Data for Cincinnati in 1973, extracted from monthly reports of the Radiochemistry and Nuclear Engineering Laboratory of the Environmental Protection Agency, Cincinnati, Ohio. The value of ^{3}H is based on an average concentration of 0.2 nCi/l in drinking water and an average monthly rainfall of 10 cm (EPA, 1973).

TABLE 15—*Radioactivity concentrations of radionuclides in soil*[a]

Radionuclide	Concentration
	pCi/g
Naturally occurring[b]	
^{40}K	10
^{87}Rb	4
^{226}Ra	0.8
^{232}Th	0.6
^{238}U	0.7
Fallout[c]	
^{90}Sr	~1
^{137}Cs	~1

[a] From Chapter 2 of NCRP (1976b).

[b] Values for the natural radionuclides are reported as world average concentrations (NCRP, 1976b), although variations as large as a factor of 4 between different locations are quite common.

[c] Values shown for fallout radionuclides are typical for surface soil (that is, the sample is taken to a depth of about 1 cm). Depth distributions for fallout ^{137}Cs in soil are given by Ragaini (1976).

Shielding alone will not necessarily eliminate some of this contribution because, for example, gaseous radionuclides such as ^{220}Rn and ^{222}Rn may be present inside the shielding, and their daughter products may be deposited upon it and the counting equipment.

Fallout activity generally has decreased since publication of the first edition of this Report. It is useful in this respect to differentiate between shorter-lived and longer-lived fission products. Thus, for example, after a nuclear test ^{99}Mo, ^{131}I, ^{140}Ba and ^{239}Np can be observed in the atmosphere for one to four weeks; ^{89}Sr, ^{95}Zr, ^{95}Nb, ^{103}Rn and ^{141}Ce for about a year; and long-lived radionuclides such as ^{3}H, ^{90}Sr, ^{106}Ru, ^{125}Sb, ^{137}Cs and ^{144}Ce for much longer. In addition some activation-product radionuclides such as ^{54}Mn and ^{60}Co can be detected.

b. *Nearby Nuclear Facilities*

The presence of a nearby nuclear facility may pose serious problems for environmental measurements. In addition to the off-gases listed in Table 16, reactors release some short-lived gaseous radionuclides that may contribute to the background count rate in the local area.

Neutron-generating facilities in the vicinity may also have an adverse effect on the detector background due to radiation arising from neutron capture. In semi-conductor detectors, for instance, Bunting

TABLE 16—*Radionuclides in effluents from typical nuclear facilities ($T_{1/2} > 1$ d)*

Nuclear power stations[a]

in gases:

^{3}H	^{85}Kr	^{133}Xe
^{14}C	^{131}I	^{133m}Xe

in airborne particles:

^{51}Cr	^{60}Co	^{134}Cs
^{54}Mn	^{89}Sr	^{137}Cs-^{137m}Ba
^{55}Fe	^{90}Sr-^{90}Y	^{140}Ba-^{140}La
^{59}Fe	^{99}Mo-^{99m}Tc	^{141}Ce
^{58}Co	^{131}I	

in liquids:

^{3}H	^{58}Co	^{124}Sb
^{14}C	^{60}Co	^{131}I
^{32}P	^{63}Ni	^{133}Xe
^{51}Cr	^{65}Zn	^{133m}Xe
^{54}Mn	^{89}Sr	^{134}Cs
^{55}Fe	^{90}Sr-^{90}Y	^{137}Cs-^{137m}Ba
^{59}Fe	^{95}Zr-^{95}Nb	^{140}Ba-^{140}La
^{57}Co	^{95}Mo-^{99m}Tc	^{141}Ce
	^{110m}Ag	^{144}Ce-^{144}Pr

Nuclear fuel reprocessing plants[b]

in gases:

^{3}H	^{85}Kr	^{129}I

in liquids:

^{3}H	^{134}Cs	^{238}U
^{14}C	^{137}Cs-^{137m}Ba	^{238}Pu
^{22}Na	^{144}Ce-^{144}Pr	^{239}Pu
^{54}Mn	^{147}Pm	^{240}Pu
^{60}Co	^{155}Eu	^{241}Pu
^{90}Sr-^{90}Y	^{233}U	^{241}Am
^{95}Zr-^{95}Nb	^{234}U	^{242}Cm
^{125}Sb	^{235}U	^{244}Cm

[a] From Kahn *et al.* (1971a, 1971b).

[b] From Magno *et al.* (1970); Cochran *et al.* (1970); Holm and Persson (1976); Cuddihy *et al.* (1976).

and Krausharr (1974) observed ^{73m}Ge, ^{75m}Ge, and ^{28}Al, the last being produced by irradiation of the detector housing. The problem is much more serious, however, for NaI(Tl) detectors (ICRU, 1972) where ^{24m}Na and ^{128}I are produced.

In addition to the effects described above, a neutron flux produces a number of high-energy electrons and photons in the surroundings that may subsequently interact within a detector.

c. Cosmic Rays

The cosmic-ray flux in the upper atmosphere consists of a "galactic" component and a "solar" component. The galactic radiation (NCRP,

1975) is composed of about 87-percent protons, 11-percent alpha particles, 1-percent heavier nuclei, and 1-percent electrons, many of which have energies in excess of 10^3 MeV. The sea-level cosmic-ray flux consists of both muonic and nucleonic components. The effects of cosmic rays on detector backgrounds have been examined by May and Marinelli (1960) and Tanaka et al. (1965). Extensive reviews are available in ICRU (1972), NCRP (1975), and Adams and Lowder (1964). The counts observed in detectors are principally due to the secondary particles generated by the interaction of these high-energy particles with matter.

d. *Instrument Noise*

The contribution to background from instrument noise is very dependent on the type of detector and energy region of interest. For instance, in scintillation detectors, spurious counts are observed at low pulse height due to noise in the electron-multiplier phototubes. Thermal noise of this nature, once a significant problem in liquid-scintillation counters, has been reduced by improved tube design (see Section 4.3.5).

Variations in the high-voltage supply may also pose a problem. All counters should be isolated from electro-mechanical equipment such as pumps, compressors, motors, and Tesla coils by an efficient filter in the main power supply in order to prevent spurious counts. A single efficient electrical ground may also be required to prevent 60-Hz noise induced between interconnected instruments at differing ground potentials. Other examples of instrument noise will be discussed in relation to specific measurements systems.

e. *Radioactivity in Detector Materials*

The contamination of the detector materials with radioactivity is a serious problem and one that is difficult to correct (Rodriguez-Pasqués, 1971). A common problem is that of ^{40}K in glass phototubes. The contribution from this source is greatly reduced by using quartz rather than glass windows. Measurements of radioactivity in materials used in low-level-counting systems have been reported by DeVoe (1961), Weller (1964), and Rodriguez-Pasqués et al. (1972); and in Section 2 of ICRU (1963).

6.1.1.6 *Interference Due to Other Radioactive Constituents.* As stated earlier, environmental samples are often mixtures of several radionuclides. In such cases, the sensitivity with which one can measure a particular constituent depends on the degree of interference from the other radionuclides in the sample. If only one species is of interest, it may be possible to effect a partial chemical separation during sampling. For instance, one might extract material of interest from several hundred liters of water using a cation-exchange column. In the laboratory, radionuclides may be separated from the bulk matrix and each other by radiochemical methods (Harley, 1972; Johns, 1975; NCRP, 1976b).

It is often possible to assay several radionuclides in the same sample by using spectrometric techniques. However, interferences may still be encountered, such as a Compton continuum caused by more energetic photons from other components in the sample. These interferences may often be minimized by instrumental techniques. Examples such as coincidence, anticoincidence, and pulse-shape discrimination will be discussed in relation to specific measurement systems in the following three subsections.

6.1.2 *Low-Level Measurement of Alpha-Particle-Emitting Nuclides*

6.1.2.1 *Introduction.* The very short ranges of alpha particles dictate that the amount of matter between the radioactive source being measured and the alpha-particle detector be minimized. Different methods for achieving this include close physical proximity of source to detector, placement of source and detector in a vacuum chamber, and mounting the source inside the detector. The detector and source holder should be constructed of materials having very low alpha-particle activity (see also references in Section 6.1.1.5e). Even materials having negligible intrinsic alpha-particle activity can indirectly contribute to the observed alpha-particle background if other background radiations cause nuclear reactions in these materials that produce alpha particles, e.g., (n,α) reactions in nylon sample planchets (Rodriguez-Pasqués *et al.*, 1972). Materials that are clean initially can become contaminated in time by exposure to atmospheres containing alpha-particle-emitting daughters of ^{222}Rn and ^{220}Rn that "plate out" on surfaces. A review of studies of radon transport through, and exhalation from, building materials has been published by Collé *et al.* (1981).

The short range of alpha particles in liquid and solid samples, less than 0.01 cm, usually requires the removal by physical and, or, chemical means of as much as possible of the inactive material originally present. For gross alpha-particle measurements, maximum sensitivity is attained when the prepared sample is at least as thick as the range of the alpha particles, i.e., so-called "infinitely thick" (Watt and Ramsden, 1964). For alpha-particle spectrometry, the source should be in a very thin layer to provide maximum resolution. If absorption corrections are to be less than 1 percent, the source thickness should be less than 1 percent of the average range of the alpha particles when counting in a 2π configuration (Lucas and Hutchinson, 1976). This criterion means that a sample of low atomic weight should have a surface density of approximately 100 $\mu g/cm^2$ or less. (Surface density is obtained by multiplying the density of the sample by its thickness.) A number of methods for preparation of alpha-particle-emitting sources has been described in NCRP (1961), by Yaffe (1962) and in references therein, and by Watt and Ramsden (1964). For a discussion of alpha-particle-emission rates from solid sources and range-energy relationships, see Evans (1955) and ICRU (1972).

Specific methods for measuring low-level alpha-particle radiation are discussed, according to type of detector, in Sections 6.1.2.2 through 6.1.2.5. The principles of some of these detectors are concisely given by Johnson et al. (1963).

6.1.2.2 *Scintillation Counting.* One of the earliest methods for detecting alpha particles used a ZnS screen and a low-power microscope to view the tiny flashes of light. Discs of polyester film coated with silver-activated ZnS and phototubes with appropriate electronic circuitry are now used. Developed by Hallden and Harley (1960), such detectors have a geometrical efficiency of approximately 50 percent. They are not suitable for spectrometric analysis. The background counting rates over the entire alpha-particle-energy range are quite low, and of the order of one event per hour for ZnS(Ag) on a thin plastic disc of 24-mm diameter. Figure 80a shows the configuration used for measuring thick sources. Thin sources can be placed between the planchet and the phosphor in the configuration shown in Figure 80b, which is also used for background measurements.

"Infinitely thick" powdered sources containing isotopes of radium and their daughters can be assayed by scintillation counting, provided the sample is sealed to prevent the loss of radon. The fraction of the total activity arising from the decay of ^{226}Ra may be found by following the ingrowth of ^{222}Rn. Biological and geological samples have been assayed in this way (Turner et al., 1958; Cherry, 1963; Inn et al., 1984).

The contributions of two of the natural radioactivity series to the total alpha-particle activity in thin sources can be measured by the so-called "fast-pairs" technique. Polonium-215 in the actinium series and ^{216}Po in the thorium series have half lives of 1.78×10^{-3} s and 0.15 s, respectively. As each of these is formed and decays, pairs of

Fig. 80. Schematic diagram of the zinc-sulfide scintillator with phototube for the detection of alpha particles. The configuration in (a) is for count-rate measurements of thick samples, while that in (b) is for thin-sample and background measurements. (From Rodriguez-Pasqués *et al.*, 1972)

nearly coincident alpha particles are emitted which can interact with the detector, giving rise to output pulses closely related in time. These fast pairs of pulses are presented to two scalers, one that records all pulses, and the other that is insensitive to any other pulse for a preset time, typically 0.3 to 0.4 s, after the arrival of the first pulse. The difference between the count rates of the two scalers gives the number of pairs in unit time. A variation of the method, that gives the proportion of pairs directly, is for the second scaler to be sensitive to the second member of the pair for a preset time after the arrival of the first. Except for very thin sources, however, a correction must be calculated to account for the differences between the ranges of the pairs of alpha particles in the sample, hence the technique is only an approximate one (Cherry, 1963, and references therein).

Other inorganic scintillators used to detect alpha particles are CsI(Tl) and CaF_2(Eu). CsI(Tl) is only very slightly, but reversibly, hygroscopic so that no encapsulation is required. Exposure to air at high relative humidity should be avoided. The resolution for 5.29-MeV alpha particles with small crystals, 5-mm diameter by 0.3-mm thick, is approximately 200-keV FWHM (Watt and Ramsden, 1964), but is poorer for larger crystals. CaF_2(Eu), is not hygroscopic, and can be cleaned with water. It is less sensitive to background gamma radiation than CsI(Tl) as it consists of lower atomic-weight elements. CaF_2(Eu) in NaI(Tl)-shielded ("phoswich") detectors have been developed by Mayhugh *et al.* (1978).

Conventional liquid-scintillation-counting systems are suitable for the assay of alpha-particle-emitting nuclides in some samples, but the background count rates may be so high as to preclude their application, in reasonable counting times, to very low-level sources. A low-background liquid-scintillation counter with a single phototube has been developed by McDowell (1975). High-detection efficiencies, that can approach 100 percent, are an advantage of liquid-scintillation counting. ^{222}Rn and its decay products have been assayed by liquid-scintillation counting (Horrocks and Studier, 1964; Rolle, 1970; Darrall et $al.$, 1973). The coupling of a multi channel analyzer to a liquid-scintillation counter with a single phototube can yield alpha-particle-spectra with an energy resolution of several percent (McDowell, 1971; Figure 81), which, however, is not satisfactory for many alpha-particle-

Fig. 81. Liquid-scintillation spectra of selected actinides using a single-phototube detector, with the peak-channel counts of each major peak preset to 40,000. (From McDowell, 1971)

spectrometry applications. Pulse-shape discrimination can reduce the beta-particle component of the spectrum (Thorngate et al., 1974).

Two scintillation-counting techniques are employed in the assay of gaseous ^{222}Rn and ^{220}Rn after separation from ^{226}Ra and ^{224}Ra, respectively. In the first, the initially positively-charged recoiling daughter atoms from the decay of radon are collected by means of an electric field, on to a ZnS (Ag)-coated electrode within the chamber containing the sample and filling gases (Bryant and Michaelis, 1952). The electrode, usually a screen, may be viewed from outside the chamber by a phototube as shown in Figure 82 (Mann and Seliger, 1958; Roberts and Davies, 1966), or removed and placed in an alpha-particle detector after a collection period (Kuchta et al., 1976). In the second technique, the gas is held in a smaller chamber and the photons, produced when the alpha-particles hit the ZnS(Ag)-coated walls, are detected by one (Lucas, 1957) or two (Kraner et al., 1964) phototubes. A regenerating activity standard of ^{222}Ra in water is being developed at the National Bureau of Standards (Hutchinson et al., 1984). Assays of drinking water for $^{224, 226, 228}$Ra have been made by McCurdy and Mellor (1981) with minimum detectable concentrations of 0.2, 0.3, and 0.5, respectively, using coincident α-γ and β-γ techniques.

6.1.2.3 *Gas-Filled Detectors.* The assay of ^{226}Ra can be accomplished by measuring its separated gaseous daughter ^{222}Rn in an ionization chamber. Ionization chambers, both direct-current and pulse, and the associated gas-handling and purification systems, were developed for the assay of radon in solution, breath and air samples, and of radium in solutions of rock and ore samples, by Evans (1935, 1937), Curtiss and Davis (1943), and Harding et al. (1958). A radon-analysis system that is similar in principle has been described by Harley (1972).

A gas-filled proportional detector (see, e.g., McDaniel et al., 1956) can be used to detect alpha particles from thick or thin solid sources. The source may be inside the detector, or outside next to a thin window. Some systems are capable of simultaneously recording alpha and beta particles by using appropriate pulse-height discrimination.

6.1.2.4 *Alpha-Particle Spectrometry.* The two major types of alpha-particle spectrometers are the pulse-ionization chamber and the semiconductor detector. Pulse-ionization chambers of many designs have been built, but nearly all can be classified as either parallel-plate or cylindrical, and each class can be divided into gridded or gridless varieties. Cylindrical chambers have the advantage in low-level measurements of providing a large surface area for the sample without building up such large capacitance between electrodes as to degrade resolution. A Frisch-type grid that shields the collecting electrode is often employed to improve resolution by reducing the pulse-widening

Fig. 82. Cross-sectional drawing of scintillation counter for radon analysis constructed at the National Bureau of Standards and based on the method of Bryant and Michaelis (1952). (From Mann and Seliger, 1958)

effect of the positive ions (see, e.g., Bunemann *et al.*, 1949). Cherry (1963) and Watt and Ramsden (1964) elaborate on the merits and construction principles of pulse-ionization chambers. For good examples of spectrometers with very large area (15,000 cm^2) and high resolution (down to 40-keV FWHM), see Hill (1961) and Osborne and Hill (1964), respectively.

Silicon surface-barrier semiconductor detectors, that were previously fragile and costly, are now probably the most widely used in alpha-particle spectrometry. A 1-cm^2 active-area detector typically has a resolution of 15- to 20-keV FWHM for 5.5-MeV alpha particles emitted from a source 1.5 detector diameters away. Placing the source closer to the detector increases the detection efficiency at the expense of degraded resolution. Compared to most pulse-ionization chambers, the larger semiconductor detectors, with an active area of about 20 cm^2, are still very small. This fact may necessitate more extensive sample preparation prior to counting than for pulse-ionization detection. Electrodeposition and evaporation are frequently used as the final step to concentrate the activity on to a small area (see Section 5.2; Yaffe, 1962; Watt and Ramsden, 1964). Contamination of silicon detectors by recoil atoms or by volatilization of ^{210}Po can raise otherwise very low background-count rates to unacceptable levels. These problems, and solutions for them, have been investigated by Sill and Olson (1970).

Mass spectrometry is another analytical method that has been applied to low-level quantitative identification of alpha-particle-emitting nuclides (NCRP, 1976b). By this method Diamond *et al.* (1960) assayed transuranic elements in thermonuclear-bomb debris, Krey and Hardy (1970) determined the isotopic composition of plutonium in contaminated soil, and Fields *et al.* (1972) assayed for several actinides in lunar materials.

Recent developments in alpha-particle spectrometry, including techniques of source preparation and the measurement of decay data, were discussed at a seminar organized by the International Committee for Radionuclide Metrology at the Central Bureau for Nuclear Measurements at Geel, in October, 1981 (ICRM, 1984b). In recent years half-life measurements of ^{239}Pu and ^{240}Pu have been carried out under the *aegis* of a U.S. Department of Energy Half-Life Evaluation Committee under the chairmanship of W.W. Strohm (Lucas and Mann, 1978 and 1984). These measurements employed the techniques of calorimetry, counting and mass spectrometry.

Horrocks (1966a) used a liquid-scintillation-spectrometer system, using a single electron-multiplier phototube, to assay solution samples of alpha-emitting nuclides, with disintegration rates in the range of 0.2 to 1.3 per minute. He was able to resolve the peaks at 4.8 and 5.5 MeV due to the alpha-particle activities of ^{233}U and ^{238}Pu, respectively, when mixed in the same solution.

6.1.2.5 *Nuclear-Track Techniques.* Special photographic emulsions, commonly referred to as nuclear emulsions, can detect alpha particles and fission fragments. These emulsions can detect very low activities provided that the exposure periods are sufficiently long, typically days or longer. The method is a form of autoradiography. A major drawback is the time-consuming counting of tracks in the developed emulsion. Elements assayed include thorium (Picciotto and Wilgain, 1954) and plutonium (Geiger *et al.*, 1961). A broad survey of the technique was provided by Yagoda (1949).

Alpha particles also leave tracks upon passing through various plastics, glasses, and micas. Although too small to be observed directly, the tracks can be enlarged by chemical etching to the point where they can be seen with a microscope. Three-dimensional recording of tracks is possible with thicker pieces of plastic. Counting of the tracks can sometimes be accomplished with a flying-spot microscope (Young and Roberts, 1951) or a spark counter (Fleischer *et al.*, 1972). This technique has applications, for example, in health physics (Becker, 1969), and the assay of uranium in ores, water, biological materials, and fossils (Fleischer *et al.*, 1972, and references therein).

The use of plastic alpha-particle-track detectors has also become the preferred method for determining radon-progeny concentrations in air in buildings, for monitoring radon exposure to the lung.

The technique of track detection has been applied to a variety of biological problems at the University of Bristol (see, for example, three papers by A. P. Fews and D. L. Henshaw in ICRM, 1984a). A review of nuclear-track techniques has been published by Ahlen *et al.* (1981).

6.1.3 *Low-Level Measurement of Beta-Particle-Emitting Nuclides*

6.1.3.1 *Introduction.* The necessity for counting beta particles lies in the fact that a number of radionuclides either emit no gamma or x rays, or emit them at so low a probability per decay as to make assay by photon detection impossible. Such radionuclides that are frequently assayed (and their maximum beta-particle energies in MeV) include ^{3}H (0.0186), ^{14}C (0.156), ^{32}P (1.71), ^{35}S (0.167), ^{45}Ca (0.257), ^{85}Kr (0.84), ^{89}Sr (1.49), ^{90}Sr (0.546), and ^{90}Y (2.28). Of these, only tritium emits beta particles of such low energy as to require special detection techniques.

Assay by beta-particle counting may be the method of choice even when low-level radionuclides emit both beta particles and gamma rays. When only one or, possibly, two such radionuclides are present in a sample, assay by beta-particle counting may be preferable to assay by gamma-ray counting, because of the higher counting efficiencies of most beta-particle detectors. Gamma-ray spectrometry, nonetheless, may also be of use in identifying radionuclides that may be present.

Because of the continuous nature of a beta-particle energy spectrum, spectrometric identification and quantification of a sample containing several beta-particle emitters are generally difficult. Thus, radiochemical separations are often necessary in the assay of such samples. An exception to this is the presence of radionuclides that emit conversion electrons in the source (Section 6.1.3.5).

For the majority of detectors, physical shielding is required primarily for the reduction of background due to gamma rays and cosmic rays. Considerable further reduction is frequently possible through the use of anticoincidence shielding. The comments made in Section 6.1.1.5e with regard to radioactivity in materials used in counter construction are also valid here.

To detect lower-energy beta particles in spectral distributions, it is generally desirable to prepare thin sources. Techniques for this are described in NCRP (1961), by Yaffe (1962), by Watt and Ramsden (1964), and in Section 5.2.

Absorption and scattering of beta particles by the sample matrix in thick sources must be determined for quantitative measurements, the corrections being greater for solid sources than for gaseous ones or liquid sources dispersed in a scintillator solution. Frequently, the detection efficiency as a function of sample surface density must be determined (see, e.g., Nervik and Stevenson, 1952), unless an "infinitely-thick" sample is utilized.

Specific methods for measuring low-level beta-particle radiation are discussed, according to type of detector, in Sections 6.1.3.2 through 6.1.3.5.

6.1.3.2 *Liquid- and Solid-Scintillation Counting.* Liquid-scintillation-counting systems are widely used for the assay of low levels of beta-particle-emitting nuclides. High-sensitivity phototubes with bi-alkali (Na-Cs-Sb) photocathodes and quartz faces are desirable. Lower backgrounds can be obtained with sample vials made of quartz, low-potassium glass, or plastic, although the last type exhibits some permeability to organic compounds. Coupling the vial optically to the phototube with fluid or a plastic light pipe will increase the light-collection efficiency, but the former technique is not practical for systems having an automatic sample changer. Degassing the sample and then sealing the vial will reduce quenching by oxygen. A technique for the quantitative transfer of volatile radioactive solutions has been developed by Garfinkel *et al.* (1965).

Extensive work has gone into the study of chemical and color quenching and the development of better scintillator mixtures, or "cocktails" (Rapkin, 1967; Parmentier and Ten Haaf, 1969; Bransome, 1970; Horrocks and Peng, 1971; Wang and Willis, 1975). Some of this effort is germane to low-level applications, even though not specifically directed toward it. (See also Section 4.3.)

All of the beta-particle emitters listed in Section 6.1.3.1 can be assayed by the method of liquid-scintillation counting. This method is frequently employed for tritium analyses (Moghissi and Carter, 1973), although considerable concentration and processing of some kinds of samples may be required to convert them to a suitable form that is soluble or miscible in, usually, water or benzene (Tamers *et al.*, 1962; Cameron, 1967). A system that can quantify very small amounts of tritium in samples as large as 100 ml has been developed by Noakes *et al.* (1973). ^{14}C can be assayed as $^{14}CO_2$ contained in an organic base, or $^{14}CO_2$ can be converted to $BaCO_3$ and counted as a suspension (Cluley, 1962; Rapkin, 1967; Section 4.3.3). The ^{3}H or ^{14}C sample can also be dispersed or dissolved directly in the scintillator solution (Fox, 1968). Inorganic compounds containing ^{35}S, ^{45}Ca, ^{89}Sr, ^{90}Sr-^{90}Y, and ^{90}Y can also be prepared and suspended in the "cocktail." For some details and references, see ICRU (1972) and Section 4.3.4. Chapters 6 and 7 in Bransome (1970) give conditions for counting some additional beta-particle emitters and advice on counting various pairs of beta-particle emitters. Assays of mixtures of three pure beta-emitting nuclides are sometimes possible with the aid of signal processing, a multi-channel analyzer and/or computer spectrum analysis (Oller and Plato, 1972; Piltingsrud and Stencel, 1972).

Solid organic scintillators, usually plastics but also organic crystals such as anthracene and trans-stilbene, are used for measuring gross beta-particle count rates and for beta-particle spectrometry. For gross counting, a thin plastic phosphor, typically with a surface density of tens of mg/cm^2, is placed between the sample and phototube as in Figure 80. With a simple mercury shield the background count rate averaged about 0.01 s^{-1} (Harley *et al.*, 1962). When the phototube and sample were fitted into the well of a reentrant-anode Geiger-Müller counter, operated in anticoincidence, the background was reduced by nearly a factor of two (HASL, 1969).

For beta-particle spectrometry, a 1- to 2-cm-thick scintillator for complete absorption of the beta-particle energies is essential. The use of plastic scintillators with shapes other than discs can reduce the scattering of beta particles out of the detector and the distortion of spectral shapes, and increase the detection efficiency (see, e.g., Gardner and Meinke, 1958). The probability of gamma rays interacting with these specially-shaped scintillators is greater because of their larger size. Hence, these scintillators have higher backgrounds than simple disc-shaped scintillators. An elegant solution to the increased background problem, shown in Figure 83, was achieved by constructing a small Geiger-Müller counter into the front side of the scintillator (Tanaka, 1961). To obtain beta-ray spectra, of medium to high energy, scintillator pulses from only those events which are coincident in the two detectors are recorded by a pulse-height analyzer. With the addition of a logarithmic amplifier to the system, Tanaka, E., et al. (1967) were able to interpret, without complicated calculations, beta-particle spectra from radionuclides with beta-particle end-point ener-

Fig. 83. Cross-section and exploded view showing the detector assembly of a beta-particle scintillation spectrometer using a coincidence method with a Geiger-Müller counter: (1) plastic scintillator; (2) lucite light pipe; (3) phototube (EMI 6097B); (4) MgO powder light reflector; (5) vinyl packing; (6) anode wire of the G-M counter; (7) window of the G-M counter (gold-plated Mylar film, 1.3 mg/cm² surface density); (8) Lucite plate; (9) counting gas inlet tube; (10) counting gas outlet tube; (11) sample inlet. (From Tanaka, 1961)

gies as low as approximately 150 keV, as for example, ^{35}S and ^{14}C.

6.1.3.3 *Čerenkov-Radiation Counting.* Čerenkov radiation is produced by a charged particle traveling through a medium at a velocity greater than that of light in the medium. The photons thus produced are usually detected by the phototubes in conventional liquid-scintillation-counting systems. Not all beta-particle emitters can be assayed by this technique because an electron must initially have sufficient energy to be relativistic in the medium. The minimum or threshold energy for electrons in water is 0.26 MeV, and photon yields are given in Table 7. However, most other solvents used in Čerenkov-radiation counting have a higher index of refraction and thus have lower threshold energies (Figure 17). Much of the radiation emitted is in the ultraviolet region and is absorbed by the glass window of the phototube and the glass vials. This drawback can be partly overcome by the use of wavelength shifters that can increase the detection efficiency several fold (Parker and Elrick, 1970).

The reader should consult Sections 2.6, 4.5.1, and 5.4.2 for general information on Čerenkov-radiation counting. Although the method is usually no more sensitive than alternative ones, an advantage lies in the relative ease of preparation for many samples. An example of the uses of the method is the measurement, by Yamada (1962), of Čerenkov radiation arising from ^{90}Y beta particles, to assay for ^{90}Sr in radioactive waste water. Randolph (1975) also used the method of consecutive Čerenkov-radiation counting and liquid-scintillation counting of a sample to assay for ^{89}Sr and ^{90}Sr, respectively.

6.1.3.4 *Gas-Multiplication Detectors.* Counters with gas-multiplication detectors may be divided into those for use with solid beta-particle-emitting sources, and those for use with gaseous sources. In the former category, the usual design comprises a small, thin-window detector that operates in the proportional or Geiger region, with a continuous flow of counting gas, at just above atmospheric pressure, through the detector, a guard detector which operates in anticoincidence with the main detector, and considerable shielding around both detectors. The detector bodies are usually made of petroleum-derived plastics, the petroleum having been in the earth long enough to be essentially free from ^{3}H and ^{14}C activities. The guard detector may be physically separated from the main detector, as shown in Figure 84a, or may be incorporated into the same body, as in Figure 84b. If separate, the guard may be constructed of other materials, such as glass or metal, because it is not necessary for the guard to be made of very low background-activity material.

The source is placed close to the window outside the detector, whose

ANTICOINCIDENCE MANTLE
(GEIGER TUBES)

8 in. STEEL

1 in. MERCURY

DETECTOR

(a)

GUARD COUNTER (+HV)

GAS
INLET

ANODE
LOOPS

GAS
OUTLET

DETECTOR
(+HV)

METALLIZED INTERIOR
SURFACES (−HV)

PLASTIC

METALLIZED PLASTIC FILM WINDOW

(b)

Fig. 84. Low-background gas-multiplication detectors for beta particles from solid sources: a. external guard detector; b. guard detector in same body. (From Johnson *et al.*, 1963) net background is generally less than 0.01 s⁻¹ with a 2.5-cm-diameter window, and less than 0.02 s⁻¹ with a 5-cm-diameter window. Windows are usually metalized polyethylene terephthalate (Mylar) films with surface densities of 0.1 to 1 mg/cm^2. Except in the case of tritium, the window does not seriously reduce the counting efficiency of the detector. Hence, windowless detectors, with the increased possibility of detector contamination, are less often used. Widely used detector-filling gases are, for proportional counters, 90 percent argon plus 10 percent methane or 100 percent methane, and, for Geiger-Müller counters, 99.05 percent helium plus 0.95 percent isobutane or 98.7 percent helium plus 1.3 percent n-butane.

Counting systems for solid sources that use this type of detector combined with automatic sample changing and data readout have been extensively developed commercially. Special laboratory-built systems are superior for certain applications, such as those described in Watt and Ramsden (1964).

Perhaps the greatest problem associated with the assay of thick, solid beta-particle sources with gas-multiplication detectors is to insure that the detection efficiency is known with an accuracy sufficient for the purpose of the measurement. In addition to the absorption and scattering in the source mentioned in Section 6.1.3.1, the effects of backscattering from the source mount, source-to-detector distance, and thickness of the source must be evaluated, and the system calibrated under known and reproducible conditions. Furthermore, these experimental conditions must be maintained as nearly constant as possible for both the calibrating and unknown sources. Discussions of these factors are found in such references as Snell (1962), Price (1964), Watt and Ramsden (1964), and Robinson (1967)

Gas-multiplication detectors with internal gaseous sources, i.e., internal gas counters (Section 3.6), are widely employed in the assay of ³H and ¹⁴C, and also for the assay of ³⁵S and radioisotopes of the noble gases. Low-level internal gas counting of ³H and ¹⁴C has been thoroughly reviewed (Watt and Ramsden, 1964; USAEC, 1965; IAEA, 1967b; Moghissi and Carter, 1973). References to internal gas counting for ³⁵S may be found in Watt and Ramsden (1964) and ICRU Report 22 (ICRU, 1972), and for radioactive noble gases in Currie (1972) and Stanley and Moghissi (1975). The gas to be analyzed may be the only gas present in the detector, but more often it is mixed with a nonradioactive counting gas.

These detectors are usually long, closed cylinders of metal or quartz with an axially centered 0.0025- to 0.0051-cm-diameter anode wire, and their volumes range from only a few milliliters to several liters.

Anti-coincidence shielding often employs separate detectors (see, e.g., Watt and Ramsden, 1964). Another design has the guard detector externally concentric to the detector in a single unit in which the two detectors may be separated only electrically by a multiwire grid or perforated, metalized plastic film (Drever et al., 1957; Houtermans and Oeschger, 1958; Section 3.6), or physically separated with a thin foil. In the last case, all the sample gas can be confined to the inner detector, and counting gas, free from activity, is used in the guard detector (Charalambus and Goebel, 1963; Schell, 1970).

The technique of pulse-shape discrimination has also been applied to reduce background (Culhane and Fabian, 1972; Sudar et al., 1973; Currie and Lindstrom, 1975).

6.1.3.5 *Semiconductor Detectors.* Semiconductor detectors can be used advantageously for the analyses of thin samples containing a radionuclide that emits conversion electrons. Background count rates under the sharp conversion-electron peaks will be quite low compared to the total background. In one instance (Folsom et al., 1965), radiochemically separated ^{137}Cs was assayed by measuring the 624-keV conversion-electron activity of its ^{137m}Ba daughter. The background in that energy region of their silicon detector was less than 4 counts per hour. Even though nearly all the conversion electrons reaching a semiconductor detector will be counted, the fraction of decays by this route for most radionuclides is so low as to preclude widespread use of this detection method.

6.1.4 *Low-Level Measurement of X-Ray- and Gamma-Ray-Emitting Nuclides*

6.1.4.1 *Introduction.* One of the most valuable techniques for low-level radioactivity measurements is gamma-ray spectrometry. The various systems, consisting of scintillation and, or, semiconductor detectors coupled to multichannel analyzers, provide for rapid simultaneous measurement of many radionuclides in the same sample. The fundamental characteristics of scintillation and semiconductor detectors are described in Sections 2.4 and 2.5, respectively, and those of gas proportional counters used in the assay of x-ray emitters in Section 2.2. Such characteristics as energy resolution and intrinsic detector efficiency are important in choosing the detector for a particular measurement (see Section 4.2). Israel et al. (1971) have compared the characteristics for several detectors in the energy range from 10 to 300 keV.

The choice of measurement system for gamma-ray emitters is also based on considerations such as sample size available, the level of activity in the sample, and time available for measurement (see also Section 6.1.1). Some of the more commonly encountered types of samples and suggested measurements systems are discussed below:

(i) *Large Sample of Low Radioactivity Concentration*—Example: Fallout ^{54}Mn in seawater. In an example such as this where great sensitivity is required, the radioelement of interest may be concentrated by radiochemical techniques, and the resultant sample counted by using a high-efficiency NaI(Tl) detector (Folsom *et al.*, 1963).

(ii) *X-Ray-Emitting Radionuclides in the Environment*—Example: Fallout ^{55}Fe in biological samples. Due to the short range of ^{55}Fe x rays, it is usually necessary to separate iron from the sample matrix prior to counting. Thus, in the assay of ^{55}Fe in serum, an electrode-posited source may be counted in an argon-methane proportional counter (Palmer and Beasley, 1967).

(iii) *Analysis of Extraterrestrial Materials*—Example: Meteorites. When a nondestructive assay of a number of radionuclides in a small sample is required, a germanium detector is clearly the best choice due to the superior resolution. (See, for example, Perkins *et al.*, 1970.)

(iv) *Effluent Monitoring at Nuclear Facilities*—Example: Mixed fission and corrosion products. Ge detectors have largely replaced NaI(Tl) in this application because of superior resolution. Time requirements usually prohibit chemical separation of individual elements in the case of routine samples. Soil samples collected in the vicinity of a nuclear-fuel-processing plant have been standardized for α, β and γ radioactivities and distributed by the National Bureau of Standards (Inn *et al.*, 1984). This standard reference material is available for the measurement of radiochemical yields in the radioactivity assay of similar soil samples.

(v) *Measurement of Gamma Rays Between 60 and 150 keV*—Example: ^{155}Eu in soil, ^{133}Xe in air. The efficiency of a typical coaxial germanium detector is difficult to measure in this energy region. Si(Li) detectors or planar Ge or Ge(Li) detectors may be most advantageous for this type of measurement.

6.1.4.2 *Inorganic Scintillation Detectors.* As mentioned previously, NaI(Tl) is the most common scintillator used in low-level assay. The following discussion of counting efficiency, standardization, data processing, and background-reduction techniques, while assuming the use of NaI(Tl) detectors with multichannel analyzers, is generally applicable to other types of scintillation counters as well.

6.1.4.2.1 *Counting Efficiency.* The overall counting efficiency may be increased by increasing either the sample size or the size of the detector. Heath (1964) has tabulated efficiencies for point sources at different distances from 7.6 x 7.6-cm to 12.7 x 12.7-cm cylindrical NaI(Tl) detectors. The effects of detector size have also been evaluated by May and Marinelli (1960).

For small samples, the efficiency is greatly increased by the use of well crystals (Hine, 1967; Lloyd et al., 1967) that give almost 4π geometry. For larger samples the source may be distributed around the detector in a Marinelli beaker (May and Marinelli, 1960). With such a geometry as this, where photopeak efficiencies are very high, the probability of summing for coincident events will also be high. This does not present a problem if one is measuring the integral count rate for a sample containing a single radionuclide for which the efficiency is known. Summing corrections must be made, however, if count rate measurements are made on a single photopeak for which the corresponding gamma ray is emitted in coincidence with other radiations.

It is difficult to predict the efficiency for the large variety of source detector arrangements used. Thus, the most common procedure is to calibrate each system by using radioactivity standards in the same geometry as the sample. (Such standards are listed in Table 9.) The calibration is usually accomplished by incorporating several radionuclides, such as ^{57}Co, ^{54}Mn, ^{137}Cs, ^{60}Co, and ^{22}Na, into a matrix similar to the sample (e.g., soil, water, or biological matrix). The results are then used to prepare photopeak-efficiency curves, such as the curve shown for point sources in Figure 85, using a NaI(Tl) detector of practically 4π geometry (Hutchinson et al., 1973b). The efficiency appropriate to the energy, or energies, of the unknown is taken from the curve. In practice, however, quantitative interpretation of NaI(Tl) scintillation-spectrometer data, for a source emitting several gamma rays, is difficult because of poor resolution (see, for example, Figure 86). The various methods used for peak-area analysis and spectrum "stripping techniques" are beyond the scope of this section. General characteristics of such methods are reviewed in Section 4.2.7 and in ICRU (1972).

6.1.4.2.2 *Background.* The major sources of background counts were discussed in Section 6.1.1.5 and will be further discussed in 6.1.4.3.3. Major techniques for reducing the contributions from these sources to the background for a scintillation counter are siting, air filtration, and physical and anticoincidence shielding.

(i) *Siting.* Siting of any low-level-measurements system is an important consideration. Counting rooms should be located below ground since the covering materials (earth, water, and concrete, for example)

Fig. 85. Photopeak-efficiency curve obtained using point-source standards in the NBS $4\pi\gamma$ 8-inch-crystal system. (From Hutchinson *et al.*, 1973b)

will provide some shielding from cosmic radiation. Wogman *et al.* (1969), for instance, describe the background of a counting room located at the base of the Grand Coulee Dam. As the authors point out, however, the covering material alone does not eliminate the need for shielding from the muon and neutron components of the cosmic radiation [see (iii) below]. When a counting room is under construction, building materials and nearby soil should be checked to insure that they do not contain excessive amounts of natural or fallout radioactivity. Siting of the room away from reactors, accelerators, and x-ray equipment is, of course, desirable.

(ii) *Ventilation.* The counting room should be equipped with air-filtration systems to remove particulates that may contain radioactivity. To avoid radon- and thoron-daughter deposition within the shielding for detectors of the NASA Lunar Sample Laboratory, Lucas (1964, 1965) designed a counting room in which fresh air was introduced

Fig. 86. Gamma-ray spectra of a 5-ml mixed-radionuclide-solution source taken with the source within a 5-inch NaI(Tl) well crystal (upper curve) and at the face of a 60-cm³ Ge(Li) detector (lower curve). The counting time in each case was 2000 s. (From measurements made at the National Bureau of Standards)

through a low-temperature charcoal-filtration system to trap the radon.

(iii) *Shielding.* For large scintillation detectors, massive shielding is required to reduce the background contribution from environmental gamma-ray emitters. Shielding requirements are less cumbersome for small NaI(Tl) wafers or CaI₂(Eu) scintillators because they have low efficiency for the more penetrating gamma rays. The most common shielding materials are lead, steel, mercury, and water. The materials used should be checked for contamination (Section 6.1.1.6). This is one advantage of using water; although it is low-Z, large quantities of pure material are easily prepared.

In the design of a shield, some thought must be given to the scattering and attenuation characteristics of the materials used. Steel, for example, is a common shielding material which serves to scatter and degrade incoming photons but has a low photoelectric cross section for photons above 100 keV. Thus, in the absence of other shielding, some degraded photons, usually in the energy region 100 to 300 keV, will interact with the detector. Also backscatter peaks may be detected in this energy region (Heath, 1957) due to source gamma rays scattered through 180°.

Another consideration in the design of a compact, efficient shield is a process known as "grading" (ICRU, 1972) in which materials of successively lower Z are used to absorb the x rays from outer layers. For instance, if lead shielding is used, then a cadmium liner is used to absorb the Pb x rays having energies between 73 and 88 keV. The Cd x rays of approximately 22 keV are then absorbed by a thin liner of copper or steel.

The design of a typical shield is described in ICRU, 1972. The outer shield is fabricated from carefully selected low-activity steel or lead. Inside this shield, one may use a neutron shield of borated paraffin wax, in which the hydrogen serves to moderate neutrons, while the boron has a high capture-cross section for neutrons at thermal energies. Within the wax shield, an anticoincidence shield, as described in (iv) below, may be used and finally, closest to the detector, a shield of selected lead, steel, or mercury.

If the outer component of the innermost shield is steel, it should be lined with about 3 mm of lead to remove the degraded photons. The lead itself must then be lined with about 1 mm each, of cadmium and copper. If lead alone is used for the shield, the cadmium and copper linings are of course still required. When no borated paraffin is used the substitution of tin for cadmium will reduce any (n, γ) background from stray thermal neutrons captured in the graded shield, *e.g*, "skyshine" from a nearby accelerator.

(iv) *Background Reduction by Instrument Techniques*. A reduction in both the background contribution from external sources and of the Compton background due to higher-energy gamma rays scattered out of the detector can be obtained by closely surrounding the main detector with an anticoincidence shield such as a plastic annulus containing p-terphenyl, or by a NaI(Tl) annulus (Perkins *et al.*, 1960).

(v) *Applications*. Examples of NaI(Tl) counting systems that incorporate most of the background-reduction features discussed here have been reported by Perkins (1965), Nielsen and Perkins (1967), Tanaka, S., *et al.* (1967), Wogman *et al.* (1969), Hutchinson *et al.* (1972, 1973b), Eldridge *et al.* (1973) and Wogman (1974). Figure 87 gives a schematic

Fig. 87. Two 8″ × 8″ NaI(Tl) well detectors surrounded by a 30″-o.d. by 20″-long plastic scintillator anti-coincidence anti-Compton shield. The detectors enclose a cavity 5″ in diameter and 1.5″ deep. (From Hutchinson *et al.*, 1973b)

of the anti-Compton, anticoincidence-shielded NaI(Tl) system at the NBS (described in Section 3.3).

One of the most sophisticated systems reported to date is a "multi-dimensional" γ-γ coincidence counter (Wogman *et al.*, 1969) that uses two NaI(Tl) detectors. The method was subsequently extended to β-γ-γ coincidence counting by Wogman and Brodzinski (1973) by employing two NaI(Tl) detectors and two gas proportional counters. In addition, each half of the main detector is surrounded by an anticoincidence shield of a plastic phosphor. Brauer *et al.* (1970) reported similar β-γ coincidence systems in which plastic phosphors were employed as the beta-particle detectors.

6.1.4.3 *Semiconductor Detectors.* The most powerful measurement technique used in low-level assays at present is gamma-ray spectrometry using semiconductor detectors. Most of the current systems utilize Ge(Li) detectors (Section 2.5) that have counting efficiencies of the order of 1 to 25 percent relative to 7.6 x 7.6-cm NaI(Tl) scintillation detectors for 1.3-MeV gamma rays. In the future, however, the drifted semiconductors may be replaced by high purity materials such as Ge, CdTe, or HgI$_2$. Comparisons of gamma-ray response by different types of detectors have been made by Israel *et al.* (1971) and Malm *et al.* (1973). Pure Ge crystals are now used to avoid continuously cooling the detector with liquid nitrogen as had been predicted by Hansen and Haller (1973). A review of detection systems currently used for low-level gamma-ray measurements has been made by Wogman (1974). Additional references may be found in Nielsen (1972).

6.1.4.3.1 *Energy Calibration.* Calibrations of analyzer channel number in terms of energy for Ge detector systems were discussed in Section 4.2.4. Such calibrations may be accomplished by counting sources that emit several gamma rays whose transition energies are well known. Unknown radionuclides may be identified by comparing observed gamma-ray energies with tabulations from Oak Ridge National Laboratory (see Appendix A). Compilations of gamma-ray spectra such as Adams and Dams (1970), Cline (1971), and Heath (1974) are also useful in the identification of unknowns.

6.1.4.3.2 *Counting Efficiency.* The principles of calibration for counting efficiency of low-level Ge gamma-ray spectrometers are the same as for systems using NaI(Tl) detectors (Section 6.1.4.2.1). Consideration must be given to the selection of the optimum counting geometry, and to the preparation of photopeak vs. energy efficiency curves.

In general, the efficiency of the detection system may be increased by using a larger detector or optimizing the sample-detector geometry (Brauer *et al.*, 1973). The size of the detector is limited by present state-of-the-art techniques in semiconductor fabrication. Near optimum counting geometries have been obtained in two ways. Latner and Sanderson (1972) have used Marinelli-type beakers in which the source surrounds the detector (see ANSI, 1978c), while Brauer and Kaye (1974) place the source between two detectors.

The photopeak efficiencies for various source-detector configurations may be predicted using Monte Carlo calculations (e.g., Coley and Frigerio, 1975), but more often the system is calibrated directly by using one or more radioactivity standards (Legrand, 1973). A single standard, such as [166m]Ho (Lavi, 1973), that emits many gamma rays of different energies may be used, or a number of standards such as ^{22}Na, ^{54}Mn, ^{57}Co, ^{60}Co, and ^{137}Cs may be individually incorporated into "blank" sample matrices (Laichter *et al.*, 1973). Systems may be calibrated for liquid and point sources with mixed radionuclide standards such as that for which the spectrum is displayed in figure 88 (Cavallo *et al.*, 1973). However, summing of cascade gamma rays can change the efficiency by 30 percent or more for small sources directly on the detector, and a direct calibration for selected radiations of each radionuclide may be required to give accurate measurements in a close geometry.

6.1.4.3.3 *Background.* As with other types of low-level assay the central problem in germanium gamma-ray spectrometry is to achieve the optimum signal-to-noise ratio in the energy region of interest. General references on the sensitivity of such systems are Cooper (1970) and Armantrout *et al.* (1972). One can discern three interrelated topics for discussion: (i) detection limits; (ii) spectrum processing; and (iii) experimental methods used to enhance the signal-to-noise ratio.

The general problem of detection limits was discussed in Section 6.1.1.3. Pasternack and Harley (1971) examined the problem of de-

CHANNEL NUMBER

Fig. 88. Gamma-ray spectrum of an NBS mixed-radionuclide gamma-ray emission-rate standard. This 50-ml solution source was counted at 5 cm from a 60-cm³ Ge(Li) detector. (From Coursey, 1976) Such calibrated sources, that are traceable to NBS, are now available from commercial suppliers. A longer-lived mixture, consisting of ^{125}Sb, ^{154}Eu, and ^{155}Eu, is available from NBS.

tecting gamma-ray photopeaks in multicomponent gamma-ray spectra, and Wood and Palms (1974) have examined gamma-ray spectrum-stripping techniques used for such spectra. Commercial multichannel analyzers with minicomputers are now available with programs for analyzing the data, listing the radionuclides found, their activities, and gamma-ray-peak energies and intensities. Users of such systems should be familiar with the detection criteria and peak-area-evaluation methods used in these computer programs. Guidance on standard methods for such analyses has been provided by the ANSI Subcommittee on Gamma-Ray Spectrometry (ANSI, 1978a). (See, also, Section 4.2.6.)

As discussed in Section 6.1.4.2.2, background reduction for gamma-ray spectrometry is generally accomplished by a combination of siting, ventilation, massive shielding and measuring-system techniques. The more sophisticated systems employ one or two main detectors of Ge surrounded by one or more plastic or NaI(Tl) anticoincidence shields (Phelps and Hamby, 1972). Less widely used are summing-Compton spectrometers (SCS) (e.g., Ivanov and Shipilov, 1974) that consist of a single germanium ingot with two different lithium-drifted regions. A common characteristic of all of these systems is that one can utilize the decay scheme of the radionuclide of interest to reduce interfering background from other sources. A typical system is shown in Figures 89 and 90. When, in the spectral analysis of monoenergetic gamma-ray-emitting radionuclides, the NaI(Tl) shield and Ge detector are operated in the anticoincidence mode, contributions from both Compton events and radionuclides emitting coincident gamma rays will be dramatically reduced. This reduction is demonstrated in Figure 91, which shows the gamma-ray spectra of a mixture of ^{137}Cs and ^{60}Co in the normal and anticoincidence modes.

6.1.4.4 *X-Ray Detection Systems.* Radionuclides of environmental interest that are detected by x-ray or low-energy gamma-ray measurements include ^{37}Ar, ^{55}Fe, ^{129}I, and ^{125}I. The assay of such low-energy-photon emitters in bulk environmental samples, without concentration, is usually complicated by attenuation of the photons within the sample.

The detectors used for x-ray measurements include Si(Li) and high purity HPGe detectors, proportional counters, gas proportional scintillation counters, and liquid-scintillation counters. The operation of these instruments is covered elsewhere in this report, for example, in Sections 2.2, 2.4.5, 2.4.6, 4.2, and 4.3.

6.1.4.4.1 *Semiconductor Detectors.* The use of semiconductor detectors in the x-ray region is discussed by Israel *et al.* (1971) and

Fig. 89. Schematic diagram of a Ge(Li)-NaI(Tl) coincidence-anticoincidence gamma-ray spectrometer. (From Cooper and Perkins, 1971)

Fig. 90. Block diagram of electronic circuitry used in anticoincidence mode of operation for a NaI(Tl) shielded Ge(Li) gamma-ray spectrometer. (From Cooper and Perkins, 1971)

Fig. 91. Gamma-ray spectra, obtained at the National Bureau of Standards, of a mixture of ^{60}Co and ^{137}Cs-^{137m}Ba (activities 8600 s^{-1} and 93 s^{-1}, respectively), contained in 50 ml of solution in a cylindrical glass bottle, using a 60-cm^3 Ge(Li) detector within a NaI(Tl) shield. The source-to-detector distance was 1 inch, and the counting time was 10^4 s for each spectrum.

Spectrum (a) is that obtained in the normal mode (i.e., with no anticoincidence or coincidence requirement between the outputs of the two detectors). Spectrum (b) was obtained with the detectors operated in the anticoincidence mode (i.e., recording only non-coincident events). The relative enhancement of the 662-keV line due to the ^{137m}Ba is clearly seen.

Swierkowski *et al.* (1974), and the calibration of such systems is reviewed by Fink (1973) and Hansen *et al.* (1973). Background-reduction techniques need not be as elaborate as for gamma-ray spectrometers because detection efficiencies are low for background gamma rays with energies greater than 0.2 MeV. Cosmic rays may, however, produce photons in surrounding material that contribute background counts in the x-ray region.

6.1.4.4.2 *Gas Proportional Counters.* Gas proportional counters are often used for the assay of x-ray emitters. Palmer and Beasley (1967) measured body burdens of ^{55}Fe in reindeer and humans by electrodepositing Fe from serum onto copper planchets and subsequently assaying the ^{55}Fe using a gas proportional counter. The activity of ^{37}Ar, an electron-capture nuclide with a 35-day half life, is normally measured by internal gas proportional counting (Sections 2.2 and 6.1.3.4). Its use as a tracer in atmospheric studies has been described by Loosli *et al.* (1975) and Currie and Lindstrom (1975). Both groups assay the argon fraction condensed from large air samples.

6.1.4.4.3 *Other Techniques Used for Analysis of X-Ray Emitters.* Some other detector systems that have been used for the analysis of x-ray emitters are

   (i) x-γ coincidence counting (Brauer *et al.*, 1970),

   (ii) liquid-scintillation counting (Horrocks, 1971),

   (iii) xenon scintillation proportional counting (Israel *et al.*, 1971), and

   (iv) neutron activation techniques (Brauer and Kaye, 1974).

It should be mentioned that in liquid-scintillation counting of electron-capture nuclides, a major fraction of the observed response may be due to the Auger electrons.

Although ^{129}I is radioactive, its specific activity is extremely low (1.57×10^7-y half life). It can, however, be readily detected by neutron activation to form ^{130}I, a gamma-ray emitter with a 12-hour half life and correspondingly higher specific activity (Edwards, 1962; Koch and Keisch, 1964; Keisch *et al.*, 1965). The ^{130}I activity is then measured by using a gamma-ray spectrometer.

6.1.5 *Further Activities in Low-Level Measurements.*

Subsequent to the publication of the first edition of this Report, five international conferences have been convened to discuss progress and techniques in low-level radioactivity measurements. Three of these were organized by the International Committee for Radionuclide Metrology (ICRM 1978, 1980, 1984a), and one by the International Atomic Energy Agency (IAEA, 1981). The fifth was held in Czechoslovakia in 1980 (Povinec, 1982).

6.2 Supply of Radioactive Materials

Radioactive materials are principally produced by neutron bombardment, usually in a reactor, or bombardment with charged particles in an accelerator. Some radioactive materials are also available as purified fission products from reactor fuels. Other radionuclides occur in nature and are isolated by appropriate methods.

Radionuclides are available in many physical forms as simple compounds, labeled compounds, radiopharmaceuticals, and radioactive sources from several commercial organizations whose catalogs should be consulted for materials and services available. The American Nuclear Society publishes an annual Buyer's Guide that contains lists of companies that supply radionuclides. The American Chemical Society publishes the Analytical Chemistry Buyer's Guide with some of the same information. Other sources of information are found in advertisements in periodicals such as Science, published by the American Association for the Advancement of Science; the Journal of Nuclear Medicine, and the Physician's Desk Reference for Radiology and Nuclear Medicine, published by the Medical Economics Company.

Standard reference materials for calibration of equipment are available from the National Bureau of Standards and many of the commercial firms that supply radioactive nuclides, sources, and pharmaceuticals. Standard reference materials are available in many geometries and as gases, liquids, or solids.

An *International Directory of Certified Radioactive Reference Materials* has been prepared jointly by the International Atomic Energy Agency and the Joint Research Center of the European Communities (Grosse and Bambynek, 1983). Reference materials are either certified by the manufacturer as to accuracy or guaranteed by catalog statement as to accuracy. *Certified* activity standards are always higher in cost, for a given radionuclide, than *guaranteed* standards. Manufacturers' labeled amounts of regular radionuclide shipments should never be used as standard reference materials since they are not packaged or calibrated for such purposes.

Before ordering reactor-produced radionuclides above certain specified levels, the prospective user must be licensed for possession of byproduct (i.e., reactor-produced) material by the U.S. Nuclear Regulatory Commission or an Agreement State. Addresses and names of regulatory personnel are available in a *Directory of Personnel Responsible for Radiological Health Programs,* issued each July by the U.S. Department of Health and Human Services, Food and Drug Administration, Center for Devices and Radiological Health, Rockville,

Maryland 20852. Other radionuclides may also require licensing and the user should check with the appropriate state agencies prior to ordering.

6.3 Users' Responsibilities for Quality Control

6.3.1 Check Procedures for Incoming Shipments

Shipments of radioactive materials received by institutions fall into four general categories, namely:
  (i) prepackaged small quantities for individual doses or tests,
  (ii) large quantities to be used as bulk solutions for dispensing in smaller quantities,
  (iii) radionuclide generators such as ^{99}Mo-^{99m}Tc generators, and
  (iv) sealed radioactive sources.
Most suppliers of radioactive materials have subjected their shipments to intensive quality control to prove identity, quantity, and quality. The quality-control procedures may be required and approved by regulatory agencies and, at the least, the competition of other suppliers will dictate some reasonable degree of reliability. However, the medical user, in particular, should take the precautions enumerated below before he uses a radionuclide as the ultimate responsibility for the accuracy of the dose administered to the patient rests with the physician.

The first step taken by all users should be to make a wipe test of the external shipping package and internal containers to be sure that no leakage has occurred and that there has been no compromise of primary container integrity. This practice is one that should be established by all laboratories using radioactive materials. In addition, it is a requirement for some packages of some radionuclides (see *Procedures for Picking up, Receiving, and Opening Packages*; Nuclear Regulatory Commission Regulations, Title 10, Part 20.205, Code of Federal Regulations).

As a minimum check of the product identity, the user should inspect the label attached to the primary container, to insure that this information indicates that it is the required radionuclide, and that the amount is correct. In the case of short-lived radionuclides, the calibration date should be checked and the necessary calculations or measurements should be made to determine the quantity at time of use. Technical data sheets and package inserts accompanying the radionuclide package should be read carefully for special instructions on stability, quality inspection, and chemistry of the product.

The next step beyond this visual inspection is to make an approximate assay of the shipment. If this assay yields results in reasonably good agreement with the value stated by the supplier, the identity of the radionuclide is unlikely to be other than expected. Gross discrepancy between the user's assay and the supplier's value should lead to the use of more accurate identification methods such as described elsewhere in this report.

If the user is using bulk solutions to prepare "in-house" pharmaceuticals or labeled materials, quality-control procedures should be instituted in order to verify activity, radionuclidic purity, and radiochemical purity by means of methods described or referenced elsewhere in this report (e.g., Section 6.4).

6.3.2 *Sterility and Pyrogenicity*

Radioactive pharmaceuticals used for *in vivo* studies must be sterile and non-pyrogenic. Materials supplied for human use by manufacturers licensed by the Food and Drug Administration are required to be sterile and non-pyrogenic and no tests for sterility and pyrogenicity are required by the user.

However, if commercial suppliers are not used, or preparation of radioactive pharmaceuticals is carried out in the hospital pharmacy, sterility and pyrogenicity testing must be done by the user for samples to be administered intravenously. For descriptions of methods one may consult the most recent edition of the U.S. Pharmacopeia (USP, 1980).

6.4 Assay of Radiopharmaceuticals

6.4.1 *Introduction*

The requirements for radioactivity assays in a clinical laboratory range from simple checks on the identity and quantity of preassayed shipments of commercial radiopharmaceuticals to complete chemical, biological, and radioactivity tests and measurements on products prepared within the laboratory. Radiopharmaceuticals fall into four general categories; (i) small single-dose quantities intended for direct administration, (ii) bulk shipments from which an aliquot may be taken for further processing or direct administration, (iii) short half-life radionuclide generators and (iv) radionuclides produced on site.

Two of the short half-life generator systems are used to prepare various chemical compounds that have different biological distributions. The ^{113}Sn-^{113m}In generator has been used to produce compounds for imaging in lung, liver and kidney (Alvarez, 1975). The ^{99}Mo-^{99m}Tc generator is, however, the source of the most often used radiopharmaceuticals in nuclear-medicine clinics. Over 80 percent of the radiopharmaceuticals contain ^{99m}Tc in various chemical forms for detecting abnormalities in the brain, kidney, liver, lung, bone, hepatobiliary tract and the vascular system. Radiopharmaceuticals in the first three categories are normally obtained from commercial producers and are supplied preassayed and presumably in conformity with pharmacopoeial standards (USP, 1980). However, when radiopharmaceuticals are prepared using nonradioactive kits and radionuclide generators, or using radionuclides produced on site, then they must be calibrated and checked by the user for both quantity and quality (Eckelman and Coursey, 1982).

However, because ultimate responsibility for the accuracy of the amount of a radiopharmaceutical administered to a patient rests with the supervising physician, good practice requires that all individual doses be checked to minimize the possibility of error. Even commercial shipments should be routinely tested for proper quantity and identity of the radionuclide.

Although large numbers of tests, such as determinations of radioactivity concentration, radionuclidic purity, radiochemical purity, chemical purity, and purely biological criteria of sterility, pyrogenicity, isotonicity, and biological affinity are normally required to determine the suitability of a radiopharmaceutical for clinical use, only those tests primarily concerned with radioacitivity measurements will be discussed here. Thus, any discussion of radiochemical-purity tests will be excluded, since such determinations rely on varied and often complex chemical separations that are beyond the scope of this report. Those interested in pursuing the subjects of radiopharmaceutical preparation and testing should consult the proceedings of several comprehensive symposia on the subject (Andrews et al., 1966; IAEA, 1970, 1971, 1973; Subramanian et al., 1975; Tubis and Wolf, 1976; Eckelman and Coursey, 1982; ICRM, 1983c).

For a detailed discussion of in vivo and in vitro radioactivity measurements and related subjects, the interested reader should consult a forthcoming NCRP Report on clinical radioactivity measurements and the references contained therein. That report also treats many of the tests for radiopharmaceutical quality not covered in this report.

The most commonly used radiopharmaceuticals are those compounded from gamma-ray-emitting nuclides, and therefore the following sections will deal mainly with the assay of such photon-emitting materials.

6.4.2 *Use of Radioactivity Standards and Reference Sources*

Radioactivity measurements in a clinical laboratory rely heavily on indirect or comparative calibration procedures. These procedures require prior calibration of instruments with calibration standards. Subsequent sample measurements must be made in geometries identical to those used for the initial calibrations in order to achieve valid comparisons. Such instrument calibrations are either performed by the user or, as is frequently the practice in nuclear medicine, by the instrument manufacturer. They should, however, be periodically checked by using such radioactivity standards as may be available.

Calibration standards are radioactive sources for which the activity of the major radionuclide and any measurable radionuclidic impurities are known within specified limits of uncertainty. These can be procured from national standardizing laboratories or commercial suppliers. When these are ordered, "certified" calibration standards should be specified that conform with the certification guidelines presented in ICRU Report 12 (ICRU, 1968). For a discussion of the assessment of uncertainties associated with the stated values of calibrated radioactive reference materials see Section 8.4.

One problem that frequently arises with such calibration standards is that they come in containers that differ from the great variety of containers used for routine measurements in individual laboratories. This problem can be overcome by quantitatively transferring the certified standard, or part of it, to a laboratory container and making up to the desired volume with appropriate carrier solution (see Section 5.1). This laboratory standard is then used to calibrate the laboratory assay system. Alternatively, if it is desired to preserve the integrity of the certified standard, working standards calibrated in the same geometry as the certified standard by relative gamma-ray measurements can be used to calibrate the variously shaped laboratory containers.

Once an instrument or counting system has been calibrated, the performance of the instrument should be monitored by carrying out periodic measurements with a reference source of a long-lived radionuclide. These reference sources need not be calibrated and are normally mounted in rugged containers to prevent leakage. ^{226}Ra in equilibrium with its decay products is frequently selected for use as a reference source, although other long-lived radionuclides, such as ^{60}Co and ^{137}Cs-^{137m}Ba, are also used for this purpose.

Other so-called "standards" occasionally used in a clinical laboratory are simulated or mock standards. These are radioactive sources prepared from long-lived radionuclides, or mixtures of radionuclides, that approximate the electromagnetic radiation characteristics of the simulated radionuclide. Thus, a simulated standard should rarely, if ever, be used as a calibration standard for any other radionuclide. Such a "standard" when prepared from *one* radionuclide can, however, serve a useful function as a reference source to monitor the response of a measurement system in the energy region where measurements are to be made. A good example of this is the use of an ^{129}I reference source to monitor the performance of a counting system used for the routine measurement of ^{125}I samples.

A recent international compilation of radioactivity standards that are available, some only at specified times, has been jointly prepared by the International Atomic Energy Agency and the Joint Research Centre of the European Communities (Grosse and Bambynek, 1983).

6.4.3 *Radioactivity Assay Methods and Instruments*

6.4.3.1 *Introduction.* Pre- and post-administration radioactivity measurements in a clinical laboratory are primarily performed to determine the total activity or concentration of the principal radionuclide or the radionuclidic purity. The majority of these measurements are made on gamma-ray-emitting radionuclides and, to a lesser extent, on pure beta emitters (Williams and Sutton, 1969; Gerrard, 1971). The assay of alpha-emitting radionuclides is seldom required and will, therefore, be neglected in this discussion.

Frequently, there exist prescribed approved standard methods for the assay of radiopharmaceuticals. The user should consult the current edition of the U.S. Pharmacopeia and any other relevant regulatory guidelines for the existence and description of recommended methods, for example, standards developed by the American National Standards Institute. Where relevant prescribed methods exist, they should be followed. It should be noted, however, that the user may substitute an alternate assay method for the recommended method. When this is done, it will be the responsibility of the user to demonstrate, if called upon to do so, that this alternate method consistently yields the same measurement results as the standard method.

6.4.3.2 *Assay of Gamma-Ray-Emitting Radionuclides.* Assays of gamma-ray-emitting radionuclides are performed to determine the total amount or concentration of the major radionuclide in a radiopharmaceutical. When the identification and assay of possible gamma-ray-emitting impurities are also required, spectrometric methods based on such detectors as NaI(Tl) or Ge crystals must be used (see Section 4.2). When assaying a gamma-ray-emitting radiopharmaceutical for radionuclidic impurities that decay without the emission of photons, such as pure alpha or beta emitters, then alternative measuring techniques are required (see Sections 3.5, 3.6, 3.7, 3.8, 6.1.2, and 6.1.3). These techniques often involve chemical separation of the impurity from the principal radionuclide, in order to avoid interference by the radiations from the principal radionuclide. For quantitative assays of samples of established purity of relatively high activities, ionization-chamber devices, that have no intrinsic energy-discrimination capabilities, are frequently selected.

6.4.3.2.1 *NaI(Tl) Detectors.* The use of NaI(Tl) detectors for the assay of gamma-ray-emitting radionuclides in clinical laboratories is well established. Three basic types or configurations of these detector systems are frequently employed in a clinical laboratory. The first and simplest is the integral-counting system employing either a solid cylindrical or a well-type NaI(Tl) crystal and it is normally used for the routine quantitative assay of radionuclidically pure samples.

This simple system can be modified to perform pulse-height spectrometry by adding a single channel or multichannel analyzer. With this additional capability, the detector system may be used for the simultaneous quantitative assays of radiopharmaceuticals containing more than one gamma-emitting radionuclide (see Sections 2.4 and 4.2). Other relevant reviews and texts should also be consulted (e.g., Price, 1964; IAEA, 1965; Hine, 1967; Wagner, 1968; Adams and Dams, 1970; Belcher and Vetter, 1971; Cradduck, 1973; Krugers, 1973; Hine and Sorenson, 1974).

NaI(Tl) well crystals are frequently used in clinical laboratories for *in vitro* measurements. If samples are small enough to be placed at the bottom of the well, the solid angles subtended by the detector approach 4π steradians and, for low-energy photons, the total detection efficiency approaches 100 percent. Care must be exercised in the control of the sample position and size, because these detectors are very sensitive to variations in either. The effects of varying detector geometries are shown in Figure 92.

Increased source-to-detector distances can be used to extend the useful range of NaI(Tl) detectors to millicurie or greater quantities of gamma-ray-emitters by performing sample measurements and calibrations at appropriate source-to-detector distances (Figure 92A).

The nearly 4π geometry achieved by using well crystals is shown in Figure 92C. When this geometry is used, the position and size of the

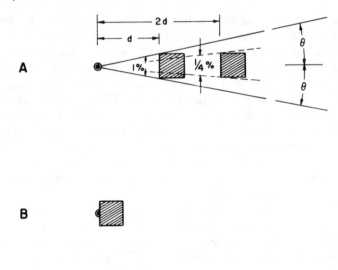

Fig. 92. Examples of different point-source-detector geometries: A, with the source positioned 2.5 times the detector diameter away, the geometric efficiency is 1 percent. At double the distance, it is 0.25 percent, using the approximate law of *inverse squares*, i.e., count rate ~1/(distance)². B, for small detector-to-source distances the inverse square law does not apply; however, the *cosine law* [geometric efficiency = ½ (1 − cos θ), with θ the half-angle of the cone subtended by the detector at the source] gives the geometric efficiency at all distances. C, with a well crystal the geometric efficiency approaches 100 percent (4π), provided that the diameter of the well is small and its depth large in relation to the source dimensions.

sample must be carefully controlled in order to achieve reproducible results as demonstrated by the change in sensitivity of a typical well-crystal counter with variations in source position and volume (Figures 93 and 94). The energy summing of two or more photons emitted in coincidence is greatly enhanced over that observed with lower-geometry detectors. For this reason, well-crystal detector systems, calibrated in terms of the photopeak efficiency as a function of photon energy, are of limited use and should be avoided for purposes of making indirect activity measurements on radionuclides that emit two or more photons in coincidence. However, this effect can frequently be used to advantage to perform direct measurements as described in Section 3.3.

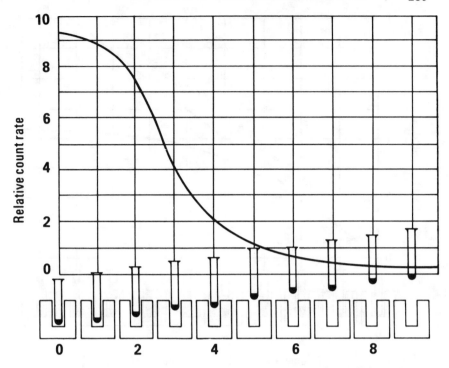

Height, in centimeters, from the bottom of a well crystal,

of a 1-ml volume of cobalt-57

Fig. 93. Response of a typical 2″ × 2″ NaI(Tl) well-crystal as a function of detector-to-source position (in one-centimeter increments) of a 1-ml-volume source. For any well-crystal-counting system, the greatest change of response with source position occurs at intermediate source-to-detector positions, as demonstrated by the 2- to 4-cm positions in this example. (After Brucer, 1970)

Several examples of the application of this method, relevant to nuclear medicine, are described by Brinkman and Aten (1963) and Eldridge and Crowther (1964) for positron-emitting radionuclides and ^{125}I, respectively.

The third configuration in which NaI(Tl) detectors are typically encountered in nuclear-medicine laboratories is in imaging devices, either scanners or Anger cameras. These systems were neither designed for, nor should they be used for, performing qualitative or quantitative radioassays. Not only will much less expensive NaI(Tl)-spectrometer systems nearly always out-perform imaging devices in terms of increased sensitivity, energy resolution, and reproducibility for routine radioactivity assays, but even the temporary removal of a

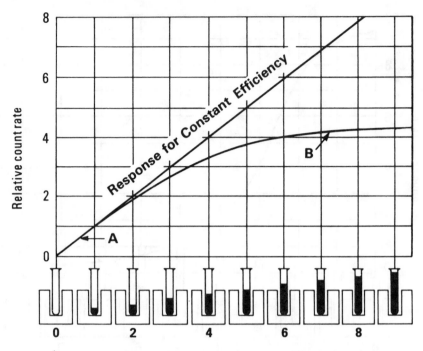

Volume, in milliliters, of a cobalt-57 solution of

constant radioactivity concentration

Fig. 94. Response of a typical 2″ × 2″ NaI(Tl) well-crystal detector to different volumes of a solution source of ^{57}Co (in one-milliliter increments) of constant radioactivity concentration. A, for a source volume <1 ml the maximum counting efficiency per milliliter of solution for the detector is approached. B, beyond approximately 7 ml in source volume, no appreciable increase in detector response is observed with increasing source volume. (After Brucer, 1970)

very expensive imaging device from service in order to perform tasks for which it is ill-suited should be avoided.

6.4.3.2.2 *Semiconductor Detectors.* The principal type of solid-state detector currently used for assaying gamma-ray-emitting radionuclides is germanium of either the earlier lithium-drifted, Ge(Li), or the modern high-purity, HPGe, type. For a review of the physics of operation of these detectors and associated assay techniques, Sections 2.5 and 4.2 of this report should be consulted.

Due largely to cost considerations, these detectors have not been widely adopted in medical laboratories. However, any laboratory that routinely performs large numbers of qualitative analyses, particularly for radionuclidic purity, on gamma-emitting radionuclides should either have, or seriously consider procuring, a high-resolution Ge-spectrometer system. Determinations of the impurities present in the ^{99m}Tc eluates from ^{99}Mo generators provide an excellent example of the advantages of using a Ge-spectrometer system for radionuclidic-purity determinations (Wood and Bowen, 1971; Barrall, 1972; Bowen and Wood, 1972; Ficken et al., 1973; Billinghurst et al., 1974). Further information relating to the use of semiconductors in medicine can be found in Hoffer et al. (1971).

Both Ge spectrometers and NaI(Tl) spectrometers must be calibrated for efficiency of response as a function of photon energy, and the efficiency for photons having energies lying between the calibration points determined from an efficiency curve as described in Sections 4.2.6 through 4.2.8. Such calibrations are normally performed in terms of the full-energy-peak efficiency, and provide the basis for one of the prescribed standard methods for determining the activities of gamma-emitting radiopharmaceuticals for which calibration standards are unavailable (USP, 1980).

6.4.3.2.3 *Dose Calibrators.* Ionization-chamber systems, reading directly in activity, have become predominantly the instruments of choice for performing quantitative preadministration assays of radiopharmaceuticals. They are generally known as "dose calibrators" in the United States. Because nuclear medicine is a multidisciplinary specialty encompassing both medicine and radiation or nuclear physics, the popularization of the term "dose calibrator" is unfortunate. The term dose has one meaning or definition when applied to radiation quantities and another when used in the medical sense. In this instance, the term dose is to be interpreted in the medical sense as the prescribed quantity of a pharmaceutical and not as the radiation quantity, absorbed dose. In the United Kingdom these instruments are called "radionuclide calibrators" and in the United States have recently been referred to as "activity (dose) calibrators" (Herrera and Paras, 1983). It is to be hoped that either "radionuclide" or "activity" may ultimately replace "dose" in common parlance, but, in the meantime, the term "dose calibrator" will be used in this Report.

One of the early forerunners of the present commercial dose calibrators is the type 1383A beta-gamma ionization chamber (Figures 72 and 95) developed for the National Physical Laboratory (Dale et al., 1961; Lewis et al., 1972). This re-entrant chamber is designed so that the sample is almost completely surrounded by the sensitive volume of the ionization chamber. This arrangement provides high geometrical efficiency that minimizes the effects of small variations in sample position and volume. The lower half of the ionization chamber has a

Fig. 95. Combined NPL beta-gamma reference ionization chamber of which a schematic diagram is shown in Fig. 72. (From Robinson, B. 1960)

thin window to admit beta particles, thereby permitting use of the chamber for the assay of beta emitters.

Most of the dose calibrators available today are variations of this basic design. Typically the beta-particle ionization chamber has been deleted and many manufacturers use pressurized chambers filled with argon up to a pressure of 20 atmospheres. With the high-pressure argon filling, the ionization current from the source is up to 20 times, while the background current is about twice, that obtained in the same chamber filled with air at atmospheric pressure.

The general acceptance of these instruments in clinical laboratories for the purpose of performing comparative or indirect radioactivity assays is primarily due to their ability to permit measurements of gamma-emitting solutions in ampoules or serum vials over a range of activities from a few microcuries to many millicuries. Because of the lack of energy selectivity in these detectors, care must be exercised to ensure that the measurements are performed on radionuclidically pure sources. Otherwise, the amount and identities of any impurities present must be determined by measurements that use energy-selective systems with the overall response corrected for any contributions due to such impurities. Sections 2.1 and 4.4 give further information on the operation and use of ionization chambers in performing comparative measurements. It should be noted, however, that recently several manufacturers have introduced dose calibrators that rely on detectors other than ionization chambers, such as plastic scintillators in a well configuration. Much of the information presented in this section will also apply to these devices.

The feature that distinguishes dose calibrators from other ionization-chamber systems is the special electronic circuitry that permits the response of the instrument to be displayed directly in terms of activity. This is accomplished by dividing the amplified output of the detector by some unique value for each radionuclide to provide a digital or analog output equivalent to the activity of the radionuclide. Thus, a dose calibrator could be considered a simple analog-computer detector system that is programmed by setting a potentiometer, or by plugging in a fixed resistor, to convert electronically the detector output into direct activity readings, usually in sub-multiples of the curie.

The air of finality imparted by a direct reading in multiples of the curie of a given radionuclide should not mislead the user and, indeed, recent reports of the accuracy of dose calibrators support this warning (Genna *et al.*, 1972; Garfinkel and Hine, 1973; Hare *et al.*, 1974; Hauser, 1974). Some confidence in a manufacturer's calibration can be gained from good, accompanying documentation, but it behooves the operator to assure himself that the calibrations and quoted estimates of uncertainty are applicable to the size, shape, and filling of containers routinely used. These calibrations should be checked by means of certified standards of the radionuclides of interest and, if necessary, the corrections applicable to containers of different geometry should be determined in the manner described in Section 6.4.2.

It may frequently happen that a dose calibrator may not have a calibration (variable-resistance-dial) setting, or plug-in resistor module, for a gamma-emitting radionuclide contained in a radiopharmaceutical that has been recently introduced into use. A calibration point for such a radiopharmaceutical can, however, be established by an interpolative method, provided that reliable decay-scheme data are available for the radionuclide in question.

Basically, for *any one radionuclide dial setting or module*, the readout is proportional to the ionization current flowing across the ionization chamber. Thus, by measuring a series of monoenergetic photon sources having known emission rates, at *any dose-calibrator setting*, a curve of reciprocal $K_{R,\gamma}$, as a function of photon energy, can be plotted as described in Section 4.4.3, and will be similar to that shown in Figure

96. If R is the instrument reading for the radionuclide of activity A, P_γ is the fractional probability of the radionuclide emitting a gamma ray of energy E_γ per decay, and R_R is the instrument reading for a long-lived reference source, then

$$K_{R,\gamma} = P_\gamma A \frac{R_R}{R},$$ (6.3)

and

$$A = \frac{K_{R,\gamma} R}{P_\gamma R_R}.$$ (6.4)

Provided that the same long-lived reference source is used, an identical curve of $1/K_{R,\gamma}$ vs. E_γ will be obtained at *all other* radionuclide settings of the dose calibrator. As stated in Section 4.4.1, the reciprocal of the K factor, or of the relative factor $K_{R,\gamma}$ (as plotted in Figure 94), is equivalent to an efficiency.

Fig. 96. The efficiency, in terms of relative response per photon per unit time, as a function of photon energy, for a typical commercial dose calibrator (see Eq. 6.3) from data obtained by R. L. Ayres (1977). (*Note:* The numerical values on the vertical scale will depend upon the particular dose calibrator and the activity of the long-lived reference standard. In some cases of non-monoenergetic photon emitters, R/R_R for photons with energies in uncalibrated regions of the curve can be calculated by subtracting from the total response the response obtainable from Eq. 6.3, due to photons with energies in regions where $K_{R,\gamma}$ is known.)

After such an efficiency curve has been established, the activity of any other monoenergetic gamma-emitting nuclide can be computed by using Eq. 6.4 and an interpolated value of $1/K_{R,\gamma}$. The activity of a nuclide emitting more than one photon can be obtained by weighting the efficiency (i.e. $1/K_{R,\gamma}$) for each photon of different energy, including x rays, by the appropriate probability per decay and by computing the activity by using Eq. 4.20, namely

$$A = \frac{R}{R_R} \cdot \sum_{i}^{n} \left(\frac{K_{R,\gamma}}{P_\gamma} \right)_i. \tag{6.5}$$

After the activity has thus been calculated, a calibration setting can be established for the "new" radiopharmaceutical. From the nature of the efficiency curve, it can be seen that serious error could arise if a preponderance of photons had energies below 200 keV, due to the difficulty in establishing the curve in this region.

One method for determining the response function of the type 1383A beta-gamma ionization chamber has been described by Dale (1961). This reference should be consulted for a description of the method itself and the physical processes that determine the observed response function. Efficiency vs. energy curves for both air and pressurized argon chambers are shown, and their use discussed, in a review by Weiss (1973) of "4π"γ ionization chamber measurements.

Two other considerations that must be taken into account in determining or checking the calibration of a dose calibrator are its linearity as a function of activity and its geometry dependence on sample size and position. The linearity of the instrument at any given setting can be checked by preparing a source of a relatively short-lived radionuclide, such as ^{99m}Tc, and following the activity of the source as a function of time. The initial activity of the source should be equal to, or greater than, the highest activity that one would routinely expect to measure, and the decay of the source should be followed down to the lowest levels of activity that would normally be measured. By following this procedure, any nonlinearities between range changes on multi-range instruments would be detected, as well as the region and magnitude of ion-recombination-induced nonlinearities. An example of both of these effects is shown in Figure 97. Over the range indicated by A, a 10-percent discrepancy is observed between the 1- and 10-millicurie-range settings. In such a case it would be necessary to return the instrument to the manufacturer for correction of the problem. For rates at or above those indicated by the range B, ion recombination produces a nonlinearity. These higher rates should either be avoided, or, alternatively, the magnitude of the nonlinearity as a function of

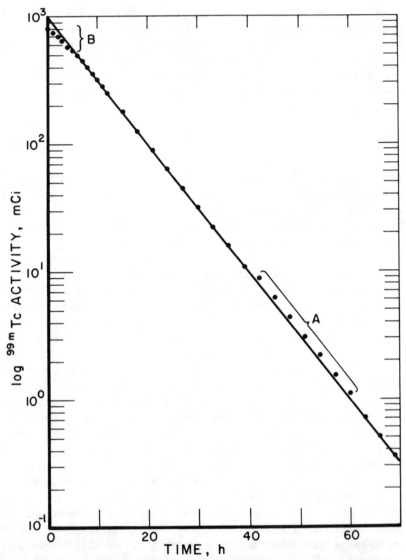

Fig. 97. Simulated linearity check of a commercial dose calibrator using ^{99m}Tc. (See Section 6.4.3.2.3 for discussion.)

activity, for a given radionuclide, can be measured and appropriate corrections applied. Test material should be free of other radionuclides in order to prevent other distortions of the half-life curve. An alternative procedure described in Section 4.4.4 is that using otherwise identical sources with different radioactivity concentrations of the same radionuclide. If measurements are made over an interval of time short compared with the half lives of any photon-emitting impurities, then the method is insensitive to these impurities.

In order to minimize sources of measurement errors, careful control of source geometry and positioning is required. The effects of varying source position and volume are similar to those presented in Figures 93 and 94, but are less severe because of the greater well depth of most dose calibrators. A discussion of the effects of varying source position in two typical reentrant ionization chambers is presented by Weiss (1973), and of varying source volume by Seliger and Schwebel (1954). A more severe geometry-dependent source of error, frequently encountered in the assay of radiopharmaceuticals, is due to the containers of varying size and composition used by commercial distributors of radiopharmaceuticals. In order to minimize such source-geometry-induced errors in the measurement, the user must either adopt a standard container and content for all of his measurements or develop individual calibrations for each source geometry. The latter option would require individual calibrations for each container and for variations in the volume of radiopharmaceutical within the container.

The importance of performing daily checks with a long-lived source cannot be over-emphasized. Without such checks, significant drift or alteration of the instrument's response could escape detection and result in measurement errors. Moreover, the practice of pairing measurements of the unknown with a long-lived reference source compensates for changes in amplifier gain—or even errors in settings. That is, the *ratio* of the response per unit activity of a given radionuclide to that of a reference source should be a constant for a given instrument. This ratio is established during the initial calibration of the instrument, and can be used thereafter to calculate the activity of that radionuclide from future paired measurements with the same reference source (corrected for decay) at any arbitrary instrument setting. For example, if the ratio of the response of 1 mCi of a calibration standard of ^{99m}Tc to that of the reference source was determined to be 0.5 and, in a later measurement of a ^{99m}Tc source of unknown activity, the ratio to the same reference source was determined to be 3.6, then the activity of the unknown ^{99m}Tc source would have been equal to 7.2 mCi.

Although a dose calibrator lacks energy selectivity, useful purity and identification checks of a radiopharmaceutical may be made by measuring the half life of a sample. Significant departures from the known value of the half life indicate either the presence of impurities or that it is the wrong radionuclide. Such a check, made immediately after receipt, can be useful in cases of short-lived radionuclide generators such as ^{99}Mo-^{99m}Tc. In addition, when pre-assayed radiopharmaceuticals are being checked, close agreement between the measured and stated activities provides some assurance that the correct radiopharmaceutical has been received.

A measure of energy discrimination can be introduced by using absorbers to selectively attenuate lower-energy gamma rays from given radionuclides and thereby enhance the relative sensitivity of the instrument for radionuclidic impurities emitting higher-energy gamma rays than those of the principal radionuclide. The primary example of the application of this procedure is the use of the "molybdenum breakthrough kit" to measure the quantity of ^{99}Mo in a solution of ^{99m}Tc. Again, due to the lack of energy selectivity in the instrument itself, gross errors have been reported when other high-energy gamma-emitting impurities, other than ^{99}Mo, have been present (Briner and Harris, 1974).

A performance standard on the use of dose-calibrators has been published by the American National Standards Institute (ANSI, 1978b).

6.4.3.3 *Assay of Beta-Particle-Emitting Radionuclides.* To obtain the activity of a pure beta-emitting radionuclide, the two methods commonly employed are gas proportional counting and liquid-scintillation counting. For a description of the operation of, and measurements using, proportional counters see Sections 2.2 and 3.5. Similarly, and for liquid-scintillation counting, see Sections 2.4 and 4.3.

Another method that is frequently used in clinical laboratories to obtain the activity of a ^{32}P source is based on the measurement of the bremsstrahlung radiation produced by the deceleration of the high-energy beta particles in the solution and source container. The bremsstrahlung radiation is measured with a photon-sensitive system such as a dose calibrator or NaI(Tl)-detector system.

In order to achieve accurate and reproducible results, great care must be exercised to minimize any variations in counting geometry. Since most of the bremsstrahlung radiation is produced through interactions of the beta particles with the solution and container, the composition and volume of the solution and the wall thickness of the container must be carefully controlled. Attempts to perform such measurements with the sample in typical molded-glass pharmaceutical serum vials will yield widely scattered and erroneous results because the wall thicknesses of these containers vary considerably. Reproducible results can be achieved only by using a "standard" container or set of containers and carefully controlling the solution volume.

Methods for the qualitative identification of pure beta emitters normally rely on measuring the half life and the maximum beta-particle energy, $E_{\beta max}$, of the radionuclide. This energy is a unique index for each beta-emitting radionuclide and is most commonly obtained by measuring the count rate of a source as a function of successive thicknesses of absorbers placed between the source and the detector. From these data the maximum range or mass absorption coefficient can be obtained and related to $E_{\beta max}$ (Friedlander *et al.*, 1964; Evans, 1955; Mann *et al.*, 1980; USP, 1980).

6.5 Laboratory Equipment for Radioactivity Measurements

6.5.1 *General-Purpose Equipment*

A well-equipped radioactivity measurements laboratory should have some or all of the following instrumentation available to it:

(1) A dose-calibrator or ionization-chamber type of detector for measuring millicurie quantities quickly and accurately in various geometries, and for providing some qualitative indication of radionuclidic purity (see Section 6.4);

(2) A multichannel-analysis system utilizing either

   (a) a NaI(Tl) detector (Sections 2.4 and 4.2) or

   (b) a semiconductor detector (Sections 2.5 and 4.2). A Ge detector will allow the user to resolve multicomponent gamma radiations from a sample for radionuclide identification.

(3) a NaI(Tl) well crystal system with either a ratemeter or digital display, and a variable discriminator circuit. This system is useful for routine counting and simple pulse-height analyses (see Sections 2.4 and 4.2). Automatic, multi-sample counting equipment of this type is also available.

(4) a liquid-scintillation-counter system for counting pure beta-emitting nuclides. This type of counter can also be used to count the x rays from electron-capture nuclides and low-energy gamma rays (and associated Auger or internal-conversion electrons). Most systems are multi-sample units with two or more channels for measurements in different energy ranges, and with some computational ability (see Sections 2.4 and 4.3).

(5) proportional counters of various types for counting alpha, beta, or gamma radiation. Single and multi-sample units are available

TABLE 17—*Types of radiation-detection instrumentation*

Type of instrument	Typical activity range	Typical sample form	Data acquisition and display
	mCi		
Gas proportional counters	10^{-7} to 10^{-3}	Film disc mount, gas	Ratemeter or scaler
Liquid-scintillation counter	10^{-7} to 10^{-3}	Up to 20 ml of liquid or gel	Digital Accessories for: background subtraction, quench correction, internal standard, sample comparison
NaI(Tl) cylindrical or well crystals	10^{-6} to 10^{-3}	Liquid, solid, or contained gas <4 ml	Ratemeter Digital Discriminators for measuring various energy regions Multichannel analyzer, or computer plus analog-to-digital converter Computational accessories for full-energy-peak identification, quantification, and spectrum stripping
Ionization chambers	10^{-2} to 10^{-3}	Liquid, solid, or contained gas (can be large in size)	Ionization-current measurement Digital (mCi) readout, as in dose calibrators (Section 6.4.3.2.3)
Solid-state detectors	10^{-2} to 10	Various	Multichannel analyzer or computer with various readout options

in a variety of configurations, some with automatic computational ability (see also Sections 2.2 and 3.6).

Table 17 indicates the instrumentation appropriate to some typical sample forms in typical ranges of activity.

The choice of equipment appropriate to a laboratory's investigations should be based on:

   (i) the types of measurements and accuracies required,

  (ii) the particular radionuclides and the radiations to be measured,

 (iii) the types of samples to be measured as well as their number and variety, and

 (iv) the funds available.

6.5.2 *Buyers' Guides*

Various publications describe the different types of equipment available to assay radioactive material. Hendee (1973) provides a basic, easy-to-read description of the operation of various types of nuclear-detection equipment.

Several Buyers' Guides published by scientific journals provide excellent sources of information concerning the variety of commercial instruments available (e.g., the "Guide to Scientific Instruments" issue of *Science*, the "Instruments Specifier" issue of *Industrial Research*, the "Buyers' Guide" issue of *Nuclear News*, and the "Physicians' Desk Reference for Radiology and Nuclear Medicine" published by the Medical Economics Co., Oradell, New Jersey).

6.5.3 *Instrument Stability*

All laboratories making radioactivity measurements utilizing the above equipment should institute an effective quality-control program for all their procedures (also see Sect 8.3). Laboratories carrying out clinical or diagnostic procedures may be required by regulation to have such quality control. Instrument performance should be checked daily with a suitable long-lived source to assure the reproducibility of daily measurements. Counting efficiencies for each radionuclide measured at a particular instrument setting, counting geometry, and sample composition should be determined and periodically checked. All detector systems should be characterized for the effect of variations of sample geometry, composition, and activity on counting efficiencies.

Records should be kept of all the measurement data, the instrument settings, and measurement geometry. Data must be checked for statistical reliability by using techniques such as the χ^2 test for "goodness of fit" (see Section 7.3). If the instrument-monitoring data, in terms of a long-lived reference source, and the instrument background count are plotted as a continuing function of time, long-term trends, reflected in small changes in slope or spread of points, can indicate a change in instrument performance or in instrument environment even before statistical evaluation of the data might do so. For nuclear-medicine technologists, a discussion on counting statistics may be found in Simmons (1970).

6.5.4 *Radiation Protection*

A radiation-protection program should be implemented to control exposure of workers to radiation and their contamination by radioactive materials, and also contamination of working areas or the environment. Some desirable elements of such a program are:

(i) availability of radiation-survey equipment for measuring levels of radiation,
(ii) availability of wipe or smear test equipment for measuring contaminant levels and identifying contaminating radionuclides
(iii) training programs for workers,
(iv) availability of personnel and area dosimeters, and
(v) radioactive-waste management facilities.

The assistance of a qualified radiation safety expert on a consultant or other ad hoc basis should be available, if competent staff assistance is not available. For a more comprehensive treatment of the subject of radiation protection, texts such as those by Cember (1983) and Shapiro (1972) should be consulted, as well as reports by the National Council on Radiation Protection and Measurements (See NCRP Publications on page 573 of this Report) and the International Commission on Radiological Protection.

6.5.5 Further Reading

Information regarding the type of equipment suitable for specific purposes and the procedures that can be used can be found in handbooks available from the Oak Ridge National Laboratory (Davis and Gupton, 1970), and the National Institutes of Health (NIH, 1973) as well as a text by Shapiro (1972). Additional general references are Lapp and Andrews (1964), Price (1964), Bertolini and Coche (1968), Adams and Dams (1970), Krugers (1973), and Ouseph (1975).

7. Statistics

7.1 Treatment of Counting Data

7.1.1. *Statistics of Pulse Counting*

The process of radioactive decay is a random sequence in time. From the law of radioactive decay (see Section 1.3) the decay rate, dN/dt, is given by

$$\frac{dN}{dt} = -\lambda N \tag{7.1}$$

where λ is the decay constant and N is the number of unstable nuclei. Let n denote the number of atoms that decay during a certain period of time T. For T small compared to the half life, n will be small compared to N, so that the decay rate $\rho (= dN/dt)$ can be treated as a constant. Then the probability of exactly k disintegrations during the period T can be approximated by the Poisson distribution:

$$\text{Probability } (n = k) = \frac{(\rho T)^k e^{-\rho T}}{k!} \tag{7.2}$$

Equation 7.2 has been derived in most treatises on probability based on either (a) axiomatic assumptions, or (b) the limit to binomial (Bernoulli) distributions (see, for example, Feller, 1968; Parzen, 1960; and Stevenson, 1966). (See also Appendix B.3.2.) The Poisson distribution is also applicable to a large number of physical phenomena in daily life, such as telephone calls and traffic accidents.

Once we accept the view that the Poisson Law adequately describes the phenomenon of radioactive decay, several useful results follow by deduction:

a. The mean and variance of the number of decays or transitions during time T are both equal to ρT. The standard deviation is $\sqrt{\rho T}$. (See Appendix B.3.5.)

b. Addition Theorem. If n_1 is the number of decays of one species of nuclide in time T with decay rate ρ_1, and n_2 the number of decays of another species of nuclide in time T with decay rate ρ_2, then the distribution of the sum $(n_1 + n_2)$ is also Poisson with mean and variance $(\rho_1 + \rho_2)T$.

c. For large numbers of decays n, the Poisson distribution is well approximated by the Gaussian, or normal, distribution, with a mean ρT and a variance ρT. Let $\xi = (n - \rho T)/\sqrt{\rho T}$ be the normalized deviate, i.e., the difference from the mean relative to the standard deviation.

Then the probability, P, of $a < \xi < b$ is

$$P(a < \xi < b) = \frac{1}{\sqrt{2\pi}} \int_a^b e^{-x^2/2} \, dx$$

d. The distribution of the time interval t between decays is the negative exponential,

$$f(t) = \rho e^{-\rho t}$$

where $0 < t < \infty$, and the probability of $t_1 < t < t_2$ is the integral

$$P(t_1 < t < t_2) = \int_{t_1}^{t_2} f(u) du.$$

The properties of Poisson and normal distributions may be found in most texts on statistics, e.g., Brownlee (1950), Haight (1967), and Cramér (1946). Tables of the individual and cumulative terms of Poisson distribution have been provided, e.g., by the General Electric Company (1962).

The decay rate ρ of a sample bears a known relationship, through the counting efficiency, to the "true" counting rate, ρ'. However, counting for a finite period of time t can only yield an estimate $r = n/t$ of the "true" counting rate ρ'. The measured r is subject to statistical fluctuations. A measure for the scatter of the measured counting rate r about the "true" counting rate ρ' can be derived from the Poisson assumption; such a measure is provided by the standard deviation.[18] The standard deviation of r is a function of t and can be expressed as

$$\sigma_r = \sqrt{\frac{\rho'}{t}}. \tag{7.3}$$

---

[18] If counting losses, or dead time, are not negligible, then the Poisson approximation does not hold. Dead time, decay, and background considerations are treated in Section 2.7.

Because ρ' is unknown, only an estimate, $\hat{\sigma}_r$, of the standard deviation, σ_r, can be computed from the measured r, namely,

$$\hat{\sigma}_r = \sqrt{\frac{r}{t}} = \frac{1}{t} \sqrt{n}. \qquad (7.4)$$

An estimate of the standard deviation of the number of counts is

$$\hat{\sigma}_n = \sqrt{n}. \qquad (7.5)$$

The standard deviation is a measure of the scatter of a set of observations around their mean value. For example, about 10,000 counts are necessary for a standard deviation in the total count of about 100 or of 1 percent. When many observations are made, approximately two-thirds of the observations ought to lie within one standard deviation from the mean and the remainder outside this limit. The following limits apply to representative multiples of the standard deviation in the theoretical distribution:

Deviation	$\pm 0.675\sigma$	$\pm\sigma$	$\pm 2\sigma$	$\pm 3\sigma$
Probability that observation lies within this deviation	0.5	0.68	0.95	0.997

The above limits allow an appraisal of the performance of a counting system to be quickly made, i.e., if a counting rate deviates more than what would be expected statistically from a previously determined rate. This is of importance when equipment is checked routinely with a performance standard. For instance, a deviation of more than 2σ from a previously established mean should be considered with suspicion, even though it may be correct, since the probability of this occurring is only about 5 percent.

The application of the general description given above can be formalized through the use of "control charts" that provide a running graphical record of the values obtained from counting a reference source or of their variability. For further details on control charts, see Natrella (1963) and Ku (1969).

7.1.2 Background Considerations

In most instances the radiation counter exhibits a background rate that is not negligible and that has to be subtracted from the gross counting rate. An estimate of the statistical error due to the background must be included in the final estimate of the error. On the basis that the source and background counts are additive, the standard deviation of the net counting rate is

$$\sigma = \sqrt{\sigma_T^2 + \sigma_B^2} = \sqrt{\frac{\rho'_T}{t_T} + \frac{\rho'_B}{t_B}}. \tag{7.6}$$

T refers to the counting of the sample plus background (total) and B to the counting of the background. In counting sources of reasonable activity, it is usually sufficient to measure the background counting rate only once (or perhaps before and after a series of samples) by counting long enough to make σ_B negligible compared to σ_T. The optimum division of available time between background and source counting has been derived by Loevinger and Berman (1951), and is given by

$$\frac{t_B}{t_T} = \sqrt{\frac{\rho'_B}{\rho'_T}}. \tag{7.7}$$

To approximate the optimum division, estimates of ρ'_B and ρ'_T have to be obtained from a preliminary run. The minimum combined time necessary to achieve a predetermined precision can be estimated for particular values of total counting rate relative to background counting rate. Curves published by Loevinger and Berman (1951) and reproduced in Figure 98 enable this to be accomplished for any practical case. As an example of the application of these curves, suppose that it is desired to measure with a standard deviation of 5 percent, the activity of a sample that generates in a given detector a count rate that is approximately half that due to background. Then k, the ratio of total to background counting rate, is 1.5, and from Figure 98, the 5-percent line intersects the oblique line for k = 1.5 at a total count of 6,600. The point of intersection lies about one-third of the distance between the background-count curves for 3,000 and 5,000. Thus, the background count should be about 3,600. These curves are valuable guides whenever the preset-count technique can be used, and are especially useful when only a few measurements are to be made. They are, however, seldom used in the case of well-established measurements systems for which "running backgrounds" are maintained, as, for example, in a low-level radioactivity measurements laboratory. A preliminary rough determination of the background and the total

Fig. 98. Relationship between random uncertainty, expressed as an estimate of the standard deviation, and the number of observed counts taking account of background. For a given desired percentage error in the measurement of sample activi^ty, the optimum distribution of counts between background and the combined background plus sample counts can be chosen so that the time spent in counting is a minimum. The ratio k is that of sample plus background counting rate to background counting rate. The curves are particularly useful when a predetermined precision is required in counting a sample of low counting rate using the pre-set count technique. (From Loevinger and Berman, 1951).

counting rates is necessary in order to establish the value of k. The preset-count technique used with these curves then enables all sample activities to be determined with approximately the same precision.

7.1.3 *Minimum Detectable Activity*

Counting intervals cannot be made indefinitely long; therefore, time and background considerations determine minimum detectable activity. In addition, the situations in which the concept of minimal detectable activity is used also tend to give rise to similar expressions but with entirely different interpretations. Currie (1968) commented on a number of such terms and proposed three of his own that constitute successive levels of activities leading from a *decision rule* that indicates detection based on the data at hand to a preselected *detection limit* such that a signal at this level is almost sure to be detected, and then to a quantitative estimate of the signal with a particular precision. Hence, in using the term "minimum detectable activity" or similar expressions, the exact definition and meaning of the term must be given at the same time.

Moreover, since these limits are based on counting statistics alone and do not include other random errors or any systematic components of error that are always inherent in the procedures, the use of these terms is limited to that of serving as guideposts only, and not as absolute levels of activity that can or cannot be detected by a counting system. Bearing these remarks in mind, definitions of decision limit and detection limit are given as two convenient and useful terms in the interpretation of minimum detectable activity in the framework of statistical hypothesis testing, as originally conceived by Neyman and Pearson (1936), and adopted by Currie (1968), and others. Treatment of statistical hypothesis testing may be found in most statistical texts, for example, Natrella (1963, Chapter 3). Ordinarily, a rule is applied to the results observed and then a decision is made between a pair of alternatives. The two alternatives may be formally stated as follows:

(i) proceed on the supposition that the observed counts (counting rate) are greater than background.

(ii) proceed on the supposition that the observed counts (counting rate) are not greater than background.

As the observations are made in a finite length of time, sometimes an erroneous decision is made. There are two ways in which a wrong action can be taken:

(i) proceed on the supposition that there is a signal when in fact there is none: Error of the first kind, or *false detection.*

(ii) proceed on the supposition that there is no signal, when a signal really exists: Error of the second kind, or *false nondetection.*

In any particular experiment, assurance that a correct decision has been made can never be absolute. The rule of decision adopted, in turn, would depend on the importance attached to each type of wrong decision, and is a compromise of (i) the time available for the experiment,[19] (ii) what proportion of false detection can be tolerated, and (iii) what proportion of false non-detection can be tolerated. A rule that works well for one laboratory, therefore, may not work at all for another laboratory with a different objective in mind.

The proportion of false detection that can be tolerated is usually

---

[19] Note that the problem is not necessarily solved even if time can be extended, since then there will be instrument drift and a host of other interfering problems.

denoted by α, and the proportion of false non-detection that can be tolerated is denoted by β. In general, both α and β should be small. A useful selection of the values of α and β is that $\alpha = \beta = 0.05$, i.e., 5 percent of false detection and 5 percent of false non-detection are acceptable in a sequence of such decisions.

If equal values of α and β have been decided upon, e.g., $\alpha = \beta = 0.05$, it can be shown that the *decision limit*, L_C and the detection limit, L_D can be expressed in terms of standardized normal deviates (normal deviates in units of σ) and the background count as follows (Currie, 1968):

$$L_C = k \sqrt{2}\sqrt{B} = 1.64 \sqrt{2}\sqrt{B} = 2.32 \sqrt{B} \qquad (7.8)$$

$$L_D = k^2 + 2L_C = 2.71 + 4.65 \sqrt{B} \qquad (7.9)$$

where:

L_C: the net number of counts (total minus background) for reaching a decision of no detection of signal if the actual net counts are less than L_C and detection if the actual net counts are larger than L_C.

L_D = the signal level such that a signal at or above this level is likely to be detected.

B = the "true" background counts.

$k = 1.64$, the value of the standardized normal deviate that is exceeded with probability $\alpha = 0.05$.

For values of L_C and L_D thus computed, the probability of a false detection at a level of L_C is approximately 5 percent and the probability of non-detection at a level L_D is approximately 5 percent.

The above formulation assumes paired observations of total and background counts. Since the "true" background count B is usually not known and an estimate is used, the computed values of L_C and L_D are approximate for given α and β. For other cases, including those where the background count can no longer be assumed to be Poisson because of interference, see Currie (1968) and Donn and Wolke (1977).

The relation of the just described procedure and definitions of two terms—the decision limit L_C and the detection limit L_D—to the minimum detectable activity of a system is indicated in the plot of β against signal level in Figure 99. This figure shows the behavior of β, the probability of not detecting a signal as a function of signal level. Note that the procedure calls for a decision of no detection if the net number of counts is less than L_C. Should the signal level corresponding to L_C be called the minimum detectable activity? Most likely not, since at that level all one needs is to flip a coin, and the results based on heads or tails is as good as the result of the measurements in the long

Fig. 99. Plot of probability of not detecting a signal (β) *vs* net signal level. The curve shown is one of a family of curves so that the location of L_C and L_D depends on α, β, and background. In this illustration for $\alpha = 0.05$, the probability of non-detection at zero signal level is 0.95.

run. This condition is unsatisfactory, because by minimum detectable activity is meant the activity level that could be detected most of the time, and not detected only occasionally.

As the signal increases from L_C toward L_D, the percentage of false non-detection continues to decrease and reaches a tolerable 5 percent error rate when the signal level is at L_D. Perhaps it could be claimed that this is the level at which the system will consistently detect the signal. However, since L_D is fixed by the background and the parameters α and β, there is no unique point corresponding to a "minimum."

In fact, fixing any two of the four parameters automatically determines the remaining two. In the above case, L_C is fixed by choosing $\alpha = 0.05$, and L_D is fixed by selecting $\beta = 0.05$ in conjunction with L_C already determined. If a low value of L_D is arbitrarily fixed, and a value for β is selected, then L_C is fixed, and the value of α corresponding to this L_C may not be acceptable for the given background counts.

Statistical hypothesis testing and hence the decision and the detection limits, is a "yes" or "no," "go" or "no go" decision making procedure. In many instances, an "interval estimate" of the signal level could be more informative than a decision process. Confidence interval for the mean is described in Section 7.3.2.

From the above discussion, it is clear that the minimum detectable activity, however defined, does not guarantee with certainty the presence or absence of a signal. These concepts have been frequently misused in the handling of data, particularly in low-level counting. For example, in a series of data to be averaged, to consider the minimum detectable activity either as zero or at the minimum detectable activity would bias the value of the average. In such a situation, all values should be used *as measured* even if some values are below the minimum detectable activity. Negative values (below background) are as valid a measurement as positive values and should be retained in any series of data.

7.1.4 *Background Equivalent Activity*

A useful concept for expressing the intrinsic sensitivity of any detector is that of the Background Equivalent Activity (BEA) (Seliger and Schwebel, 1954; Mann and Seliger, 1958). This quantity, for any given instrument and radioactive nuclide, is, as its name implies, the activity of the nuclide that will, in the particular geometry used, produce a response of the instrument that is equal to its background reading.

7.1.5 *Statistics of Ionization Chambers*

Statistical variation in the number of particles produced in a radioactive-decay process is reflected in the variation of total charge released in an ionization chamber (or "mean-level" system; see Price, 1958). The total charge, Q, collected in a given time in the chamber will be equal to the product qn, where q is the average charge released per interacting particle and n is the number of particles interacting within the chamber in that time. Therefore the standard deviation in Q will be q times as large as the standard deviation in n, or $\sigma_Q = q \sqrt{n}$. Substituting $Q = qn$ one obtains

$$\sigma_Q = \sqrt{qQ}. \tag{7.10}$$

This treatment neglects the fluctuations in the charge released by the individual particles. Price (1958) gives the treatment for rate-type ionization-chamber instruments. For certain applications (*e.g.*, semiconductor detectors, Section 2.5), the Fano factor (Fano, 1947) is used to correct σ_Q to agree with experimental results.

7.2 Statistical Models of a Measurement Process

The previous section has dealt with counting errors, the variability due to the decay process. The counting error is the fundamental, inescapable statistical fluctuation associated with the counting process. In addition to these counting errors, however, other sources of error may also be considered as "random" in nature. These errors in a sense are also inescapable because no two samples are exactly alike and no physical quantities can be measured without error. The standard deviation computed from counting errors can be used as a measure to judge if these other sources of error are in fact significant or negligible.

The use of a factor to correct for "geometry" of the counting instrument may introduce a systematic error in a series of results; the use of a specific calibration curve or half-life value may bias results from those of another laboratory using a different curve or a different value. Systematic errors may be the consequences of a particular measurement procedure or of particular instruments. The use of referee methods and standard reference samples helps to coordinate results of different laboratories to the same benchmark.

Another source of bias, which is important in the statistical sense, is the reduction and summary of data. The weighting of observations, including the rejection of outlying values (outliers), is forever a problem. Should the weights represent counting error alone or should the weights also include a component representing measurement errors in addition to counting error? Should a line be fitted to the data with a model that goes through the origin? These and a number of similar problems do not have readily available answers, since the appropriate technique may depend on the way data are generated, the theoretical basis for their relationship, and their distributional properties.

The measurement process and its error structure can usually be approximated by statistical models, shorthand expressions to account for and to focus attention on the variability of results of measurements. In a statistical model, a measured value is usually expressed as a function of parameters on which the measured value depends plus the errors that may be associated with various factors involved in the measurement process, including assumptions made with regard to the distribution of these errors.

For illustration, consider a set of measurements of ^{90}Sr activities in each of the soil samples collected in duplicate at five sites near a reactor. The measured activity, for example, of the second sample from the third site may be represented by the following statistical model:

Measured activity of the second sample of the third site = (grand mean of samples from five sites at date of collection)
+ (effect of third site)
+ (activity difference of second sample)
+ (possible instrument drift between tests)
+ (errors in sample preparation)
+ (counting error, including background).

The activity is also influenced by other possible errors, *e.g.*, in the:
(geometry factor),
(efficiency factor used for instrument),
(half-life value used), and
(chemical yield).

The last four items may introduce errors that are systematic in nature and that would affect equally the measured activity at different times, and thus would cancel out if the measured activity of the soil at one time period was compared to that measured at another time period by the same laboratory. These systematic errors could, however, be the main sources of discrepancy when results from one laboratory were compared to results from another.

Errors due to sample preparation and instrument drift may be assumed to be random and their standard deviations may be estimated from duplicate determinations, but the counting-error component has to be considered. The effects of different sites may or may not be a subject of interest. If the average of the five sites selected is defined to be the criterion for safe levels of radioactivity surrounding a reactor, then the sites by themselves are of no interest; but if the five sites are selected as a sample of all sites surrounding the reactor, then variability of activity from site to site may contribute substantially to the uncertainty of the reported value.[20]

The formalization of an "error account" in a statistical model such as above, for a particular measurement process, serves at least two purposes:

    i. to list exhaustively the possible sources of errors that are relevant to the interpretation and use of the results, and

    ii. to help in planning or designing the experiment such that the results may be interpreted as unambiguously as possible and the errors due to various factors can be estimated.

In the following section a selection of statistical procedures is presented to help in explaining some of the concepts and techniques commonly used. For additional references see Jarrett (1946), Brownlee (1960), Youden (1951, 1963), Natrella (1963), Stevenson (1966), and Ku (1969).

7.3 Statistical Terms and Formulae Useful in Radioactivity Measurements

7.3.1 *Population, Mean, and Standard Deviation*

One of the basic concepts of statistics is the notion of a *population* of "things" with fixed properties, and the use of observations on a *sample* from the population to estimate such properties. Greek characters are usually used to denote the parameters of a population, and Roman characters for estimators of these parameters, or estimates of these parameters computed from a series of observations say, x_1, x_2, ..., x_n. Some of the most commonly used parameters and their estimators are given in Table 18.

The mean is a measure of location (center) of the distribution of the population, and the standard deviation is a measure of the spread (scatter) of the distribution about the mean. The square of the standard deviation, σ^2, is called the variance.

---

[20] The above example, on soil samples taken about a reactor, points to another area in which more research and standardization are needed, namely, as to how many samples (where, when, and how obtained) should be taken in order to ensure that the radioactive effluent from a reactor conforms with prescribed environmental limits. Statistical sampling methods deal with this type of problem, but a general plan for selecting sampling sites cannot be given because different situations arise in each individual case. For further information on this subject, see NCRP Report No. 50 (NCRP, 1976b).

TABLE 18—*Population parameters and corresponding estimators*

Population parameters	Corresponding estimators of parameters
μ_x (mean or first moment)	$\bar{x} = \dfrac{1}{n} \sum\limits_{i=1}^{n} x_i$
σ_x^2 (variance or second central moment)	$s_x^2 = \dfrac{1}{n-1} \sum\limits_{i=1}^{n} (x_i - \bar{x})^2$
$\sigma_{xy} = \sigma_{yx}$ (covariance)	$s_{xy} = s_{yx} = \dfrac{1}{n-1} \sum\limits_{i=1}^{n} (x_i - \bar{x})(y_i - \bar{y})$
σ_{xy} (correlation coefficient)	$r_{xy} = r_{yx} = \dfrac{s_{xy}}{s_x s_y}$
σ_x (standard deviation of x about μ_x)	s_x
$\sigma_{\bar{x}} = \dfrac{\sigma_x}{\sqrt{n}}$ (standard deviation of the average $\bar{x}$, or standard error)	$s_{\bar{x}} = \dfrac{s_x}{\sqrt{n}}$
$100 \dfrac{\sigma}{\mu}$ (coefficient of variation or relative standard deviation, expressed in per cent)	$v_x = 100 \dfrac{s_x}{\bar{x}}$

Since repeated counts for equal intervals of time follow the Poisson distribution, which has the property that the variance is equal to the mean, this property can be used to test for the presence of extraneous variability in addition to the inherent variability of the counts. Let c_1, $c_2, \ldots, c_k$ be the results of repeated counts. Theoretically an estimate of the variance of each of these counts is expected to be equal to $\bar{c}$ (Appendix B3.5). An experimental estimate of the variance is given by:

$$s_c^2 = \frac{\sum\limits_{i=1}^{k} (c_i - \bar{c})^2}{k - 1}. \qquad (7.11)$$

In the absence of other random errors the ratio of these two quantities, $s_c^2/\bar{c}$ would be expected to be about unity. Large values of this ratio indicate excessive variation among the repeated counts, perhaps due to instrument drift or sample placements; values significantly less than one are equally suspicious and may indicate large dead-time losses, or other de-randomizing effects, that invalidate the assumption of Poisson statistics.

The quantity $\sum(c_i - \bar{c})^2/\bar{c}$ follows approximately the Chi-squared (χ^2) distribution with $(k-1)$ degrees of freedom. A table of percentiles of the Chi-squared distribution is readily available (e.g., Natrella, 1963), and the value for $\chi_{0.95}^2$ with $k - 1$ degrees of freedom can be used as a criterion to decide if the variance computed from the k counts is larger than that expected purely due to chance. The degree of freedom is the number of independent measurements used in the estimate of the variance.

Results of experiments, however, are not always expressed in counts, or counts per unit time, but are reduced to numbers such as activities per gram, etc. These numerical values are often treated as if they are distributed normally, with mean μ and standard deviation σ. Such an assumption affords mathematical convenience, and can often be shown to be valid to a satisfactory degree of approximation.

A normal (Gaussian) distribution is completely specified by its mean and standard deviation. For distributions other than normal, other additional parameters may be needed to specify properties of the distribution, e.g., skewness (third central moment) and peakedness (fourth central moment). While it is true that, in many applications, the normal distribution can be assumed and used as an adequate approximation, caution must be exercised to check for departure of data from normality. A simple, graphical test for normality can be made using normal-probability paper (Dixon and Massey, 1957).

The median of a set of n observations can also be used as a measure of location, and is defined as the value occupying the $[(n + 1)/2]$th position for n odd if the n values are arranged in an ascending sequence with $x_1 \leq x_2 \leq x_3 \ldots \leq x_n$. For n even, the median can be defined as the value midway between the $(n/2)$th position and the $[(n/2) + 1]$th position. A large difference between median and the mean indicates nonnormality, or the presence of outlying observations.

For small sets of normally distributed data, an estimate of standard deviation can be obtained by multiplying the range R (difference between largest and smallest) of these measurements by a constant (approximately equal to $1/\sqrt{n}$). Table 19 lists these constants corresponding to the number n of measurements in the set. Clearly the use

TABLE 19—*Estimate of σ from the range*[a]

n	Multiplying factor	$1/\sqrt{n}$
2	0.886	.707
3	0.591	.577
4	0.486	.500
5	0.430	.447
6	0.395	.408
7	0.370	.378
8	0.351	.354
9	0.337	.333
10	0.325	.316
12	0.307	.289
16	0.283	.250

[a]Adapted from Table 2-1, Natrella (1963).

of one set of three or four measurements gives a very uncertain estimate of the standard deviation. If there are k sets of n measurements each, the average range $\bar{R}$ can be used.

In summarizing a set of data, the following three quantities should be computed and presented:

1. The average, $\bar{x}$—the quantity of interest;
2. An estimate of the standard deviation, s—a measure of the precision of the measurements; and
3. The number of measurements, n—an indication of how well $\bar{x}$ and s are known as estimates of μ and σ.

These three quantities provide a useful summary of the information in the data, unless there are very discrepant values, usually called "outliers". "Outliers" may provide information on the cause of lack of control in the measurement procedure and should never be discarded, i.e., expunged from the record, because a later review of outliers may reveal the cause of their production. An outlier should not be excluded from computation of the mean without adequate physical explanation. Outliers that recur frequently should always be included in the computation of the standard deviation to avoid an over-optimistic evaluation of the precision of the measurements.

In calculating a grand average, $\bar{X}$, of a number of groups of measurements of the same quantity, each group having a mean $\bar{x}_i$ and a variance of the mean, $\mathrm{var}(\bar{x}_i)$, it may be desirable to take a weighted average when the variances of the mean of the different groups differ greatly. The reciprocal of the estimated variance is the proper weighting factor, so that the weighted average and variance are given by (Stevenson, 1966):

$$\bar{X} = \frac{\sum\limits_{i} \dfrac{\bar{x}_i}{\mathrm{var}(\bar{x}_i)}}{\sum\limits_{i} \dfrac{1}{\mathrm{var}(\bar{x}_i)}}, \qquad \mathrm{var}\,\bar{X} = \frac{1}{\sum\limits_{i} \dfrac{1}{\mathrm{var}(\bar{x}_i)}}. \tag{7.12}$$

This variance of $\bar{X}$ is the "internal" one, referring only to the variance of the sample due to fluctuations in decay. Stevenson (1966, p. 98) also shows how an "external" variance, the variability associated with errors other than counting, can be incorporated into a weighted average.

If it is assumed that the data are normally distributed to a close approximation, more can be said about the results. The confidence intervals for the mean and the statistical tolerance interval for a fixed percentage of the population are discussed in the next two sections.

7.3.2 *Confidence Interval for the Mean*

In practice, the exact values of the population parameters are never known; only sample estimates are available computed from one or more sets of data. It is postulated, however, that the larger the sample, the closer these estimates will approach to the true parameter values, provided there are no systematic errors in the sample values. For the population mean in particular, an interval about the observed $\bar{x}$ can be constructed that will cover μ_x with a specified probability, 0.95 or 0.99 for instance. This interval is called the *confidence interval for the mean*, and is often used as the term expressing limits to random measurement errors in the uncertainty statement of a reported value.

To construct a confidence interval *for the mean*, three quantities must be calculated; and to be sure of the probability level, i.e., the percentage of coverage in the long run when the same experiment is repeated a large number of times, two assumptions must be verified. The three quantities are:

$\bar{x}_n$: an average computed from n observations.

s_ν: a standard deviation based on ν degrees of freedom. If the same n observations are used to compute the standard deviation, the $\nu = n - 1$. If the standard deviation is pooled from k sets of data each with n observations, then $\nu = k(n - 1)$.

$t_{(1-\alpha),\nu}$: a tabulated value of the "t" statistic with entries ν and $(1 - \alpha)$, where $(1 - \alpha)$ is the confidence coefficient (or probability level), 0.99 for example, or a confidence level of 99 percent.

The two assumptions are:

(1) The observations $x_1, x_2, \ldots, x_n$ are independent in the statistical sense. A minimum requirement would be that a plot of these values in some meaningful order (e.g., run numbers, days, instruments, operators) does not show pattern, trend, or grouping.

(2) The observations are approximately normally distributed. A convenient way to check the validity of this assumption is to plot the observations on normal-probability paper and to see if these observations fall approximately on a straight line. (If the distribution is skewed, i.e., not symmetrical with respect to the mean, the probability level, $(1 - \alpha)$, could be seriously in error.)

The confidence interval *for the mean* is computed as:

$$\bar{x}_n - t_{(1-\alpha),\nu}\left(\frac{s_\nu}{\sqrt{n}}\right) \text{ to } \bar{x}_n + t_{(1-\alpha),\nu}\left(\frac{s_\nu}{\sqrt{n}}\right). \tag{7.13}$$

Sometimes $\bar{x}_n \pm t_{(1-\alpha),\nu}(s_\nu/\sqrt{n})$ is used to express this interval and the endpoints are termed confidence limits.

From the above expression it is seen that "n" is associated with $\bar{x}_n$ and the denominator $\sqrt{n}$, but "ν" is associated with $t_{(1-\alpha),\nu}$ and s_ν. Since various sets of measurements could have the same precision, the standard deviations calculated from these sets can be pooled together to yield a better estimate of σ.[21] The pooled standard deviation, s_p, could be used with any average to form the confidence interval provided the average is computed from a set of data derived from the same measuring process.

If duplicate determinations are made on each of k ampoules of different but similar activities, then pooled standard deviations from the k pairs of duplicates can be computed by a simplified formula as:

$$s_p^2 = \frac{1}{2k} \sum_1^k d_i^2 \tag{7.14}$$

where d_i is the difference of the duplicate readings of the ith ampoule, and s_p has k degrees of freedom. Then the confidence interval for the mean activity of the ith ampoule based on the duplicate determinations is:

$$\bar{x}_i - t_{(1-\alpha),k}\left(\frac{s_p}{\sqrt{2}}\right) \text{ to } \bar{x}_i + t_{(1-\alpha),k}\left(\frac{s_p}{\sqrt{2}}\right) \tag{7.15}$$

where $\nu = k$, but $n = 2$.

Since t is a constant and s is a number that converges to σ as a limit, the confidence interval would have zero width with n sufficiently large. That is to say, for a large number of measurements, $\bar{x}$ converges to μ, and the mean of the distribution is known almost exactly.

Many an experimenter feels uneasy with this situation where the confidence interval of his value shrinks toward an arbitrarily small interval while, in fact, he knows that his uncertainty is probably larger than these limits. This illusion, however, is probably due to any or all of the following causes:

(1) In precision measurements, it is typical that variability between sequences of measurements made on different occasions is larger than variability within a sequence of measurements. A sequence of measurements, $x_1, x_2, \ldots, x_n$, could depend on various envi-

---

[21] In general, the pooled standard deviation can be computed from k sets of data as:

$$s_p^2 = (\nu_1 s_1^2 + \nu_2 s_2^2 + \ldots + \nu_k s_k^2)/(\nu_1 + \nu_2 + \ldots + \nu_k)$$

with $\nu = \nu_1 + \nu_2 + \ldots + \nu_k$ degrees of freedom.

ronmental conditions, *e.g.*, time, temperature, barometric pressure, the particular instrument setting, etc. It follows, then, that the first assumption has been violated because of correlation within sets of measurements resulting in (a) an underestimate of σ, and more important, (b) an erroneous use of the divisor $\sqrt{n}$, since n is the number of independent measurements.[22]

(2) The confidence interval relates only to random errors, and not to systematic errors. Some systematic errors may be small when compared to the standard deviation of a single measurement, s_x, and hence cannot be detected, but may show up when averages of a large number of measurements are compared. These systematic errors would then be a dominant part of the uncertainty.

(3) The homogeneity of samples measured is always a factor that must be considered, since any minor differences between samples can only add to the uncertainty of the value reported. Limits to the inhomogeneity that may be inherent in the samples can be constructed using the concept of statistical tolerance limits discussed below.

7.3.3 *Statistical Tolerance Interval for a Fixed Percentage of Population*

Frequently, a large number of units (ampoules, bottles) is prepared, of which only a fraction can be measured. From this sample of units examined, it is hoped to find out, first, whether these units are sufficiently homogeneous for the purpose intended, and secondly, if these tested units are acceptable, what can one say about the expected variability within the entire lot (population) of units.

For simplicity, assume that the measurement error is negligible in comparison with the variability of the lot of units, and that the number of units is large. Then, under the same two basic assumptions listed

---

[22] Two cases are of particular interest:

a. If σ_b^2 is the component of variance corresponding to between sequences of measurements, and σ_w^2 that within a sequence of measurements, then,

$$\sigma_{\bar{x}}^2 = \sigma_b^2 + \frac{\sigma_w^2}{n}.$$

b. If successive measurements are correlated with average coefficient of correlation r, then,

$$\sigma_{\bar{x}}^2 = \frac{\sigma^2}{n}\{1 + (n-1)r\}.$$

(See page 58 of Brownlee, 1960.)

in Section 7.3.2, i.e., independence and normality, an interval can be constructed to cover at least p percent of all the units in the population with the desired probability $\gamma = (1 - \alpha)$. This statistical tolerance interval is computed from a random sample of n units as:

$$\bar{x}_n - Ks \text{ to } \bar{x}_n + Ks \qquad (7.16)$$

where K is the tabulated value of the two-sided tolerance factor for the specified p and γ. Note that, when n is large and s approaches the standard deviation of the population, the tolerance interval approaches to an interval of constant width, as expected.

Some selected values of "t" and "K" for $\gamma = 1 - \alpha = 0.95$ and 0.99 are given in Table 20 for ready reference.

If both analytical (measurement) error and material variability are present, then standard deviations corresponding to each of the two sources of error have to be estimated from the data. In the simplest case, duplicate determinations can be made on n units selected at

TABLE 20—*t factors for confidence interval and K factors for tolerance interval*[a]

Degrees of freedom	Confidence interval, t Factor[b] $1 - \alpha$		Tolerance interval, K Factor[b]			
			$\gamma = 0.95$		$\gamma = 0.99$	
	0.95	0.99	p = 0.95	p = 0.99	p = 0.95	p = 0.99
1	12.706	63.657	37.674	48.430	188.491	242.300
2	4.303	9.925	9.916	12.861	22.401	29.055
3	3.182	5.841	6.370	8.299	11.150	14.527
4	2.776	4.604	5.079	6.634	7.855	10.260
5	2.571	4.032	4.414	5.775	6.345	8.301
6	2.447	3.707	4.007	5.248	5.488	7.187
7	2.365	3.499	3.732	4.891	4.936	6.468
8	2.306	3.355	3.532	4.631	4.550	5.966
9	2.262	3.250	3.379	4.433	4.265	5.594
10	2.228	3.169	3.259	4.277	4.045	5.308
15	2.131	2.947	2.903	3.812	3.421	4.492
20	2.086	2.845	2.723	3.577	3.121	4.100
25	2.060	2.787	2.612	3.432	2.941	3.865
30	2.042	2.750	2.536	3.332	2.820	3.706
60	2.000	2.660	2.329	3.061	2.500	3.285
∞	1.960	2.576	1.960	2.576	1.960	2.576

[a] Adapted from Tables A-4 and A-6, Natrella (1963).

[b] The t and K factors are for "two-sided" limits based on a normal distribution. The term "two-sided" means that both over and under limits from the average are of interest. Some tables give "one-sided" factors, in which case the use of $\gamma/2$ instead of α yields the same result. (Such t-factor tables are sometimes called "Percentiles of the t distribution.")

random from the lot. The standard deviation s_a due to analytical error can be computed from the replicate measurements as:

$$s_a^2 = \frac{1}{2n} \sum_1^n d_i^2 \qquad (7.17)$$

where d_i represents the difference between duplicate measurements of the ith pair, and that for the material variability, s_m, as:

$$s_m^2 = s^2 - \frac{s_a^2}{2} \qquad (7.18)$$

where s is the standard deviation computed from the n averages of duplicates.

The value of $\bar{x}_n$ may be used in two distinct ways:
(a) to characterize the true mean of the whole lot, or
(b) to characterize the bulk (e.g., 95 or 99 percent) of measured values of units from the lot.

In the first case, the concept of confidence interval applies, i.e., if n samples are taken at random repeatedly from the lot, are measured, and $\bar{x}_n$ computed, $\bar{x}_n \pm ts/\sqrt{n}$ is expected to include the true mean of the lot at a specified confidence level, for example, 95 percent.

In the second case, the concept of tolerance interval applies, i.e., if all units were measured, then the interval $\bar{x}_n \pm Ks$ is expected to include at least p percent of all such units, say 99 percent, with a confidence level of 95 percent. In either case, the number of degrees of freedom associated with s is $n - 1$, since s is computed from n values.

It is to be noted that by measuring more units, the confidence interval will decrease in size proportional to $\sqrt{n}$, but the tolerance limits will approach a constant width, representing the variability inherent in the units within the lot.

The separation of variability into its components corresponding to the various sources of possible error (e.g., the model used for ^{90}Sr activities given in Section 7.2) demands careful planning and arrangement of the measurements schedule, and can be facilitated by statistical design of experiments.

7.3.4 *Formulae for the Propagation of Errors*

Frequently one cannot measure the quantity of interest directly, but must derive the value of this quantity as a function of other measurable quantities. If W is the quantity of interest, and X and Y are the quantities to be measured, then the problem is to determine the error in the computed value w, of W, that is due to the measurement errors in $\bar{x}$ and $\bar{y}$.

The results in the measurements of X and Y can be expressed by their averages $\bar{x}$ and $\bar{y}$, and standard deviations of these averages $s_{\bar{x}}$ and $s_{\bar{y}}$, respectively. If the estimate of W is computed from

$$\hat{W} = F(\bar{x}, \bar{y}), \tag{7.19}$$

then the variance of $\hat{W}$ is approximately:

$$\text{var } \hat{W} = \left[\frac{\partial F}{\partial X}\right]^2 \frac{\sigma_{\bar{x}}^2}{n} + \left[\frac{\partial F}{\partial Y}\right]^2 \frac{\sigma_{\bar{y}}^2}{n} + 2\left[\frac{\partial F}{\partial X}\right]\left[\frac{\partial F}{\partial Y}\right]\frac{\sigma_{\bar{x}\bar{y}}}{n} \tag{7.20}$$

where the partial derivatives in square brackets are to be evaluated at $\bar{x}$ and $\bar{y}$. If $\bar{x}$ and $\bar{y}$ are statistically independent, $\sigma_{\bar{x}\bar{y}} = 0$, and, therefore the last term equals zero. If $\bar{x}$ and $\bar{y}$ are measured in pairs, $s_{\bar{x}\bar{y}}$ can be used as an estimate of $\sigma_{\bar{x}\bar{y}}$. Also $\sigma_{\bar{x}}^2$ and $\sigma_{\bar{y}}^2$ are not known in practice and $s_{\bar{x}}^2$ and $s_{\bar{y}}^2$ are usually used in the formula in their place.

The above formula is called the propagation of error formula and is based on the first order Taylor series approximation of the function form. Formulae for some simple functions are shown in Table 21.

In these formulae, if
(i) the partial derivatives when evaluated at the averages are small, and
(ii) $\sigma_{\bar{x}}$ and $\sigma_{\bar{y}}$ are small compared to $\bar{x}$ and $\bar{y}$,
then the approximations are good and $\hat{W}$ tends to be distributed normally with var $\hat{W}$ given by Eq. 7.20.

7.3.5   Linear Relationships and Calibration Curves

By use of the arithmetic mean of n measurements as an estimate of the population mean, in fact, a constant has been fitted to the data by the method of least squares, i.e., a value $\hat{m}$ has been selected for μ such that

$$\sum_1^n (y_i - \hat{m}) = \sum_1^n d_i^2 \tag{7.21}$$

is a minimum. The solution is $\hat{m} = \bar{y}$. The deviations $d_i = y_i - \hat{m} = y_i - \bar{y}$ are called residuals.

The measurements can be expressed in the form of a statistical model

$$y = \mu + \epsilon, \tag{7.22}$$

TABLE 21—*Propagation of error formulae for some simple functions*[a]

Function form of $\hat{w}$[b]	Approx. formula for var ($\hat{w}$) (x and y are assumed to be statistically independent)	Term to be added if x and y are correlated, and a reliable estimate of $\sigma_{\bar{x}\bar{y}}$, $s_{\bar{x}\bar{y}}$, can be assumed
$A\bar{x} + B\bar{y}$	$A^2 s_{\bar{x}}^2 + B^2 s_{\bar{y}}^2$	$2AB s_{\bar{x}\bar{y}}$
$\left(\dfrac{\bar{x}}{\sigma_{\bar{x}}^2} + \dfrac{\bar{y}}{\sigma_{\bar{y}}^2}\right) \Big/ \left(\dfrac{1}{\sigma_{\bar{x}}^2} + \dfrac{1}{\sigma_{\bar{y}}^2}\right)^c$	$1 \Big/ \left(\dfrac{1}{\sigma_{\bar{x}}^2} + \dfrac{1}{\sigma_{\bar{y}}^2}\right)$	
$\dfrac{\bar{x}}{\bar{y}}$	$\left(\dfrac{\bar{x}}{\bar{y}}\right)^2 \left(\dfrac{s_{\bar{x}}^2}{\bar{x}^2} + \dfrac{s_{\bar{y}}^2}{\bar{y}^2}\right)$	$\left(\dfrac{\bar{x}}{\bar{y}}\right)^2 \left(-2\dfrac{s_{\bar{x}\bar{y}}}{\bar{x}\bar{y}}\right)$
$\dfrac{1}{\bar{y}}$	$\dfrac{s_{\bar{y}}^2}{\bar{y}^4}$	
$\dfrac{\bar{x}}{\bar{x}+\bar{y}}$	$\left(\dfrac{\hat{w}}{\bar{x}}\right)^4 (\bar{y}^2 s_{\bar{x}}^2 + \bar{x}^2 s_{\bar{y}}^2)$	$\left(\dfrac{\hat{w}}{\bar{x}}\right)^4 (-2\bar{x}\bar{y}s_{\bar{x}\bar{y}})$
$\dfrac{\bar{x}}{1+\bar{x}}$	$\dfrac{s_{\bar{x}}^2}{(1+\bar{x})^4}$	
$\bar{x}\bar{y}$	$(\bar{x}\bar{y})^2 \left(\dfrac{s_{\bar{x}}^2}{\bar{x}^2} + \dfrac{s_{\bar{y}}^2}{\bar{y}^2}\right)$	$(\bar{x}\bar{y})^2 \left(2\dfrac{s_{\bar{x}\bar{y}}}{\bar{x}\bar{y}}\right)$
$\bar{x}^2$	$4\bar{x}^2 s_{\bar{x}}^2$	
$\sqrt{\bar{x}}$	$\dfrac{1}{4}\dfrac{s_{\bar{x}}^2}{\bar{x}}$	
$\ln \bar{x}$	$\dfrac{s_{\bar{x}}^2}{\bar{x}^2}$	
$k\bar{x}^a\bar{y}^b$	$\hat{w}^2 \left(a^2 \dfrac{s_{\bar{x}}^2}{\bar{x}^2} + b^2 \dfrac{s_{\bar{y}}^2}{\bar{y}^2}\right)$	$(\hat{w})^2 \left(2ab\dfrac{s_{\bar{x}\bar{y}}}{\bar{x}\bar{y}}\right)$
$e^{\bar{x}d}$	$e^{2\bar{x}}s_{\bar{x}}^2$	
$\dfrac{\sin\left(\dfrac{\bar{x}+\bar{y}}{2}\right)}{\sin\dfrac{\bar{x}}{2}}$	$\dfrac{1}{4}\left\{ \dfrac{\cos^2\left(\dfrac{\bar{x}+\bar{y}}{2}\right)}{\sin^2\dfrac{\bar{x}}{2}}s_{\bar{y}}^2 + \dfrac{\sin^2\dfrac{\bar{y}}{2}}{\sin^4\dfrac{\bar{x}}{2}}s_{\bar{x}}^2 \right\}$	$-\dfrac{\sin\dfrac{\bar{y}}{2}\cos\left(\dfrac{\bar{x}+\bar{y}}{2}\right)}{2\sin^3\dfrac{\bar{x}}{2}}s_{\bar{x}\bar{y}}$
	($s_{\bar{x}}$ and $s_{\bar{y}}$ in radians)	
$100\dfrac{s_{\bar{x}}}{\bar{x}}$ (= coefficient of variation in percent)	$\dfrac{\hat{w}^2}{2(n-1)}$ (not directly derived from the formulae)[e]	

[a] From Ku (1966).

[b] It is assumed that the value of $\hat{w}$ is finite and real, e.g., $\bar{y} \neq 0$ for ratios with $\bar{y}$ as denominator, $\bar{x} > 0$ for $\sqrt{\bar{x}}$ and $\ln \bar{x}$.

[c] Weighted mean as a special case of $A\bar{x} + B\bar{y}$, with σ_x and σ_y considered known.

[d] Distribution of $\hat{w}$ is highly skewed and normal approximation could be seriously in error for small n.

[e] See, for example, *Statistical Theory with Engineering Applications*, p. 301, by A. Hald (John Wiley & Sons, New York, N.Y., 1952).

where y stands for the observed values, μ the population mean (a constant), and ϵ the random error (normal) of measurement with a population mean, μ_ϵ, equal to zero and with a standard deviation σ. It follows that

$$\mu_y = \mu + \mu_\epsilon = \mu \tag{7.23}$$

and

$$\sigma_y^2 = \sigma^2.$$

The method of least squares requires us to use that estimator $\hat{m}$ for μ such that the sum of squares of the residuals is a minimum. As a corollary, the method also states that the sum of squares of residuals divided by the number of measurements, n, less the number of fitting constants, k, will give us an estimate of σ^2, i.e.,

$$s^2 = \frac{\Sigma(y_i - \hat{m})^2}{n - k}. \tag{7.24}$$

It is seen that for $k = 1$, the above agrees with the definition of s^2.

Suppose Y, the quantity measured, exhibits a linear functional relationship with a variable that can be controlled accurately, then a model can be written as:

$$y = a + bX + \epsilon, \tag{7.25}$$

where, as before, y is the quantity measured, a (the intercept) and b (the slope) are two constants to be estimated, and ϵ the measurement error with a mean equal to zero, and variance σ^2 not depending on X. X is set at x_i, and y_i is observed. For example, y_i might be the natural logarithm of counting rates observed in a series of measurements on n successive days designated as x_i. In this case the quantity of interest is the decay constant, λ, which is the slope of the line.

For any estimates of a and b, say $\hat{a}$ and $\hat{b}$, a value $\hat{y}_i$ can be computed for each x_i, thus,

$$\hat{y}_i = \hat{a} + \hat{b}x_i.$$

If the sum of squares of the residuals

$$\sum_{i=1}^{n} (y_i - \hat{y}_i)^2$$

are to be a minimum, then it can be shown that

$$\hat{b} = \frac{\displaystyle\sum_{i=1}^{n} (x_i - \bar{x})(y_i - \bar{y})}{\displaystyle\sum_{i=1}^{n} (x_i - \bar{x})^2} \tag{7.26}$$

and

$$\hat{a} = \bar{y} - \hat{b}\bar{x}. \tag{7.27}$$

The variance of Y can be estimated by

$$s^2 = \frac{\Sigma(y_i - \hat{y}_i)^2}{n - 2}, \tag{7.28}$$

with $n - 2$ degrees of freedom since two constants have been estimated from the data.

The standard errors of $\hat{b}$ and $\hat{a}$ are respectively estimated by $s_{\hat{b}}$ and $s_{\hat{a}}$, where

$$s_{\hat{b}}^2 = \frac{s^2}{\Sigma(x_i - \bar{x})^2}, \tag{7.29}$$

$$s_{\hat{a}}^2 = s^2 \left[\frac{1}{n} + \frac{\bar{x}^2}{\Sigma(x_i - \bar{x})^2} \right]. \tag{7.30}$$

With these estimates and the degrees of freedom associated with s^2, confidence limits can be computed for $\hat{a}$ and $\hat{b}$ for the confidence coefficient selected if it is assumed that errors are normally distributed.

Thus, the lower and upper limits of a and b, respectively, are:

$$\hat{a} - ts_{\hat{a}}, \quad \hat{a} + ts_{\hat{a}} \tag{7.31}$$
$$\hat{b} - ts_{\hat{b}}, \quad \hat{b} + ts_{\hat{b}}$$

for the value of t corresponding to the degree of freedom and the selected confidence coefficient.

There are many variations to the problem of fitting a function to a set of data, viz., curvilinear relationships, weighted least squares, nonlinear least squares, etc. For a comprehensive treatment on these subjects, Draper and Smith (1968) is one of the best references available.

7.4 Statistical Experiment Design[23]

The definition of an experiment as "a considered course of action aimed at answering one or more carefully framed questions" was coined by the late W. J. Youden, an expert in the theory and practice of the design of experiments. Eleven of his papers on designs useful in physical sciences, including inter-laboratory tests, are reprinted in Ku (1969).

[23] See also Chapter 11 of Natrella (1963).

There are certain characteristics an experiment must have in order to accomplish the defined objective. These essential requisites are:

(1) As far as possible, the effects of the factors selected for study should not be obscured by other variables. The use of an appropriate experimental pattern helps to free the comparisons of interest from the effects of variables not under experimental study, and simplifies the analysis of results.

(2) As far as possible, the experiment should be free from bias. The effects of known variables may be balanced out by planned grouping and effects of unsuspected variables may be given a chance to play their parts through randomization.

(3) An experiment should provide a measure of precision. Replicated measurements provide a measure of precision. The standard deviation computed from replicated results may be compared to that of counting statistics to find out whether the precision of experiment can be improved. Planned grouping may also aid to trace the source of undesired or excess variability.

Since each experiment has its own individual characteristics and constraints, consultation with an experienced statistician who understands the problem is perhaps the prudent way to achieve an optimal design.

The Youden Diagram, or Plot, is given below as an illustration of a simple, but highly effective, procedure used in inter-laboratory tests. The name "Youden Plot" is used universally to specify not only the plotting technique, but also the experimental procedure for sampling the respective performance of the laboratories through the results obtained on paired test items. Suppose the accuracy of radiation measurement detectors used in standard laboratories or hospitals is to be checked; two samples, A and B, are prepared at different levels of activity but reasonably close to the level of interest. Identical samples of A and B are sent to each laboratory and the activities of the pair are to be measured as unknowns. The results reported by the participating laboratories are used to prepare a graph.

In Figure 100, the value of A is plotted in the horizontal direction and that of B in the vertical direction. Thus each laboratory is represented by one point. The known values of A and B are used as the origin, and lines through the origin parallel to the x and y axes divide the graph into four quadrants as shown. The points that fall in the four quadrants are, respectively,

Q_1: high in A, high in B
Q_2: high in A, low in B
Q_3: low in A, low in B
Q_4: low in A, high in B.

Fig. 100. A Youden Diagram. Each point represents values of samples A and B obtained by a laboratory. The results of 15 laboratories are shown here, divided into four quadrants.

If there are no systematic biases among the laboratories, all points should scatter about the center and divide equally among the four quadrants. In practice, however, this ideal picture is never realized on the first try, but can be achieved approximately after various causes of troubles are removed. A majority of points usually clusters about the 45° line drawn through the origin, and for extremely precise measurements, such as those on the voltage of standard cells, the points fall almost exactly on the line. An ellipse with the major axis on the 45° line can be constructed to enclose practically all the points in most cases, indicating the existence of some systematic errors among laboratories.

Laboratory L_1 will receive a copy of the diagram with a circle around its point. Immediately, this laboratory knows that its values are unreasonably high as compared with the standards, and that corrective measures have to be taken. L_3 is assured that its detector is in order, whereas L_2 has to improve its measurement procedure and locate the causes of its inconsistent value.

The paper and rubber industries have adopted this procedure and have used it since 1969 for routine surveillance of their laboratory measurements procedures. The samples are prepared at NBS and computerized reports are distributed monthly to more than 100 laboratories. For simplicity in concept and execution, and for ready and meaningful interpretation, this inter-laboratory test design cannot be surpassed. The original paper by Youden was published in *Industrial Quality Control*, 1959, and reprinted as paper 3.1 in Ku (1969).

Many of Youden's other designs are presented in papers reprinted in the same NBS Special Publication (Ku, 1969), including one on "Comparison of four national radium standards" (original references: Davenport *et al.*, 1954; Connor and Youden, 1954), and one on "Design and statistical procedures for the evaluation of an automatic gamma-ray point-source calibration" with W. B. Mann and S. B. Garfinkel (original reference: Garfinkel *et al.*, 1966). A quick and economical procedure in locating sources of trouble, e.g., to check on seven sources in eight determinations, has been described by Youden (1963).

7.5 Precision, Systematic Error, and Uncertainty

In the above sections, the terms precision and systematic error have been used, without specific definitions as to their meanings. This section brings into focus the meanings of these two terms and clarifies their connotations in usage.

The term precision refers to the closeness of a set of observations, (data) among themselves, being a sample of a large sequence of such observations that can be generated by the same process. Individual values of any of the observations cannot be predicted, yet collectively the scatter of the set can be numerically characterized by a measure of precision (strictly, of imprecision), e.g., the standard deviation σ if known, or its estimate s.

The calculated value of s, as an estimate of the standard deviation σ, is a function of the sample values $x_1, x_2, \ldots, x_n$, and therefore is a statistic. For repeated sets of data, each containing n values, these calculated standard deviations will not usually be the same, but will vary as individuals from a population of such standard deviations so generated, and their properties may be predicted.

Systematic error, on the other hand, cannot be estimated or calculated from the data themselves, as each and every datum point is affected by the systematic error in the same way. The magnitude and direction of the systematic error remain the same in the average of these data points as in each individual point, and are not reduced by taking the average, as is the case with standard deviations, where the standard deviation of the average is $1/\sqrt{n}$ times the standard deviation of individual values.

If the instrument used is suspected to be a source of systematic error, the experiment may be repeated with another instrument. If the confidence intervals constructed about the two results do not overlap, the presence of a systematic error in one instrument with respect to the other is indicated.

If the second instrument used is a standard and known to be accurate, then an estimate can be made for the systematic error of the first instrument as the difference between the two averages. If the two instruments are claimed to be equally good, the further course of action is not clear. The instruments may have to be recalibrated to estimate the systematic uncertainty in each instrument.

Suppose a number of instruments is available. Then the results obtained by using these instruments may be construed as a sample from a population of such results obtainable with the use of such instruments. Accordingly, a standard deviation can be computed from these results and a statistical tolerance interval used to give bounds to the uncertainty in the result of any one of the instruments.

What has been said about instruments applies equally well to other factors involved in any experiment: operators, reagents, time between sampling and analysis, size of sample, etc. Obviously a definitive program to establish the effects of all these factors is out of the question for most laboratories in view of the time and cost involved. Nevertheless, statistical designs are available that use a minimum number of experimental observations to check on the gross effects of these factors. Preferably, these experiments are incorporated in the course of regular routine so that if all of these effects are negligible, no effort is wasted. Youden (1963) gives an example in which a design is suggested to check seven factors each at two levels using only eight determinations.

An immediate extension of the above example would be to different laboratories, then to different test methods, to different national standards, and finally to fundamental physical constants. It follows then that the use of the terms "precision" and "systematic error" must be conditioned by the frame of reference, and these terms cannot be defined in a single way applicable to all situations. The precision of a test method [called "reproducibility" by some ASTM committees (American Society for Testing Materials)] used by many laboratories must include in its measure of variability a component due to differences among laboratories, as well as the component due to factors within each individual laboratory (called repeatability by some ASTM committees). Thus, a systematic error, peculiar to a laboratory, behaves as if it were a random error when the results of two test methods used by all of a large number of qualified laboratories are under investigation. When there are only a few laboratories, systematic differences among them should be treated as systematic error while an effort is made to remove the cause.

Proceeding in the opposite direction, the random measurement uncertainty on a particular standard calibrated by a national laboratory becomes a systematic error for the laboratory receiving the standard, to be used as if the assigned value were exact and correct. Since all related measurements will inherit the same deviation attached to that standard, the uncertainty given for the standard is a systematic error common to all the laboratory's results, although perhaps of negligible magnitude.

If a measure of the magnitude of the systematic error can be obtained or verified by an experiment, this should always be done as indicated in Section 7.4. If time and cost prohibit such an experiment, then some reasonable bounds to the systematic error will have to be estimated through experience or judgment. Since different experimenters will use different criteria, there is no way to give guidelines for such estimates. Moreover, the combined effect of two or more such sources of error may exceed the sum total of individual effects. For a detailed discussion on systematic errors, see Eisenhart (1963), reprinted in Ku (1969).

The uncertainty of a reported value refers to its likely inaccuracy in terms of credible limits, and combines both (1) components based on data that are amenable to statistical treatments and (2) components due to systematic errors that cannot be treated statistically. Depending on the circumstance in which the value is likely to be used, the uncertainty statement should be phrased accordingly so that it will be understandable and useful to the user. For example, in ICRU Report 12 (ICRU, 1968) the overall uncertainty is recommended to be expressed as $\pm (ts_{\bar{x}} + \delta)$ where $ts_{\bar{x}}$ is the confidence interval for the mean at a specified probability level, and δ the linear sum of estimated maximum values of conceivable systematic errors. Eisenhart (1968) provides further guidelines on the expression of the uncertainties of final results, reprinted in Ku (1969). The question of the statement of uncertainty is further discussed in Section 8.4.

8. Measurements Assurance, Standards, Traceability and the Statement of Uncertainty

8.1 General

In a sense, this is the motherhood and apple-pie (MAP) section of this report. But here it is called Measurements Assurance Programs. The ability to reproduce, whether with the help of genes or gauges, is important in all fields of human endeavor, but uniformity is as often mistaken for accuracy as it is for excellence in apple pie. To create and reproduce and then, from that base, further to create is the key to all human progress, and measurement is the foundation on which all scientific progress to the full comprehension of life and the universe is built. Cooperative endeavor and progress in practically every field of science depend on the ability to communicate experience in terms of common measure and mutual comprehension between scientific research groups throughout the world. Diverse systems of units can impede such a free exchange of ideas and information. Before metrication was enacted in Canada, the gallon changed in size as one went there from the United States. This could lead to some embarrassment, as anyone who has poured an imperial half pint of milk into (and overflowing) a U.S. half-pint glass will, perhaps, remember. To remove impediments to world trade, standards of length and mass were defined, often different in different parts of the world, but eventually related to each other by intercomparative measurement. In science, the metric system, based on the defined meter, kilogram, and second, became the common denominator for the normalization of scientific data. And because these were defined standards, they were also absolute because they, in themselves, had no error and errors arose only in comparing other lengths, masses, or intervals of time with them.

On the economic side of the world's trade, money became the measure of effort that went into the production of a kilogram of butter, or a square meter of woven cloth. It is not just by chance that in many countries the standardizing laboratories were set up under the aegis of departments or ministries of trade or commerce.

In the case of radioactivity, early standards consisted of masses of pure radium in a salt of that element, often radium chloride or radium bromide, with the mass of radium ranging from about 1 to 100 mg. The international radium standards were kept at the Bureau International des Poids et Mesures (BIPM) at Sèvres. In 1959, however, it was agreed that the relative values of a set of radium standards prepared by O. Hönigschmid in 1934 were so well known that the whole system of Hönigschmid radium standards distributed in many countries of the world could be considered to be the international standard.

In 1910 the *curie* was defined as the quantity of radon that is in equilibrium with one gram of radium. A curie of radon consists of about 0.65 μl of gas at 0 °C and 760 torr (760 mmHg, 0.1013 MPa). Subsequently, measurements were made to obtain the alpha-particle activity of a curie of radon and the first measurements yielded a result of 3.7×10^{10} s^{-1}. Later values tended to give a value for the curie of around 3.608×10^{10} s^{-1}. The best current value is 3.658×10^{10} s^{-1}.

With the advent of nuclear energy, it became necessary to have an activity standard applicable to any radionuclide. Accordingly in 1950, at meetings of the Joint Commission on Standards, Units, and Constants of Radioactivity of the International Unions of Pure and Applied Chemistry and Physics and at the Sixth International Congress of Radiology, it was decided to redefine the curie to be applicable to all radionuclides at a value of exactly 3.700×10^{10} transitions per second. The millicurie and the microcurie are, respectively, 3.700×10^7 and 3.700×10^4 s^{-1}.

In 1975 a new unit of radioactivity was introduced to be consistent with the Système International d'Unités (SI) (BIPM, 1973; NBS, 1977), in which the unit of choice is in terms of one kilogram, one meter, or one second. In the case of radioactive transitions, the unit of activity was named the becquerel (Bq), which is an activity of one transition per second.

In many branches of technology, however, the curie, or multiples thereof, has become so well established in the last thirty years that it is unlikely to be immediately replaced. Thus, in medicine, the millicurie is as commonplace to the nuclear-medicine practitioner as a cup of sugar is to the housewife. And in the environment, picocurie per liter seems to be becoming as well known to the lay public as it is to those who measure radioactivity around nuclear-power plants or nuclear-fuel-processing plants. There is also a considerable literature based on the curie that has been produced over many years in several fields of science.

8.2 The Role of Standards in the Measurement of Radioactivity

The primary role of a radioactivity standard, or of any standard, is to enable the result of a measurement to be communicated as a quantity, in terms of a number and a unit, to users or to other workers in the field. A radioactivity standard is a suitable piece of radioactive material, the quantity of any given radionuclide which is certified in terms of the average number of decays per second, or becquerels, to within stated limits of uncertainty. The standard must be accompanied by a certificate giving the radionuclide activity, or activities if more than one radionuclide, to within *stated* limits of uncertainty, and characterizing all other important parameters, for example, chemical and physical form, mass, and chemical, radiochemical and radionuclide impurities (and their amounts, if any).

Such standards can then be used to calibrate instruments in order to use these instruments to quantify the activities of the same radionuclide (or sometimes others), in suitable configurations.

Such was the role of radioactivity standards described in the opening scientific paper at the Herceg Novi, Yugoslavia, International Summer School (Cavallo *et al.*, 1973), as follows:

"As in other fields of scientific endeavor, radioactivity standards are needed to relate observed effects in terms of a well-defined and known stimulus. Thus, if a known flux density of alpha particles, defined in terms of the standards of length and time, of known energy, defined in terms of the standards of mass, length, and time, bombard a surface of beryllium, how many neutrons are emitted in unit time per unit volume? For a given alpha-particle energy, and knowing the number of beryllium atoms per unit volume, this experiment reduces to a cross-section measurement for the reaction $^9Be(\alpha,n)^{12}C$. Through the use of standards, a cross section obtained in Vinča can be compared with one measured in Chalk River, without the necessity of moving equipment, or people, from one place to another. Biological and physiological effects in response to a given applied radioactivity can also be related, one to another, through standards of that activity that are used to calibrate the measuring equipment used by the different investigators. But all of this implies a continuing availability of radioactivity standards both nationally and internationally. By and large, standards of most radionuclides of interest can be produced, but to keep a continuous supply available at all times, including standards of the shorter-lived activities is beyond the production capacity of the national and commercial laboratories in most countries. Therefore, throughout the world, the need for radioactivity standards has been met by calibrated instruments of greater or lesser degrees of complexity, all of which are normalized to the national or international radioactivity measurements systems by calibrations, by means of radioactivity standards, over a wide spectrum of radionuclides with widely different modes of decay. Thus, the national standard of ^{131}I in the U.S.A. is, for the periods between recalibrations, a "4π"γ ionization chamber and a sealed radium reference source at NBS, and any laboratory that wishes to check the calibration of its own measuring equipment for ^{131}I can, at any time, send a solution of that radionuclide to NBS for assay. In turn, NBS has assayed ^{131}I distributed internationally in order to be assured that the national system conforms with the international system."

It is of interest to note that, in the early days of investigating the properties of, and effects caused by, newly discovered radionuclides at the Radiation Laboratory at Berkeley, the most popular instrument used for measuring relative radioactivity in the late 1930's was the Lauritsen electroscope. Thus, the biological effects observed when a radionuclide was administered to experimental animals were often related to an activity measured as a drift rate of the electroscope, in divisions per second, for the given radionuclide in a specified geometry relative to the electroscope. The half lives of the newly discovered radionuclides were invariably measured with such instruments. The Lauritsen electroscope, the principle of which is incorporated into many pen-type dosimeters, is still a relatively inexpensive, accurate, and rugged asset to a radioactivity measurements laboratory (see Price, 1964; Mann *et al.*, 1980).

Radioactivity standards are ephemeral; unlike many other standards, they are not calibrated masses of, say, platinum that can be stored in some safe place for future use. If a calibration error is made in the assay of, say, ^{99m}Tc or ^{198}Au, administered to a patient, there may be no evidence of this months later, in the event of legal or regulatory proceedings. It is therefore of the greatest importance to have the highest assurance of continuous quality in the measuring processes and procedures at all and every stage. The steps taken to achieve this are described in the next section.

8.3 Measurements Assurance, Quality Control and Traceability

The role of Measurements Assurance Programs at the National Bureau of Standards and the concept of Traceability were briefly described in the introduction to a paper given at an International Atomic Energy Agency (IAEA) seminar in 1981 (Mann *et al.*, 1981), as follows:

"One of the earliest, if not the earliest, formal Measurement Assurance Program carried out by the National Bureau of Standards (NBS) is that described by Paul Pontius in 1966, in an NBS Technical Note entitled "The Measurement Philosophy of the Pilot Program for Mass Measurement" (Pontius, 1966). This program was carried out in conjunction with three military laboratories, namely the Calibration Laboratory of the White Sands Missile Range, the Navy Eastern Standards Laboratory, and the Calibration Laboratory of Redstone Arsenal, in order to establish consistency between their mass measurements and the mass standards of the National Bureau of Standards. In order to achieve this, NBS supplied each laboratory with two calibrated one-kilogram reference standards. Each laboratory was then asked to calibrate one set of weights that were multiples of the kilogram (e.g. 1 kg to 20 kg) and one set of weights that were sub-multiples of the kilogram (e.g. 1 kg to 10 g), in terms of the two NBS reference standards, and to submit these two sets to NBS for confirmatory calibration. These exercises could then be repeated at such intervals of time as were deemed necessary to achieve satisfactory quality control of each laboratory's measurements of mass. Pontius, in his 1966 Technical Note makes two cogent statements, namely, (i) that "It is necessary to know that a particular process is capable of operating in a state of control, and that, at any given time, its operation remains in a state of control," and (ii) that "Facilities that can demonstrate a continuous 'in control' operation through the use of the Pilot Program are, in essence, extensions of the NBS facilities and, as such, require only minimal calibration support."

Briefly it may therefore be stated that the purpose of "Quality Assurance" in any production process is to provide the necessary assurance that, within certain well specified limits, that process shall remain in control.

Such is the purpose of the so-called traceability programs provided by NBS in the various fields of radioactivity measurements, as, for example, in the practice of nuclear medicine, in the production of nuclear power, and in the monitoring of the environment. Traceability to NBS in radioactivity measurements is, however, less simple to achieve in practice than in the field of mass where two calibrated one-kilogram standards suffice for long periods of time, and calibrations of the subsets of masses may be carried out over a period of four or five months. In radioactivity, NBS provides standards or calibrations for some seventy different radionuclides, and intercomparative measurements with other laboratories may have to be carried out in the order of a few days, as, for example, in the case of the widely used radiopharmaceutical ^{99m}Tc that has a half life of 6 hours....

The term "traceability" is not yet wholly accepted for use in the vocabulary of science. But it is not too inappropriate a term when one considers the use of the word "trace" in "tracing one's ancestry" or "tracing our footprints back along their track." The literature of science has often adapted everyday words to its own special use. ... [In the discipline of radionuclide metrology we] use this term to describe a demonstrable hierarchical relationship to the national and international fountain-heads[2] of the standards of measurement.

The NBS traceability programs in radioactivity, at levels of activity ranging from environmental up to multimillicurie levels such as are common in radiopharmaceuticals, are possibly even more pragmatic than those designed for traceability in measurements of mass. In demonstrating traceability to NBS in its radioactivity-measurement processes we do not require that a laboratory use NBS standards to calibrate its measuring equipment; it is merely required to show that it can independently assign a value of activity to an amount of radionuclide in a sample, which value shall not differ from that obtained by NBS for the same sample by more than a specified range of, say, five or ten percent. The reason for not requiring the use of NBS standards is that (i) such standards are not always available from NBS, given the many species of radionuclides in common use, (ii) the laboratory's measuring systems should be maintained "in control" over large periods of time by the use of long-lived radioactive reference sources as monitors, (iii) many laboratories have their own in-house capability of performing direct measurements of activity by, for example, coincidence counting, and (iv) many good-quality standards of radioactivity are available from commercial suppliers that are themselves traceable to NBS through appropriate quality-assurance programs."

[2] To trace an error to its fountain-head is to refute it (Jeremy Bentham, 1748–1832).

In the United States scientists are trying to achieve a nationwide system of radioactivity measurements that are in agreement with the national standards maintained by the National Bureau of Standards (NBS). For this purpose, laboratory standards prepared both by the Las Vegas Environmental Monitoring Systems Laboratory (EMSL) of the Environmental Protection Agency (EPA), and by the Idaho Falls Radiological and Environmental Services Laboratory (RESL) of the Department of Energy (DOE), under contract to the Nuclear Regulatory Commission (NRC), have been measured and certified by NBS as being in agreement with the NBS measurements to within a stated percent. The laboratory in question can then state that its measurements were traceable to NBS within that percent for that particular radionuclide at that time.

The basis for the most widely used form of traceability exercise in the United States is for NBS or other monitoring laboratory, such as EMSL (EPA) or RESL (DOE/NRC), to distribute calibrated radioactivity sources of undisclosed value ("blinds") for quantification by the laboratories that are seeking accreditation. The first such exercise was carried out by NBS at the request of the College of American Pathologists (CAP) in 1970, to check the quality of activity measurements at some twenty hospital and medical laboratories. Then in 1971 the U.S. Atomic Energy Commission (AEC) asked NBS to distribute "blinds" of a mixture of gamma-ray-emitting radionuclides in solution to seven laboratories that were monitoring for possible radioactive effluents at the start-up of a nuclear-power plant (Cavallo et al., 1973). In this exercise these laboratories, including the health-physics laboratory of the plant itself, had both to identify and quantify the six parent radionuclides in the mixture. The interesting results of the next CAP traceability exercise carried out in 1972 are shown in Figure 101. In this exercise none of the submitted results lay within the limits of ±5 percent stipulated for ^{125}I by the U.S. Pharmacopoeia. The results of early CAP "round robins" have been summarized and discussed by Hauser (1974).

The next important request to NBS for traceability testing came from the Committees on Radioisotope Production and on Radiopharmaceuticals of the Atomic Industrial Forum (AIF) in 1972. As a result of this request an NBS-AIF research-associate program was established at NBS to distribute, initially every month, standards or "blinds" at microcurie and millicurie levels of activity (Mann et al., 1981). This program continues under the guidance of a joint NBS-AIF committee, but recently two months in every year have been designated as "open months" in which the commercial suppliers of radioactivity standards and radiopharmaceuticals can submit calibrated sources of radionuclides of their own choice to NBS for verification. These are submitted with their certified values and any relevant documentation and then measured by NBS, and an appropriate NBS certificate issued to each participant stating the degree of traceability achieved.

These traceability programs have been described by Coursey et al. (1975), Collé (1976), and Golas and Calhoun (1983). Soon after the

Fig. 101. Results of assays, using different dose calibrators (Section 6.4.3.2.3), of "unknown" NBS samples of ^{125}I (each having an activity of 75.10 μCi), distributed to various hospitals and medical laboratories in 1972. It is interesting to note that none of the results lies in the range of ±5 percent required by the U.S. Pharmacopoeia; that the measurements range from 2.5 to 294.0 μCi; and that results, obtained by 10 different laboratories using dose calibrators "N," are distributed around a value some 100 percent too high, which indicates an error in the "radionuclide" dial setting.

inception of the NBS-AIF program, the U.S. Food and Drug Administration became a participant; thus both regulator and regulatee were assured of being on the same scale of values.

The central criterion for the traceability of a laboratory to the national laboratory (or to a quality-control or regulatory laboratory, as shown in Figure 102) is that the lower-echelon laboratory should be able to measure, within required limits of uncertainty, the activity of a "blind" sample supplied by the higher-echelon laboratory; and, if necessary, identify the principle radionuclide, or radionuclides, and any impurities.

If a laboratory seeking such traceability uses, for example, NBS standards to calibrate its measuring systems, this cannot in itself constitute traceability, because traceability must also include the performance of the measuring equipment and the performance of the operating staff. Only when that laboratory can quantify or identify a "blind" sample and obtain a value agreeing with the NBS value, or that of another quality-control laboratory, within a specified range, can traceability be achieved. And this can be achieved using a national or commercial standard, or a directly calibrated "in-house" standard.

International and National Radioactivity Measurements "Traceability Tree"

Fig. 102. The international and national radioactivity measurements traceability "tree." Most of the acronyms used in this figure have been explained in the text, but at the national standardizing laboratory level, NRC refers to the National Research Council of Canada. On the regulatory level, NRC refers to the U.S. Nuclear Regulatory Commission.

There are three separate methods by which NBS checks laboratories in the United States for traceability to NBS. These are:

(i) Quantification, and/or identification of "blind" samples specially prepared and distributed for that purpose by NBS;

(ii) Essentially the same as (i) but using radioactivity standards that may be available at the time as NBS Standard Reference Materials (SRM's). These can be sent to client laboratories without certification or documentation—in other words as "blinds." On receiving that laboratory's value, NBS sends it the certificate of the standard and a certificate of traceability for its competence to measure the activity of that radionuclide, and, possibly, even to identify the radionuclide in question. The values of such SRM's can be adjusted by NBS, if necessary, by gravimetric dilution with the appropriate carrier solution; and

(iii) The client laboratory sends to NBS a radioactive sample that the client has assayed, together with the certification and full documentation. NBS checks the calibrated sample and then certifies the degree of its traceability to NBS.

At the conclusion of any traceability exercise each participant receives an NBS certificate stating the NBS calibration value and the percentage difference between it and the participant's submitted value, in the form $100(X - N)/N$, where X is the participant's value and N is the NBS value. In this way the participant is certified as being traceable to NBS to within that stated percentage. The NBS certified value can then also be used to effect any required corrections to the calibrations of the participant's instruments.

The kind of traceability discussed above is usually *implicit*, insofar as it only establishes the ability of a laboratory to assay any given radionuclide *at any given time*. In the long term it is the competence and reliability of a laboratory and the skill of its staff that comprise its traceability.

An *explicit* form of traceability to NBS can be established when a distributing laboratory prepares a batch of radioactivity standards from one master solution, and submits two or more randomly selected samples to NBS for verification.

On the other hand, it is incumbent upon the national standardizing laboratories to maintain agreement among themselves and this is attempted in a small way by international "round robins" organized by the BIPM or by bilateral intercomparisons between two or more national laboratories. BIPM and IAEA have also installed "4π"γ pressure ionization chambers for the purpose of monitoring the agreement between gamma-ray-emitting solution standards of activity prepared by different national laboratories and the consistency, over the years, of such standards produced by a given laboratory. Standards can be submitted to BIPM and could, until recently, be submitted to IAEA for monitoring, in standard glass ampoules (Section 4.4.7). In the process, BIPM and IAEA have been able to build up very excellent calibrations, for photon-emission-rate response as a function of energy for their own "4π"γ chambers that could enable them, in turn, to calibrate solution standards for countries requiring a standard of any particular gamma-ray-emitting radionuclide. Details of the BIPM program, the Système International de Référence (SIR) have recently been given by Rytz (1983).

In addition to distributing "blind" samples in the activity range of 0.05 to 5 μCi to EMSL (EPA) and RESL (DOE/NRC) for traceability measurements, NBS has also carried out intercomparative measurements, with RESL, of several of the NBS natural-matrix environmental reference materials. These environmental materials have been rather fully described in Hutchinson and Mann (1978, 1980), and

Mann *et al.* (1981). More recently the preparation of one of these standards has been described in detail by Inn *et al.* (1984).

Environmental materials are also distributed for international intercomparison by the IAEA's Analytical Quality Control Service. These samples are prepared by the Agency's Seibersdorf Laboratory and the Laboratory of Marine Radioactivity in Monaco (Suschny *et al.*, 1978).

Figure 102 illustrates how the national and international radioactivity measurements systems can be envisaged as being traceable to and consistent with each other. The national laboratories maintain traceability links with BIPM and with each other, while each of the national laboratories will presumably seek to maintain traceability with the radioactivity measurements laboratories in its own jurisdiction. In the United States the NBS, in the second generation of the genealogical tree, will seek to establish traceability between itself and key laboratories in the third generation, as shown in Figure 102, namely NRC, EPA, and so forth. With an annual industrial production of radionuclides, including radiopharmaceuticals, of the order of $100 million in the United States, to say nothing of radioactive waste and effluents from the use of nuclear energy, it is clearly quite impossible for traceability to be maintained, in the measurement of radioactivity, *directly* between NBS and every laboratory measuring radionuclides. NBS can only attempt to maintain traceability to the third-generation organizations, but would expect occasionally to penetrate to the fourth generation of measurements laboratories, and beyond if necessary, to confirm that traceability continues to be maintained across the generation gaps.

The National Bureau of Standards seeks, therefore, to maintain strong links of traceability, renewed as frequently as possible, with every one of the key laboratories of the "Quality Control and Regulatory Bodies" shown in the third-generation level of Figure 102.

8.4 The Statement of Uncertainty

It has been suggested in Section 1.5.1 that a realistic assessment and statement of the estimate of the uncertainty of a standard is far more important than categorizing it as "primary," or "secondary," etc. The seven defined SI base units (metre, kilogram, second, etc.) are absolute standards and in a sense "primary," although many *national* standards are designated as primary. In fact, however, many national standards of α-, β-, or γ-emitting nuclides, especially of short-lived radionuclides, are "secondary" in the sense that they have been calibrated by means of an ionization chamber that has been previously calibrated for the particular radionuclide in terms of a ^{226}Ra reference source. The additional uncertainty introduced by using the ionization chamber and radium source to preserve the initial activity calibration of any given radionuclide is expected to be no worse than 0.1 percent, and they are in every sense the *national standards*.

In the conclusion to the previous chapter, Section 7.5, attention is focused on the existence of experimental uncertainties due to the possibilities of random error and systematic error. Random error gives rise to imprecision, or irreproducibility, in any scientific experiment, but in the measurement of radioactivity its influence is compounded by the fact that radioactivity itself is a random, *i.e.*, stochastic, phenomenon. Therefore, even with the most precise instrument and observer, any two consecutive readings of the activity of a radionuclide (allowing for radioactive decay, if necessary) will probably not be the same (see Appendix B 3.1). The potential for systematic error[24] arises both from uncertainties associated with the measurements of ancillary data such as efficiencies and dead times, needed to calculate the activity from a simple count of events in a given time, and from instrumental and operator bias.

In Chapter 7 of this Report a clear distinction has been drawn between *population parameters* and their *estimators* (Section 7.3.1). The true *mean*, μ_x, and standard deviation, σ_x, of the population can not be measured experimentally but the *average* $\bar{x}$ of a group of n measurements can be, and the estimated standard deviation of the average value, $s_{\bar{x}}$, gives an indication of how well $\bar{x}$ represents μ_x.

A clear distinction must also be drawn between the terms *confidence level* and *confidence interval* (or *limits*). The former states the probability of a given parameter lying within the latter.

When it comes to the systematic component of the overall uncertainty there is no general method for estimating it. Random error or imprecision gives rise to a range of final results distributed around an average value; systematic error, however, represents a fixed or varying unknown difference from a "true" reading of an instrument or the "true" value of the quantity being measured. As the "true" value of the quantity being measured is *ipso facto* unknown, it is impossible to assign a magnitude or, usually, a sign to the systematic error.[25]

---

[24] Throughout this report the terms "systematic error" and "systematic uncertainty" are used with the connotation given in Section 7.5, and by Webster (systematic error: An error that is not determined by chance but by a bias).

[25] Great care has been taken in this report to distinguish between the terms "error" and "uncertainty." *By chance* an experimental result might coincide with the "true" value of the quantity being measured. Unknown to the observer, the error would then be zero but the uncertainty would not be changed by this (unknown) coincidence.

As is suggested in Section 7.5, if the quantity in question can be measured by each of a number of different instruments then "the results obtained by using these instruments may be construed as a sample from the population" thereby giving an insight into the magnitude of systematic uncertainty. In radioactivity measurements it is often necessary to prepare several sources for measurement in different instruments, and this again generates the possibility of further random and systematic error (*e.g.*, weighing and pipetting in the dilution and deposition of the source).

Frequently a good assessment of systematic uncertainty is derived from an experimenter's extensive previous experience with a given instrument, for example, a microbalance. An estimate of possible systematic uncertainty is only a guess, but a guess by an *experienced* experimenter may be the only reasonable estimate available. Where practical, sources of systematic error should be investigated in separate experiments.

As explained in Section 7.3 of this Report, there are two common population parameters concerned with the assessment of the confidence interval for random uncertainty, namely σ_x, and $\sigma_{\bar{x}}$.

The corresponding estimators of these population parameters are s_x, and $s_{\bar{x}}$ (Table 18), where $s_{\bar{x}} = s_x/\sqrt{n}$ for n independent measurements. Another statistical estimate discussed in Section 7.3.2 is the confidence interval for the mean μ, given by $\pm ts_{\bar{x}}$, where t is the student-t factor for n independent measurements and a stated confidence level, such as 99 percent. In other words, $ts_{\bar{x}}$ expresses the limits around a measured average value $\bar{x}$ within which, at a stated level of confidence, the "true" mean, μ_x, may be expected to lie.

A fundamental difference between random uncertainty of an average value and systematic uncertainty is that the former can be reduced by increasing the number n of observations, but the latter will not necessarily be affected. In radioactivity measurements at the National Bureau of Standards, the statement of systematic uncertainty has, in the past and in accordance with the recommendations of ICRU Report 12, started with a realistic attempt to estimate the maximum conceivable systematic error—"maximum" to match the statistical range in which the true mean μ_x has a 99-percent probability of being found, and "conceivable" because some source of possible systematic error could have a finite probability of being overlooked, or underestimated.

One option that has been used in practice was to consider the "maximum conceivable" systematic uncertainty as being roughly equivalent to the random-uncertainty interval, $\pm ts_{\bar{x}}$, in which the true mean μ_x, may have a 99-percent probability of being found. From

Table 20 it is seen that for n equal to between six and a very large number of experimental observations (*i.e.*, for degrees of freedom, ν, between five and infinity) t has values between three and four, and predominantly three. Analogously therefore to dividing by about three to derive the estimated standard deviation, $s_{\bar{x}}$, of the overall average value from the confidence interval, $ts_{\bar{x}}$, of the mean for a confidence level of 99 percent, it is possible to divide the estimate of the maximum conceivable systematic error by three to give a quantity that may be considered to be equivalent to one estimated standard deviation of the systematic uncertainty. This has been combined, linearly (ICRU, 1968[26]) or in quadrature (but *which, must* be stated), with the estimated random standard deviation of the average value. The sum of the two was then treated, or accepted, *as if* it were *equivalent* to a *total (random plus systematic) estimated standard deviation of an average value.*

Recent recommendations by the Bureau International des Poids et Mesures (Giacomo, 1981) raise the question of how to assign a value equivalent to a "standard deviation" for error components that cannot be evaluated by statistical methods. These recommendations, which were adopted by the Comité International des Poids et Mesures (CIPM) in October 1981, were essentially based on a paper by Müller (1979) entitled "Some Second Thoughts on Error Statements." Quoting from Giacomo (1981) the recommendations, which are five in number (with the first two paragraphs of "1", here omitted, constituting a preamble), are as follows:

"1... The uncertainty in the result of a measurement generally consists of several components which may be grouped into two categories according to the way in which their numerical value is estimated:

A. those which are evaluated by applying statistical methods to a series of repeated determinations,

B. those which are evaluated by other means.

There is not always a simple correspondence between the classification into categories A or B and the previously used classification into "random" and "systematic" uncertainties. The term "systematic uncertainty" can be misleading and should be avoided.

Any detailed report of the uncertainty should consist of a complete list of the components, specifying for each the method used to obtain its numerical value.

2. The components in category A are characterized by the estimated

---

[26] ICRU Report No. 12 was intended as a guide to obtaining conservative limits for total uncertainty, and was used for many years by many laboratories.

variances, s_i^2, (or the estimated standard deviations s_i) and the number of degrees of freedom, ν_i. Where appropriate, the covariances should be given.

3. The components in category B should be characterized by quantities u_j^2, which may be considered as approximations to the corresponding variances, the existence of which is assumed. The quantities u_j^2 may be treated like variances and the quantities u_j like standard deviations.

4. The combined uncertainty should be characterized by the numerical value obtained by applying the usual method for the combination of variances. The combined uncertainty and its components should be expressed in the form of "standard deviations."

5. If, for particular applications, it is necessary to multiply the combined uncertainty by a factor to obtain an overall uncertainty, the multiplying factor used must always be stated."

Giacomo then goes on to state that it is the intention of BIPM "to encourage the application of the principles outlined in the above Draft Recommendation in international comparisons" that take place under their auspices. A BIPM Working Group on the Statement of Uncertainties which drafted the recommendations, states in the conclusion of its Report, that "It is realized that the proposed recommendation may not give all the information needed for implementation of the new guidelines into laboratory work." Therefore, pending the dissemination of this further information contemplated by the Working Group as being possibly needed, it may be premature for the great majority of users of radioactive materials in the United States to adopt these recommendations at the present time. Indeed while the CIPM Recommendations provide guidelines for a uniform format for statements of uncertainty, they do not, nor does any other method, provide a mathematical basis for the treatment of so-called systematic error. It is also very important to recognize that the CIPM Recommendations do not suggest that the components of uncertainty in category B are variances but that, in order to achieve an international uniformity in reporting data, they *could* be thought of, and treated, *as if* they were variances.

On reflection, it would seem that there is more a difference in style rather than in substance between the BIPM guidelines and the everyday usage of the national laboratories and the users of their standard reference materials. Consider the following passage from Müller (1979):

"However, experience shows that some users might still hesitate to adopt this position and for them the question seems to be an important one, as can be judged by the length of the discussions devoted to this problem. Essentially, there are two main suggestions. One of them is to use *"maximum possible limits"** which should practically never be exceeded, as implied by the word maximum; and the other asks the experimenter to make an attempt at estimating an uncertainty which would correspond as closely as possible *to something like** a "standard deviation." Since both proposals have some merits and drawbacks, the choice is not quite obvious and calls for some second thoughts."

Clearly maximum possible limits are $\pm\infty$, but in practical terms many experienced experimenters would be thinking in terms of intervals at confidence levels of 99.0 or 99.5 percent. From such intervals it is not difficult to proceed to a confidence level of 67 percent or "something like a standard deviation," or *vice versa*. Or as Müller states:

"Thus, one could for instance demand that the range covered by the interval $\pm s$ should contain the "true" value in about 70% of the cases, i.e. with a probability approaching 2/3. If the estimation of a 50% confidence level is considered to be more practical, such a "probable error" will have to be multiplied by a factor 3/2 to obtain an approximate value of the corresponding standard deviation. The theoretically "exact" values would be 1.48 for a normal and 1.39 for a triangular density, for example.

Apparently, some people are not very happy with a suggestion of this kind and a few even claim that an experimenter is unable to indicate such a quantity, stating that a "maximum limit" (as discussed above) is all he can actually estimate. However, a look at the literature shows that experienced people in fact can produce such estimates."

Anyone wishing to understand the philosophical background to the CIPM recommendations should most certainly read Müller (1979).

In current everyday practice at the "user level" (Fig. 102), in many countries, three components of uncertainty inherent in the average value of any measured quantity are generally recognized. They are (i) the random component associated with the measurements themselves, including, in the case of radioactivity, the randomness of radioactive decay itself that makes it improbable that consecutively measured values of the activity of a given source will be the same; (ii) the random component associated with the measurements of preliminary ancillary data, such as weighings, dead times of electronic circuits, and so on, that are incorporated into the computation of the final measured

* Emphasis added.

average value, and whose uncertainties can be considered to be random or systematic depending upon the experimental circumstances and the predilection of the experimenter (but *which, must* be stated); and (iii) the systematic uncertainty, or the uncertainty due to a fixed or variable but unknown bias in the readings of an instrument or of the observer of that instrument's output.

Components (i) and (ii) have long been discussed in texts on statistics and in the scientific literature. But the third component remains, and will remain, a major obstacle to the estimation of overall uncertainty. As a simple everyday example how does one estimate the uncertainty associated with the digital output readings of a "dose calibrator?" If a medical technician takes readings on a radiopharmaceutical and obtains results equal to, for example, 1.42, 1.43 and 1.44 μCi, but mostly 1.43 μCi and unequal numbers of readings at 1.42 and 1.44 μCi what is the random uncertainty, let alone the instrument bias—especially if there is only time to take two or three readings? Most experimenters develop, with time and experience, an intuitive awareness of the measuring capabilities of their various pieces of equipment and will be able to allot to each of them an order-of-magnitude estimate of the uncertainties in the data they produce; and they should, where possible, characterize each of the components of such uncertainties. Such estimates can be fortified by frequent quality-control, or traceability, exercises in which they quantify "unknown" sources of radioactivity distributed by national laboratories, or by laboratories whose traceabilities to the national laboratory have been established (Section 8.3).

A striking example of the demonstration of the possible existence of systematic bias emerged from an international comparison of mass-spectrometric measurements of the differences in ^{12}C and ^{13}C abundances in the old (1958) and new (1983) NBS contemporary ^{14}C standards of oxalic acid. These measurements were made by thirteen leading radiocarbon-dating laboratories in China, Europe, South Africa and the United States. When each laboratory's values for a parameter δ^{13}C, which reflects the divergence of the ^{13}C:^{12}C ratio from the natural abundance, were plotted in a histogram for the old standard and a histogram for the new, neither histogram showed any resemblance to a normal distribution. But on plotting differences in each laboratory's values for the old and new standards (δ^{13}C for the old less δ^{13}C for the new for each laboratory) in a histogram, a very good approximation to a normal distribution emerged (see Mann, 1983). This indicated that each laboratory had a given systematic bias that was the same for its δ^{13}C measurements for both the old and the new standards.

In the same intercomparison the method of Paule and Mandel (1982) was used to derive the estimated within-laboratory and between-laboratory standard deviations of the grand weighted average of the results submitted by all laboratories for the ratio of activity concentrations in the new and old standards.

Whatever the method adopted to arrive at an estimate of all components of the uncertainty, it should be fully and clearly described. The evaluation of experimentally measured nuclear-decay parameters is also impossible without an estimate of overall uncertainty and a clear statement of every component of uncertainty, and how they have been derived in terms, as appropriate, of s_x, $s_{\bar{x}}$, or $ts_{\bar{x}}$, ν (the number of degrees of freedom), the estimate of systematic bias, and the methods used to combine or propagate these estimates.

Many of the problems associated with the estimation of uncertainty have been reviewed by Churchill Eisenhart and Ronald Collé in "Postscripts," written in 1980 and reproduced in Eisenhart et al. (1983).

This section can only be regarded as a guide to the treatment of a somewhat nebulous subject in which the concepts are still not wholly clear and the literature and recommendations conflicting. The only criteria of importance are to present clearly the experimental data and to list each of the estimated components of their uncertainties, so that future researchers in the field can understand and, if necessary, unscramble them.

As the techniques available for measuring a given quantity become more precise, systematic error in earlier measurements may be revealed.[27]

[27] An interesting example is presented by the historical vagaries of the recommended and measured values of the Avogadro constant during the last two decades. These, together with the quoted values of their uncertainty at a level of one estimated standard deviation, are as follows:

$6.02252 \pm 0.00009 \times 10^{23}$ mol^{-1}	(NBS, 1963; Cohen and DuMond, 1965)
6.022169 ± 0.000040	(Taylor et al., 1969)
6.022045 ± 0.000031	(Cohen and Taylor, 1973; CODATA, 1973)
6.0220978 ± 0.0000063	(Deslattes et al., 1974; Deslattes et al., 1976; Deslattes, 1980)

Each of the three changes is greater than the sum of the standard deviations of the respective values involved.

Appendix A

Nuclear-Decay Data for Selected Radionuclides

A.1 Description of the Tabulated Information[28]

A.1.1 *Introduction*

Contained in this Appendix are tabulations of the atomic and nuclear radiations emitted by some 200 radioactive nuclides.[29] The nuclei included comprise most of those currently of interest in medical practice or research, health physics, industry, nuclear power, environmental impact studies, and as reference standards. Listed in tabular form are recommended values for half-lives, energies, intensities (probabilities per decay), and equilibrium absorbed-dose constants (discussed in Section A.1.4.4) for each of the atomic and nuclear radiations emitted by these radioactive atoms.

Data on which evaluations were completed between August 1983 and April 1984 have been incorporated in this Appendix. These data supersede those in earlier publications by Martin and Blichert-Toft (1970), Martin (1973, 1976), Martin (1978), Kocher (1981) and Martin (1982). The latter two references, however, contain data on more nuclei than are included here.

Space limitations do not permit the inclusion of references for the nuclear data appearing in this table. If documentation or the most

---

[28] Data provided by M. J. Martin, Nuclear Data Project, Oak Ridge National Laboratory, from the Evaluated Nuclear Structure Data File and the Nuclear Structure Reference File maintained by the National Nuclear Data Center at Brookhaven National Laboratory. Oak Ridge National Laboratory is operated by Martin Marietta Energy Systems, Inc. under contract DE-AC05-840R21400 with the U.S. Department of Energy.

[29] Many of the symbols in this Appendix, including those in the table, are nonstandard (thus D rather than d for day, etc.). Their form is dictated by the symbols available on the computer print train used to prepare the table. See also Appendix A.1.4.3.

recent nuclear data are required, they can be obtained by writing to the Nuclear Data Project.[30]

A.1.2 Review of Radiation Processes

A.1.2.1 *Alpha-Particle Transitions (α).* In α decay, an atom, atomic number Z and mass number A, emits an α particle (^{4}He nucleus, $Z = 2$, $A = 4$) and decays to an atom with atomic number $Z - 2$ and mass number $A - 4$, and two surplus electrons.

For α decay between ground states of parent and daughter atoms, the maximum energy available for the α particle is,

$$E_\alpha = Q_\alpha - E_R$$

where Q_α is the difference, in energy units, between the mass of the parent atom and the sum of the masses of the daughter atom and a helium atom. E_R is the recoil energy of the daughter atom which, if M and M_α are the masses of the daughter atom and the α particle, respectively, is given by

$$E_R \approx Q_\alpha \frac{M_\alpha}{M_\alpha + M}$$

The energy available will be increased or decreased, respectively, if α decay is from an excited state of the parent or to an excited state of the daughter nucleus.

A.1.2.2 Beta-Particle Transitions (β).

A.1.2.2.1 β^- *Decay.* In β^- decay, an antineutrino ($\bar{\nu}$) and a negative electron (β^-) are emitted from the nucleus as a result of the process $n \rightarrow p + \beta^- + \bar{\nu}$. The decay increases the nuclear charge by one unit.

The energy released in a single β transition is divided between the β particle and the antineutrino in a statistical manner so that, when a large number of transitions is considered, both the antineutrinos and β particles have energy distributions extending from zero up to some maximum value. For decay to a particular energy level, E_L, in the daughter nucleus, the maximum energy available is $E_{max} = Q^- - E_L$, where Q^- is the atomic mass difference, expressed in energy units, between ground states of parent and daughter nuclides. The average energy of a β particle in this transition is given by

$$E_{av} = \int_0^{E_{max}} EN(E)dE \left/ \int_0^{E_{max}} N(E)dE \right.$$

---

[30] Oak Ridge National Laboratory, Post Office Box X, Oak Ridge, Tennessee 37830.

where $N(E)$ is the number of β particles with energy between E and $E + dE$. Thus, the β^- particles from a particular transition i, denoted in the main table by β^-i, constitute a "group" having a continuous distribution of energies with a definite E_{max} and E_{av}.

A.1.2.2.2 *β^+ Decay.* In β^+ decay, a neutrino (ν) and a positive electron (β^+) (positron) are emitted from the nucleus as a result of the process p $\rightarrow$ n + β^+ + ν. This decay decreases the nuclear charge by one unit.

As in β^- decay (see above), the β^+ particles emitted in a transition to a particular level in the daughter nucleus have a continuous distribution of energies with a definite E_{max} and E_{av}. For positron decay, $E_{max} = Q^+ - 2mc^2$, where Q^+ is the atomic mass difference, expressed in energy units, between ground states of parent and daughter nuclides. Note that β^+ decay to a particular energy level, E_L, cannot occur unless $Q^+ - E_L > 2mc^2$ ($2mc^2 = 1022$ keV).

A.1.2.2.3 *Electron-Capture Decay (ϵ).* In electron-capture decay, an atomic electron is captured by a nucleus and a neutrino is emitted as a result of the process p + e$^-$ $\rightarrow$ n + ν. This decay decreases the nuclear charge by one unit and leaves the daughter nucleus with a vacancy in one of its atomic shells. K-shell electron capture, for example, refers to a capture process where the final-state vacancy is in the K-shell. For K-shell electron-capture to a particular energy level, E_L, the energy released is given by $Q^+ - E_L - E_K$, where E_K is the K-shell binding energy in the daughter atom.

The electron-capture process always competes with β^+ decay, but also can occur when the transition energy is too small to allow β^+ emission, that is, when $Q^+ - E_L < 1022$ keV.

A.1.2.3 *Gamma-Ray Transitions (γ).* Electromagnetic radiation is emitted by a nucleus in a transition from a higher- to a lower-energy state. A numeric index on γ, as in γ1, denotes a γ ray emitted in the transition between a particular pair of nuclear levels. The energy of this γ ray, $E(\gamma 1)$, is equal to the energy difference between the two levels (except for a usually negligible amount of nuclear recoil energy, E_r (in keV) $\simeq 5.4 \times 10^{-7} E^2(\gamma)/A$, where A is the mass number and $E(\gamma)$ is in keV).

A.1.2.4. *Internal-Conversion-Electron Transitions (ce).* An atomic electron can be emitted as an alternative to γ-ray emission in the transition of a nucleus from a higher- to a lower-energy state. In the internal-conversion process, the energy difference between these states is transferred directly to a bound atomic electron which is ejected from the atom. A letter hyphenated to "ce" refers to the shell from which the atomic electron is ejected. Thus, ce-K denotes K-shell conversion and ce-MNO denotes conversion in the M, N, and O shells combined.

A numeric index, as in ce-K-1, has a meaning analogous to that used above for γ rays. For a transition with energy $E(\gamma 1)$, the K-shell internal-conversion electron is emitted with energy $E(\text{ce-}K\text{-1}) = E(\gamma 1) - E_K$, where E_K is the K-shell binding energy. The energy of an electron converted in one of the other shells is given by this same expression with the replacement of E_K by the binding energy appropriate to that shell.

For a particular transition, the ratio of the probability for emission of a K-conversion electron to that for emission of a γ ray, $I(\text{ce-}K)/I(\gamma)$, is called the K-shell conversion coefficient for that transition. Conversion coefficients for the other shells are defined in an analogous manner.

A.1.2.5 *X-Ray and Auger-Electron Transitions (X, e_A).* Whenever a vacancy is produced in an inner electron shell of an atom, the filling of this vacancy is accompanied by the emission of either an x ray (X) or Auger electron (e_A). Vacancies created by the filling of the initial vacancy will in turn produce further x rays or Auger electrons. This cascade of radiations continues until all vacancies have been transferred to the outermost electron shell. Inner-shell vacancies are always produced in two types of nuclear decay, electron capture and internal conversion. Other processes that produce electron vacancies following a nuclear decay, such as electron shakeoff (the ejection of one or more atomic electrons due to a sudden change in nuclear charge) or the ejection of atomic electrons by escaping α or β particles, are not discussed here because of the low probability of their occurrence.

A.1.2.6 *Fluorescence Yield (ω).* The fluorescence yield for a particular atomic shell (ω_K, ω_L, etc.) is defined as the probability that a vacancy in that shell will result in the emission of an x ray. If n_K is the number of K-shell vacancies per transition, the number of K x rays will be $n_K \omega_K$, and the number of K Auger electrons will be $n_K(1 - \omega_K)$.

The most recent and thorough review of fluorescence yields is that of Bambynek *et al.* (1972). The K-shell fluorescence yields, ω_K, adopted in the present work are the "fitted values" from Table III·IV of the above reference. The average L-shell fluorescence yields, $\bar\omega_L$, with uncertainties, have been estimated by Martin (1976) from Fig. 4–34 of this same review (Bambynek *et al.*, 1972).

A.1.2.7 *X Rays*

A.1.2.7.1 X_K. An x ray emitted as the result of the filling of a K-shell vacancy by an electron from a higher shell, for example the Y

shell, has an energy $E_K - E_Y$, where E_K and E_Y are the electron binding energies in the K and Y shells, respectively. This transition can be denoted by $K - Y$. In order of decreasing intensity, the most important transitions are

$$
\begin{aligned}
K_{\alpha 1} &= K - L_3 & K_{\beta_3} &= K - M_2 \\
K_{\alpha 2} &= K - L_2 & K_{\beta_4} &= K - N_2 \\
K_{\beta_1} &= K - M_3 & K_{\beta_5} &= K - M_4. \\
K_{\beta_2} &= K - N_3 &
\end{aligned}
$$

Energies and intensities for the $K_{\alpha 1}$, $K_{\alpha 2}$ and K_β x rays, the last summed over $K_{\beta i}$, are given explicitly. The x-ray energies are taken from Bearden (1967), and the intensity ratios K_β/K_α and $K_{\alpha 2}/K_{\alpha 1}$ are taken from Rao et al. (1972).

As already mentioned, the number of K x rays per disintegration is $n_K \omega_K$. Thus, the number of $K_{\alpha 1}$ x rays is $n_K \omega_K/(1 + K_\beta/K_\alpha)(1 + K_{\alpha 2}/K_{\alpha 1})$ with similar expressions for $K_{\alpha 2}$ and K_β. The number of K-shell vacancies per disintegration is the sum of the vacancies produced by K capture and those produced by internal conversion in the K shell. Thus, $n_K = \epsilon_K + I(\text{ce-}K)$, where ϵ_K is the total number of K-capture events per disintegration, and $I(\text{ce-}K)$ is the total number of K-shell internal-conversion electrons per disintegration.

A.1.2.7.2 X_L. As in the case of the K shell, many transitions contribute to the L x-ray spectrum. However, since the relative intensities of these transitions are not known for all Z values, and since the energy differences between the strong transitions are small (≤ 3 keV for $Z \lesssim 92$), the total L x radiation is treated here as a single group having the energy of the strongest transition, $L_{\alpha 1} = L_3 - M_5$. For $Z < 37$, the M_5 transitions have not been resolved experimentally from the M_4 transitions. In these cases, the energy given is that for the transition $L_{\alpha 1, \alpha 2} = L_3 - M_4, M_5$. Figures 4-8 to 4-11 of Bambynek et al. (1972) show examples of L x-ray spectra.

The calculation of the number of L x rays per disintegration, $n_L \omega_L$, is similar to that for K x rays except that, in addition to vacancies produced by direct L-shell capture and by L-shell conversion, those created by the transfer of L-shell electrons to K-shell vacancies must also be included. Thus,

$$n_L = \epsilon_L + I(\text{ce-}L) + n_{KL} n_K,$$

where n_{KL} is the number of L-shell vacancies created per K-shell vacancy, and the other symbols have meanings analogous to those used above for the K shell. Values of n_{KL} are taken from Bambynek et al. (1972).

A.1.2.8 *Auger Electrons* (e_A). The Auger process competes with the emission of x rays as a means of carrying off the energy released in the filling of an inner-shell vacancy by an electron from an outer shell. In the Auger process, the filling of an inner-shell vacancy is accompanied by the simultaneous ejection to the continuum of an outer-shell electron. The resulting atom is left with two vacancies. The Auger-electron yield per disintegration of a particular atomic shell is $n_K(1 - \omega_K)$, $n_L(1 - \bar{\omega}_L)$, etc. If the original vacancy is in the K shell and if this vacancy is filled by an electron from shell X with the ejection of an electron from shell Y, the transition is denoted by KXY. The energy of the ejected electron is $E_K - E_X - E'_Y$ where E_K and E_X are neutral atom K- and X-shell binding energies, respectively, and E'_Y is the binding energy of a Y-shell electron in an atom containing a vacancy in the X shell. The most intense K-Auger transitions are of the type KLL. Since the relative intensities of the electrons in the KLL group are not accurately known for all Z values and since the energy difference between the transitions is small (≤ 5 keV for $Z \lesssim 92$), the K-Auger electrons are treated here as a single group having the energy of the strong transition, KL_2L_3. The energy of this transition as a function of atomic number is taken from Table 1 of Bergstrom and Nordling (1965). A K-Auger-electron spectrum is shown in Fig. 3-5 of Bambynek *et al.* (1972).

In the case of the L-Auger process, very little is known about the energies or relative intensities of the individual L-Auger electrons. For this compilation, the L-Auger electrons are treated as a single group having the energy appropriate to an $L_3M_4M_5$ transition. An example of an L-Auger electron spectrum is shown in Fig. 4-13 of Bambynek *et al.* (1972).

A.1.2.9 *Bremsstrahlung.* In addition to any monoenergetic x rays or γ rays that may be present in radioactive decay, every β or electron-capture decay produces continuous electromagnetic radiation called bremsstrahlung. Two processes contribute to this continuous spectrum:

1. "External" bremsstrahlung is produced by collisions between β's or conversion electrons and the atoms of the material surrounding the radiating atoms. Since the intensity of the external bremsstrahlung depends on the atomic number, Z, of the surrounding material, this radiation is not included in the present tables. In many cases this radiation cannot be neglected in dose calculations. An approximate value for the average energy of the external bremsstrahlung associated with a β group of maximum energy E_β (in keV) is given, for the thick-target approximation, by Evans (1955) as

$$E_{av} \approx 1.4 \times 10^{-7} \, ZE_\beta^2 \text{ keV per } \beta.$$

In the decay of ^{32}P, where $E_\beta = 1711$ keV, the average energy per disintegration of the external bremsstrahlung is $\approx 0.4 \, Z$ keV;

2. "Internal" bremsstrahlung, originating within a decaying atom, is produced by the sudden change of nuclear charge which occurs in β^+, β^-, or ϵ decay. This radiation is not included here because of its low intensity and low average energy. The average energy of the internal bremsstrahlung associated with a β group of maximum energy E_β (in keV) is given, for $E_\beta \gg mc^2$ (511 keV) by Schopper (1966, pp. 76–84) as

$$E_{av} \approx 1.5 \times 10^{-3} \, E_\beta \log(0.004 \, E_\beta - 2.2)\text{keV per } \beta.$$

In electron capture, the corresponding expression for a transition with energy E_ϵ (in keV) is (Schopper, 1966, pp. 76–84)

$$E_{av} \approx 1.5 \times 10^{-7} \, E_\epsilon^2 \text{ keV per capture.}$$

In the decay of ^{32}P, where $E_\beta = 1711$ keV, the average internal bremsstrahlung energy per disintegration is ≈ 1.7 keV.

A.1.3 Computer Program, MEDLIST[31]

Based on the radioactive decay data sets in the Evaluated Nuclear Structure Data File and computerized tabulations of the relevant Z-dependent constants (ω_K, n_{KL}, etc.) discussed in the preceding section, the computer program, MEDLIST, calculates the energies and intensities of the atomic radiations (x-ray and Auger-electron transitions). The program then combines these radiations with the nuclear radiations, sorts them according to radiation type (i.e., α, β, γ, ce ...), and, within each type, arranges and numerically labels them in order of increasing energy.

Uncertainties in all experimental quantities, including the atomic constants, are consistently carried through the calculations. A 3-percent uncertainty is assigned to all theoretical conversion coefficients and is propagated, within MEDLIST, along with the experimental uncertainties.

A variable low-intensity cutoff limit is built into MEDLIST. For the present tables this intensity limit is 0.1 percent. Immediately following the listing of each of the radiation types α, β, and γ,

---

[31] Written by W. B. Ewbank, Nuclear Data Project, and M. J. Kowalski, Great Lakes College Association (Oak Ridge Science Semester participant from Ohio Wesleyan University).

MEDLIST prints a comment giving the number and the summed intensity of the radiations omitted because of the chosen cutoff limit, provided that this summed intensity is greater than 0.01 percent.

It should be noted that the numerical labels on the γ radiations are based on all the radiations contained in the original data set on which MEDLIST operates. These labels are maintained throughout the MEDLIST calculations and are carried into the output. In particular, when low-intensity radiations are omitted, the remaining radiations are not, in the γ-ray listing, relabeled. Thus, in the γ-ray listing, there may appear the sequence $\gamma 1$, $\gamma 2$, $\gamma 4$, ..., which tells the reader that $\gamma 3$, with energy between that of $\gamma 2$ and $\gamma 4$, has been omitted because its intensity is below the cutoff limit. Note that there could still appear a conversion line associated with $\gamma 3$, say ce-K-3, if its conversion coefficient is large enough such that the intensity of ce-K-3 exceeds the cutoff limit.

For those nuclei that decay by positron emission, MEDLIST prints a comment at the end of the γ-ray list, giving the maximum possible observable intensity of the annihilation radiation ($\gamma\pm$), calculated from $I(\gamma\pm) = 2 \sum_i I(\beta_i^+)$. The actual fraction of these events observed in practice depends upon source attenuation and the detector efficiency.

A.1.4   Explanation of Table of Nuclear-Decay Data

A.1.4.1   Policies Followed. This table contains data on the following atomic and nuclear radiations: Auger (K, L shells); x ray (L, $K_{\alpha 1}$, $K_{\alpha 2}$, K_β); β^-; β^+; α; γ; and ce (K, L, ... shells) emitted with intensity >0.1 per 100 disintegrations of the parent nucleus (i.e., >0.1 percent). Following the listing of α, β, and γ radiations for each data set is a comment stating the number and total intensity of the observed radiations omitted because of this intensity cutoff.

Auger electrons and x rays resulting from the filling of vacancies in the M, N, ... shells, although sometimes occurring with large intensities, are not included. These transitions have energies ranging from $\lesssim 2$ keV in ^{197}Hg decay to $\lesssim 0.6$ keV in ^{113}Sn decay. Their intensities, however, are not, in general, calculable since the contribution to the total number of, for example, M-shell vacancies due to initial K- or L-shell vacancies has not been measured. However, theoretical estimates of some of these quantities are available for limited Z regions. For ^{83m}Kr, based on theoretical values of the required atomic constants (Bambynek *et al.*, 1972), the yield of M-shell Auger transitions (with energy ≈0.2 keV) is estimated to be ≈300 percent. As an additional example, for ^{125}I decay, where M-shell vacancies are produced both in electron capture decay and the subsequent internal conversion process, the yield of M-shell Auger transitions is estimated to be ≈590 percent. The total energy emitted per decay by these low-energy Auger transitions (≈0.5 keV) is ≈3 keV.

The radiations are listed in the first column by type. For an explanation of the notation employed see SYMBOLS AND DEFINITIONS (Section A.1.4.3). Particle radiations (Auger, ce, α, β) are given first, followed by the electromagnetic radiations (x ray, γ). In the case of β decay, whenever more than one group is present, a separate entry (designated "total β") is made for the composite spectrum.

The intensity given for the "total β" entry is the sum of the intensities of the individual groups. For β^- emitters with no alternate mode of decay, this intensity must be exactly 100 percent. However, it should be noted that since the intensities of the individual β^- groups are usually determined from the γ-ray intensities, which in turn are usually determined by the requirement that the total $(\beta^- + \gamma)$ feeding of the daughter ground state be 100 percent, the intensity of the total β^- contribution, as calculated from the individual β^- intensities, does not necessarily always come to exactly 100 percent. This summed value could be replaced by 100 percent, but this has not been done in the tables.

Columns 2 and 3 contain, respectively, the energy (in keV) and intensity (particles or photons per 100 disintegrations of the parent nucleus). For the β groups, both the maximum and the average energies are given.

In column 4, Δ, the mean energy emitted per nuclear decay, is tabulated in the units g rad/μCi h. For a discussion of Δ and its units, see Appendix A.1.4.4.

The nuclei are listed in order of increasing mass number, A. For nuclei with the same A, the order is by increasing Z.

Each decay mode of a given radionuclide is listed separately. In those cases where a nucleus decays by more than one mode, a cross reference is given to the alternate mode or modes. For example, ^{64}Cu decays by both β^- and β^+ branches. A separate listing is given for each of these decay modes. Following the heading "64CU B− DECAY" is the comment "SEE ALSO 64CU B+ DECAY" and following the heading "64CU B+ DECAY" is the common "SEE ALSO 64CU B− DECAY". As another example, 154-d ^{121}Te decays by both electron capture (EC) and isomeric transition (IT) branches. Following the heading "121TE EC DECAY" is the comment "SEE ALSO 121TE IT DECAY," and conversely. In such cases, the radiations in each data list should simply be combined to obtain all the radiations from the parent under consideration. In those cases where a parent decays to one or more radioactive descendants, the parent and each descendant are treated separately. For example, there is a separate listing for ^{28}Mg β^- decay and for the radioactive daughter, ^{28}Al β^- decay. If one of the daughters is an isomer, then for that isomer data listing, a comment is made giving the percentage feeding from the immediate parent. For example, ^{52}Fe decays to ^{52m}Mn. Following the heading in Table A.3 "52FE EC DECAY" is the comment "SEE ALSO 52MN EC, IT DECAYS (21.1 M)" and following the headings "52MN EC DECAY (21.1 M)" and "52MN IT DECAY (21.1 M)" is the comment "FEEDING OF 21.1-M 52MN IN 52FE EC DECAY = 100%." To include correctly all the contributions of such radioactive chains, the user must combine the radiations from each member of the chain using the standard equations for parent-daughter activity relations discussed in Section A.1.4.2 and the feeding of each member of the chain (given in the table only for ground states with feeding not equal to 100% and for isomers; otherwise, feeding is 100%). In a few cases where there is a single daughter and that daughter is an isomer with a half life much shorter than that of the parent (see Section A.1.4.2), the parent and daughter radiations at equilibrium have been combined. An example is "99MO B−, 99TC IT DECAYS." Note that in this case a separate listing of ^{99m}Tc is given since it is of importance in itself as a source.

No separate entry is made in the γ-ray listing for the 511-keV annihilation radiation ($\gamma\pm$) accompanying positron decay. Instead, the intensity of this radiation, calculated from $I(\gamma\pm) = 2 \times \sum_i I(\beta_i^+)$, is given in a comment at the end of each data set. This intensity is presented separately, and as a "maximum" value, since the annihilation process takes place external to the decaying atom in the medium surrounding the atom. The fraction of the maximum that will be of importance in a given application therefore depends upon the extent of the region surrounding the decaying atom that is being considered in that application.

The following policy on the use of "B+" and "EC" in decay type headings should be noted. "EC" as in ". . . EC DECAY" is used to denote either EC or EC + B+ (if positrons are energetically allowed). "B+" as in ". . . B+ DECAY" is also sometimes used, especially if the positron decay dominates electron capture, or if "B+ DECAY" is a more common descriptor of a certain nucleus. Thus, the use of "EC" as a heading does not preclude the possibility of positron decay.

A.1.4.2 *Parent-Daughter Activity Ratios.* A thorough discussion of the general case of activity relations among members of a radioactive decay chain is contained in Evans (1955). For the special case of a radioactive parent with one radioactive daughter, fed in a fraction, f, of the parent decays, the ratio of daughter activity to that of the parent (both at time t) is given by

$$\frac{f\lambda_d}{\lambda_d - \lambda_p} [1 - e^{-(\lambda_d - \lambda_p)t}], \tag{A.1}$$

where λ_p and λ_d are the decay constants ($\lambda = \ln 2/T_{1/2}$) of the parent and daughter nuclides, respectively. The activity of the daughter is taken to be zero at time $t = 0$ (i.e., the parent source is initially pure).

A common case occurring in these tables is that in which the daughter half life is shorter than the parent half life. In such cases, the ratio of daughter-to-parent activities first increases with time and then approaches a constant value (see Figure 2 of this report). For time, t, large compared with $1/(\lambda_d - \lambda_p)$, Eq. A.1 reduces to

$$\frac{f\lambda_d}{\lambda_d - \lambda_p} = \frac{fT_{1/2,p}}{T_{1/2,p} - T_{1/2,d}} \tag{A.2}$$

An example of this is the case of the decay of ^{52}Fe to ^{52m}Mn where $T_{1/2,p} = 8.275 \pm 0.008$ h, $T_{1/2,d} = 21.1 \pm 0.2$ m, and f = 1. For $t \gtrsim 10$ $T_{1/2,d}$, i.e., $\gtrsim 3.5$ h, the ratios of ^{52m}Mn activity to that of ^{52}Fe is, from Eq. A.2, (1) (1.0444 ± 0.0004) = 1.0444 ± 0.0004. Thus, to correctly account for the ^{52m}Mn radiations from a ^{52}Fe source, the radiations from ^{52m}Mn should be multiplied by 1.0444 ± 0.0004 and combined with those from ^{52}Fe. Note furthermore that since ^{52m}Mn decays by a 1.75 percent branch to the ^{52}Mn ground state, which itself is radioactive with a half life of 5.591 d, a generalization of Eq. A.1 to the case of two daughters is needed (with the factor f in this case equal to 0.0175 ± 0.0005) to account for the radiations from 5.591-d ^{52}Mn in a ^{52}Fe source.

Each nuclide entry in the table of nuclear-decay data (Appendix A.3) is characterized by atomic mass number, chemical symbol, and half life. The "m" which usually accompanies the mass number for decays that do not originate from the ground state, as in ^{99m}Tc and ^{110m}Ag, does not appear in the table.

A.1.4.3 *Symbols and Definitions.*

Auger-K, Auger-L	*K*-shell Auger electron, *L*-shell Auger electron
avg	average

ce-K-1, ce-L-2, etc.	K-shell conversion electron for transition 1, L-shell conversion electron for transition 2, etc.
Ci	curie
D	day
H	hour
$K_{\alpha 1}$, $K_{\alpha 2}$, $K\beta$; L	K x rays; L x ray
M	minute
max	maximum
rad	unit of absorbed dose, equal to 6.25 × 10^7 MeV/g
S	second
Y	year (365.242 days)
$\alpha 1$, $\alpha 2$, . . .	alpha particle corresponding to transition 1, 2, etc.
$\beta 1$, $\beta 2$, . . .	beta particle corresponding to transition 1, 2, etc.
$\gamma 1$, $\gamma 2$, . . .	gamma ray corresponding to transition 1, 2, etc.
$\gamma\pm$	annihilation radiation
Δ	mean energy emitted per time integral of activity (See Appendix A.1.4.4)
μCi-h	microcurie-hour
3.624 12	3.624 ± 0.012
3.6 H 12	3.6 ± 1.2 hours
2.1 E5 Y 2	(2.1 ± 0.2) × 10^5 years

A.1.4.4 *Mean Energy Emitted per Time Integral of Activity (Δ).* When a radiopharmaceutical is administered to an individual, it is possible to estimate the absorbed dose to various organs from a knowledge of the administered activity, A_o, the kinetics of the administered material, the anatomy of the organs of interest, and the radiations emitted in the nuclear transition.[32] The activity administered to an individual will be distributed in the organs and tissues of the body in some more or less complicated manner. In most organs of the body the activity will be zero at the time of administration, will rise to some maximum value at a later time, and will fall to zero at a sufficiently long time. The activity in the organ of interest, considered here to be the source organ, can be represented as $A_s(t) = A_o$ f(t),

[32] At the time of writing (January 1984), the Medical Internal Radiation Dose (MIRD) Committee of the Society of Nuclear Medicine is preparing for publication a detailed introduction to the MIRD method of calculating the internal dose from radiopharmaceuticals. It will be entitled "MIRD Primer and Collected Dose Estimate Reports," to be published by the Society of Nuclear Medicine.

where A_o is the activity administered at time $t = 0$, and $f(t)$ is the activity in the source organ expressed as a fraction of the administered activity. The time integral of the activity in the source organ is

$$\int_0^\infty A_s(t) \, dt = A_o \int_0^\infty f(t) \, dt = A_o\tau \qquad \text{(A.3)}$$

where the time integral of the fraction $f(t)$ is represented by the symbol τ, and is called the *residence time*.

The mean absorbed dose per unit administered activity, $\bar{D}/A_o$, to another organ (the target organ), due to the activity in the source organ, can be given in the form (Loevinger and Berman, 1976; ICRU, 1979b; ICRP, 1975):

$$D(t \leftarrow s)/A_o = \tau \sum_i \Delta_i \phi_i(t \leftarrow s)/m_t. \qquad \text{(A.4)}$$

In this equation $\phi_i(t \leftarrow s)$ is the fraction of the energy of the ith radiation type, from the radionuclide in the source organ, that is imparted to the target organ, and is called the *absorbed fraction*. The mass of the target organ is m_t. (The quotient $\phi_i(t \leftarrow s)/m_t = \Phi_i(t \leftarrow s)$ is called the *specific absorbed fraction*.)

Δ_i is the mean energy of the ith radiation type emitted per unit time integral of activity, and can be expressed as

$$\Delta_i = kP_iE_i \qquad \text{(A.5)}$$

where P_i is the probability of emission of an i-type particle per nuclear transition, and E_i is the mean energy of an i-type particle. (Photons are considered to be particles, for this purpose.) The value of k is dependent upon the system of units used, and is equal to unity for the Système International d'Unités (SI). Then using SI units

$$\Delta_i = P_iE_i \qquad \text{(A.6)}$$

with the units J/Bq s. Since 1 Bq s = 1 nuclear transition, Δ_i can also be considered the mean energy of the ith radiation type emitted per nuclear transition, when using SI units.

Using traditional units, namely the microcurie for the unit of activity, the rad for absorbed dose, the kilo-electron volt for the mean energy E_i, and the gram and hour for mass and time, respectively, Δ_i is given by

$$\Delta_i = 0.00213 \, P_iE_i \qquad \text{(A.7)}$$

with the units rad g/μCi h. It should be noted that, while these conventional units are dimensionally correct, they are conceptually in error, since Δ_i is the energy *emitted* per time integral of activity and the rad is a unit of energy *imparted* per unit mass. Nevertheless, these traditional units are widely used in current practice, and are the units used in Appendix A.3 for tabulating Δ_i.

The total mean energy emitted per time integral of activity is

$$\Delta = \sum_i \Delta_i. \tag{A.8}$$

The SI units of Δ are J/Bq s, as in Eq. A.6, and Δ can then properly be considered the total mean energy emitted per nuclear transition. The traditional units of Δ are rad g/μCi h, as in Eq. A.7, and Δ must then be considered the total mean energy emitted for a certain number of nuclear transitions.

Within an infinite homogeneous medium that contains a uniform distribution of activity, the energy emitted per unit mass must just equal the energy absorbed per unit mass. Under these conditions, Δ is the dose rate from unit activity per unit mass (in any consistent system of units). For this reason, Δ was formerly called the *equilibrium absorbed-dose constant*.

Appendix A.2 Selected Radionuclides of biomedical importance

A.2 List of Radionuclides

3H B- DECAY (12.35 Y 1)	51CR EC DECAY (27.702 D 4)	76AS B- DECAY (26.32 H 7)
7BE EC DECAY (53.284 D 4)	52MN EC DECAY (5.591 D 3)	77AS B- DECAY (38.8 H 3)
11C B+ DECAY (20.38 M 2)	52MN EC DECAY (21.1 M 2)	77BR EC DECAY (56 H 2)
13N B+ DECAY (9.965 M 1)	52MN IT DECAY (21.1 M 2)	79KR B+ DECAY (35.04 H 10)
14C B- DECAY (5730 Y 40)	52FE EC DECAY (8.275 H 8)	81KR EC DECAY (2.1E5 Y 2)
15O B+ DECAY (122.24 S 14)	54MN EC DECAY (312.14 D 5)	81KR IT DECAY (13 S 1)
16N B- DECAY (7.12 S 2)	55FE EC DECAY (2.72 Y 2)	82BR B- DECAY (35.34 H 2)
18F B+ DECAY (109.71 M 2)	56MN B- DECAY (2.577 H 1)	82RB EC DECAY (1.273 M 2)
22NA B+ DECAY (2.602 Y 1)	56CO EC DECAY (77.9 D 12)	82SR EC DECAY (25.55 D 15)
24NA B- DECAY (14.959 H 4)	57CO EC DECAY (271.7 D 2)	84RB B+ DECAY (32.77 D 4)
26AL B+ DECAY (7.2E5 Y 3)	57NI B+ DECAY (35.9 H 2)	84RB B- DECAY (32.77 D 4)
27MG B- DECAY (9.458 M 12)	58CO B+ DECAY (70.82 D 3)	85KR B- DECAY (10.72 Y 1)
28MG B- DECAY (20.91 H 3)	59FE B- DECAY (44.51 D 2)	85KR IT DECAY (4.480 H 8)
28AL B- DECAY (2.244 M 3)	60CO B- DECAY (5.2714 Y 5)	85KR B- DECAY (4.480 H 8)
32SI B- DECAY (104 Y 13)	60CU B+ DECAY (23.2 M 3)	85SR EC DECAY (64.854 D 3)
32P B- DECAY (14.29 D 2)	61CU EC DECAY (3.37 H 4)	86RB B- DECAY (18.64 D 2)
33P B- DECAY (25.34 D 12)	62CU B+ DECAY (9.74 M 2)	86Y B+ DECAY (14.74 H 2)
35S B- DECAY (87.44 D 7)	63NI B- DECAY (96 Y 4)	86ZR EC DECAY (16.5 H 1)
36CL EC DECAY (3.01E5 Y 2)	63ZN B+ DECAY (38.47 M 5)	87KR B- DECAY (76.3 M 5)
36CL B- DECAY (3.01E5 Y 2)	64CU B+ DECAY (12.701 H 2)	87SR EC DECAY (2.805 H 3)
38CL B- DECAY (36.51 M 4)	64CU B- DECAY (12.701 H 2)	87SR IT DECAY (2.805 H 3)
39AR B- DECAY (269 Y 3)	65ZN EC DECAY (243.9 D 1)	87Y EC DECAY (80.3 H 3)
41AR B- DECAY (1.827 H 7)	66GA B+ DECAY (9.40 H 7)	88KR B- DECAY (2.84 H 2)
42K B- DECAY (12.36 H 1)	67CU B- DECAY (2.575 D 3)	88RB B- DECAY (17.8 M 1)
43K B- DECAY (22.3 H 1)	67GA EC DECAY (3.261 D 1)	88Y EC DECAY (106.61 D 2)
44SC EC DECAY (3.927 H 8)	68GA B+ DECAY (68.0 M 2)	89RB B- DECAY (15.2 M 1)
44SC EC DECAY (58.6 H 1)	68GE EC DECAY (270.8 D 3)	89SR B- DECAY (50.5 D 1)
44SC IT DECAY (58.6 H 1)	69ZN B- DECAY (57 M 1)	90SR B- DECAY (28.5 Y 2)
44TI EC DECAY (47.3 Y 12)	69ZN IT DECAY (13.76 H 3)	90Y B- DECAY (64.0 H 1)
45CA B- DECAY (163 D 1)	72GA B- DECAY (14.10 H 1)	94NB B- DECAY (2.03E4 Y 16)
46SC EC DECAY (83.79 D 2)	72AS EC DECAY (26.0 H 1)	94NB IT DECAY (6.26 M 1)
46SC IT DECAY (18.70 S 5)	73AS EC DECAY (80.30 D 6)	94NB B- DECAY (6.26 M 1)
47CA B- DECAY (4.536 D 2)	73SE EC DECAY (7.15 H 8)	97ZR B- DECAY (16.90 H 5)
47SC B- DECAY (3.351 D 2)	74AS B+ DECAY (17.78 D 2)	97NB B- DECAY (72.1 M 7)
48SC B- DECAY (43.7 H 1)	74AS B- DECAY (17.78 D 3)	97NB IT DECAY (60 S 1)
48V EC DECAY (15.974 D 3)	75SE EC DECAY (119.8 D 1)	97RU EC DECAY (2.9 D 1)

```
99MO B-,99TC IT DECAY (65.92 H 2)
99TC IT DECAY      (6.007 H 1)
103RU B- DECAY     (39.26 D 2)
103RH IT DECAY     (56.114 M 6)
103PD EC DECAY     (16.97 D 2)
108AG EC DECAY     (2.37 M 1)
108AG B- DECAY     (2.37 M 1)
108AG EC DECAY     (127 Y 21)
108AG IT DECAY     (127 Y 21)
109PD B- DECAY     (13.427 H 14)
109CD EC DECAY     (462.6 D 4)
110AG EC DECAY     (24.6 S 2)
110AG B- DECAY     (24.6 S 2)
110AG IT DECAY     (249.8 D 1)
110AG B- DECAY     (249.8 D 1)
111AG B- DECAY     (7.45 D 1)
111IN EC DECAY     (2.805 D 1)
113IN IT DECAY     (1.658 H 1)
113SN EC DECAY     (115.08 D 3)
115IN IT DECAY     (4.486 H 4)
115IN B- DECAY     (4.486 H 4)
116IN B- DECAY     (54.15 M 6)
117SN IT DECAY     (13.61 D 4)
117SB EC DECAY     (2.80 H 1)
119SN IT DECAY     (293.0 D 13)
121SN IT DECAY     (55 Y 5)
121SN B- DECAY     (55 Y 5)
121TE EC DECAY     (17 D 1)
121TE EC DECAY     (154 D 7)
121TE IT DECAY     (154 D 7)
122SB EC DECAY     (2.70 D 1)
122SB B- DECAY     (2.70 D 1)
123I EC DECAY      (13.221 H 3)
124SB B- DECAY     (60.20 D 3)
124I B+ DECAY      (4.18 D 2)
125SB B- DECAY     (2.758 Y 1)
125TE IT DECAY     (57.40 D 5)
125I EC DECAY      (59.6 D 2)
126I EC DECAY      (13.02 D 7)

140LA B- DECAY     (40.28 H 1)
141CE B- DECAY     (32.50 D 1)
142PR B- DECAY     (19.13 H 4)
143CE B- DECAY     (33.0 H 2)
143PR B- DECAY     (13.58 D 3)
144CE B- DECAY     (285.0 D 1)
144PR B- DECAY     (17.28 M 5)
145PM EC DECAY     (17.7 Y 4)
147ND B- DECAY     (10.98 D 1)
147PM B- DECAY     (2.6234 Y 2)
148PM B- DECAY     (5.37 D 1)
148PM IT DECAY     (41.3 D 1)
148PM B- DECAY     (41.3 D 1)
149ND B- DECAY     (1.73 H 1)
149PM B- DECAY     (53.08 H 5)
152EU EC DECAY     (13.4 Y 1)
152EU B- DECAY     (13.4 Y 1)
152EU EC DECAY     (9.274 H 3)
152EU B- DECAY     (9.274 H 3)
153SM B- DECAY     (46.7 H 1)
153GD EC DECAY     (242 D 1)
154EU B- DECAY     (8.55 Y 6)
155EU B- DECAY     (4.72 Y 2)
157DY EC DECAY     (8.1 H 1)
157TB EC DECAY     (99 Y 10)
166HO B- DECAY     (1200 Y 180)
169YB EC DECAY     (32.03 D 1)
170TM B- DECAY     (128.6 D 3)
171ER B- DECAY     (7.52 H 3)
171TM B- DECAY     (1.92 Y 1)
175YB B- DECAY     (4.19 D 1)
177LU B- DECAY     (6.71 D 1)
178TA EC DECAY     (2.45 H 5)
180TA EC DECAY     (8.152 H 6)
180TA B- DECAY     (8.152 H 6)
181HF B- DECAY     (42.4 D 1)
181W EC DECAY      (121.2 D 3)
181HF B- DECAY     (42.4 D 1)
181W EC DECAY      (121.2 D 3)
```

126I B- DECAY (13.02 D 7)	
127XE EC DECAY (36.4 D 1)	
128CS EC DECAY (3.62 M 2)	
128BA EC DECAY (2.43 D 5)	
129TE B- DECAY (69.6 M 2)	
129TE IT DECAY (33.6 D 1)	
129TE B- DECAY (33.6 D 1)	
129I B- DECAY (1.57E7 Y 4)	
129XE IT DECAY (8.0 D 2)	
129CS EC DECAY (32.06 H 6)	
130I B- DECAY (12.36 H 1)	
131I B- DECAY (8.021 D 1)	
131XE IT DECAY (11.9 D 1)	
131CS EC DECAY (9.69 D 1)	
131BA EC DECAY (11.8 D 2)	
132TE B- DECAY (76.3 H 2)	
132I B- DECAY (2.30 H 3)	
132CS B+ DECAY (6.475 D 10)	
132CS B- DECAY (6.475 D 10)	
133I B- DECAY (20.8 H 1)	
133XE B- DECAY (5.243 D 1)	
133XE IT DECAY (2.19 D 1)	
133BA EC DECAY (10.5 Y 1)	
133BA IT DECAY (38.9 H 1)	
134TE B- DECAY (41.8 M 8)	
134I B- DECAY (52.6 M 5)	
134CS B- DECAY (2.065 Y 1)	
135I B- DECAY (6.61 H 1)	
135XE B- DECAY (9.09 H 1)	
135XE IT DECAY (15.29 M 3)	
135BA IT DECAY (28.7 H 2)	
136CS B- DECAY (13.1 D 1)	
137CS B- DECAY (30.0 Y 2)	
137BA IT DECAY (2.553 M 1)	
138XE B- DECAY (14.08 M 8)	
138CS B- DECAY (32.2 M 1)	
139BA B- DECAY (82.8 M 2)	
139CE EC DECAY (137.64 D 2)	
140BA B- DECAY (12.76 D 1)	

182TA B- DECAY (114.43 D 4)	
185W B- DECAY (75.1 D 3)	
187W B- DECAY (23.9 H 1)	
188W B- DECAY (69.4 D 5)	
188RE B- DECAY (16.98 H 2)	
191OS B- DECAY (15.4 D 2)	
191IR IT DECAY (4.94 S 3)	
192IR EC DECAY (73.83 D 1)	
192IR B- DECAY (73.83 D 18)	
194IR B- DECAY (19.15 H 3)	
195PT IT DECAY (4.02 D 1)	
195AU EC DECAY (183 D 2)	
195AU IT DECAY (30.5 S 2)	
197PT B- DECAY (18.3 H 3)	
197HG EC DECAY (64.1 H 1)	
198AU B- DECAY (2.696 D 2)	
199AU B- DECAY (3.139 D 7)	
201TL EC DECAY (72.91 H 2)	
203HG B- DECAY (46.60 D 1)	
203PB EC DECAY (51.88 H 1)	
206BI EC DECAY (6.243 D 3)	
207BI EC DECAY (32.2 Y 9)	
226RA A DECAY (1600 Y 7)	(Ra)
222RN A DECAY (3.8235 D 3)	(Rn)
218PO A DECAY (3.05 M)	(RaA)
214PB B- DECAY (26.8 M)	(RaB)
214BI B- DECAY (19.9 M 4)	(RaC)
214PO A DECAY (164.3 US 18)	(RaC')
210TL B- DECAY (1.30 M 3)	(RaC")
210PB B- DECAY (22.3 Y 2)	(RaD)
210BI B- DECAY (5.013 D 5)	(RaE)
210PO A DECAY (138.376 D 2)	(RaF)
232TH A DECAY (1.405E10 Y 6)	(Th)
228RA B- DECAY (5.75 Y 3)	(MsTh1)
228AC B- DECAY (6.13 H)	(MsTh2)
228TH A DECAY (1.913 Y 2)	(RdTh)
224RA A DECAY (3.66 D 4)	(ThX)
220RN A DECAY (55.6 S 1)	(Tn)
216PO A DECAY (0.145 S 2)	(ThA)

212PB B- DECAY (10.64 H 1)	(ThB)
212BI B- DECAY (60.55 M 6)	(ThC)
212BI A DECAY (60.55 M 6)	(ThC)
212PO A DECAY (0.298 US 3)	(ThC')
208TL B- DECAY (3.053 M 4)	(ThC")
238PU A DECAY (87.74 Y 4)	
239NP B- DECAY (2.355 D 4)	
240PU A DECAY (6564 Y 11)	
241AM A DECAY (432.2 Y 5)	
241PU B- DECAY (14.4 Y 2)	
242PU A DECAY (3.76E5 Y 2)	
242CM A DECAY (162.8 D 4)	
243AM A DECAY (7380 Y 40)	

Appendix A.3   Table of Nuclear-Decay Data (as of April 1984)

3H B- DECAY (12.35 Y 1)   I(MIN) = 0.10%

Radiation Type	Energy (keV)	Intensity (%)	Δ(g-rad/μCi-h)
β^- 1 max	18.600 10		
avg	5.685 4	100	0.0121

7BE EC DECAY (53.284 D 4)   I(MIN) = 0.10%

Radiation Type	Energy (keV)	Intensity (%)	Δ(g-rad/μCi-h)
γ 1	477.606 3	10.32 4	0.105

14C B- DECAY (5730 Y 40)   I(MIN) = 0.10%

Radiation Type	Energy (keV)	Intensity (%)	Δ(g-rad/μCi-h)
β^- 1 max	156.475 9		
avg	49.469 3	100	0.105

150 B+ DECAY (122.24 S 14)   I(MIN) = 0.10%

Radiation Type	Energy (keV)	Intensity (%)	Δ(g-rad/μCi-h)
Auger-K	0.38	0.107 6	≈0
β^+ 1 max	1732.0 7		
avg	735.3 4	99.900 6	1.56

Maximum γ± intensity = 199.80%

11C B+ DECAY (20.38 M 2) I(MIN)= 0.10%

Radiation Type	Energy (keV)	Intensity (%)	Δ(g-rad/μCi-h)
Auger-K	0.17	0.230 10	≈0
β⁺ 1 max	960.1 11		
avg	385.6 5	99.760 10	0.819

Maximum γ± intensity =199.52%

13N B+ DECAY (9.965 M 4) I(MIN)= 0.10%

Radiation Type	Energy (keV)	Intensity (%)	Δ(g-rad/μCi-h)
Auger-K	0.26	0.170 10	≈0
β⁺ 1 max	1198.5 3		
avg	491.80 20	99.800 10	1.05

Maximum γ± intensity =199.60%

16N B- DECAY (7.12 S 2) I(MIN)= 0.10%

Radiation Type	Energy (keV)	Intensity (%)	Δ(g-rad/μCi-h)
β⁻ 1 max	1547.0 24		
avg	630.5 11	1.00 20	0.0134
β⁻ 2 max	3302.1 23		
avg	1461.6 12	4.9 4	0.153
β⁻ 3 max	4288.5 23		
avg	1941.3 12	68.0 20	2.81
β⁻ 4 max	10418.9 23		
avg	4979.3 12	26.0 20	2.76
Total β⁻ avg	2695.2 15	100 3	5.74

1 weak β's omitted (ΣI_β = 0.01%)

γ 3	1754.9 5	0.13 4	0.0049
γ 5	2741.2 5	0.76 15	0.0444
γ 8	6129.17 5	69.0 20	9.01
γ 10	7115.15 14	5.0 4	0.758

7 weak γ's omitted (ΣI_γ = 0.16%)

18F B+ DECAY (109.71 M 2)   I(MIN)= 0.10%

Radiation Type	Energy (keV)	Intensity (%)	Δ(g-rad/μCi-h)
Auger-K	0.52	3.08 10	≈0
β⁺ 1 max	633.2 7		
avg	249.7 3	96.90 17	0.515

Maximum γ± intensity =193.80%

22NA B+ DECAY (2.602 Y 1)   I(MIN)= 0.10%

Radiation Type	Energy (keV)	Intensity (%)	Δ(g-rad/μCi-h)
Auger-K	0.82	9.20 5	0.0002
β⁺ 1 max	545.4 5		
avg	215.50 21	89.84 11	0.412
	1 weak β's omitted (ΣIβ = 0.06%)		
K X-ray 1	0.84	0.12 4	≈0
γ	1274.542 7	99.940 20	2.71

Maximum γ± intensity =179.80%

27MG B- DECAY (9.458 M 12)   I(MIN)= 0.10%

Radiation Type	Energy (keV)	Intensity (%)	Δ(g-rad/μCi-h)
β⁻ 1 max	1595.7 11		
avg	646.1 5	29.0 4	0.399
β⁻ 2 max	1766.3 11		
avg	724.7 5	71.0 4	1.10
Total β⁻ avg	701.9 5	100.0 6	1.50
γ 1	170.686 15	0.84 3	0.0031
γ 2	843.76 3	71.8 4	1.29
γ 3	1014.44 4	28.2 4	0.609

28MG B- DECAY (20.91 H 3)   I(MIN)= 0.10%

Radiation Type	Energy (keV)	Intensity (%)	Δ(g-rad/μCi-h)
Auger-K	1.39	26 6	0.0008
ce-K-1	29.080 20	27 7	0.0168
ce-L-1	30.522 20	2.6 7	0.0017
β⁻ 1 max	211.8 20		
avg	65.2 7	4.60 20	0.0064
β⁻ 2 max	458.9 20		
avg	155.9 8	94.0 20	0.312
β⁻ 3 max	859.6 20		
avg	319.3 9	0.21 12	0.0014
Total β⁻ avg	152.0 9	98.8 21	0.320

Radiation Type	Energy (keV)	Intensity (%)	Δ(g-rad/μCi-h)
K X-ray	1.48	0.97 24	≈0
γ 1	30.640 20	66. 4	0.0431
γ 2	400.690 20	36.6 10	0.312
γ 5	941.45 3	38.3 10	0.768
γ 8	1342.25 3	52.6 16	1.50
γ 9	1372.89 6	4.70 20	0.137
γ 10	1589.36 3	4.20 20	0.142
γ 11	1620.00 15	0.30 10	0.0104

4 weak γ's omitted (ΣIγ = 0.13%)

28AL B- DECAY (2.244 M 3)   I(MIN)= 0.10%

Radiation Type	Energy (keV)	Intensity (%)	Δ(g-rad/μCi-h)
β⁻ 1 max	2862.8 3		
avg	1241.60 20	100	2.64
γ 1	1778.987 15	100	3.79

24NA B- DECAY (14.959 H 4)   I(MIN)= 0.10%

Radiation Type	Energy (keV)	Intensity (%)	Δ(g-rad/μCi-h)
β⁻ 1 max	1390.8 7		
avg	554.2 4	99.935 4	1.18

3 weak β's omitted (ΣIβ = 0.07%)

γ 2	1368.633 6	100	2.92
γ 3	2754.030 14	99.863 5	5.86

4 weak γ's omitted (ΣIγ = 0.06%)

26AL B+ DECAY (7.2E5 Y 3)   I(MIN)= 0.10%

Radiation Type	Energy (keV)	Intensity (%)	Δ(g-rad/μCi-h)
Auger-K	1.18	16.21 19	0.0004
β⁺ 1 max	1174.0 4		
avg	543.80 20	81.90 20	0.947
K X-ray	1.25	0.44 6	≈0
γ 1	1129.67 10	2.50 20	0.0602
γ 2	1808.65 7	99.76 4	3.84
γ 3	2938.24 13	0.240 20	0.0150

Maximum γ± intensity =163.60%

32SI B- DECAY (104 Y 13)   I(MIN)= 0.10%

Radiation Type	Energy (keV)	Intensity (%)	Δ(g-rad/μCi-h)
β⁻ 1 max	226.8 12		
avg	69.4 5	100	0.148

32P B- DECAY (14.29 D 2)   I(MIN)= 0.10%

Radiation Type	Energy (keV)	Intensity (%)	Δ(g-rad/μCi-h)
β⁻ 1 max	1710.3 6		
avg	694.8 3	100	1.48

33P B- DECAY (25.34 D 12)   I(MIN)= 0.10%

Radiation Type	Energy (keV)	Intensity (%)	Δ(g-rad/μCi-h)
β⁻ 1 max	249.0 20		
avg	76.6 7	100	0.163

36CL B- DECAY (3.01E+5 Y 2)   I(MIN)= 0.10%
%B-=98.1 1
SEE ALSO 36CL EC DECAY

Radiation Type	Energy (keV)	Intensity (%)	Δ(g-rad/μCi-h)
β⁻ 1 max	709.3 3		
avg	251.23 13	98.10 10	0.525

38CL B- DECAY (36.51 M 4)   I(MIN)= 0.10%

Radiation Type	Energy (keV)	Intensity (%)	Δ(g-rad/μCi-h)
β⁻ 1 max	1106.5 9		
avg	420.1 4	32.5 6	0.291
β⁻ 2 max	2748.9 9		
avg	1181.2 5	11.4 8	0.287
β⁻ 3 max	4916.5 9		
avg	2243.9 5	56.0 5	2.68
Total β⁻			
avg	1529.3 8	99.9 12	3.25
γ 1	1642.42 6	32.5 6	1.14
γ 2	2167.51 5	44.0 5	2.03

1 weak γ's omitted (ΣIγ = 0.03%)

35S B- DECAY (87.44 D 7)   I(MIN) = 0.10%

Radiation Type	Energy (keV)	Intensity (%)	Δ(g-rad/µCi-h)
β⁻ 1 max	166.74 23		
avg	48.60 8	100	0.104

36CL EC DECAY (3.01E+5 Y 2)   I(MIN) = 0.10%
%EC=1.9 1
SEE ALSO 36CL B- DECAY

Radiation Type	Energy (keV)	Intensity (%)	Δ(g-rad/µCi-h)
Auger-K	2.1	1.58 9	≈0
K X-ray	2.30	0.130 19	≈0

39AR B- DECAY (269 Y 3)   I(MIN) = 0.10%

Radiation Type	Energy (keV)	Intensity (%)	Δ(g-rad/µCi-h)
β⁻ 1 max	565 5		
avg	218.8 21	100	0.466

41AR B- DECAY (1.827 H 7)   I(MIN) = 0.10%

Radiation Type	Energy (keV)	Intensity (%)	Δ(g-rad/µCi-h)
β⁻ 1 max	1198.5 8		
avg	459.4 4	99.170 20	0.970
β⁻ 2 max	2492.2 8		
avg	1076.8 4	0.780 20	0.0179
Total β⁻ avg	464.1 4	100.00 3	0.989

1 weak β's omitted (ΣIβ = 0.05%)

γ 1	1293.64 4	99.160 20	2.73

1 weak γ's omitted (ΣIγ = 0.05%)

42K B- DECAY (12.36 H 1)   I(MIN)= 0.10%

Radiation Type	Energy (keV)	Intensity (%)	Δ(g-rad/μCi-h)
β⁻ 1 max	1685.7 17		
avg	701.8 8	0.330 20	0.0049
β⁻ 2 max	1998.4 17		
avg	823.3 8	18.1 5	0.317
β⁻ 3 max	3523.1 17		
avg	1564.8 8	81.5 5	2.72
Total β⁻			
avg	1426.3 9	100.0 7	3.04

2 weak β's omitted (ΣIβ = 0.11%)

γ 1	312.75 3	0.329 17	0.0022
γ 6	1524.665 20	18.5 4	0.601

6 weak γ's omitted (ΣIγ = 0.15%)

44SC EC DECAY (3.927 H 8)   I(MIN)= 0.10%

Radiation Type	Energy (keV)	Intensity (%)	Δ(g-rad/μCi-h)
Auger-L	0.3	8.55 16	≈0
Auger-K	3.3	4.22 9	0.0003
β⁺ 1 max	1476 3		
avg	632.7 13	94.37 5	1.27
X-ray Kα₂	3.68809 4	0.244 25	≈0
X-ray Kα₁	3.69168 4	0.48 5	≈0
γ 1	1157.025 11	99.88 4	2.46
γ 2	1499.450 13	0.912 20	0.0291
γ 5	2656.47 3	0.112 4	0.0064

Maximum γ± intensity =188.74%

44SC EC DECAY (58.6 H 1)   I(MIN)= 0.10%
%EC=1.38 2
SEE ALSO 44SC IT DECAY (58.6 H)

Radiation Type	Energy (keV)	Intensity (%)	Δ(g-rad/μCi-h)
Auger-L	0.3	2.10 5	≈0
Auger-K	3.3	1.034 25	≈0
X-ray Kα₁	3.69168 4	0.119 12	≈0
γ 1	1001.877 17	1.38 3	0.0294
γ 2	1126.082 18	1.38 3	0.0331
γ 3	1157.025 11	1.38 3	0.0340

44SC IT DECAY (58.6 H 1)   I(MIN)= 0.10%
%IT=98.62 2
SEE ALSO 44SC EC DECAY (58.6 H)

Radiation Type	Energy (keV)	Intensity (%)	Δ(g-rad/μCi-h)
Auger-L	0.37	17.6 6	0.0001
Auger-K	3.64	8.6 3	0.0007
ce-K-1	266.752 9	10.6 4	0.0600
ce-L-1	270.745 9	1.07 4	0.0061
ce-MNO-1	271.191 9	0.3550 13	0.0021

(continued on next page)

43K B- DECAY (22.3 H 1)   I(MIN)= 0.10%

Radiation Type	Energy (keV)	Intensity (%)	Δ(g-rad/μCi-h)
β- 1 max	422 10		
avg	137 4	2.24 9	0.0065
β- 2 max	827 10		
avg	298 5	92.1 13	0.585
β- 3 max	1224 10		
avg	469 5	3.7 4	0.0370
β- 4 max	1444 10		
avg	568 5	0.8 13	0.0097
β- 5 max	1817 10		
avg	762 5	1.3 3	0.0211
Total β-			
avg	309 6	100.1 19	0.659
γ 1	184.00 20	0.27 6	0.0011
γ 2	220.608 18	4.11 22	0.0193
γ 3	372.763 15	87.3 5	0.693
γ 4	396.870 20	11.43 12	0.0966
γ 5	404.30 20	0.109 8	0.0009
γ 6	593.40 8	11.0 3	0.139
γ 7	617.494 25	80.5 14	1.06
γ 8	800.8 10	0.147 10	0.0025
γ 9	990.25 20	0.33 7	0.0069
γ 10	1015.1 10	0.16 7	0.0034
γ 11	1021.79 18	1.88 8	0.0409
γ 12	1394.2 7	0.102 12	0.0030

44SC IT DECAY (58.6 H 1) (CONTINUED)

Radiation Type	Energy (keV)	Intensity (%)	Δ(g-rad/μCi-h)
X-ray Kα₂	4.08610 20	0.59 6	≈0
X-ray Kα₁	4.09060 20	1.18 11	0.0001
X-ray Kβ	4.46	0.232 22	≈0
γ 1	271.245 9	86.6 3	0.500

44TI EC DECAY (47.3 Y 12)   I(MIN)= 0.10%

Radiation Type	Energy (keV)	Intensity (%)	Δ(g-rad/μCi-h)
Auger-L	0.37	172 6	0.0014
Auger-K	3.64	83 3	0.0064
ce-K-1	63.36 3	10.8 21	0.0146
ce-L-1	67.35 3	1.1 3	0.0015
ce-MNO-1	67.80 3	0.35 7	0.0005
ce-K-2	73.91 6	2.9 5	0.0046
ce-L-2	77.90 6	0.28 5	0.0005
X-ray L	0.4	0.34 12	≈0
X-ray Kα₂	4.08610 20	5.8 6	0.0005
X-ray Kα₁	4.09060 20	11.5 11	0.0010
X-ray Kβ	4.46	2.26 21	0.0002
γ 1	67.85 3	87.7 15	0.127

46SC IT DECAY (18.70 S 5)   I(MIN)= 0.10%

Radiation Type	Energy (keV)	Intensity (%)	Δ(g-rad/μCi-h)
Auger-L	0.37	54.7 14	0.0004
Auger-K	3.64	26.6 8	0.0021
ce-K-1	138.035 3	32.8 6	0.0964
ce-L-1	142.028 3	3.34 11	0.0101
ce-MNO-1	142.474 3	1.10 4	0.0033
X-ray L	0.4	0.11 4	≈0
X-ray Kα₂	4.08610 20	1.84 16	0.0002
X-ray Kα₁	4.09060 20	3.7 4	0.0003
X-ray Kβ	4.46	0.72 7	≈0
γ 1	142.528 3	62.7 7	0.190

47CA B- DECAY (4.536 D 2)   I(MIN)= 0.10%

Radiation Type	Energy (keV)	Intensity (%)	Δ(g-rad/μCi-h)
β⁻ 1 max	691 4		
avg	241.0 17	82.0 20	0.421
β⁻ 2 max	1988 4		

γ 2	78.40 6	94.7 15	0.158
γ 3	147.0 15	0.10 3	0.0003

45CA B- DECAY (163 D 1)   I(MIN)= 0.10%

Radiation Type	Energy (keV)	Intensity (%)	Δ(g-rad/ μCi-h)
β⁻ 1 max	256.7 10		
avg	77.1 4	100	0.164

46SC B- DECAY (83.79 D 2)   I(MIN)= 0.10%

Radiation Type	Energy (keV)	Intensity (%)	Δ(g-rad/ μCi-h)
β⁻ 1 max	356.8 7		
avg	111.8 3	99.9964 7	0.238
γ 1	889.277 3	99.9840 10	1.89
γ 2	1120.545 4	99.9870 10	2.39

avg	817.0 19	18.0 20	0.313
Total β⁻ avg	344.7 22	100 3	0.735

2 weak β's omitted (ΣIβ = 0.11%)

γ 2	489.23 10	6.7 3	0.0702
γ 3	530.4 3	0.105 16	0.0012
γ 4	767.0 3	0.195 16	0.0032
γ 5	807.86 10	6.9 3	0.119
γ 6	1297.09 10	74.9 19	2.07

2 weak γ's omitted (ΣIγ = 0.03%)

47SC B- DECAY (3.351 D 2)   I(MIN)= 0.10%

Radiation Type	Energy (keV)	Intensity (%)	Δ(g-rad/ μCi-h)
Auger-L	0.42	0.347 18	≈0
Auger-K	4	0.175 9	≈0
ce-K-1	154.415 15	0.224 10	0.0007
β⁻ 1 max	441.2 20		
avg	142.7 8	68.0 20	0.207

(continued on next page)

48V EC DECAY (15.974 D 3) (CONTINUED)

Radiation Type	Energy (keV)	Intensity (%)	Δ(g-rad/μCi-h)
γ 7	2240.395 8	2.41 4	0.115

2 weak γ's omitted (ΣIγ = 0.02%)
Maximum γ± intensity =100.77%

51CR EC DECAY (27.702 D 4)   I(MIN)= 0.10%

Radiation Type	Energy (keV)	Intensity (%)	Δ(g-rad/μCi-h)
Auger-L	0.47	144.7 12	0.0014
Auger-K	4.38	66.9 7	0.0062
X-ray L	0.5	0.33 12	≈0
X-ray $K\alpha_2$	4.94464 4	6.59 21	0.0007
X-ray $K\alpha_1$	4.95220 4	13.1 4	0.0014
X-ray $K\beta$	5.43	2.62 9	0.0003
γ 1	320.0842 9	9.85 9	0.0672

47SC B- DECAY (3.351 D 2) (CONTINUED)

Radiation Type	Energy (keV)	Intensity (%)	Δ(g-rad/μCi-h)
β^- 2 max	600.6 20		
avg	204.1 8	32.0 20	0.139
Total β^- avg	162.3 9	100 3	0.346
γ 1	159.381 15	68.0 20	0.231

48SC B- DECAY (43.7 H 1)   I(MIN)= 0.10%

Radiation Type	Energy (keV)	Intensity (%)	Δ(g-rad/μCi-h)
β^- 1 max	481 6		
avg	157.9 23	9.85 8	0.0331
β^- 2 max	657 6		
avg	226.5 25	90.0 5	0.434
Total β^- avg	220 3	99.8 5	0.467
γ 1	175.361 5	7.47 7	0.0279
γ 2	983.526 12	100.0 3	2.09
γ 3	1037.522 12	97.5 5	2.15

	Energy (keV)	Intensity (%)	Δ(g-rad/μCi-h)
γ 4	1212.880 17	2.38 3	0.0615
γ 5	1312.120 12	100.0 5	2.79

48V EC DECAY (15.974 D 3)   I(MIN)= 0.10%

Radiation Type	Energy (keV)	Intensity (%)	Δ(g-rad/μCi-h)
Auger-L	0.42	73.6 19	0.0007
Auger-K	4	34.8 9	0.0030
β⁺ 1 max	697 3		
avg	291.4 13	50.30 20	0.312

1 weak β's omitted (ΣIβ = 0.09%)

Radiation Type	Energy (keV)	Intensity (%)	Δ(g-rad/μCi-h)
X-ray L	0.45	0.15 6	≈0
X-ray Kα₂	4.50486 4	2.89 25	0.0003
X-ray Kα₁	4.51084 4	5.7 5	0.0006
X-ray Kβ	5	1.15 10	0.0001
γ 1	803.23 8	0.150 20	0.0026
γ 2	928.327 5	0.77 5	0.0152
γ 3	944.132 5	7.76 9	0.156
γ 4	983.521 7	99.9940 20	2.09
γ 5	1312.096 6	97.49 20	2.72
γ 6	1436.80 10	0.120 20	0.0037

(continued on next column)

52MN EC DECAY (5.591 D 3)   I(MIN)= 0.10%

Radiation Type	Energy (keV)	Intensity (%)	Δ(g-rad/μCi-h)
Auger-L	0.54	100.2 10	0.0012
Auger-K	4.78	45.1 5	0.0046
β⁺ 1 max	576.0 21		
avg	241.9 10	29.70 10	0.153
X-ray L	0.57	0.26 9	≈0
X-ray Kα₂	5.40551 5	5.25 15	0.0006
X-ray Kα₁	5.41472 5	10.4 3	0.0012
X-ray Kβ	6	2.08 7	0.0003
γ 2	346.03 3	1.000 20	0.0074
γ 4	399.56 5	0.18 8	0.0015
γ 5	502.05 5	0.220 20	0.0024
γ 6	600.18 4	0.390 10	0.0050
γ 7	647.50 4	0.410 20	0.0057
γ 8	744.214 11	90.8 3	1.44
γ 9	848.13 3	3.38 3	0.0611
γ 11	935.520 20	95.0 3	1.89
γ 13	1246.246 12	4.27 7	0.113
γ 14	1247.85 9	0.38 4	0.0101
γ 15	1333.615 16	5.07 3	0.144
γ 16	1434.056 16	100	3.05

9 weak γ's omitted (ΣIγ = 0.37%)

Maximum γ± intensity = 59.40%

54MN EC DECAY (312.14 D 5)   I(MIN)= 0.10%

Radiation Type	Energy (keV)	Intensity (%)	Δ(g-rad/μCi-h)
Auger-L	0.54	142.1 12	0.0016
Auger-K	4.78	63.9 7	0.0065
X-ray L	0.57	0.37 13	≈0
X-ray Kα₂	5.40551 5	7.43 21	0.0009
X-ray Kα₁	5.41472 5	14.7 4	0.0017
X-ray Kβ	5.947 6	2.95 10	0.0004
γ 1	834.843 6	99.9760 10	1.78

55FE EC DECAY (2.72 Y 2)   I(MIN)= 0.10%

Radiation Type	Energy (keV)	Intensity (%)	Δ(g-rad/μCi-h)
Auger-L	0.6	139 4	0.0018
Auger-K	5.19	60.7 21	0.0067
X-ray L	0.64	0.42 14	≈0
X-ray Kα₂	5.88765 3	8.2 7	0.0010
X-ray Kα₁	5.89875 3	16.3 12	0.0020
X-ray Kβ	6.49	3.29 25	0.0005

52MN EC DECAY (21.1 M 2)   I(MIN)= 0.10%

%(EC+B+)=98.32 4
SEE ALSO 52MN IT DECAY (21.1 M).
FEEDING OF 21.1-M 52MN IN 52FE B+ DECAY=100%

Radiation Type	Energy (keV)	Intensity (%)	Δ(g-rad/μCi-h)
Auger-L	0.54	2.37 5	≈0
Auger-K	4.78	1.067 24	0.0001
β⁺ 1 max	905.9 21		
avg	383.3 10	0.164 8	0.0013
β⁺ 2 max	2633.5 21		
avg	1174.1 10	96.5 20	2.41
Total β⁺			
avg	1172.4 10	96.7 20	2.42

3 weak β's omitted (ΣIβ = 0.05%)

X-ray Kα₂	5.40551 5	0.124 5	≈0
X-ray Kα₁	5.41472 5	0.246 8	≈0
γ 4	1434.060 10	98.29 4	3.00
γ 6	1727.53 7	0.216 10	0.0080

12 weak γ's omitted (ΣIγ = 0.17%)
Maximum γ± intensity =193.43%

56MN B- DECAY (2.577 H 1)   I(MIN)= 0.10%

Radiation Type	Energy (keV)	Intensity (%)	Δ(g-rad/μCi-h)
β⁻ 1 max	326.0 9		
avg	99.2 4	1.16 3	0.0025
β⁻ 2 max	735.8 9		
avg	255.3 4	14.6 4	0.0794
β⁻ 3 max	1038.2 9		
avg	382.0 4	27.8 8	0.226
β⁻ 4 max	2848.9 9		
avg	1216.9 5	56.3 9	1.46
Total β⁻			
avg	830.4 7	100.0 13	1.77

3 weak β's omitted (ΣIβ = 0.12%)

γ 1	846.764 6	98.9 3	1.78
γ 4	1810.72 4	27.2 8	1.05
γ 5	2113.05 4	14.3 4	0.645
γ 6	2522.88 6	0.99 3	0.0531
γ 8	2657.45 5	0.653 20	0.0369
γ 9	2959.77 6	0.306 10	0.0193
γ 10	3369.60 7	0.168 10	0.0121

3 weak γ's omitted (ΣIγ = 0.16%)

52MN IT DECAY (21.1 M 2)   I(MIN)= 0.10%
%IT=1.68 4
SEE ALSO 52MN EC DECAY (21.1 M)
FEEDING OF 21.1-M 52MN IN 52FE B+ DECAY=100

Radiation Type	Energy (keV)	Intensity (%)	Δ(g-rad/μCi-h)
γ 1	377.738 11	1.68 5	0.0135

52FE EC DECAY (8.275 H 8)   I(MIN)= 0.10%
SEE ALSO 52MN EC,IT DECAYS (21.1 M)

Radiation Type	Energy (keV)	Intensity (%)	Δ(g-rad/μCi-h)
Auger-L	0.6	62.6 23	0.0008
Auger-K	5.19	27.4 12	0.0030
ce-K-1	162.145 11	0.705 22	0.0024
β⁺ 1 max	804 12		
avg	340 6	56.0 13	0.406
X-ray L	0.64	0.19 7	≈0
X-ray Kα₂	5.88765 3	3.7 3	0.0005
X-ray Kα₁	5.89875 3	7.3 6	0.0009
X-ray Kβ	6.49	1.48 12	0.0002
γ 1	168.684 11	99.200 20	0.356
γ 2	377.738 11	1.68 5	0.0135

Maximum γ± intensity =112.00%

56CO EC DECAY (77.9 D 12)   I(MIN)= 0.10%

Radiation Type	Energy (keV)	Intensity (%)	Δ(g-rad/ μCi-h)
Auger-L	0.67	109.1 12	0.0016
Auger-K	5.62	46.2 6	0.0055
β⁺ 1 max	421.2 20		
avg	178.8 9	1.045 20	0.0040
β⁺ 2 max	1459.0 20		
avg	631.2 9	17.9 4	0.241
Total β⁺			
avg	606.2 10	18.9 4	0.245
X-ray L	0.7	0.34 11	≈0
X-ray Kα₂	6.39084 3	7.28 20	0.0010
X-ray Kα₁	6.40384 3	14.4 4	0.0020
X-ray Kβ	7	2.90 9	0.0004
γ 6	733.72 15	0.193 3	0.0030
γ 7	787.88 7	0.313 5	0.0053
γ 8	846.764 6	99.9900 10	1.80
γ 10	977.49 6	1.420 10	0.0296
γ 11	997.33 16	0.108 6	0.0023
γ 12	1037.844 4	14.06 5	0.311
γ 14	1140.28 10	0.1300 20	0.0032
γ 16	1175.099 8	2.250 20	0.0563
γ 18	1238.287 6	66.7 5	1.76
γ 20	1335.56 8	0.1230 20	0.0035
γ 21	1360.206 6	4.265 17	0.124
γ 22	1442.75 8	0.177 5	0.0054
γ 25	1771.350 15	15.50 7	0.585
γ 26	1810.722 17	0.645 6	0.0249
γ 27	1963.714 12	0.717 6	0.0300
γ 28	2015.179 11	3.030 20	0.130
γ 29	2034.759 11	7.77 5	0.337
γ 30	2113.107 12	0.380 5	0.0171
γ 31	2212.921 10	0.391 6	0.0184

57CO EC DECAY (271.7 D 2)   (CONTINUED)

Radiation Type	Energy (keV)	Intensity (%)	Δ(g-rad/ μCi-h)
γ 1	14.4127 4	9.54 14	0.0029
γ 2	122.0614 3	85.6 7	0.223
γ 3	136.4743 5	10.58 8	0.0308
γ 9	692.00 3	0.160 5	0.0024

6 weak γ's omitted (ΣIγ = 0.03%)

57NI B+ DECAY (35.9 H 2)   I(MIN)= 0.10%

Radiation Type	Energy (keV)	Intensity (%)	Δ(g-rad/ μCi-h)
Auger-L	0.75	77 3	0.0012
Auger-K	6	31.4 14	0.0041
β⁺ 1 max	324 3		
avg	139.0 13	0.49	0.0015
β⁺ 2 max	485 3		
avg	206.4 13	0.86	0.0038
β⁺ 3 max	738 3		
avg	313.7 13	6.7	0.0448
β⁺ 4 max	865 3		
avg	368.7 13	35	0.275
Total β⁺			
avg	354.3 14	43	0.325

γ	Energy		Intensity		Δ
32	2276.36	16	0.124	4	0.0060
35	2598.460	10	16.98	7	0.940
36	3009.596	17	1.010	10	0.0647
37	3201.954	14	3.04	3	0.207
38	3253.417	14	7.46	10	0.517
39	3272.998	14	1.780	20	0.124
41	3451.154	13	0.890	20	0.0654
42	3547.92	6	0.185	5	0.0140

17 weak γ's omitted (ΣIγ = 0.78%)
Maximum γ± intensity = 37.89%

57CO EC DECAY (271.7 D 2)   I(MIN)= 0.10%

Radiation Type	Energy (keV)		Intensity (%)		Δ(g-rad/μCi-h)
Auger-L	0.67		249	3	0.0036
Auger-K	5.62		105.5	14	0.0126
ce-K-1	7.3007	10	69.6	5	0.0108
ce-L-1	13.5666	6	7.79	22	0.0023
ce-MNO-1	14.3198	10	1.15	7	0.0004
ce-K-2	114.9494	10	1.87	10	0.0046
ce-L-2	121.2153	5	0.183	10	0.0005
ce-K-3	129.3623	11	1.40	12	0.0038
ce-L-3	135.6282	7	0.140	12	0.0004
X-ray L	0.7		0.8	3	≈0
X-ray Kα2	6.39084	3	16.6	5	0.0023
X-ray Kα1	6.40384	3	32.8	8	0.0045
X-ray Kβ	7		6.63	21	0.0010

(continued on next column)

	Energy		Intensity		Δ
X-ray L	0.78		0.27	10	≈0
X-ray Kα2	6.91530	4	5.7	5	0.0008
X-ray Kα1	6.93032	4	11.3	8	0.0017
X-ray Kβ	7.65		2.30	17	0.0004
γ 1	127.19	3	16.2	4	0.0440
γ 8	1046.40	20	0.128	8	0.0029
γ 11	1377.59	4	80.0	4	2.35
γ 14	1757.48	8	6.10	17	0.228
γ 16	1919.43	8	13.6	4	0.556
γ 19	2803.90	20	0.136	8	0.0081

14 weak γ's omitted (ΣIγ = 0.51%)
Maximum γ± intensity = 86.10%

58CO B+ DECAY (70.82 D 3)   I(MIN)= 0.10%

Radiation Type	Energy (keV)		Intensity (%)		Δ(g-rad/μCi-h)
Auger-L	0.67		116.5	13	0.0017
Auger-K	5.62		49.4	6	0.0059
β+ 1 max	475.1	12			
avg	201.3	5	14.950	20	0.0641
X-ray L	0.7		0.36	12	≈0
X-ray Kα2	6.39084	3	7.78	21	0.0011
X-ray Kα1	6.40384	3	15.4	4	0.0021
X-ray Kβ	7		3.10	10	0.0005
γ 1	810.772	14	99.4	3	1.72
γ 2	863.935	18	0.676	10	0.0124
γ 3	1674.68	4	0.517	10	0.0184

Maximum γ± intensity = 29.90%

60CU EC DECAY (23.2 M 3) (CONTINUED)

Radiation Type		Energy (keV)	Intensity (%)	Δ(g-rad/μCi-h)
β+ 9	avg	748.8 8	0.86 7	0.0137
	max	1835.4 21	4.51 19	0.0774
β+10	avg	805.8 8		
	max	1910.7 21	11.5 3	0.206
β+11	avg	840.5 8		
	max	1980.7 21	48.9 22	0.909
β+12	avg	872.9 8		
	max	2478.6 21	2.8 3	0.0659
β+13	avg	1105.1 8		
	max	2945.9 21	15.1 5	0.426
β+14	avg	1325.9 8		
	max	3772.3 21	5.2 24	0.191
Total β+	avg	1721.0 8	93 9	1.93

8 weak β's omitted (ΣIβ = 0.20%)

Radiation Type	Energy (keV)	Intensity (%)	Δ(g-rad/μCi-h)
X-ray Kα₂	7.46089 4	0.84 7	0.0001
X-ray Kα₁	7.47815 4	1.66 13	0.0003
X-ray Kβ	8.26	0.34 3	≈0
γ 1	120.5 3	0.193 18	0.0005
γ 2	467.30 20	3.52 18	0.0350
γ 3	497.90 20	1.67 9	0.0177
γ 4	643.2 3	0.97 6	0.0133
γ 8	826.40 20	21.7 11	0.382
γ 9	839.2 4	0.46 7	0.0082
γ 10	896.1 5	0.13 6	0.0025
γ 11	909.20 20	2.02 9	0.0392
γ 12	952.40 20	2.73 18	0.0553
γ 13	965.2 3	0.30 7	0.0061

59FE B- DECAY (44.51 D 2)   I(MIN)= 0.10%

Radiation Type		Energy (keV)	Intensity (%)	Δ(g-rad/μCi-h)
β- 1	max	130.6 7		
	avg	35.70 20	1.27 4	0.0010
β- 2	max	273.3 7		
	avg	80.9 3	45.7 8	0.0787
β- 3	max	465.6 7		
	avg	149.2 3	53.0 14	0.168
Total β-	avg	116.5 4	100.0 17	0.248

1 weak β's omitted (ΣIβ = 0.08%)

	Energy (keV)	Intensity (%)	Δ(g-rad/μCi-h)
γ 1	142.6520 20	1.00 4	0.0030
γ 2	192.349 5	3.01 8	0.0123
γ 3	334.80 20	0.271 11	0.0019
γ 5	1099.251 4	56.3 14	1.32
γ 6	1291.596 7	43.7 10	1.20

2 weak γ's omitted (ΣIγ = 0.08%)

60CO B- DECAY (5.2714 Y 5)   I(MIN)= 0.10%

Radiation Type		Energy (keV)	Intensity (%)	Δ(g-rad/μCi-h)
β- 1	max	317.92 11		
	avg	95.80 4	99.920 20	0.204

	Energy	Intensity	Δ
γ 17	1035.20 20	3.69 18	0.0814
γ 18	1110.5 4	1.06 18	0.0250
γ 19	1173.2	0.26 9	0.0066
γ 20	1234.2 7	0.11 5	0.0028
γ 21	1293.70 20	1.85 18	0.0509
γ 22	1307.1 6	0.11 3	0.0029
γ 23	1332.5	87.9 3	2.50
γ 24	1420.1 5	0.114 18	0.0035
γ 26	1451.4 5	0.17 3	0.0052
γ 31	1791.6 3	45.4 23	1.73
γ 32	1861.6 3	4.7 3	0.188
γ 33	1919.7 4	0.69 7	0.0284
γ 34	1936.9 3	2.20 9	0.0907
γ 35	2061.0 3	0.79 5	0.0347
γ 36	2158.90 20	3.34 18	0.154
γ 38	2263.6 8	0.11 4	0.0055
γ 41	2389.6 10	0.12 4	0.0063
γ 42	2403.3 6	0.77 8	0.0396
γ 44	2675.3 8	0.13 3	0.0075
γ 45	2687.9 3	0.44 7	0.0252
γ 46	2746.1 3	1.06 9	0.0617
γ 48	2986.3 5	0.123 18	0.0078
γ 49	3124.1 3	4.7 3	0.316
γ 50	3160.8 3	0.58 3	0.0391
γ 51	3194.1 3	2.02 9	0.138
γ 55	3269.4 3	0.77 5	0.0539
γ 63	4020.4 4	0.77 8	0.0655

35 weak γ's omitted (ΣI$_\gamma$ = 1.47%)
Maximum γ± intensity =185.25%

2 weak β's omitted (ΣI$_\beta$ = 0.08%)

	Energy	Intensity	
γ 3	1173.238 4	99.900 20	2.50
γ 4	1332.501 5	99.9824 5	2.84

4 weak γ's omitted (ΣI$_\gamma$ = 0.02%)

60CU EC DECAY (23.2 M 3)   I(MIN)= 0.10%

Radiation Type	Energy (keV)	Intensity (%)	Δ(g-rad/μCi-h)
Auger-L	0.84	10.1 5	0.0002
Auger-K	6.54	4.01 22	0.0006
β⁺ 1 max	611.1 22		
avg	261.2 7	0.190 20	0.0011
β⁺ 2 max	785.9 22		
avg	336.1 7	0.10 10	0.0007
β⁺ 3 max	1026.1 21		
avg	440.8 7	1.17 7	0.0110
β⁺ 4 max	1084.4 22		
avg	466.5 8	0.96 9	0.0095
β⁺ 5 max	1097.2 22		
avg	472.2 8	0.15 3	0.0015
β⁺ 6 max	1179.2 22		
avg	508.5 8	0.24 5	0.0026
β⁺ 7 max	1369.3 22		
avg	593.6 8	0.74 8	0.0094
β⁺ 8 max	1711.2 21		

(continued on next column)

61CU EC DECAY (3.37 H 4)   I(MIN)= 0.10%

Radiation Type	Energy (keV)	Intensity (%)	Δ(g-rad/μCi-h)
Auger-L	0.84	52 5	0.0009
Auger-K	6.54	20.7 19	0.0029
ce-K-1	59.079 3	0.48 9	0.0006
β+ 1 max	560.7 15		
avg	238.8 6	2.7 6	0.0137
β+ 2 max	933.7 15		
avg	399.3 7	5.6 11	0.0476
β+ 3 max	1149.3 15		
avg	494.2 7	2.0 5	0.0211
β+ 4 max	1216.7 15		
avg	524.2 7	51 5	0.569
Total β+ avg	499.0 8	61 6	0.652

1 weak β's omitted (ΣIβ = 0.04%)

Radiation Type	Energy (keV)	Intensity (%)	Δ(g-rad/μCi-h)
X-ray L	0.85	0.20 7	≈0
X-ray Kα2	7.46089 4	4.3 5	0.0007
X-ray Kα1	7.47815 4	8.5 9	0.0014
X-ray Kβ	8.26	1.74 19	0.0003
γ 1	67.412 3	3.9 8	0.0057
γ 5	282.9560 20	12.5 23	0.0753
γ 6	373.050 5	2.1 4	0.0171
γ 8	529.169 22	0.41 11	0.0046
γ 9	588.605 9	1.20 23	0.0150
γ 12	656.008 4	10.7 20	0.149
γ 14	816.692 13	0.36 7	0.0063
γ 16	841.211 17	0.25 5	0.0044
γ 18	908.631 17	1.21 23	0.0235
γ 22	1099.560 19	0.28 6	0.0066
γ 25	1185.234 15	3.7 7	0.0931

63NI B- DECAY (96 Y 4)   I(MIN)= 0.10%

Radiation Type	Energy (keV)	Intensity (%)	Δ(g-rad/μCi-h)
β- 1 max	65.88 15		
avg	17.13 4	100	0.0365

63ZN B+ DECAY (38.47 M 5)   I(MIN)= 0.10%

Radiation Type	Energy (keV)	Intensity (%)	Δ(g-rad/μCi-h)
Auger-L	0.92	9.32 19	0.0002
Auger-K	7	3.56 8	0.0005
β+ 1 max	933.5 18		
avg	400.3 8	0.51 4	0.0043
β+ 2 max	1383.4 18		
avg	600.0 9	4.9 4	0.0626
β+ 3 max	1675.8 18		
avg	732.7 9	7.2 4	0.112
β+ 4 max	2345.5 18		
avg	1042.5 9	80.2 8	1.78
Total β+ avg	991.3 10	92.9 10	1.96

5 weak β's omitted (ΣIβ = 0.05%)

22 weak γ's omitted (ΣIγ = 0.84%)
Maximum γ± intensity =122.67%

62CU B+ DECAY (9.74 M 2)   I(MIN)= 0.10%

Radiation Type	Energy (keV)	Intensity (%)	Δ(g-rad/μCi-h)
Auger-L	0.84	2.88 10	≈0
Auger-K	6.54	1.14 6	0.0002
β+ 1 max	1754 5		
avg	767.5 23	0.135 9	0.0022
β+ 2 max	2927 5		
avg	1316.0 24	97.200 20	2.72
Total β+			
avg	1314.3 24	97.430 23	2.73

2 weak β's omitted (ΣIβ = 0.10%)

X-ray Kα2	7.46089 4	0.239 17	≈0
X-ray Kα1	7.47815 4	0.47 4	≈0
γ 3	875.71 7	0.150 10	0.0028
γ 6	1173.02 10	0.342 14	0.0085

14 weak γ's omitted (ΣIγ = 0.10%)
Maximum γ± intensity =194.86%

X-ray Kα2	8.027830 10	0.848 24	0.0001
X-ray Kα1	8.047780 10	1.67 5	0.0003
X-ray Kβ	9	0.340 11	≈0
γ 5	449.93 5	0.242 21	0.0023
γ 11	669.62 5	8.4 4	0.120
γ 20	962.06 4	6.4 4	0.132
γ 23	1123.72 7	0.113 13	0.0027
γ 34	1412.08 5	0.75 4	0.0226
γ 37	1547.04 6	0.125 8	0.0041

61 weak γ's omitted (ΣIγ = 1.04%)
Maximum γ± intensity =185.71%

64CU B+ DECAY (12.701 H 2)   I(MIN)= 0.10%
%(EC+B+)=62.7 4
SEE ALSO 64CU B- DECAY

Radiation Type	Energy (keV)	Intensity (%)	Δ(g-rad/μCi-h)
Auger-L	0.84	59.1 22	0.0011
Auger-K	6.54	23.3 12	0.0033
β+ 1 max	652.9 9		
avg	278.1 4	17.9 3	0.106
X-ray L	0.85	0.23 8	≈0
X-ray Kα2	7.46089 4	4.9 4	0.0008
X-ray Kα1	7.47815 4	9.6 7	0.0015
X-ray Kβ	8.26	1.96 14	0.0003
γ 1	1345.88 12	0.472 18	0.0135

Maximum γ± intensity = 35.80%

66GA B+ DECAY (9.40 H 7) (CONTINUED)

Radiation Type	Energy (keV)	Intensity (%)	A(g-rad/µCi-h)
γ 22	1356.2	0.38 4	0.0110
γ 23	1356.6	0.128 20	0.0037
γ 24	1357	0.19 8	0.0056
γ 25	1418.88 10	0.654 20	0.0198
γ 26	1459.2 3	0.101 5	0.0031
γ 27	1508.33 10	0.591 21	0.0190
γ 29	1899.18 10	0.440 17	0.0178
γ 30	1918.64 10	2.20 7	0.0898
γ 32	2173.90 20	0.12 3	0.0056
γ 33	2190.00 20	5.83 19	0.272
γ 34	2213.60 20	0.144 9	0.0068
γ 36	2393.30 20	0.257 11	0.0131
γ 37	2422.70 10	2.00 6	0.103
γ 40	2752.10 20	23.8 7	1.39
γ 41	2780.50 20	0.130 6	0.0077
γ 42	2934.30 20	0.222 8	0.0139
γ 46	3229.26 10	1.52 5	0.105
γ 47	3256.57 15	0.101 12	0.0070
γ 48	3381.32 10	1.45 5	0.105
γ 49	3422.64 15	0.84 3	0.0615
γ 50	3433.00 15	0.288 9	0.0210
γ 53	3767.40 20	0.144 9	0.0115
γ 54	3791.47 10	1.04 4	0.0839
γ 58	4086.36 10	1.16 5	0.101
γ 59	4295.70 20	3.57 11	0.327
γ 60	4462.01 15	0.727 22	0.0691
γ 61	4806.59 15	1.51 5	0.155

25 weak γ's omitted (ΣIγ = 1.06%)
Maximum γ± intensity =110.16%

64CU B- DECAY (12.701 H 2)   I(MIN)= 0.10%
%B-=37.3 4
SEE ALSO 64CU B+ DECAY

Radiation Type	Energy (keV)	Intensity (%)	A(g-rad/µCi-h)
β⁻ 1 max	577.9 15		
avg	190.1 6	37.7 5	0.153

65ZN EC DECAY (243.9 D 1)   I(MIN)= 0.10%

Radiation Type	Energy (keV)	Intensity (%)	A(g-rad/µCi-h)
Auger-L	0.92	126.7 18	0.0025
Auger-K	7	48.3 8	0.0072
β⁺ 1 max	329.9 11		
avg	143.0 5	1.42 3	0.0043
X-ray L	0.93	0.57 20	≈0
X-ray Kα₂	8.027830 10	11.5 3	0.0020
X-ray Kα₁	8.047780 10	22.6 5	0.0039
X-ray Kβ	9	4.61 13	0.0009
γ 3	1115.546 4	50.70 13	1.20

Maximum γ± intensity = 2.84%

66GA B+ DECAY (9.40 H 7)   I(MIN)= 0.10%

Radiation Type	Energy (keV)	Intensity (%)	Δ(g-rad/μCi-h)
Auger-L	0.99	57 3	0.0012
Auger-K	7.53	20.7 13	0.0033
β+ 1 max	361 3		
avg	156.8 13	0.98 5	0.0033
β+ 2 max	720 3		
avg	308.7 13	0.170 10	0.0011
β+ 3 max	772 3		
avg	330.9 13	0.70 10	0.0049
β+ 4 max	924 3		
avg	397.0 14	3.67 12	0.0310
β+ 5 max	1780 3		
avg	781.5 14	0.36 3	0.0060
β+ 6 max	4153 3		
avg	1904.1 15	49.2 15	2.00
Total β+ avg	1740.3 21	55.1 15	2.04
X-ray L	1	0.28 12	≈0
X-ray Kα2	8.61578 5	5.7 4	0.0010
X-ray Kα1	8.63886 5	11.1 8	0.0020
X-ray Kβ	9.57	2.26 16	0.0005
γ 4	448.90 10	0.113 5	0.0011
γ 7	686.28 10	0.265 11	0.0039
γ 8	833.56 10	6.20 19	0.110
γ 10	856.70 10	0.124 6	0.0023
γ 11	907.0 3	0.117 9	0.0023
γ 15	1039.29 10	38.9 11	0.861
γ 18	1190.36 10	0.136 13	0.0035
γ 20	1232.9	0.545 22	0.0143
γ 21	1333.20 20	1.27 4	0.0360

(continued on next column)

67CU B- DECAY (2.575 D 3)   I(MIN)= 0.10%

Radiation Type	Energy (keV)	Intensity (%)	Δ(g-rad/μCi-h)
Auger-L	0.99	19.4 13	0.0004
Auger-K	7.53	7.0 6	0.0011
ce-K-1	81.607 17	0.51 3	0.0009
ce-K-2	83.656 13	12.1 8	0.0215
ce-L-2	92.121 13	1.48 9	0.0029
ce-M-2	93.179 13	0.218 14	0.0004
ce-K-3	174.918 20	0.93 11	0.0034
β- 1 max	182 8		
avg	51.0 25	1.13 5	0.0012
β- 2 max	391 8		
avg	121 8	57 3	0.147
β- 3 max	483 8		
avg	154 3	21.6 13	0.0709
β- 4 max	576 8		
avg	189 3	20 4	0.0805
Total β- avg	141 4	100 6	0.299
X-ray Kα2	8.61578 5	1.93 17	0.0004
X-ray Kα1	8.63886 5	3.8 4	0.0007
X-ray Kβ	9.57	0.77 7	0.0002
γ 1	91.266 17	7.0 4	0.0136
γ 2	93.315 13	16.1 9	0.0320
γ 3	184.577 20	49 3	0.191
γ 4	208.951 20	0.115 8	0.0005
γ 5	300.219 25	0.80 5	0.0051
γ 6	393.529 20	0.220 14	0.0018

67GA EC DECAY (3.261 D 1)   I(MIN)= 0.10%

Radiation Type	Energy (keV)	Intensity (%)	Δ(g-rad/ μCi-h)
Auger-L	0.99	168 9	0.0035
Auger-K	7.53	61 4	0.0098
ce-K-1	81.607 11	0.222 8	0.0004
ce-K-2	83.652 11	28.4 11	0.0507
ce-L-2	92.117 11	3.49 14	0.0068
ce-M-2	93.175 11	0.512 20	0.0010
ce-K-3	174.918 14	0.40 5	0.0015
X-ray L	1	0.8 4	≈0
X-ray Kα₂	8.61578 5	16.7 12	0.0031
X-ray Kα₁	8.63886 5	32.8 23	0.0060
X-ray Kβ	9.57	6.7 5	0.0014
γ 1	91.266 11	3.04 7	0.0059
γ 2	93.311 11	37.9 10	0.0754
γ 3	184.577 14	20.9 5	0.0821
γ 4	208.951 14	2.37 6	0.0105
γ 5	300.219 14	16.8 4	0.107
γ 6	393.529 14	4.66 11	0.0391
γ 10	887.693 18	0.147 5	0.0028

3 weak γ's omitted (ΣIγ = 0.13%)

68GE EC DECAY (270.8 D 3)   (CONTINUED)

Radiation Type	Energy (keV)	Intensity (%)	Δ(g-rad/ μCi-h)
Auger-K	8	42.4 7	0.0073
X-ray L	1	0.67 19	≈0
X-ray Kα₂	9.22482 7	13.1 3	0.0026
X-ray Kα₁	9.25174 7	25.6 5	0.0050
X-ray Kβ	10.3	5.45 14	0.0012

69ZN B- DECAY (57 M 1)   I(MIN)= 0.10%

Radiation Type	Energy (keV)	Intensity (%)	Δ(g-rad/ μCi-h)
β⁻ 1 max	906 3		
avg	321.4 13	100	0.685

69ZN IT DECAY (13.76 H 3)   I(MIN)= 0.10%

Radiation Type	Energy (keV)	Intensity (%)	Δ(g-rad/ μCi-h)
Auger-L	0.99	6.3 4	0.0001
Auger-K	7.53	2.28 15	0.0004

68GA B+ DECAY (68.0 M 2)   I(MIN)= 0.10%

Radiation Type	Energy (keV)	Intensity (%)	Δ(g-rad/μCi-h)
Auger-L	0.99	14.0 6	0.0003
Auger-K	7.53	5.1 3	0.0008
β⁺ 1 max	821.7 12		
avg	352.6 6	1.20 6	0.0090
β⁺ 2 max	1899.1 12		
avg	836.0 6	87.80 20	1.56
Total β⁺			
avg	829.5 6	89.00 21	1.57
X-ray Kα₂	8.61578 5	1.40 9	0.0003
X-ray Kα₁	8.63886 5	2.75 18	0.0005
X-ray Kβ	9.57	0.56 4	0.0001
γ 3	1077.40 10	3.24 16	0.0744
γ 6	1883.20 10	0.140 8	0.0056

6 weak γ's omitted (ΣIγ = 0.24%)
Maximum γ± intensity =178.00%

ce-K-1	428.975 18	4.38 14	0.0401
ce-L-1	437.440 18	0.512 19	0.0048
ce-MNO-1	438.498 18	0.171 10	0.0016
X-ray Kα₂	8.61578 5	0.62 5	0.0001
X-ray Kα₁	8.63886 5	1.23 9	0.0002
X-ray Kβ	9.57	0.250 18	≈0
γ 1	438.634 18	94.90 16	0.887

72GA B- DECAY (14.10 H 1)   I(MIN)= 0.10%

Radiation Type	Energy (keV)	Intensity (%)	Δ(g-rad/μCi-h)
Auger-L	1.19	0.62 6	≈0
Auger-K	8.56	0.228 24	≈0
ce-K-20	680.097 7	0.45 5	0.0065
β⁻ 1 max	236 3		
avg	67.5 10	0.150 10	0.0002
β⁻ 2 max	315 3		
avg	93.6 10	0.78 4	0.0016
β⁻ 3 max	427 3		
avg	132.9 11	0.230 10	0.0007
β⁻ 4 max	538 3		
avg	173.8 11	0.340 10	0.0013
β⁻ 5 max	554 3		
avg	179.8 11	0.300 20	0.0011
β⁻ 6 max	651 3		
avg	217.5 12	15.00 20	0.0695
β⁻ 7 max	668 3		

(continued on next page)

68GE EC DECAY (270.8 D 3)   I(MIN)= 0.10%

Radiation Type	Energy (keV)	Intensity (%)	Δ(g-rad/μCi-h)
Auger-L	1	121.4 18	0.0028

(continued on next column)

72GA B- DECAY (14.10 H 1) (CONTINUED)

Radiation Type	Energy (keV)	Intensity (%)	Δ(g-rad/μCi-h)
avg	224.1 12	21.7 3	0.104
β⁻ 8 max	957 3		
avg	342.3 13	27.7 5	0.202
β⁻ 9 max	1049 3		
avg	381.4 13	1.87 4	0.0152
β⁻10 max	1478 3		
avg	569.5 13	8.90 15	0.108
β⁻11 max	1529 3		
avg	592.4 13	0.15 3	0.0019
β⁻12 max	1591 3		
avg	620.2 14	0.310 20	0.0041
β⁻13 max	1928 3		
avg	774.6 14	2.99 12	0.0493
β⁻14 max	2265 3		
avg	931.2 14	0.80 20	0.0159
β⁻15 max	2529 3		
avg	1055.5 14	8.5 6	0.191
β⁻16 max	3159 3		
avg	1354.9 14	10.3 8	0.297
Total β⁻ avg	498.1 19	100.2 12	1.06

8 weak β's omitted (ΣIβ = 0.21%)

Radiation Type	Energy (keV)	Intensity (%)	Δ(g-rad/μCi-h)
X-ray Kα₁	9.88642 7	0.154 16	≈0
γ 2	112.52 3	0.118 14	0.0003
γ 6	289.31 7	0.191 14	0.0012
γ 9	336.63 4	0.107 3	0.0008
γ 10	381.24 8	0.268 10	0.0022
γ 12	428.42 18	0.198 19	0.0018
γ 17	587.44 24	0.121 7	0.0015
γ 18	600.95 3	5.54 11	0.0709
γ 19	629.96 4	24.8 5	0.332

72AS EC DECAY (26.0 H 1)   I(MIN)= 0.10%

Radiation Type	Energy (keV)	Intensity (%)	Δ(g-rad/μCi-h)
Auger-L	1.19	16.2 8	0.0004
Auger-K	8.56	5.5 4	0.0010
ce-K-7	679.897 7	1.4 3	0.0207
β⁺ 1 max	818 4		
avg	353.0 20	0.464 18	0.0035
β⁺ 2 max	931 4		
avg	402.0 20	0.166 9	0.0014
β⁺ 3 max	1869 4		
avg	824.0 20	5.82 18	0.102
β⁺ 4 max	2499 4		
avg	1117.0 20	64.2 15	1.53
β⁺ 5 max	2642 4		
avg	1205.0 20	0.7 4	0.0180
β⁺ 6 max	3333 4		
avg	1528.0 20	16.3 17	0.531
Total β⁺ avg	1167.8 21	87.8 23	2.18

5 weak β's omitted (ΣIβ = 0.17%)

Radiation Type	Energy (keV)	Intensity (%)	Δ(g-rad/μCi-h)
X-ray Kα₂	9.85532 5	1.91 11	0.0004
X-ray Kα₁	9.88642 5	3.74 21	0.0008
X-ray Kβ	11	0.84 5	0.0002
γ 5	600.92 5	0.315 11	0.0040
γ 6	629.92 5	7.92 22	0.106
γ 9	786.2 4	0.448 21	0.0075
γ 10	833.99 3	79.5 17	1.41
γ 12	894.26 3	0.775 21	0.0148
γ 21	1050.75 4	0.98 3	0.0220
γ 25	1215.14 5	0.208 10	0.0054
γ 28	1390.44 5	0.238 10	0.0071
γ 29	1464.00 6	1.11 3	0.0345

γ	21	735.75 15	0.367 7	0.0058
γ	24	786.44 8	3.20 6	0.0537
γ	25	810.20 9	2.01 4	0.0347
γ	26	834.03 3	95.63 7	1.70
γ	27	861.11 5	0.913 20	0.0168
γ	29	894.25 10	9.88 17	0.188
γ	30	924.22 18	0.142 4	0.0028
γ	32	939.36 7	0.259 7	0.0052
γ	34	970.55 6	1.104 17	0.0228
γ	36	999.86 6	0.799 15	0.0170
γ	39	1050.69 5	6.91 12	0.155
γ	43	1215.15 7	0.789 14	0.0204
γ	44	1230.86 7	1.454 20	0.0381
γ	46	1260.10 7	1.13 3	0.0303
γ	47	1276.76 7	1.565 17	0.0425
γ	50	1464.00 7	3.55 6	0.111
γ	54	1568.10 20	0.199 7	0.0066
γ	55	1571.60 14	0.818 20	0.0274
γ	56	1596.68 8	4.24 9	0.144
γ	59	1680.77 7	0.90 5	0.0322
γ	60	1710.90 14	0.388 12	0.0141
γ	62	1837.6 3	0.209 12	0.0082
γ	63	1861.09 6	5.25 8	0.208
γ	64	1877.90 21	0.231 6	0.0093
γ	65	1920.20 17	0.158 5	0.0065
γ	66	1991.3 3	0.112 3	0.0047
γ	67	2029.4 5	0.123 6	0.0053
γ	68	2109.50 9	1.042 20	0.0468
γ	69	2201.66 7	25.9 5	1.22
γ	70	2214.3 3	0.178 14	0.0084
γ	73	2490.98 7	7.68 23	0.407
γ	74	2507.79 7	12.78 23	0.682
γ	75	2515.0 5	0.250 9	0.0134
γ	78	2621.06 24	0.132 4	0.0074
γ	81	2844.00 14	0.43 3	0.0261

46 weak γ's omitted (ΣIγ = 1.45%)

γ	30	1475.90 6	0.506 15	0.0159
γ	32	1568.20 7	0.131 7	0.0044
γ	36	1680.72 9	0.117 4	0.0042
γ	37	1710.90 7	0.248 13	0.0090
γ	41	1991.15 8	0.350 22	0.0148
γ	43	2105.90 17	0.64 3	0.0285
γ	44	2109.59 17	0.266 14	0.0119
γ	46	2201.72 8	0.484 19	0.0227
γ	48	2248.50 10	0.316 14	0.0151
γ	51	2402.86 12	0.101 5	0.0052
γ	53	2507.86 9	0.323 11	0.0172
γ	58	2621.46 12	0.386 12	0.0216
γ	64	2940.07 12	0.297 20	0.0186
γ	67	3094.1 3	0.106 20	0.0070
γ	77	3803.40 21	0.100 6	0.0081

55 weak γ's omitted (ΣIγ = 1.36%)
Maximum γ± intensity =175.64%

73AS EC DECAY (80.30 D 6)   I(MIN)= 0.10%

Radiation Type	Energy (keV)	Intensity (%)	Δ(g-rad/µCi-h)
Auger-L	1.19	321 10	0.0081
ce-K-1	2.160 17	27.8 6	0.0013
Auger-K	8.56	88 5	0.0161
ce-L-1	11.849 15	60.3 7	0.0152
ce-M-1	13.083 15	8.9 3	0.0025
ce-NOP-1	13.263 15	3.00 10	0.0008
ce-K-2	42.334 12	76.1 6	0.0686

(continued on next page)

73AS EC DECAY (80.30 D 6)   (CONTINUED)

Radiation Type	Energy (keV)	Intensity (%)	Δ(g-rad/µCi-h)
ce-L-2	52.023 9	11.1 4	0.0123
ce-M-2	53.257 9	1.73 5	0.0020
ce-NOP-2	53.437 9	0.580 10	0.0007
X-ray L	1.19	1.9 7	≈0
X-ray Kα₂	9.85532 7	30.5 16	0.0064
X-ray Kα₁	9.88642 7	60. 3	0.0125
X-ray Kβ	11	13.3 7	0.0031
γ 2	53.437 9	10.5 3	0.0120

1 weak γ's omitted (ΣIγ = 0.09%)

73SE EC DECAY (7.15 H 8)   I(MIN)= 0.10%

Radiation Type	Energy (keV)	Intensity (%)	Δ(g-rad/µCi-h)
Auger-L	1.24	67 4	0.0018
Auger-K	9.1	21.8 16	0.0042
ce-K-1	55.20 10	18.69 17	0.0220
ce-L-1	65.54 10	2.043 22	0.0029
ce-M-1	66.87 10	0.318 4	0.0005
ce-K-2	349.3 3	1.13 4	0.0084
ce-L-2	359.7 3	0.124 4	0.0009

74AS B+ DECAY (17.78 D 2)   I(MIN)= 0.10%
%(EC+B+)=65.8 16
SEE ALSO 74AS B- DECAY

Radiation Type	Energy (keV)	Intensity (%)	Δ(g-rad/µCi-h)
Auger-L	1.19	43.5 19	0.0011
Auger-K	8.56	14.7 9	0.0027
β+ 1 max	944.3 18		
avg	407.9 8	25.7 7	0.223
β+ 2 max	1540.1 18		
avg	700.9 9	3.4 8	0.0508
Total β+ avg	442.1 9	29.1 11	0.274
X-ray L	1.19	0.26 9	≈0
X-ray Kα₂	9.85532 8	5.1 3	0.0011
X-ray Kα₁	9.88642 7	9.9 6	0.0021
X-ray Kβ	11	2.22 13	0.0005
γ 1	595.80 8	59.2 15	0.752
γ 2	608.39 8	0.53 12	0.0069
γ 9	1204.34 8	0.290 14	0.0074

8 weak γ's omitted (ΣIγ = 0.09%)
Maximum γ± intensity = 58.20%

74AS B- DECAY (17.78 D 3)   I(MIN)= 0.10%

%B-=34.2 16
SEE ALSO 74AS B+ DECAY

Radiation Type	Energy (keV)	Intensity (%)	Δ(g-rad/μCi-h)
β- 1 max	719.7 24		
avg	243.4 10	15.4 3	0.0798
β- 2 max	1354.5 24		
avg	531.6 11	18.8 16	0.213
Total β-			
avg	401.4 12	34.2 17	0.293

1 weak β's omitted (ΣIβ = 0.04%)

γ 1	634.78 8	15.4 8	0.208
γ 2	635.0 20	0.142 7	0.0019

75SE EC DECAY (119.8 D 1)   I(MIN)= 0.10%

Radiation Type	Energy (keV)	Intensity (%)	Δ(g-rad/μCi-h)
Auger-L	1.24	131 7	0.0035
Auger-K	9.1	42 3	0.0082
ce-K-1	12.53 20	4.3 4	0.0011
ce-L-1	22.87 20	0.84 7	0.0004
ce-M-1	24.20 20	0.135 11	≈0
ce-K-2	54.193 7	0.40 3	0.0005
ce-K-4	84.8673 22	2.73 10	0.0049

(continued on next page)

	Energy (keV)	Intensity (%)	Δ(g-rad/μCi-h)
β+ 1 max	1290 10		
avg	562 5	65.0 7	0.778
β+ 2 max	1651 10		
avg	751 5	0.6 3	0.0096
Total β+			
avg	564 5	65.6 8	0.788

6 weak β's omitted (ΣIβ = 0.04%)

X-ray L	1.28	0.47 14	≈0
X-ray Kα2	10.50800 10	8.4 5	0.0019
X-ray Kα1	10.54370 10	16.4 10	0.0037
X-ray Kβ	11.7	3.79 23	0.0009
γ 1	67.07 10	76.8 6	0.110
γ 2	361.2 3	97.4 5	0.749
γ 5	510	0.267 8	0.0029
γ 7	575.9 5	0.146 7	0.0018
γ 12	765.07 12	0.1276 21	0.0021
γ 19	865.09 12	0.526 17	0.0097
γ 22	900.73 10	0.1354 21	0.0026
γ 31	1110.64 6	0.2016 22	0.0048
γ 44	1422.68 7	0.136 5	0.0041

54 weak γ's omitted (ΣIγ = 1.03%)
Maximum γ± intensity =131.28%

75SE EC DECAY (119.8 D 1)   (CONTINUED)

Radiation Type	Energy (keV)	Intensity (%)	Δ(g-rad/μCi-h)
ce-L-4	95.2075 22	0.364 14	0.0007
ce-K-5	109.252 3	0.650 21	0.0015
ce-K-6	124.135 3	1.56 5	0.0041
ce-L-6	134.475 4	0.162 5	0.0005
ce-K-10	252.789 4	0.381 12	0.0021
ce-K-11	267.671 3	0.180 13	0.0010
X-ray L	1.28	0.9 3	≈0
X-ray $K\alpha_2$	10.50800 10	16.3 10	0.0037
X-ray $K\alpha_1$	10.54370 10	31.8 18	0.0071
X-ray $K\beta$	11.7	7.4 5	0.0018
γ 2	66.060 7	1.142 25	0.0016
γ 4	96.7340 20	3.50 8	0.0072
γ 5	121.119 3	17.43 21	0.0450
γ 6	136.002 3	59.4 6	0.172
γ 7	198.596 6	1.49 3	0.0063
γ 10	264.656 4	59.5 6	0.335
γ 11	279.538 3	25.3 3	0.151
γ 13	303.924 3	1.339 14	0.0087
γ 16	400.6570 20	11.60 14	0.0990

15 weak γ's omitted (ΣI$_γ$ = 0.09%)

76AS B- DECAY (26.32 H 7)   (CONTINUED)

Radiation Type	Energy (keV)	Intensity (%)	Δ(g-rad/μCi-h)
γ 36	1216.04 5	3.45 16	0.0893
γ 37	1228.51 6	1.19 6	0.0312
γ 38	1439.12 8	0.277 13	0.0085
γ 39	1453.64 6	0.112 7	0.0035
γ 45	1787.67 6	0.282 15	0.0107
γ 49	2096.32 10	0.559 25	0.0249
γ 50	2110.80 11	0.317 17	0.0143

37 weak γ's omitted (ΣI$_γ$ = 0.60%)

77AS B- DECAY (38.8 H 3)   I(MIN)= 0.10%

Radiation Type	Energy (keV)	Intensity (%)	Δ(g-rad/μCi-h)
Auger-L	1.32	0.11 3	≈0
β⁻ 1 max	166 4		
avg	45.7 12	0.70 20	0.0007
β⁻ 2 max	437 4		
avg	135.9 15	0.70 20	0.0020
β⁻ 3 max	448 4		
avg	139.8 15	1.6 4	0.0048
β⁻ 4 max	687 4		
avg	230.5 16	97.1 8	0.477
Total β⁻ avg	227.1 17	100.1 10	0.484

76AS B- DECAY (26.32 H 7)   I(MIN)= 0.10%

Radiation Type	Energy (keV)	Intensity (%)	Δ(g-rad/μCi-h)
β- 1 max	294.0 12		
avg	86.2 4	0.53 3	0.0010
β- 2 max	308.5 12		
avg	91.1 4	1.04 5	0.0020
β- 3 max	534.8 12		
avg	171.8 5	1.70 10	0.0062
β- 4 max	1176.2 12		
avg	434.2 6	1.74 9	0.0161
β- 5 max	1747.7 12		
avg	689.3 6	7.4 4	0.109
β- 6 max	1841.5 12		
avg	747.0 6	0.79 6	0.0126
β- 7 max	2404.7 12		
avg	994.0 6	35.1 15	0.743
β- 8 max	2963.8 12		
avg	1264.6 6	51.5 21	1.39
Total β-			
avg	1069.1 8	100 3	2.28

11 weak β's omitted (ΣIβ = 0.28%)

γ 7	559.08 4	44.7 19	0.532
γ 8	563.17 5	1.18 6	0.0142
γ 9	571.53 6	0.143 8	0.0017
γ 12	657.03 4	6.1 3	0.0851
γ 14	665.31 4	0.36 4	0.0051
γ 17	740.11 6	0.116 7	0.0018
γ 19	771.77 6	0.116 7	0.0019
γ 24	867.65 6	0.130 7	0.0024
γ 33	1129.84 5	0.112 15	0.0027
γ 35	1212.88 7	1.46 9	0.0376

(continued on next column)

2 weak β's omitted (ΣIβ = 0.03%)

γ 3	87.876 20	0.21 6	0.0004
γ 5	161.933 20	0.13 4	0.0004
γ 7	238.999 20	1.6 4	0.0081
γ 8	249.790 20	0.43 11	0.0023
γ 12	520.652 20	0.62 16	0.0069

7 weak γ's omitted (ΣIγ = 0.10%)

77BR EC DECAY (56 H 2)   I(MIN)= 0.10%

Radiation Type	Energy (keV)	Intensity (%)	Δ(g-rad/μCi-h)
Auger-L	1.32	115 7	0.0032
Auger-K	9.67	36 3	0.0074
ce-K-3	75.14 10	0.146 6	0.0002
ce-K-8	149.17 8	0.79 4	0.0025
ce-L-8	160.18 8	0.134 6	0.0005
ce-K-14	226.32 7	0.213 8	0.0010
β+ 1 max	343 3		
avg	151.7 12	0.73 3	0.0024
X-ray L	1.38	1.05 24	≈0
X-ray Kα2	11.18140 20	15.4 10	0.0037
X-ray Kα1	11.22240 20	30.0 18	0.0072
X-ray Kβ	12.5	7.2 5	0.0019
γ 3	87.80 10	1.40 4	0.0026
γ 5	138.95 9	0.129 6	0.0004

(continued on next page)

77BR EC DECAY (56 H 2)   (CONTINUED)

Radiation Type	Energy (keV)	Intensity (%)	Δ(g-rad/μCi-h)
γ 8	161.83 8	1.10 3	0.0038
γ 9	180.68 7	0.284 10	0.0011
γ 12	200.40 6	1.21 6	0.0052
γ 14	238.98 7	23.1 5	0.118
γ 16	249.77 7	2.98 10	0.0159
γ 17	270.83 7	0.321 14	0.0019
γ 19	281.65 7	2.29 7	0.0137
γ 20	297.23 8	4.16 21	0.0263
γ 21	303.76 9	1.18 4	0.0076
γ 27	384.99 8	0.84 3	0.0069
γ 32	439.47 6	1.56 5	0.0146
γ 34	484.57 7	1.00 4	0.0103
γ 36	517.9 4	0.16 5	0.0018
γ 37	520.69 6	22.4 6	0.249
γ 39	565.91 19	0.427 17	0.0052
γ 40	567.90 8	0.86 3	0.0104
γ 41	574.64 8	1.19 4	0.0145
γ 42	578.91 7	2.96 10	0.0365
γ 43	585.48 7	1.57 5	0.0196
γ 48	755.35 5	1.67 5	0.0268
γ 51	817.79 6	2.08 7	0.0362
γ 59	1005.05 6	0.92 3	0.0198

37 weak γ's omitted (ΣIγ = 0.80%)
Maximum γ± intensity = 1.46%

79KR B+ DECAY (35.04 H 10)   (CONTINUED)

Radiation Type	Energy (keV)	Intensity (%)	Δ(g-rad/μCi-h)
γ 25	832.04 10	1.26 6	0.0223
γ 30	934.81 15	0.126 7	0.0025
γ 31	1025.70 10	0.156 9	0.0034
γ 35	1115.1 3	0.370 20	0.0088
γ 39	1332.13 10	0.44 3	0.0125

22 weak γ's omitted (ΣIγ = 0.52%)
Maximum γ± intensity = 14.21%

81KR EC DECAY (2.1E5 Y 2)   I(MIN)= 0.10%

Radiation Type	Energy (keV)	Intensity (%)	Δ(g-rad/μCi-h)
Auger-L	1.4	110 5	0.0033
Auger-K	10.2	31 3	0.0067
X-ray L	1.48	1.1 4	≈0
X-ray Kα₂	11.87760 20	14.8 8	0.0038
X-ray Kα₁	11.92420 20	28.8 15	0.0073
X-ray Kβ	13.3	7.1 4	0.0020
γ 1	275.990 11	3.5 4	0.0206

81KR IT DECAY (13 S 1) I(MIN)= 0.10%

Radiation Type	Energy (keV)	Intensity (%)	Δ(g-rad/μCi-h)
Auger-L	1.5	34.4 18	0.0011
Auger-K	10.8	9.3 9	0.0021
ce-K-1	175.97 3	26.4 6	0.0990
ce-L-1	188.38 3	4.50 14	0.0181
ce-MNO-1	190.01 3	1.50 5	0.0061
X-ray L	1.59	0.45 18	≈0
X-ray Kα2	12.5980 20	5.0 3	0.0013
X-ray Kα1	12.6490 20	9.6 5	0.0026
X-ray Kβ	14	2.45 14	0.0007
γ 1	190.30 3	67.7 7	0.274

82BR B- DECAY (35.34 H 2) I(MIN)= 0.10%

Radiation Type	Energy (keV)	Intensity (%)	Δ(g-rad/μCi-h)
β- 1 max	264.4 15		
avg	76.2 5	1.40 5	0.0023
β- 2 max	444.1 15		
avg	137.8 5	98.0 21	0.288
Total β-			
avg	136.9 5	99.4 21	0.290
γ 1	92.190 16	0.75 4	0.0015
γ 4	137.40 6	0.150 17	0.0004

(continued on next page)

79KR B+ DECAY (35.04 H 10) I(MIN)= 0.10%

Radiation Type	Energy (keV)	Intensity (%)	Δ(g-rad/μCi-h)
Auger-L	1.4	105 7	0.0032
Auger-K	10.2	31 3	0.0067
ce-K-1	30.7 4	0.214 22	0.0001
β+ 1 max	348 8		
avg	154 4	0.180 17	0.0006
β+ 2 max	609 8		
avg	265 4	6.9 7	0.0389
Total β+			
avg	262 4	7.1 7	0.0396

2 weak β's omitted (ΣIβ = 0.03%)

X-ray L	1.48	1.1 4	≈0
X-ray Kα2	11.87760 20	14.9 11	0.0038
X-ray Kα1	11.92420 20	29.0 22	0.0074
X-ray Kβ	13.3	7.2 6	0.0020
γ 1	44.2 4	0.210 20	0.0002
γ 2	135.99 10	1.00 10	0.0029
γ 3	180.21 15	0.10 5	0.0004
γ 4	208.45 10	0.78 4	0.0035
γ 5	217.02 10	2.40 10	0.0111
γ 6	261.26 10	12.7 4	0.0707
γ 8	299.51 10	1.57 7	0.0100
γ 9	306.31 15	2.60 10	0.0170
γ 10	344.70 10	0.240 10	0.0018
γ 11	389.00 10	1.52 7	0.0126
γ 12	397.56 10	9.5 3	0.0804
γ 16	522.98 20	0.250 10	0.0028
γ 17	525.32 15	0.430 20	0.0048
γ 19	606.07 10	8.10 20	0.105

(continued on next column)

82BR B- DECAY (35.34 H 2) (CONTINUED)

Radiation Type	Energy (keV)	Intensity (%)	Δ(g-rad/μCi-h)
γ 6	221.4800 20	2.27 9	0.0107
γ 7	273.480 20	0.84 5	0.0049
γ 12	554.348 3	70.9 20	0.837
γ 14	606.370 20	1.21 5	0.0156
γ 15	619.107 4	43.1 10	0.569
γ 16	698.374 4	28.7 10	0.426
γ 18	776.517 3	83.6 17	1.38
γ 19	827.827 5	24.0 8	0.423
γ 21	952.02 3	0.368 17	0.0075
γ 22	1007.59 3	1.31 5	0.0282
γ 23	1044.002 5	27.4 8	0.609
γ 25	1081.29 5	0.62 4	0.0142
γ 28	1317.476 6	26.9 8	0.755
γ 29	1426.00 10	0.11 5	0.0033
γ 30	1474.884 6	16.6 5	0.522
γ 31	1650.339 8	0.794 17	0.0279
γ 32	1779.66 3	0.117 17	0.0044

15 weak γ's omitted (ΣI$_\gamma$ = 0.71%)

82RB EC DECAY (1.273 M 2)   I(MIN)= 0.10%

Radiation Type	Energy (keV)	Intensity (%)	Δ(g-rad/μCi-h)
Auger-L	1.5	4.98 25	0.0002
Auger-K	10.8	1.41 13	0.0003

82SR EC DECAY (25.55 D 15) (CONTINUED)

Radiation Type	Energy (keV)	Intensity (%)	Δ(g-rad/μCi-h)
X-ray L	1.69	1.6 6	≈0
X-ray Kα₂	13.33580 20	16.8 4	0.0048
X-ray Kα₁	13.39530 20	32.4 5	0.0092
X-ray Kβ	15	8.51 20	0.0027

84RB B+ DECAY (32.77 D 4)   I(MIN)= 0.10%
%(EC+B+)=97.0 3
SEE ALSO 84RB B- DECAY

Radiation Type	Energy (keV)	Intensity (%)	Δ(g-rad/μCi-h)
Auger-L	1.5	78 7	0.0025
Auger-K	10.8	22 3	0.0051
β+ 1 max	778 3		
avg	339.2 12	11.2 11	0.0809
β+ 2 max	1660 3		
avg	757.1 8	10.2 6	0.164
Total β+			
avg	538.4 13	21.4 13	0.245
X-ray L	1.59	1.0 4	≈0
X-ray Kα₂	12.5980 20	11.8 11	0.0032
X-ray Kα₁	12.6490 20	22.8 21	0.0061
X-ray Kβ	14	5.8 6	0.0017

	Energy (keV)	Intensity (%)	Δ(g-rad/μCi-h)
γ 1	881.610 3	74.0 9	1.39
γ 2	1016.162 13	0.374 12	0.0081
γ 3	1897.761 14	0.792 21	0.0320

Maximum γ± intensity = 42.80%

84RB B- DECAY (32.77 D 4)   I(MIN)= 0.10%

%B-=3.0 3

SEE ALSO 84RB B+ DECAY

Radiation Type	Energy (keV)	Intensity (%)	Δ(g-rad/μCi-h)
β- 1 max	893 4		
avg	332.3 14	3.0 3	0.0212

85KR B- DECAY (10.72 Y 1)   I(MIN)= 0.10%

Radiation Type	Energy (keV)	Intensity (%)	Δ(g-rad/μCi-h)
β- 1 max	173.5 20		
avg	47.7 6	0.434 10	0.0004
β- 2 max	687.5 20		
avg	251.6 8	99.570 10	0.534
Total β-			
avg	250.7 8	100.004 15	0.534
γ 1	514.009 12	0.434 10	0.0048

	Energy (keV)	Intensity (%)	Δ(g-rad/μCi-h)
β+ 1 max	1190 18		
avg	520 8	0.278 13	0.0031
β+ 2 max	1887 18		
avg	835 9	0.106 15	0.0019
β+ 3 max	2585 18		
avg	1160 9	11.6 5	0.287
β+ 4 max	3362 18		
avg	1526 9	83.4 5	2.71
Total β+			
avg	1477 9	95.5 7	3.00

9 weak β's omitted (ΣIβ = 0.08%)

X-ray Kα2	12.5980 20	0.75 4	0.0002
X-ray Kα1	12.6490 20	1.45 8	0.0004
X-ray Kβ	14	0.370 20	0.0001
γ 4	698.37 5	0.133 7	0.0020
γ 6	776.520 10	13.4 5	0.222
γ 18	1395.14 3	0.470 18	0.0140

43 weak γ's omitted (ΣIγ = 0.44%)

Maximum γ± intensity =190.93%

82SR EC DECAY (25.55 D 15)   I(MIN)= 0.10%

Radiation Type	Energy (keV)	Intensity (%)	Δ(g-rad/μCi-h)
Auger-L	1.68	107.1 18	0.0038
Auger-K	11.4	28.5 7	0.0069

(continued on next column)

85KR IT DECAY (4.480 H 8) I(MIN)= 0.10%
%IT=21.3 5
SEE ALSO 85KR B- DECAY (4.48 H)

Radiation Type	Energy (keV)	Intensity (%)	Δ(g-rad/μCi-h)
Auger-L	1.5	7.8 5	0.0002
Auger-K	10.8	2.15 21	0.0005
ce-K-1	290.544 20	6.1 3	0.0376
ce-L-1	302.949 20	0.90 4	0.0058
ce-MNO-1	304.582 20	0.183 6	0.0012
X-ray L	1.59	0.10 4	≈0
X-ray Kα₂	12.5980 20	1.15 8	0.0003
X-ray Kα₁	12.6490 20	2.22 15	0.0006
X-ray Kβ	14	0.57 4	0.0002
γ 1	304.870 20	14.1 5	0.0914

85KR B- DECAY (4.480 H 8) I(MIN)= 0.10%
%B-=78.7 5
SEE ALSO 85KR IT DECAY (4.48 H)

Radiation Type	Energy (keV)	Intensity (%)	Δ(g-rad/μCi-h)
Auger-L	1.68	3.69 12	0.0001
Auger-K	11.4	1.00 4	0.0002
ce-K-2	135.99 3	3.01 9	0.0087
ce-L-2	149.12 3	0.339 17	0.0011

86RB B- DECAY (18.64 D 2) I(MIN)= 0.10%

Radiation Type	Energy (keV)	Intensity (%)	Δ(g-rad/μCi-h)
β⁻ 1 max	697.3 19		
avg	232.4 8	8.67 4	0.0429
β⁻ 2 max	1773.9 19		
avg	709.1 9	91.32 4	1.38
Total β⁻ avg	667.8 10	99.99 6	1.42
γ 1	1076.63 10	8.67 4	0.199

86Y B+ DECAY (14.74 H 2) I(MIN)= 0.10%

Radiation Type	Energy (keV)	Intensity (%)	Δ(g-rad/μCi-h)
Auger-L	1.79	71 4	0.0027
Auger-K	12	18.4 17	0.0047
β⁺ 1 max	387 20		
avg	173 9	0.22 6	0.0008
β⁺ 2 max	452 20		
avg	201 9	0.26 5	0.0011
β⁺ 3 max	573 20		
avg	252 9	0.31 4	0.0017
β⁺ 4 max	856 20		
avg	375 9	0.180 18	0.0014

Radiation Type		Energy (keV)	Intensity (%)	Δ(g-rad/μCi-h)
β+ 5	max	900 20		
	avg	394 9	1.15 22	0.0097
β+ 6	max	1033 20		
	avg	452 9	1.9 5	0.0183
β+ 7	max	1162 20		
	avg	509 9	1.36 14	0.0147
β+ 8	max	1221 20		
	avg	535 9	11.9 6	0.136
β+ 9	max	1340 20		
	avg	589 9	0.66 17	0.0083
β+ 10	max	1545 20		
	avg	681 9	5.5 5	0.0798
β+ 11	max	1736 10		
	avg	768 10	1.7 10	0.0278
β+ 12	max	1988 20		
	avg	883 10	3.6 10	0.0677
β+ 13	max	2364 20		
	avg	1078 10	0.9 10	0.0207
β+ 14	max	3141 20		
	avg	1437 10	1.8 12	0.0551
Total β+	avg	659 11	31.7 24	0.445

9 weak β's omitted (ΣIβ = 0.26%)

Radiation Type	Energy (keV)	Intensity (%)	Δ(g-rad/μCi-h)
X-ray L	1.8	1.2 4	≈0
X-ray Kα2	14.09790 20	11.9 6	0.0036
X-ray Kα1	14.16500 20	23.0 11	0.0069
X-ray Kβ	15.8	6.2 4	0.0021
γ 1	132.34 10	0.165 9	0.0005
γ 3	182.34 20	0.11 4	0.0004
γ 4	187.87 13	1.26 5	0.0051
γ 5	190.80 13	1.01 4	0.0041
γ 6	209.80 23	0.396 17	0.0018
γ 7	235.37 23	0.396 17	0.0020
γ 8	237.9 3	0.132 25	0.0007
γ 9	252.05 13	0.371 17	0.0020

(continued on next page)

Radiation Type		Energy (keV)	Intensity (%)	Δ(g-rad/μCi-h)
β- 1	max	711.3 20		
	avg	238.4 8	0.290 10	0.0015
β- 2	max	841.2 20		
	avg	290.6 9	78.7 5	0.487
Total β-	avg	290.4 9	79.0 5	0.489

1 weak β's omitted (ΣIβ = 0.02%)

Radiation Type	Energy (keV)	Intensity (%)	Δ(g-rad/μCi-h)
X-ray Kα2	13.33580 20	0.585 21	0.0002
X-ray Kα1	13.39530 20	1.13 4	0.0003
X-ray Kβ	15	0.297 11	≈0
γ 1	129.810 20	0.301 8	0.0008
γ 2	151.19 3	75.3 16	0.242

2 weak γ's omitted (ΣIγ = 0.02%)

85SR EC DECAY (64.854 D 3)   I(MIN)= 0.10%

Radiation Type	Energy (keV)	Intensity (%)	Δ(g-rad/μCi-h)
Auger-L	1.68	108.2 23	0.0039
Auger-K	11.4	29.1 9	0.0071
ce-K-1	498.809 12	0.617 20	0.0066
X-ray L	1.69	1.6 6	≈0
X-ray Kα2	13.33580 20	17.1 4	0.0048
X-ray Kα1	13.39530 20	33.0 7	0.0094
X-ray Kβ	15	8.66 24	0.0028
γ 1	514.009 12	98.0 10	1.07

1 weak γ's omitted (ΣIγ = 0.01%)

86Y B+ DECAY (14.74 H 2) (CONTINUED)

Radiation Type	Energy (keV)		Intensity (%)		Δ(g-rad/μCi-h)
γ 11	264.53	13	0.536	25	0.0030
γ 12	307.00	10	3.46	9	0.0227
γ 13	331.08	23	0.83	3	0.0059
γ 15	370.28	17	0.82	5	0.0065
γ 16	380.4	3	0.45	4	0.0037
γ 17	382.86	23	3.63	12	0.0296
γ 18	425.97	23	0.305	17	0.0028
γ 19	439.5	3	0.20	7	0.0019
γ 20	443.13	10	16.9	5	0.160
γ 21	444.18	23	0.64	17	0.0061
γ 23	469.24	25	0.297	25	0.0030
γ 26	515.18	20	4.89	15	0.0537
γ 27	580.57	10	4.78	15	0.0592
γ 28	608.29	10	2.01	15	0.0261
γ 29	618.2	4	0.21	4	0.0028
γ 30	627.72	10	32.6	10	0.436
γ 32	644.82		2.2	4	0.0300
γ 33	645.87		9.2	11	0.126
γ 35	689.29	25	0.17	4	0.0025
γ 36	702.2	6	0.25	9	0.0037
γ 37	703.33	10	15.4	5	0.231
γ 38	709.90	10	2.62	8	0.0397
γ 39	719.17	23	0.22	4	0.0034
γ 40	740.81	13	1.36	5	0.0215
γ 41	767.63	13	2.4	4	0.0391
γ 42	768.25		0.32	11	0.0053
γ 43	777.37	10	22.4	6	0.372
γ 44	783.6	3	0.26	4	0.0044
γ 45	826.02	13	3.30	9	0.0581
γ 46	833.72		1.5	4	0.0264
γ 47	835.67		4.4	6	0.0778
γ 48	882.96	17	0.25	9	0.0047

86Y B+ DECAY (14.74 H 2) (CONTINUED)

Radiation Type	Energy (keV)		Intensity (%)		Δ(g-rad/μCi-h)
γ 99	2865.9	3	0.38	7	0.0232
γ 101	3069.7	4	0.115	17	0.0076
γ 102	3334.0	5	0.124	17	0.0088

23 weak γ's omitted (Σiγ = 1.03%)
Maximum γ± intensity = 63.40%

86ZR EC DECAY (16.5 H 1)   I(MIN)= 0.10%

Radiation Type	Energy (keV)		Intensity (%)		Δ(g-rad/μCi-h)
Auger-L	2		189	10	0.0077
ce-K-1	11.0	5	69.9	4	0.0163
Auger-K	12.7		46	5	0.0126
ce-L-1	25.6	5	8.7	3	0.0047
ce-MNO-1	27.6	5	2.00	10	0.0012
ce-K-2	226.0	10	3.55	11	0.0171
ce-L-2	240.6	10	0.440	10	0.0023
ce-MNO-2	242.6	10	0.18		0.0009
X-ray L	2		4.2	14	0.0002
X-ray Kα2	14.88290	20	33.0	16	0.0105
X-ray Kα1	14.95840	20	64	3	0.0203
X-ray Kβ	16.7		17.5	9	0.0062

	Energy (keV)	Intensity (%)	Δ(g-rad/μCi-h)
γ 1	28.0 5	19.4 6	0.0116
γ 2	243.0 20	95.80 10	0.496
γ 3	612.0 20	5.2 8	0.0678

87KR B- DECAY (76.3 M 5)   I(MIN)= 0.10%

Radiation Type	Energy (keV)	Intensity (%)	Δ(g-rad/μCi-h)
Auger-L	1.68	0.218 10	≈0
ce-K-2	387.378 20	0.178 8	0.0015
β⁻ 1 max	577 5		
avg	187.0 20	0.50 3	0.0020
β⁻ 2 max	831 5		
avg	286.0 20	0.110 10	0.0007
β⁻ 3 max	925 5		
avg	325.0 20	4.4 3	0.0305
β⁻ 4 max	1075 5		
avg	388 3	0.58 3	0.0048
β⁻ 5 max	1331 5		
avg	499 3	9.5 3	0.101
β⁻ 6 max	1472 5		
avg	561 3	5.60 20	0.0669
β⁻ 7 max	1508 5		
avg	577 3	0.42 6	0.0052
β⁻ 8 max	2145 5		
avg	869 3	0.67 10	0.0124
β⁻ 9 max	2308 5		
avg	944 3	0.16 6	0.0032
β⁻10 max	2496 5		

(continued on next page)

	Energy (keV)	Intensity (%)	Δ(g-rad/μCi-h)
γ 49	887.40 17	0.44 5	0.0083
γ 50	955.35 20	1.04 5	0.0212
γ 51	971.43 18	0.27 4	0.0056
γ 52	1017.93 23	0.18 12	0.0039
γ 53	1024.04 10	3.79 17	0.0828
γ 54	1076.63 10	82.5 4	1.89
γ 56	1092.68 13	0.69 5	0.0161
γ 57	1102.02 23	0.198 25	0.0046
γ 58	1133.3 10	0.297 25	0.0072
γ 61	1153.05 10	30.5 10	0.750
γ 63	1163.03 10	1.18 5	0.0292
γ 64	1253.11 10	1.53 5	0.0410
γ 65	1270.16 13	0.65 10	0.0176
γ 66	1283.96 13	0.29 11	0.0079
γ 67	1294.9 3	0.29 9	0.0080
γ 68	1296.03 23	0.54 4	0.0150
γ 70	1349.15 10	2.95 10	0.0846
γ 71	1404.8 4	0.18 5	0.0054
γ 72	1415.20 23	0.33 9	0.0099
γ 73	1507.86 10	0.35 5	0.0114
γ 74	1533.19 13	0.22 4	0.0073
γ 75	1535.67 13	0.12 4	0.0038
γ 76	1564.4 5	0.18 5	0.0060
γ 77	1696.25 13	0.635 17	0.0230
γ 78	1711.6 7	0.17 4	0.0063
γ 79	1724.15 10	0.55 5	0.0203
γ 80	1790.90 10	1.00 5	0.0381
γ 81	1801.70 10	1.65 5	0.0633
γ 82	1854.38 13	17.2 5	0.678
γ 83	1920.72 13	20.8 7	0.851
γ 85	2017.1 6	0.132 17	0.0057
γ 86	2088.09 25	0.247 25	0.0110
γ 89	2291.8 5	0.124 9	0.0060
γ 90	2482.08 17	0.115 9	0.0061
γ 92	2567.97 18	2.25 11	0.123
γ 93	2610.11 20	1.24 8	0.0688
γ 94	2641.9 4	0.16 5	0.0093
γ 96	2794.9 4	0.206 17	0.0123

(continued on next column)

87KR B- DECAY (76.3 M 5) (CONTINUED)

Radiation Type	Energy (keV)	Intensity (%)	Δ(g-rad/μCi-h)
β-11 max avg	1032 3	0.120 10	0.0026
β-11 max	3041 5		
avg	1293 3	6.9 4	0.190
β-12 max	3483 5		
avg	1501 3	40.8 14	1.30
β-13 max	3886 5		
avg	1694 3	30.5 22	1.10
Total β- avg	1322 4	100 3	72.82
γ 2	402.578 20	49.5 16	0.424
γ 6	673.87 4	1.91 10	0.0274
γ 7	814.25 7	0.168 12	0.0029
γ 8	836.37 6	0.75 4	0.0134
γ 9	845.43 4	7.3 4	0.131
γ 12	946.64 15	0.139 11	0.0028
γ 15	1175.40 8	1.12 6	0.0281
γ 16	1337.96 8	0.65 4	0.0185
γ 17	1382.53 7	0.287 18	0.0085
γ 18	1389.91 16	0.124 11	0.0037
γ 20	1531.2 4	0.236 6	0.0116
γ 21	1577.99 14	0.129 11	0.0043
γ 22	1611.16 16	0.104 20	0.0036
γ 23	1740.52 8	2.05 10	0.0760
γ 24	1842.61 24	0.139 11	0.0054
γ 25	2011.88 12	2.90 14	0.124
γ 27	2408.50 20	0.213 17	0.0109
γ 28	2554.80 20	9.3 6	0.506
γ 29	2558.10 20	3.9 3	0.213
γ 31	2811.40 20	0.317 23	0.0190
γ 34	3308.50 20	0.450 25	0.0317

13 weak γ's omitted (ΣIγ = 0.65%)

87Y EC DECAY (80.3 H 3)   I(MIN)= 0.10%
SEE ALSO 87SR IT DECAY (2.805 H)
SEE ALSO 87SR EC DECAY (2.805 H)

Radiation Type	Energy (keV)	Intensity (%)	Δ(g-rad/μCi-h)
Auger-L	1.79	105 6	0.0040
Auger-K	12	27.0 23	0.0070
ce-K-2	468.60 20	0.240 8	0.0024
β+ 1 max	450.6 14		
avg	200.2 6	0.160 20	0.0007
X-ray L	1.8	1.7 6	≈0
X-ray Kα2	14.09790 20	17.5 8	0.0053
X-ray Kα1	14.16500 20	33.8 14	0.0102
X-ray Kβ	15.8	9.1 4	0.0031
γ 2	484.70 20	92.2 10	0.952

Maximum γ± intensity = 0.32%

88KR B- DECAY (2.84 H 2)   I(MIN)= 0.10%

Radiation Type	Energy (keV)	Intensity (%)	Δ(g-rad/μCi-h)
Auger-L	1.68	14.8 7	0.0005
Auger-K	11.4	4.00 20	0.0010
ce-K-1	12.313 14	10.7 6	0.0028
ce-L-1	25.448 14	1.23 8	0.0007
ce-MNO-1	27.191 14	0.214 18	0.0001

Radiation Type	Energy (keV)		Intensity (%)		A(g-rad/µCi-h)
ce-K-4	150.78	4	0.21	3	0.0007
ce-K-7	181.120	15	1.14	12	0.0044
ce-L-7	194.255	15	0.135	17	0.0006
β- 1 max	140	15			
avg	38	5	0.350	20	0.0003
β- 2 max	363	15			
avg	109	6	2.65	16	0.0062
β- 3 max	519	15			
avg	165	6	67	3	0.235
β- 4 max	679	15			
avg	226	6	9.1	5	0.0438
β- 5 max	822	15			
avg	283	7	0.14	3	0.0008
β- 6 max	995	15			
avg	354	7	0.200	20	0.0015
β- 7 max	1196	15			
avg	440	7	1.92	11	0.0180
β- 8 max	1250	15			
avg	463	7	0.23	4	0.0023
β- 9 max	1469	15			
avg	560	7	0.22	3	0.0026
β-10 max	1729	15			
avg	677	7	0.90	6	0.0130
β-11 max	1770	15			
avg	696	7	0.10	6	0.0015
β-12 max	2049	15			
avg	824	7	1.31	25	0.0230
β-13 max	2549	15			
avg	1057	7	0.14	14	0.0032
β-14 max	2715	15			
avg	1135	7	1.80	25	0.0435
β-15 max	2911	15			
avg	1232	7	14	4	0.367
Total β- avg	358	11	100	5	0.765

(continued on next page)

87SR EC DECAY (2.805 H 3)   I(MIN)= 0.10%
%EC=0.30 8
FEEDING OF 2.805-H 87SR IN 87Y DECAY=100%
SEE ALSO 87SR IT DECAY (2.805 H)

Radiation Type	Energy (keV)	Intensity (%)	A(g-rad/µCi-h)
Auger-L	1.68	0.32 8	≈0

87SR IT DECAY (2.805 H 3)   I(MIN)= 0.10%
%IT=99.70 8
FEEDING OF 2.805-H 87SR IN 87Y DECAY=100%
SEE ALSO 87SR EC DECAY (2.805 H)

Radiation Type	Energy (keV)	Intensity (%)	A(g-rad/µCi-h)
Auger-L	1.79	18.1 10	0.0007
Auger-K	12	4.5 4	0.0012
ce-K-1	372.30 8	14.6 5	0.115
ce-L-1	386.18 8	2.39 8	0.0196
ce-MNO-1	388.04 8	0.4113 9	0.0034
X-ray L	1.8	0.29 10	≈0
X-ray Kα2	14.09790 20	2.92 15	0.0009
X-ray Kα1	14.16500 20	5.6 3	0.0017
X-ray Kβ	15.8	1.51 8	0.0005
γ 1	388.40 8	82.26 17	0.681

88KR B- DECAY (2.84 H 2) (CONTINUED)

Radiation Type	Energy (keV)	Intensity (%)	Δ(g-rad/μCi-h)
6 weak β's omitted (ΣIβ = 0.25%)			
X-ray L	1.69	0.23 8	≈0
X-ray $K\alpha_2$	13.33580 20	2.35 12	0.0007
X-ray $K\alpha_1$	13.39530 20	4.54 21	0.0013
X-ray $K\beta$	15	1.19 6	0.0004
γ 1	27.513 14	1.94 17	0.0011
γ 3	122.27 6	0.197 12	0.0005
γ 4	165.98 4	3.10 15	0.0110
γ 7	196.320 15	26.0 13	0.109
γ 8	240.71 4	0.253 14	0.0013
γ 9	311.69 3	0.107 9	0.0007
γ 10	334.71 3	0.145 10	0.0010
γ 12	362.226 13	2.25 12	0.0174
γ 14	390.543 11	0.64 6	0.0054
γ 16	421.70 18	0.128 25	0.0011
γ 17	471.80 3	0.73 4	0.0073
γ 25	677.34 5	0.235 18	0.0034
γ 30	788.28 4	0.53 3	0.0089
γ 31	790.32 7	0.125 12	0.0021
γ 34	834.830 3	13.0 7	0.231
γ 35	850.34 5	0.173 14	0.0031
γ 36	862.327 19	0.67 4	0.0123
γ 39	944.92 4	0.294 20	0.0059
γ 42	985.780 16	1.31 7	0.0276
γ 43	990.09 9	0.142 19	0.0030
γ 44	1039.59 3	0.48 3	0.0107
γ 45	1049.48 12	0.142 13	0.0032
γ 48	1141.33 6	1.28 7	0.0312
γ 49	1179.51 3	1.00 5	0.0250
γ 50	1184.95 4	0.69 5	0.0174
γ 51	1209.84 8	0.14 3	0.0037
γ 52	1212.73 17	0.14 5	0.0036
γ 53	1245.22 4	0.363 25	0.0096

88RB B- DECAY (17.8 M 1) (CONTINUED)

Radiation Type	Energy (keV)	Intensity (%)	Δ(g-rad/μCi-h)
β⁻ 2 max	473 4		
avg	147.4 15	0.370 20	0.0012
β⁻ 3 max	575 4		
avg	185.4 16	0.144 10	0.0006
β⁻ 4 max	804 4		
avg	274.9 17	2.13 11	0.0125
β⁻ 5 max	904 4		
avg	315.8 17	0.211 17	0.0014
β⁻ 6 max	2100 4		
avg	845.9 19	0.98 6	0.0177
β⁻ 7 max	2584 4		
avg	1072.0 19	13.3 7	0.304
β⁻ 8 max	3482 4		
avg	1497.9 19	4.2 3	0.134
β⁻ 9 max	5318 4		
avg	2373.9 20	77.4 14	3.91
Total β⁻ avg	2071 3	99.5 16	4.39
4 weak β's omitted (ΣIβ = 0.06%)			
γ 5	898.03 4	14.0 8	0.269
γ 8	1366.26 12	0.103 14	0.0030
γ 9	1382.39 5	0.74 5	0.0219
γ 11	1779.83 7	0.216 18	0.0082
γ 13	1836.00 5	21.4 13	0.837
γ 14	2111.22 12	0.118 13	0.0053
γ 17	2577.72 6	0.180 14	0.0099
γ 18	2677.86 5	1.96 12	0.112
γ 19	2734.03 7	0.109 9	0.0064
γ 20	3009.43 7	0.244 14	0.0156
γ 22	3218.48 8	0.214 14	0.0147
γ 23	3486.46 9	0.131 9	0.0097

γ	54	1250.67 4	1.12 6	0.0299
γ	57	1324.98 4	0.16 4	0.0045
γ	59	1352.32 11	0.159 22	0.0046
γ	60	1369.50 20	1.48 9	0.0431
γ	61	1406.94 10	0.218 20	0.0065
γ	62	1464.84 9	0.114 15	0.0036
γ	63	1518.39 3	2.15 12	0.0696
γ	64	1529.77 3	10.9 6	0.356
γ	65	1603.79 5	0.46 4	0.0156
γ	68	1685.6 4	0.66 8	0.0239
γ	72	1892.76 13	0.14 3	0.0056
γ	73	1908.7 4	0.100 15	0.0041
γ	74	2029.84 3	4.53 23	0.196
γ	75	2035.411 18	3.74 21	0.162
γ	76	2186.5 3	0.29 6	0.0134
γ	77	2195.842 7	13.2 7	0.617
γ	78	2231.772 21	3.39 17	0.161
γ	80	2352.08 4	0.73 4	0.0366
γ	82	2392.11 4	34.6 16	1.76
γ	83	2408.91 7	0.104 12	0.0053
γ	85	2548.40 3	0.62 3	0.0338
γ	86	2771.02 5	0.149 10	0.0088

36 weak γ's omitted (ΣIγ = 1.88%)

88RB B- DECAY (17.8 M 1)   I(MIN)= 0.10%

Radiation Type	Energy (keV)	Intensity (%)	Δ(g-rad/μCi-h)
β⁻ 1 max	465 4		
avg	144.7 15	0.67 4	0.0021

(continued on next column)

γ	26	4742.69 11	0.143 11	0.0145

14 weak γ's omitted (ΣIγ = 0.38%)

88Y EC DECAY (106.61 D 2)   I(MIN)= 0.10%

Radiation Type	Energy (keV)	Intensity (%)	Δ(g-rad/μCi-h)
Auger-L	1.79	105 6	0.0040
Auger-K	12	26.9 23	0.0069
β⁺ 1 max	764.5 15	0.200 10	0.0015
avg	359.5 7		
X-ray L	1.8	1.7 6	≈0
X-ray Kα₂	14.09790 20	17.5 7	0.0052
X-ray Kα₁	14.16500 20	33.7 13	0.0102
X-ray Kβ	15.8	9.1 4	0.0030
γ	2	898.042 4	93.7 3	1.79
γ	4	1836.063 13	99.3 3	3.89
γ	5	2734.03 7	0.64 4	0.0370

3 weak γ's omitted (ΣIγ = 0.10%)
Maximum γ± intensity = 0.40%

89RB B- DECAY (15.2 M 1)   I(MIN)= 0.10%

Radiation Type	Energy (keV)	Intensity (%)	Δ(g-rad/μCi-h)
β⁻ 1 max	992 8		
avg	352 4	1.52 13	0.0114
β⁻ 2 max	1273 8		
avg	473 4	33 3	0.332
β⁻ 3 max	1794 8		
avg	718 4	2.19 13	0.0335
β⁻ 4 max	1931 8		
avg	768 4	3.0 9	0.0491
β⁻ 5 max	2221 8		
avg	902 4	34 4	0.653
β⁻ 6 max	2444 8		
avg	1006 4	0.49 13	0.0105
β⁻ 7 max	2493 8		
avg	1029 4	0.46 21	0.0101
β⁻ 8 max	2561 8		
avg	1061 4	0.22 4	0.0050
β⁻ 9 max	3028 8		
avg	1284 4	0.25 5	0.0068
β⁻10 max	3469 8		
avg	1492 4	1 4	0.0318
β⁻11 max	4501 8		
avg	1986 4	25 5	1.06
Total β⁻			
avg	1019 6	102 9	2.20

7 weak β's omitted (ΣIβ = 0.38%)

γ 3	272.45 10	1.42 12	0.0082
γ 4	289.76 10	0.54 10	0.0033
γ 8	657.71 7	10.0 9	0.140
γ 10	766.79 15	0.162 21	0.0027
γ 14	947.69 7	9.2 8	0.186
γ 16	1025.3 5	0.23 9	0.0049

90SR B- DECAY (28.5 Y 2)   I(MIN)= 0.10%

Radiation Type	Energy (keV)	Intensity (%)	Δ(g-rad/μCi-h)
β⁻ 1 max	546.2 20		
avg	195.8 8	100	0.417

90Y B- DECAY (64.0 H 1)   I(MIN)= 0.10%

Radiation Type	Energy (keV)	Intensity (%)	Δ(g-rad/μCi-h)
β⁻ 1 max	2283 3		
avg	934.3 13	99.984 3	1.99

1 weak β's omitted (ΣIβ = 0.02%)

94NB B- DECAY (2.03E4 Y 16)   I(MIN)= 0.10%

Radiation Type	Energy (keV)	Intensity (%)	Δ(g-rad/μCi-h)
β⁻ 1 max	471.3 23		
avg	145.8 9	100	0.311

| γ | 1 | 702.645 6 | 100 | 1.50 |
| γ | 2 | 871.119 4 | 100 | 1.86 |

94NB IT DECAY (6.26 M 1)   I(MIN)= 0.10%
%IT=99.52 10
SEE ALSO 94NB B- DECAY (6.26 M)

Radiation Type	Energy (keV)	Intensity (%)	Δ(g-rad/μCi-h)
Auger-L	2.15	91 4	0.0042
Auger-K	14	14.5 19	0.0043
ce-K-1	22.5 5	57.62 6	0.0276
ce-L-1	38.8 5	33.34 4	0.0276
ce-M-1	41.0 5	6.369 7	0.0056
ce-NOP-1	41.4 5	2.0899 21	0.0018
X-ray L	2.17	2.8 10	0.0001
X-ray $K\alpha_2$	16.52100 20	12.4 6	0.0044
X-ray $K\alpha_1$	16.61510 20	23.8 11	0.0084
X-ray $K\beta$	18.6	6.9 4	0.0027

1 weak γ's omitted (ΣIγ = 0.08%)

γ	17	1031.88 7	58 5	1.27
γ	23	1220.32 10	0.220 24	0.0057
γ	24	1228.40 15	0.122 20	0.0032
γ	26	1248.10 7	43 4	1.13
γ	29	1473.22 20	0.35 4	0.0111
γ	30	1501.07 20	0.197 22	0.0063
γ	31	1538.08 10	2.55 25	0.0836
γ	35	1940.2 3	0.33 4	0.0137
γ	37	2007.54 10	2.38 24	0.102
γ	38	2058.0 11	0.23 9	0.0102
γ	40	2196.00 15	13.3 13	0.624
γ	42	2280.06 10	0.180 22	0.0087
γ	45	2570.14 10	9.9 9	0.540
γ	48	2707.20 10	2.03 19	0.117
γ	57	3508.84 25	1.15 11	0.0858

41 weak γ's omitted (ΣIγ = 1.18%)

89SR B- DECAY (50.5 D 1)   I(MIN)= 0.10%

Radiation Type	Energy (keV)	Intensity (%)	Δ(g-rad/μCi-h)
β^- 1 max	1492 4		
avg	583.3 8	100	1.24

94NB B- DECAY (6.26 M 1)   I(MIN)= 0.10%
%B-=0.48 10
SEE ALSO 94NB IT DECAY (6.26 M)

Radiation Type	Energy (keV)	Intensity (%)	Δ(g-rad/µCi-h)
β- 1 max	1214.9 23		
avg	443.9 10	0.48 10	0.0045
γ 2	871.119 4	0.48 10	0.0089

97ZR B- DECAY (16.90 H 5)   I(MIN)= 0.10%

Radiation Type	Energy (keV)	Intensity (%)	Δ(g-rad/µCi-h)
Auger-L	2.15	1.96 12	≈0
Auger-K	14	0.43 6	0.0001
ce-K-17	724.37 10	1.65 5	0.0254
ce-L-17	740.66 10	0.213 7	0.0034
β- 1 max	410.8 19		
avg	124.6 7	0.41 4	0.0011
β- 2 max	552.4 19		
avg	175.7 7	5.5 3	0.0206
β- 3 max	893.9 19		
avg	309.6 8	2.10 20	0.0138
β- 4 max	907.8 19		
avg	315.3 8	0.4 3	0.0027
β- 5 max	1005.5 20		
avg	355.7 8	0.120 20	0.0009

97ZR B- DECAY (16.90 H 5)   (CONTINUED)

Radiation Type	Energy (keV)	Intensity (%)	Δ(g-rad/µCi-h)
γ 26	1362.66 10	1.34 14	0.0390
γ 27	1750.46 10	1.34 14	0.0501
γ 28	1851.55 10	0.35 4	0.0139

4 weak γ's omitted (ΣIγ = 0.21%)

97NB B- DECAY (72.1 M 7)   I(MIN)= 0.10%

Radiation Type	Energy (keV)	Intensity (%)	Δ(g-rad/µCi-h)
Auger-L	2.27	0.180 13	≈0
ce-K-5	637.92 10	0.182 6	0.0025
β- 1 max	304.8 19		
avg	88.5 7	0.120 20	0.0002
β- 2 max	418.2 19		
avg	126.9 7	0.170 20	0.0005
β- 3 max	665.2 19		
avg	218.0 8	0.210 20	0.0010
β- 4 max	909.3 20		
avg	315.2 8	1.10 10	0.0074
β- 5 max	1275.8 19		
avg	470.2 9	98.30 10	0.984
Total β- avg	466.8 9	99.99 15	0.994

1 weak β's omitted (ΣIβ = 0.09%)

		Energy (keV)	Intensity (%)	Δ(g-rad/μCi-h)
γ	3	480.9	0.148 20	0.0015
γ	5	657.92 10	98.34 11	1.38
γ	12	1268.63 10	0.157 20	0.0043
γ	13	1515.64 20	0.118 20	0.0038

10 weak γ's omitted (ΣIγ = 0.49%)

97NB IT DECAY (60 S 1)   I(MIN)= 0.10%

Radiation Type	Energy (keV)	Intensity (%)	Δ(g-rad/μCi-h)
Auger-L	2.15	1.99 12	≈0
Auger-K	14	0.44 6	0.0001
ce-K-1	724.37 10	1.74 5	0.0268
ce-L-1	740.66 10	0.225 7	0.0035
X-ray Kα₂	16.52100 20	0.375 20	0.0001
X-ray Kα₁	16.61510 20	0.72 4	0.0003
X-ray Kβ	18.6	0.207 12	≈0
γ 1	743.36 10	97.95 6	1.55

		Energy (keV)	Intensity (%)	Δ(g-rad/μCi-h)
β⁻ 6	max	1110.0 19		
	avg	399.8 9	0.45 7	0.0038
β⁻ 7	max	1110.5 19		
	avg	400.0 8	0.65 7	0.0055
β⁻ 8	max	1382.2 19		
	avg	517.6 9	0.21 11	0.0023
β⁻ 9	max	1407.3 19		
	avg	528.7 9	4.1 5	0.0462
β⁻ 10	max	1915.0 19		
	avg	757.0 9	86.0 7	1.39
Total β⁻	avg	697.0 10	99.9 10	1.48
X-ray Kα₂		16.52100 20	0.370 20	0.0001
X-ray Kα₁		16.61510 20	0.71 4	0.0003
X-ray Kβ		18.6	0.204 12	≈0
γ	3	218.87 20	0.176 19	0.0008
γ	4	254.15 20	1.25 14	0.0068
γ	5	272.27 20	0.25 4	0.0015
γ	8	330.43 20	0.11 3	0.0008
γ	9	355.39 10	2.27 24	0.0172
γ	10	400.39 20	0.32 5	0.0028
γ	11	507.63 10	5.1 6	0.0547
γ	12	513.38 20	0.56 10	0.0061
γ	13	602.41 20	1.39 14	0.0179
γ	14	690.63 20	0.25 4	0.0037
γ	15	699.2 3	0.121 19	0.0018
γ	16	703.80 10	0.93 10	0.0139
γ	17	743.36 10	92.7 4	1.47
γ	18	804.53 10	0.65 7	0.0111
γ	19	829.80 10	0.223 19	0.0039
γ	20	854.90 10	0.33 4	0.0061
γ	21	971.39 10	0.29 3	0.0059
γ	22	1021.3 3	1.34 14	0.0293
γ	23	1110.45 20	0.111 19	0.0026
γ	24	1147.95 10	2.6 3	0.0646
γ	25	1276.09 10	0.97 10	0.0265

(continued on next column)

97RU EC DECAY (2.9 D 1) I(MIN)= 0.10%

Radiation Type	Energy (keV)	Intensity (%)	Δ(g-rad/μCi-h)
Auger-L	2.17	97 6	0.0045
Auger-K	15.5	20 3	0.0065
ce-K-5	194.64 4	2.94 8	0.0122
ce-L-5	212.64 4	0.361 18	0.0016
ce-K-6	303.51 7	0.173 9	0.0011
X-ray L	2.42	4.6 16	0.0002
X-ray Kα₂	18.2508 8	20.0 9	0.0078
X-ray Kα₁	18.3671 8	38.3 17	0.0150
X-ray Kβ	20.6	11.5 6	0.0050
γ 2	108.80 4	0.108 12	0.0003
γ 5	215.68 4	86.0 4	0.395
γ 6	324.55 7	10.2 4	0.0705
γ 7	460.55 5	0.117 5	0.0011
γ 11	569.27 5	0.89 4	0.0108

14 weak γ's omitted (ΣIγ = 0.34%)

99TC IT DECAY (6.007 H 1) I(MIN)= 0.10%

Radiation Type	Energy (keV)	Intensity (%)	Δ(g-rad/μCi-h)
ce-M-1	1.630 5	86.84 16	0.0030
ce-NOP-1	2.106 5	12.18 20	0.0005
Auger-L	2.17	10.4 7	0.0005
Auger-K	15.5	2.1 4	0.0007
ce-K-2	119.4670 13	8.8 4	0.0224
ce-K-3	121.63 3	0.69 4	0.0018
ce-L-2	137.4685 11	1.06 4	0.0031
ce-L-3	139.63 3	0.215 14	0.0006
ce-M-2	139.9670 15	0.198 10	0.0006
X-ray L	2.42	0.49 17	≈0
X-ray Kα₂	18.2508 8	2.12 13	0.0008
X-ray Kα₁	18.3671 8	4.06 24	0.0016
X-ray Kβ	20.6	1.22 8	0.0005
γ 2	140.5110 10	88.97 24	0.266

2 weak γ's omitted (ΣIγ = 0.02%)

99MO B-,99TC IT DECAY (65.92 H 2) I(MIN)=0.10%

Radiation Type	Energy (keV)	Intensity (%)	Δ(g-rad/μCi-h)
ce-M-1	1.630 5	83.6 7	0.0029
ce-NOP-1	2.106 5	11.73 22	0.0005
Auger-L	2.17	15.5 9	0.0007

103RU B- DECAY (39.26 D 2) I(MIN)= 0.10%

Radiation Type	Energy (keV)	Intensity (%)	Δ(g-rad/μCi-h)
Auger-L	2.39	76.9 15	0.0039
ce-K-1	16.535 12	9.4 3	0.0033
Auger-K	17	2.0 4	0.0007

Auger-K	15.5		0.0010
ce-K-1	19.5395 8	3.1 5	0.0016
ce-L-2	37.5410 5	0.458 15	0.0004
ce-MNO-2	40.0395 11	0.110 6	≈0
ce-K-4	119.4670 13	9.0 3	0.0228
ce-K-5	121.63 3	0.67 3	0.0017
ce-L-5	137.4685 11	1.09 4	0.0032
ce-L-5	139.63 3	0.207 9	0.0006
ce-M-4	139.9670 15	0.199 7	0.0006
ce-K-8	160.013 15	0.765 24	0.0026
ce-L-8	178.014 15	0.114 4	0.0004
β⁻ 1 max	215.2 10		
avg	59.9 3	0.111 3	0.0001
β⁻ 2 max	353.0 10		
avg	104.3 4	0.134 4	0.0003
β⁻ 3 max	436.4 10		
avg	133.1 4	16.55 7	0.0469
β⁻ 4 max	847.9 10		
avg	289.7 4	1.17 3	0.0072
β⁻ 5 max	1214.3 10		
avg	442.8 5	81.93 10	0.773
Total β⁻			
avg	388.8 6	99.91 13	0.827

2 weak β's omitted (ΣI$_\beta$ = 0.01%)

X-ray L	2.42	0.73 25	≈0
X-ray Kα₂	18.2508 8	3.17 16	0.0012
X-ray Kα₁	18.3671 8	6.1 3	0.0024
X-ray Kβ	20.6	1.82 9	0.0008
γ 2	40.58347 17	1.16 3	0.0010
γ 4	140.5110 10	90.6 7	0.271
γ 8	181.057 15	6.07 5	0.0234
γ 11	366.421 15	1.203 15	0.0094
γ 25	739.500 15	12.223 12	0.193
γ 27	777.921 20	4.31 4	0.0715
γ 28	822.972 15	0.133 4	0.0023

25 weak γ's omitted (ΣI$_\gamma$ = 0.29%)

ce-K-3	30.055 10	0.70 4	0.0004
ce-L-1	36.343 12	70.50 21	0.0546
ce-M-1	39.128 12	14.2 4	0.0119
ce-NOP-1	39.674 12	4.65 20	0.0039
ce-K-14	473.864 6	0.416 13	0.0042
β⁻ 1 max	116 4		
avg	30.8 10	6.64 19	0.0044
β⁻ 2 max	230 4		
avg	64.3 11	92 3	0.126
β⁻ 3 max	472 4		
avg	145.1 13	0.245 9	0.0008
β⁻ 4 max	767 4		
avg	256.1 14	0.87 5	0.0047
Total β⁻			
avg	63.9 12	100 3	0.136

1 weak β's omitted (ΣI$_\beta$ = 0.09%)

X-ray L	2.7	4.0 13	0.0002
X-ray Kα₂	20.07370 20	2.44 13	0.0010
X-ray Kα₁	20.21610 20	4.63 23	0.0020
X-ray Kβ	22.7	1.44 8	0.0007
γ 3	53.275 10	0.384 19	0.0004
γ 9	294.980 20	0.256 9	0.0016
γ 13	443.800 20	0.329 10	0.0031
γ 14	497.084 6	91 3	0.968
γ 16	557.040 20	0.8500 19	0.0101
γ 18	610.330 20	5.76 19	0.0749

14 weak γ's omitted (ΣI$_\gamma$ = 0.20%)

108AG B- DECAY (2.37 M 1)   I(MIN)= 0.10%

%B=97.15 20
SEE ALSO 108AG EC DECAY (2.37 M)

Radiation Type	Energy (keV)	Intensity (%)	A(g-rad/μCi-h)
β⁻ 1 max	1017 8		
avg	356 4	1.76 10	0.0133
β⁻ 2 max	1650 8		
avg	629 4	95.4 3	1.28
Total β⁻			
avg	624 4	97.2 4	1.29
γ 1	632.98 5	1.76 10	0.0237

108AG EC DECAY (127 Y 21)   I(MIN)= 0.10%

%EC=91.3 6
SEE ALSO 108AG IT DECAY (127 Y)

Radiation Type	Energy (keV)	Intensity (%)	A(g-rad/μCi-h)
Auger-L	2.5	82 5	0.0044
Auger-K	17.7	14.2 24	0.0053
ce-K-1	409.586 4	0.715 22	0.0062
ce-L-1	430.332 4	0.145 5	0.0013
ce-K-2	589.931 4	0.261 10	0.0033
ce-K-3	698.579 4	0.173 7	0.0026
X-ray L	2.84	4.9 17	0.0003
X-ray Kα₂	21.02010 20	18.3 8	0.0082

103RH IT DECAY (56.114 M 6)   I(MIN)= 0.10%

Radiation Type	Energy (keV)	Intensity (%)	A(g-rad/μCi-h)
Auger-L	2.39	76.6 15	0.0039
ce-K-1	16.535 12	9.5 3	0.0033
Auger-K	17	1.8 3	0.0007
ce-L-1	36.343 12	71.290 10	0.0552
ce-M-1	39.128 12	14.4 4	0.0120
ce-NOP-1	39.674 12	4.70 20	0.0040
X-ray L	2.7	4.0 13	0.0002
X-ray Kα₂	20.07370 20	2.19 12	0.0009
X-ray Kα₁	20.21610 20	4.17 21	0.0018
X-ray Kβ	22.7	1.30 7	0.0006

1 weak γ's omitted (ΣIγ = 0.07%)

103PD EC DECAY (16.97 D 2)   I(MIN)= 0.10%

Radiation Type	Energy (keV)	Intensity (%)	A(g-rad/μCi-h)
Auger-L	2.39	167 7	0.0085
ce-K-1	16.535 12	9.5 4	0.0033
Auger-K	17	18 3	0.0067
ce-L-1	36.343 12	71 3	0.0552
ce-M-1	39.128 12	14.4 6	0.0120
ce-NOP-1	39.674 12	4.72 20	0.0040

X-ray	Energy (keV)	Intensity (%)	Δ(g-rad/μCi-h)
X-ray L	2.7	9 3	0.0005
X-ray Kα2	20.07370 20	22.1 9	0.0094
X-ray Kα1	20.21610 20	41.9 17	0.0181
X-ray Kβ	22.7	13.1 6	0.0063

9 weak γ's omitted (ΣIγ = 0.10%)

108AG EC DECAY (2.37 M 1)   I(MIN)= 0.10%

%EC+%B+=2.85 20

SEE ALSO 108AG B- DECAY (2.37 M)

Radiation Type	Energy (keV)	Intensity (%)	Δ(g-rad/μCi-h)
Auger-L	2.5		0.0001
Auger-K	17.7	2.28 18	0.0002
		0.40 7	
β+ 1 max	893 7		
avg	398 3	0.280 20	0.0024
X-ray L	2.84	0.13 5	≈0
X-ray Kα2	21.02010 20	0.52 4	0.0002
X-ray Kα1	21.17710 20	0.99 7	0.0004
X-ray Kβ	23.8	0.313 22	0.0002
γ 3	433.936 4	0.50 4	0.0046
γ 6	618.86 5	0.261 23	0.0034

10 weak γ's omitted (ΣIγ = 0.03%)
Maximum γ± intensity = 0.57%

X-ray	Energy (keV)	Intensity (%)	Δ(g-rad/μCi-h)
X-ray Kα1	21.17710 20	34.8 13	0.0157
X-ray Kβ	23.8	11.0 5	0.0056
γ 1	433.936 4	90.5 6	0.836
γ 2	614.281 4	89.8 19	1.18
γ 3	722.929 4	90.8 19	1.40

108AG IT DECAY (127 Y 21)   I(MIN)= 0.10%

%IT=8.7 6

SEE ALSO 108AG EC DECAY (127 Y)

Radiation Type	Energy (keV)	Intensity (%)	Δ(g-rad/μCi-h)
Auger-L	2.6	8.0 5	0.0004
ce-K-1	4.87 6	0.226 17	≈0
Auger-K	18.5	0.34 6	0.0001
ce-L-1	26.57 6	6.4 5	0.0036
ce-M-1	29.66 6	1.59 12	0.0010
ce-NOP-1	30.28 6	0.52 4	0.0003
ce-K-2	53.617 3	1.80 14	0.0021
ce-L-2	75.325 3	0.223 17	0.0004
X-ray L	3	0.54 19	≈0
X-ray Kα2	21.9903 3	0.48 4	0.0002
X-ray Kα1	22.16290 10	0.91 7	0.0004
X-ray Kβ	24.9	0.293 23	0.0002
γ 2	79.131 3	6.6 5	0.0112

109PD B- DECAY (13.427 H 14)   I(MIN)= 0.10%

Radiation Type	Energy (keV)	Intensity (%)	Δ(g-rad/μCi-h)
Auger-L	2.6	76 6	0.0042
Auger-K	18.5	6.9 13	0.0027
ce-K-2	62.5201 12	40 4	0.0538
ce-L-2	84.2283 12	43 5	0.0764
ce-M-2	87.3166 12	8.7 9	0.0162
ce-NOP-2	87.9389 12	1.55 16	0.0029
β- 1 max	1027.9 20		
avg	361.0 9	99.900 10	0.768

12 weak β's omitted (ΣIβ = 0.11%)

Radiation Type	Energy (keV)	Intensity (%)	Δ(g-rad/μCi-h)
X-ray L	3	5.1 18	0.0003
X-ray Kα₂	21.9903 3	9.6 10	0.0045
X-ray Kα₁	22.16290 10	18.1 19	0.0086
X-ray Kβ	24.9	5.8 7	0.0031
γ 2	88.0341 11	3.6 4	0.0068

35 weak γ's omitted (ΣIγ = 0.13%)

109CD EC DECAY (462.6 D 4)   I(MIN)= 0.10%

Radiation Type	Energy (keV)	Intensity (%)	Δ(g-rad/μCi-h)
Auger-L	2.6	166 7	0.0092
Auger-K	18.5	21 3	0.0082

110AG B- DECAY (24.6 S 2)   I(MIN)= 0.10%
%B-=99.70 6
SEE ALSO 110AG EC DECAY (24.6 S)

Radiation Type	Energy (keV)	Intensity (%)	Δ(g-rad/μCi-h)
β- 1 max	2235.6 20		
avg	894.4 10	4.4 3	0.0838
β- 2 max	2893.4 20		
avg	1199.6 10	95.2 3	2.43
Total β- avg	1185.4 10	99.7 5	2.52

9 weak β's omitted (ΣIβ = 0.09%)

γ 2	657.749 10	4.50 23	0.0630

12 weak γ's omitted (ΣIγ = 0.10%)

110AG IT DECAY (249.8 D 1)   I(MIN)= 0.10%
%IT=1.31 7
SEE ALSO 110AG B- DECAY (249.8 D)

Radiation Type	Energy (keV)	Intensity (%)	Δ(g-rad/μCi-h)
ce-M-1	0.56 10	0.98 6	≈0
ce-NOP-1	1.18 10	0.327 18	≈0
Auger-L	2.6	1.10 7	≈0
Auger-K	18.5	0.139 22	≈0

	Energy (keV)	Intensity (%)	Δ(g-rad/ μCi-h)
ce-K-1	62.5201 12	41.9	0.0558
ce-L-1	84.2283 12	44	0.0789
ce-M-1	87.3166 12	9.0 3	0.0167
ce-NOP-1	87.9389 12	1.60 5	0.0030
X-ray L	3	11 4	0.0007
X-ray Kα₂	21.9903 3	29.1 10	0.0136
X-ray Kα₁	22.16290 10	55.1 18	0.0260
X-ray Kβ	24.9	17.8 7	0.0094
γ 1	88.0341 11	3.72 10	0.0070

110AG EC DECAY (24.6 S 2)   I(MIN)= 0.10%
%EC=0.30 6
SEE ALSO 110AG B- DECAY (24.6 S)

Radiation Type	Energy (keV)	Intensity (%)	Δ(g-rad/ μCi-h)
Auger-L	2.5	0.23 5	≈0
X-ray Kα₁	21.17710 20	0.115 24	≈0

	Energy (keV)	Intensity (%)	Δ(g-rad/ μCi-h)
ce-K-2	90.97 5	0.81 5	0.0016
ce-L-2	112.67 5	0.388 23	0.0009
X-ray Kα₂	21.9903 3	0.193 13	≈0
X-ray Kα₁	22.16290 10	0.365 23	0.0002
X-ray Kβ	24.9	0.118 8	≈0

110AG B- DECAY (249.8 D 1)   I(MIN)= 0.10%
%B=98.69 7
SEE ALSO 110AG IT DECAY (249.8 D)

Radiation Type	Energy (keV)	Intensity (%)	Δ(g-rad/ μCi-h)
Auger-L	2.72	0.336 25	≈0
ce-K-29	631.0510 22	0.249 10	0.0033
β⁻ 1 max	84.4 20		
avg	22.0 6	67.27 17	0.0315
β⁻ 2 max	134.3 20		
avg	35.9 6	0.408 10	0.0003
β⁻ 3 max	531.2 20		
avg	165.8 8	30.57 11	0.108
Total β⁻ avg	66.9 14	98.38 21	0.140

4 weak β's omitted (ΣIβ = 0.13%)

	Energy (keV)	Intensity (%)	Δ(g-rad/ μCi-h)
X-ray Kα₁	23.17360 20	0.171 8	≈0
γ 18	446.811 3	3.70 3	0.0352

(continued on next

110AG EC DECAY (24.6 S 2) (CONTINUED)

Radiation Type	Energy (keV)	Intensity (%)	Δ(g-rad/μCi-h)
γ 24	620.360 3	2.787 19	0.0368
γ 25	626.262 10	0.232 7	0.0031
γ 29	657.7622 21	94.38 21	1.32
γ 31	676.58 10	0.142 10	0.0020
γ 32	677.6227 24	10.55 8	0.152
γ 33	687.010 4	6.44 6	0.0943
γ 34	706.682 3	16.60 10	0.250
γ 35	708.133 20	0.274 19	0.0041
γ 37	744.277 3	4.70 3	0.0746
γ 38	763.944 3	22.28 9	0.362
γ 40	818.031 4	7.32 4	0.128
γ 42	884.685 3	72.6 3	1.37
γ 44	937.493 4	34.29 10	0.685
γ 46	997.258 15	0.129 5	0.0027
γ 57	1334.341 17	0.139 6	0.0039
γ 58	1384.300 4	24.28 10	0.716
γ 61	1475.788 6	3.987 20	0.125
γ 62	1505.040 5	13.05 5	0.418
γ 63	1562.302 5	1.181 13	0.0393

51 weak γ's omitted (ΣIγ = 1.06%)

111IN EC DECAY (2.805 D 1)   I

Radiation Type	Energy (keV)	Intensity (%)	Δ
Auger-L	2.72	100 6	0.0065
Auger-K	19.3	16 3	0.0243
ce-K-1	144.57 3	7.89 23	0.0034
ce-L-1	167.26 3	0.96 4	0.0008
ce-MNO-1	170.51 3	0.222 11	0.0008
ce-K-2	218.68 3	5.03 14	0.0234
ce-L-2	241.37 3	0.785 23	0.0040
ce-MNO-2	244.62 3	0.182 6	0.0009
X-ray L	3.13	7.0 24	0.0005
X-ray Kα2	22.98410 20	23.5 9	0.0115
X-ray Kα1	23.17360 20	44.4 16	0.0219
X-ray Kβ	26	14.5 6	0.0081
γ 1	171.28 3	90.93 22	0.332
γ 2	245.39 3	94.00 17	0.491

113IN IT DECAY (1.658 H 1)   I(MIN)= 0.10%

Radiation Type	Energy (keV)	Intensity (%)	Δ(g-rad/μCi-h)
Auger-L	2.84	29.7 19	0.0018
Auger-K	20	4.2 9	0.0018
ce-K-1	363.762 4	28.2 6	0.218
ce-L-1	387.464 4	5.48 16	0.0452
ce-M-1	390.876 4	1.100 10	0.0092
ce-NOP-1	391.580 4	0.245 3	0.0020

111AG B- DECAY (7.45 D 1)   I(MIN) = 0.10%

Radiation Type	Energy (keV)	Intensity (%)	Δ(g-rad/µCi-h)
Auger-L	2.72	0.200 17	≈0
ce-K-4	315.419 20	0.105 8	0.0007
β⁻ 1 max	694.9 20		
avg	227.0 8	6.3 16	0.0305
β⁻ 2 max	791.6 20		
avg	282.4 8	1.0 3	0.0060
β⁻ 3 max	1037.0 20		
avg	364.1 9	92.7 16	0.719
Total β⁻ avg	354.5 9	100.0 23	0.755

3 weak β's omitted (ΣIβ = 0.04%)

Radiation Type	Energy (keV)	Intensity (%)	Δ(g-rad/µCi-h)
X-ray Kα₁	23.17360 20	0.102 6	≈0
γ 1	96.750 20	0.121 13	0.0002
γ 2	245.400 20	1.23 6	0.0064
γ 4	342.130 20	6.7 4	0.0488

7 weak γ's omitted (ΣIγ = 0.05%)

Radiation Type	Energy (keV)	Intensity (%)	Δ(g-rad/µCi-h)
X-ray L	3.29	2.3 8	0.0002
X-ray Kα₂	24.00200 20	6.8 3	0.0035
X-ray Kα₁	24.20970 20	12.9 6	0.0066
X-ray Kβ	27.3	4.27 19	0.0025
γ 1	391.702 4	64.90 20	0.541

113SN EC DECAY (115.08 D 3)   I(MIN) = 0.10%

Radiation Type	Energy (keV)	Intensity (%)	Δ(g-rad/µCi-h)
Auger-L	2.84	115 7	0.0070
Auger-K	20	17 4	0.0073
ce-K-2	363.762 4	28.2 6	0.218
ce-L-2	387.464 4	5.48 16	0.0452
ce-M-2	390.876 4	1.100 10	0.0092
ce-NOP-2	391.580 4	0.245 3	0.0020
X-ray L	3.29	9 3	0.0006
X-ray Kα₂	24.00200 20	27.5 11	0.0141
X-ray Kα₁	24.20970 20	51.9 19	0.0267
X-ray Kβ	27.3	17.2 7	0.0100
γ 1	255.115 15	1.85 6	0.0101
γ 2	391.702 4	64.90 20	0.541

115IN IT DECAY (4.486 H 4)   I(MIN)= 0.10%

%IT=95.0 7
SEE ALSO 115IN B- DECAY (4.486 H)

Radiation Type	Energy (keV)	Intensity (%)	Δ(g-rad/μCi-h)
Auger-L	2.84	41 3	0.0025
Auger-K	20	5.8 12	0.0025
ce-K-1	308.30 3	38.8 9	0.255
ce-L-1	332.00 3	7.95 24	0.0562
ce-M-1	335.42 3	1.66 5	0.0119
ce-NOP-1	336.12 3	0.370 10	0.0027
X-ray L	3.29	3.2 11	0.0002
X-ray Kα₂	24.00200 20	9.4 4	0.0048
X-ray Kα₁	24.20970 20	17.7 8	0.0091
X-ray Kβ	27.3	5.9 3	0.0034
γ 1	336.241 25	45.8 5	0.328

115IM B- DECAY (4.486 M 4)   I(MIN)= 0.10%

%B-=5.0 7
SEE ALSO 115IN IT DECAY (4.486 H)

Radiation Type	Energy (keV)	Intensity (%)	Δ(g-rad/μCi-h)
β⁻ 1 max	862 4		
avg	291.7 16	5.0 7	0.0311

1 weak β's omitted (ΣIβ = 0.05%)

116IN B- DECAY (54.15 M 6)   (CONT)

Radiation Type	Energy (keV)	Intensity (%)	
γ 29	818.70 20	11.5 5	0.200
γ 32	972.55 3	0.454 17	0.0094
γ 34	1097.30 20	56.2 12	1.31
γ 37	1293.54 4	84.4 18	2.33
γ 38	1507.40 20	10.0 4	0.320
γ 40	1753.8 6	2.46 8	0.0917
γ 42	2112.1 4	15.5 5	0.699

25 weak γ's omitted (ΣIγ = 1.01%)

117SN IT DECAY (13.61 D 4)   I(MIN)= 0.10%

Radiation Type	Energy (keV)	Intensity (%)	Δ(g-rad/μCi-h)
Auger-L	3	91 5	0.0057
Auger-K	21	10.8 22	0.0048
ce-K-1	126.82 3	64.9 20	0.175
ce-K-2	129.360 20	11.7 4	0.0322
ce-L-1	151.56 3	26.2 8	0.0845
ce-L-2	154.095 20	1.48 5	0.0048
ce-M-1	155.14 3	5.64 18	0.0186
ce-NOP-1	155.88 3	1.35 5	0.0045
ce-M-2	157.676 20	0.289 9	0.0010

116IN B- DECAY (54.15 M 6)   I(MIN)=0.10%

Radiation Type	Energy (keV)	Intensity (%)	Δ(g-rad/μCi-h)
Auger-L	3	0.73 10	≈0
Auger-K	21	0.12 3	≈0
ce-K-4	109.126 8	0.86 11	0.0020
β⁻ 1 max	305 4		
avg	87.5 13	0.33 4	0.0006
β⁻ 2 max	355 4		
avg	104.0 20	2.71 10	0.0060
β⁻ 3 max	600 4		
avg	190.0 20	10.2 4	0.0413
β⁻ 4 max	872 4		
avg	296.0 20	33.6 15	0.212
β⁻ 5 max	1010 4		
avg	352.0 20	51.8 11	0.388
Total β⁻			
avg	308.5 22	98.6 19	0.648
X-ray Kα₂	25.04400 20	0.21 3	0.0001
X-ray Kα₁	25.27130 20	0.39 5	0.0002
X-ray Kβ	28.5	0.133 17	≈0
γ 4	138.326 8	3.29 12	0.0097
γ 8	262.95 8	0.12 3	0.0007
γ 10	278.49 8	0.143 17	0.0009
γ 11	303.80 7	0.118 17	0.0008
γ 13	355.36 4	0.83 5	0.0063
γ 14	416.86 3	29.2 15	0.259
γ 17	463.14 12	0.83 5	0.0082
γ 22	655.7 4	0.11 5	0.0015
γ 24	689.0 3	0.16 3	0.0024
γ 25	705.7 3	0.17 3	0.0025
γ 28	781.1 8	0.110 21	0.0018

(continued on next column)

	Energy (keV)	Intensity (%)	Δ(g-rad/μCi-h)
X-ray L	3.44	8 3	0.0006
X-ray Kα₂	25.04400 20	18.8 9	0.0100
X-ray Kα₁	25.27130 20	35.2 15	0.0189
X-ray Kβ	28.5	11.9 6	0.0072
γ 1	156.02 3	2.113 12	0.0070
γ 2	158.560 20	86.4 4	0.292

117SB EC DECAY (2.80 H 1)   I(MIN)= 0.10%

Radiation Type	Energy (keV)	Intensity (%)	Δ(g-rad/μCi-h)
Auger-L	3	94 6	0.0059
Auger-K	21	14 3	0.0061
ce-K-1	129.362 15	11.7 4	0.0322
ce-L-1	154.097 15	1.46 5	0.0048
ce-MNO-1	157.678 15	0.4896 23	0.0016
β⁺ 1 max	575 9		
avg	262 4	1.83 12	0.0102
X-ray L	3.44	8 3	0.0006
X-ray Kα₂	25.04400 20	23.5 9	0.0125
X-ray Kα₁	25.27130 20	44.1 16	0.0237
X-ray Kβ	28.5	14.9 6	0.0090
γ 1	158.562 15	85.9 4	0.290
γ 4	861.35 5	0.31 4	0.0057
γ 5	1004.51 15	0.21 3	0.0044
γ 6	1020.6 5	0.103 18	0.0022
γ 7	1021.0 5	0.112 18	0.0024

7 weak γ's omitted (Σiγ = 0.26%)
Maximum γ± intensity = 3.66%

119SN IT DECAY (293.0 D 13) I(MIN)= 0.10%

Radiation Type	Energy (keV)	Intensity (%)	Δ(g-rad/μCi-h)
Auger-L	3	138 5	0.0087
ce-L-1	19.405 8	67.4 7	0.0279
Auger-K	21	4.5 9	0.0020
ce-M-1	22.986 8	13.2 4	0.0065
ce-NOP-1	23.733 8	3.40 10	0.0017
ce-K-2	36.460 10	32.2 7	0.0250
ce-L-2	61.195 10	52.3 8	0.0682
ce-M-2	64.776 10	12.4 3	0.0171
ce-NOP-2	65.523 10	3.10 10	0.0043
X-ray L	3.44	12 4	0.0009
γ 1	23.870 8	16.1 4	0.0082
X-ray Kα2	25.04400 20	7.9 4	0.0042
X-ray Kα1	25.27130 20	14.8 6	0.0080
X-ray Kβ	28.5	4.99 22	0.0030

1 weak γ's omitted (ΣIγ = 0.02%)

121TE EC DECAY (17 D 1) I(MIN)= 0.10%

Radiation Type	Energy (keV)	Intensity (%)	Δ(g-rad/μCi-h)
Auger-L	3	84 5	0.0055
ce-K-1	6.6478 21	1.120 12	0.0002
Auger-K	21.8	11.5 25	0.0054
ce-L-1	32.4407 21	0.145 5	0.0001
ce-K-2	35.059 8	0.49 3	0.0004
ce-K-4	477.113 5	0.127 6	0.0013
ce-K-5	542.662 4	0.369 14	0.0043
X-ray L	3.6	8 3	0.0006
X-ray Kα2	26.11080 20	21.4 8	0.0119
X-ray Kα1	26.35910 20	40.1 14	0.0225
X-ray Kβ	29.7	13.7 6	0.0087
γ 1	37.1390 20	0.117 4	0.0004
γ 2	65.550 8	0.259 9	0.0004
γ 3	470.484 8	1.41 4	0.0141
γ 4	507.604 5	17.7 5	0.191
γ 5	573.153 4	80.3 18	0.981

121SN IT DECAY (55 Y 5) I(MIN)= 0.10%
%IT=77.6 20
SEE ALSO 121SN B- DECAY (55 Y)

Radiation Type	Energy (keV)	Intensity (%)	Δ(g-rad/μCi-h)
ce-L-1	1.83 8	48.1 14	0.0019
Auger-L	3	44.3 18	0.0028

121TE EC DECAY (154 D 7) I(MIN)= 0.10%
%EC=11.4 11
SEE ALSO 121TE IT DECAY (154 D)

Radiation Type	Energy (keV)	Intensity (%)	Δ(g-rad/μCi-h)
Auger-L	3	18.0 17	0.0012
ce-K-1	6.647 10	9.0 13	0.0013

Radiation Type	Energy (keV)	Intensity (%)	Δ(g-rad/μCi-h)
ce-M-1	5.41 8	22.2 8	0.0026
ce-NOP-1	6.15 8	7.3 3	0.0010
X-ray L	3.44	3.8 13	0.0003

121SN B- DECAY (55 Y 5)   I(MIN)= 0.10%
%B-=22.4 20
SEE ALSO 121SN IT DECAY (55 Y)

Radiation Type	Energy (keV)	Intensity (%)	Δ(g-rad/μCi-h)
Auger-L	3	17.0 18	0.0011
ce-K-1	6.6478 21	17.7 18	0.0025
Auger-K	21.8	2.3 6	0.0011
ce-L-1	32.4407 21	2.29 23	0.0016
ce-M-1	36.1953 21	0.45 5	0.0003
ce-NOP-1	36.9870 21	0.150 15	0.0001
β^- 1 max	357 3		
avg	121.0 8	22.4	0.0577
X-ray L	3.6	1.6 6	0.0001
X-ray $K\alpha_2$	26.11080 20	4.4 5	0.0024
X-ray $K\alpha_1$	26.35910 20	8.2 9	0.0046
X-ray $K\beta$	29.7	2.8 3	0.0018
γ 1	37.1390 20	1.85 18	0.0015

Radiation Type	Energy (keV)	Intensity (%)	Δ(g-rad/μCi-h)
Auger-K	21.8	2.5 6	0.0011
ce-L-1	32.440 10	1.17 17	0.0008
ce-M-1	36.194 10	0.23 4	0.0002
X-ray L	3.6	1.7 6	0.0001
X-ray $K\alpha_2$	26.11080 20	4.6 5	0.0025
X-ray $K\alpha_1$	26.35910 20	8.6 8	0.0048
X-ray $K\beta$	29.7	2.9 3	0.0019
γ 1	37.138 10	0.94 14	0.0007
γ 8	1102.149 18	2.5 3	0.0586

8 weak γ's omitted (ΣI_γ = 0.16%)

121TE IT DECAY (154 D 7)   I(MIN)= 0.10%
%IT=88.6 11
SEE ALSO 121TE EC DECAY (154 D)

Radiation Type	Energy (keV)	Intensity (%)	Δ(g-rad/μCi-h)
Auger-L	3.19	72 4	0.0049
Auger-K	22.7	5.1 12	0.0025
ce-K-1	49.976 15	34.5 8	0.0367
ce-L-1	76.851 15	41.6 9	0.0680
ce-M-1	80.784 15	9.9 3	0.0171
ce-NOP-1	81.622 15	2.65 9	0.0046
ce-K-2	180.38 3	6.12 19	0.0235
ce-L-2	207.25 3	0.81 3	0.0036
ce-MNO-2	211.18 3	0.198 7	0.0009

(continued on next page)

121TE IT DECAY (154 D 7) (CONTINUED)

Radiation Type	Energy (keV)	Intensity (%)		Δ(g-rad/μCi-h)
X-ray L	3.77	7.1	24	0.0006
X-ray Kα₂	27.20170 20	10.1	4	0.0059
X-ray Kα₁	27.47230 20	18.9	8	0.0110
X-ray Kβ	31	6.5	3	0.0043
γ 2	212.19 3	81.5	11	0.368

1 weak γ's omitted (ΣIγ = 0.05%)

122SB EC DECAY (2.70 D 1)   I(MIN)= 0.10%
%(EC+β+)=2.38 13
SEE ALSO 122SB B- DECAY

Radiation Type	Energy (keV)	Intensity (%)		Δ(g-rad/μCi-h)
Auger-L	3	2.00	15	0.0001
Auger-K	21	0.29	6	0.0001
X-ray L	3.44	0.17	6	≈0
X-ray Kα₂	25.04400 20	0.50	4	0.0003
X-ray Kα₁	25.27130 20	0.93	6	0.0005
X-ray Kβ	28.5	0.315	21	0.0002
γ 1	1140.67 4	0.74	6	0.0180

123I EC DECAY (13.221 H 3)   I(MIN)= 0.10%

Radiation Type	Energy (keV)	Intensity (%)		Δ(g-rad/μCi-h)
Auger-L	3.19	94	6	0.0064
Auger-K	22.7	12	3	0.0060
ce-K-1	127.16 5	13.61	7	0.0369
ce-L-1	154.03 5	1.773	10	0.0058
ce-M-1	157.96 5	0.352	11	0.0012
X-ray L	3.77	9	4	0.0007
X-ray Kα₂	27.20170 20	24.7	9	0.0143
X-ray Kα₁	27.47230 20	46.0	16	0.0269
X-ray Kβ	31	16.0	6	0.0105
γ 1	158.97 5	83.3	4	0.282
γ 20	346.35 5	0.126	5	0.0009
γ 23	440.02 5	0.428	15	0.0040
γ 25	505.33 5	0.316	11	0.0034
γ 26	528.96 5	1.39	5	0.0157
γ 27	538.54 5	0.382	13	0.0044

39 weak γ's omitted (ΣIγ = 0.48%)

124SB B- DECAY (60.20 D 3)   I(MIN)= 0.10%

Radiation Type	Energy (keV)	Intensity (%)	Δ(g-rad/μCi-h)
Auger-L	3.19	0.341 24	≈0
ce-K-9	570.916 3	0.411 13	0.0050
β⁻ 1 max	130.1 19		
avg	34.6 6	0.51 3	0.0004
β⁻ 2 max	194.2 19		
avg	53.2 6	0.10 3	0.0001
β⁻ 3 max	203.2 19		
avg	55.9 6	0.550 20	0.0007
β⁻ 4 max	211.3 19		
avg	58.3 6	8.80 5	0.0109
β⁻ 5 max	421.7 19		
avg	126.3 7	0.324 16	0.0009
β⁻ 6 max	611.3 19		
avg	194.0 7	51.24 17	0.212
β⁻ 7 max	722.4 19		
avg	235.9 8	0.490 20	0.0025
β⁻ 8 max	813.4 19		
avg	271.3 8	0.670 20	0.0039
β⁻ 9 max	865.7 19		
avg	292.0 8	4.06 3	0.0253
β⁻ 10 max	947.1 19		
avg	324.7 8	2.10 4	0.0145
β⁻ 11 max	1579.5 19		
avg	593.4 9	4.81 7	0.0608
β⁻ 12 max	1656.4 19		
avg	627.3 9	2.67 5	0.0357
β⁻ 13 max	2302.3 19		
avg	918.6 9	23.53 18	0.460
Total β⁻			
avg	388.6 14	100.1 3	0.829

(continued on next page)

122SB B- DECAY (2.70 D 1)   I(MIN)= 0.10%
%B-=97.62 13
SEE ALSO 122SB EC DECAY

Radiation Type	Energy (keV)	Intensity (%)	Δ(g-rad/μCi-h)
Auger-L	3.19	0.290 21	≈0
ce-K-1	532.43 4	0.350 11	0.0040
β⁻ 1 max	725 4		
avg	236.8 14	4.64 13	0.0234
β⁻ 2 max	1417 4		
avg	522.5 16	66.6 3	0.741
β⁻ 3 max	1982 4		
avg	772.3 16	26.40 20	0.434
Total β⁻			
avg	576.3 17	97.7 4	1.20

3 weak β's omitted (ΣIβ = 0.04%)

X-ray Kα₁	27.47230 20	0.163 8	≈0
γ 1	564.24 4	70.04 22	0.842
γ 3	692.65 4	3.82 13	0.0563
γ 6	1256.93 4	0.81 5	0.0216

4 weak γ's omitted (ΣIγ = 0.04%)

124I B+ DECAY (4.18 D 2) (CONTINUED)

Radiation Type	Energy (keV)		Intensity (%)		Δ(g-rad/μCi-h)
γ 49	1509.49	4	3.01	25	0.0969
γ 50	1559.80	20	0.16	3	0.0055
γ 52	1637.7	5	0.195	21	0.0068
γ 54	1675.8	4	0.11	3	0.0039
γ 55	1691.02	4	10.5	9	0.378
γ 56	1720.37	14	0.171	19	0.0063
γ 58	1851.4	4	0.21	3	0.0082
γ 59	1918.58	4	0.159	23	0.0065
γ 60	2038.3	3	0.34	3	0.0149
γ 61	2078.86	7	0.35	4	0.0154
γ 62	2091.00	10	0.57	5	0.0255
γ 64	2099.09	9	0.140	13	0.0063
γ 65	2144.320	10	0.110	11	0.0050
γ 66	2232.25	7	0.57	5	0.0273
γ 67	2283.25	8	0.66	7	0.0323
γ 75	2746.90	10	0.46	5	0.0271

48 weak γ's omitted (ΣIγ = 1.24%)
Maximum γ± intensity = 45.79%

125SB B- DECAY (2.758 Y 1)   I(MIN)= 0.10%

Radiation Type	Energy (keV)	Intensity (%)		Δ(g-rad/μCi-h)
Auger-L	3.19	51	4	0.0034

124SB B- DECAY (60.20 D 3) (CONTINUED)

Radiation Type	Energy (keV)		Intensity (%)		Δ(g-rad/μCi-h)
	8 weak β's omitted (ΣIβ = 0.25%)				
X-ray Kα2	27.20170	20	0.102	5	≈0
X-ray Kα1	27.47230	20	0.191	9	0.0001
γ 4	400.01	6	0.126	16	0.0011
γ 5	443.96	5	0.201	10	0.0019
γ 8	525.40	10	0.115	12	0.0013
γ 9	602.730	3	97.92	5	1.26
γ 10	632.39	5	0.112	6	0.0015
γ 11	645.8550	20	7.45	3	0.103
γ 13	709.320	13	1.370	11	0.0207
γ 14	713.781	5	2.289	12	0.0348
γ 15	722.786	4	10.79	4	0.166
γ 16	735.74	3	0.130	9	0.0020
γ 19	790.712	7	0.742	9	0.0125
γ 24	968.201	4	1.879	13	0.0388
γ 27	1045.131	4	1.825	17	0.0406
γ 36	1325.512	6	1.551	22	0.0438
γ 37	1355.175	22	1.02	3	0.0294
γ 38	1368.164	7	2.61	3	0.0762
γ 39	1376.15	10	0.491	20	0.0144
γ 41	1436.563	7	1.200	24	0.0367
γ 42	1445.06	4	0.351	17	0.0108
γ 44	1488.886	24	0.665	19	0.0211
γ 45	1526.38	10	0.401	24	0.0131
γ 46	1579.90	20	0.424	12	0.0143
γ 48	1690.980	6	47.57	16	1.71
γ 56	2090.942	8	5.473	22	0.244

46 weak γ's omitted (ΣIγ = 1.25%)

124I B+ DECAY (4.18 D 2)   I(MIN)= 0.10%

Radiation Type	Energy (keV)	Intensity (%)	Δ(g-rad/μCi-h)
Auger-L	3.19	64 4	0.0043
Auger-K	22.7	8.4 19	0.0040
ce-K-9	570.91 4	0.256 23	0.0031
β+ 1 max	810 4		
avg	365.7 18	0.295 17	0.0023
β+ 2 max	1532 4		
avg	685.9 18	11.3 6	0.165
β+ 3 max	2135 4		
avg	973.6 18	11.3 6	0.234
Total β+			
avg	823.7 19	22.9 9	0.402
X-ray L	3.77	6.3 22	0.0005
X-ray Kα2	27.20170 20	16.7 7	0.0097
X-ray Kα1	27.47230 20	31.1 13	0.0182
X-ray Kβ	31	10.8 5	0.0071
γ 6	541.20 10	0.189 17	0.0022
γ 9	602.72 4	61 5	0.783
γ 11	645.82 4	0.95 9	0.0131
γ 13	695.0 10	0.122 10	0.0018
γ 16	713.80 20	0.110 16	0.0017
γ 17	722.78 4	10.1 9	0.155
γ 27	968.22 8	0.41 4	0.0086
γ 28	976.32 14	0.104 15	0.0022
γ 31	1045.00 10	0.43 5	0.0095
γ 32	1054.00 20	0.122 12	0.0027
γ 40	1325.50 4	1.45 12	0.0408
γ 42	1368.20 6	0.29 3	0.0084
γ 43	1376.00 10	1.68 14	0.0492
γ 48	1488.90 10	0.183 17	0.0058

(continued on next column)

Radiation Type	Energy (keV)	Intensity (%)	Δ(g-rad/μCi-h)
ce-K-2	3.690 15	51.3 23	0.0040
ce-L-1	14.94 15	0.18 9	≈0
Auger-K	22.7	6.6 15	0.0032
ce-L-2	30.565 15	6.87 22	0.0045
ce-M-2	34.498 15	1.37 5	0.0010
ce-K-7	144.520 11	0.938 18	0.0029
ce-L-7	171.395 11	0.148 6	0.0005
ce-K-17	396.075 15	0.347 11	0.0029
β- 1 max	95.4 20		
avg	24.9 6	13.60 20	0.0072
β- 2 max	124.6 20		
avg	33.1 6	5.79 6	0.0041
β- 3 max	130.8 20		
avg	34.8 6	18.10 20	0.0134
β- 4 max	241.6 20		
avg	67.5 7	1.620 20	0.0023
β- 5 max	303.4 20		
avg	87.0 7	40.2 5	0.0745
β- 6 max	445.7 20		
avg	134.5 7	7.14 4	0.0205
β- 7 max	622.0 20		
avg	215.5 8	13.50 20	0.0620
Total β-			
avg	86.4 11	100.0 7	0.184

1 weak β's omitted (ΣIβ = 0.09%)

Radiation Type	Energy (keV)	Intensity (%)	Δ(g-rad/μCi-h)
X-ray L	3.77	5.0 17	0.0004
X-ray Kα2	27.20170 20	13.2 8	0.0076
X-ray Kα1	27.47230 20	24.6 14	0.0144
X-ray Kβ	31	8.5 5	0.0056
γ 2	35.504 15	4.28 14	0.0032
γ 4	116.952 11	0.255 5	0.0006
γ 6	172.615 15	0.182 4	0.0007
γ 7	176.334 11	6.79 7	0.0255
γ 10	204.129 25	0.323 6	0.0014

(continued on next page)

125SB B- DECAY (2.758 Y 1) (CONTINUED)

Radiation Type	Energy (keV)	Intensity (%)	Δ(g-rad/μCi-h)
γ 11	208.088 25	0.236 5	0.0010
γ 12	227.91	0.132 5	0.0006
γ 14	321.03 4	0.410 6	0.0028
γ 15	380.435 20	1.520 19	0.0123
γ 16	408.01 4	0.183 7	0.0016
γ 17	427.889 15	29.4 3	0.268
γ 18	443.50 4	0.303 8	0.0029
γ 19	463.383 15	10.45 11	0.103
γ 21	600.557 18	17.78 18	0.227
γ 22	606.641 19	5.02 6	0.0649
γ 23	635.895 18	11.32 12	0.153
γ 24	671.409 20	1.80 5	0.0257

7 weak γ's omitted (ΣIγ = 0.07%)

125TE IT DECAY (57.40 D 5)   I(MIN)= 0.10%
FEEDING OF 57.4-D 125TE IN 125SB B- DECAY=23.0% 2

Radiation Type	Energy (keV)	Intensity (%)	Δ(g-rad/μCi-h)
Auger-L	3.19	153 9	0.0104
ce-K-1	3.65 3	80 4	0.0062
Auger-K	22.7	16 4	0.0080
ce-L-1	30.52 3	10.7 3	0.0070
ce-M-1	34.45 3	2.14 6	0.0016
ce-K-2	77.456 20	51.7 7	0.0853
ce-L-2	104.331 20	37.2 6	0.0827

126I EC DECAY (13.02 D 7)   I(MIN)= 0.10%
%(EC+B+)=56.3 20
SEE ALSO 126I B- DECAY

Radiation Type	Energy (keV)	Intensity (%)	Δ(g-rad/μCi-h)
Auger-L	3.19	43 3	0.0029
Auger-K	22.7	5.7 13	0.0027
ce-K-1	634.517 12	0.108 9	0.0015
β+ 1 max	1134 5		
avg	508.4 23	3.34 22	0.0362
X-ray L	3.77	4.3 15	0.0003
X-ray Kα2	27.20170 20	11.3 7	0.0065
X-ray Kα1	27.47230 20	21.1 13	0.0123
X-ray Kβ	31	7.3 5	0.0048
γ 1	666.331 12	33.1 25	0.470
γ 3	753.819 13	4.2 4	0.0669
γ 6	1420.19 3	0.295 23	0.0089

Maximum γ± intensity = 6.68%

126I B- DECAY (13.02 D 7)   I(MIN)= 0.10%
%B-=43.7 20
SEE ALSO 126I EC DECAY

Radiation Type	Energy (keV)	Intensity (%)	Δ(g-rad/μCi-h)
Auger-L	3.43	0.51 5	≈0
ce-K-1	354.072 11	0.53 5	0.0040

Radiation Type	Energy (keV)	Intensity (%)	Δ(g-rad/μCi-h)
ce-M-2	108.264 20	8.56 23	0.0197
ce-NOP-2	109.102 20	2.24 7	0.0052
X-ray L	3.77	15 5	0.0012
X-ray Kα2	27.20170 20	32.8 14	0.0190
X-ray Kα1	27.47230 20	61 3	0.0358
X-ray Kβ	31	21.2 10	0.0140
γ 1	35.46 3	6.67 18	0.0050
γ 2	109.270 20	0.291 9	0.0007

125I EC DECAY (59.6 D 2)   I(MIN)= 0.10%

Radiation Type	Energy (keV)	Intensity (%)	Δ(g-rad/μCi-h)
Auger-L	3.19	156 9	0.0106
ce-K-1	3.6781 6	79.8 6	0.0063
Auger-K	22.7	20 5	0.0096
ce-L-1	30.5527 6	10.7 3	0.0070
ce-M-1	34.4859 6	2.10 20	0.0015
ce-NOP-1	35.3236 6	0.70 20	0.0005
X-ray L	3.77	15 6	0.0012
X-ray Kα2	27.20170 20	39.7 14	0.0230
X-ray Kα1	27.47230 20	74.1 25	0.0434
X-ray Kβ	31	25.7 10	0.0170
γ 1	35.4919 5	6.67 22	0.0050

Radiation Type	Energy (keV)	Intensity (%)	Δ(g-rad/μCi-h)
β- 1 max	371 5		
β- 1 avg	108.9 17	3.6 3	0.0084
β- 2 max	862 5		
β- 2 avg	289.7 20	32 3	0.197
β- 3 max	1251 5		
β- 3 avg	449.5 22	8 3	0.0766
Total β- avg	304.1 22	44 5	0.282
X-ray Kα2	29.4580 10	0.134 12	≈0
X-ray Kα1	29.7790 10	0.249 22	0.0002
γ 1	388.633 11	34 3	0.282
γ 2	491.243 11	2.85 22	0.0298
γ 3	879.876 13	0.75 6	0.0141

127XE EC DECAY (36.4 D 1)   I(MIN)= 0.10%

Radiation Type	Energy (keV)	Intensity (%)	Δ(g-rad/μCi-h)
Auger-L	3.3	96 6	0.0068
Auger-K	23.6	12 3	0.0059
ce-K-1	24.441 20	4.28 18	0.0022
ce-L-1	52.422 20	0.61 3	0.0007
ce-M-1	56.538 20	0.123 6	0.0001
ce-K-2	112.083 10	1.54 7	0.0037
ce-K-3	138.963 10	3.65 16	0.0108
ce-L-2	140.064 10	0.391 17	0.0012
ce-L-3	166.944 10	0.475 15	0.0017
ce-K-4	169.691 10	6.63 7	0.0240
ce-L-4	197.672 10	0.98 3	0.0041

(continued on next page)

127XE EC DECAY (36.4 D 1) (CONTINUED)

Radiation Type	Energy (keV)	Intensity (%)	Δ(g-rad/μCi-h)
ce-M-4	201.788 10	0.198 3	0.0009
ce-K-5	341.822 12	0.289 13	0.0021
X-ray L	4	10 4	0.0008
X-ray Kα₂	28.3172 4	25.1 9	0.0152
X-ray Kα₁	28.6120 3	46.8 16	0.0285
X-ray Kβ	32.3	16.4 6	0.0113
γ 1	57.610 20	1.33 6	0.0016
γ 2	145.252 10	4.29 14	0.0133
γ 3	172.132 10	25.5 8	0.0937
γ 4	202.860 10	68.3 4	0.295
γ 5	374.991 12	17.2 6	0.137

1 weak γ's omitted (ΣI_γ = 0.01%)

128CS EC DECAY (3.62 M 2)   I(MIN)= 0.10%

Radiation Type	Energy (keV)	Intensity (%)	Δ(g-rad/μCi-h)
Auger-L	3.43	24.5 14	0.0018
Auger-K	24.6	2.9 6	0.0015
ce-K-1	408.340 10	0.287 17	0.0025
β⁺ 1 max	1322 6		
avg	593 3	0.291 16	0.0037
β⁺ 2 max	1936 6		

128BA EC DECAY (2.43 D 5) (CONTINUED)

Radiation Type	Energy (keV)	Intensity (%)	Δ(g-rad/μCi-h)
X-ray L	4.29	10 3	0.0009
X-ray Kα₂	30.6251 3	21.3 5	0.0139
X-ray Kα₁	30.9728 3	39.4 8	0.0260
X-ray Kβ	35	14.1 4	0.0105
γ 1	229.50 6	0.106 10	0.0005
γ 2	273.440 10	14.5 7	0.0845
γ 3	374.990 20	0.309 21	0.0025

129TE B- DECAY (69.6 M 2)   I(MIN)= 0.10%

Radiation Type	Energy (keV)	Intensity (%)	Δ(g-rad/μCi-h)
Auger-L	3.3	59 7	0.0041
ce-L-1	22.62 4	65 7	0.0313
ce-M-1	26.74 4	13.1 14	0.0075
ce-NOP-1	27.62 4	4.4 5	0.0026
β⁻ 1 max	387 4		
avg	114.4 14	0.84 3	0.0020
β⁻ 2 max	669 4		
avg	215.3 15	0.204 9	0.0009
β⁻ 3 max	939 4		
avg	321.0 17	0.240 10	0.0016
β⁻ 4 max	1012 4		
avg	350.4 17	8.9 4	0.0664
β⁻ 5 max	1221 4		

avg	869 3	0.89 5	0.0165
β+ 3 max	2462 6		
avg	1110 3	14.3 8	0.338
β+ 4 max	2905 6		
avg	1315 3	53.1 11	1.49
Total β+			
avg	1263 3	68.6 14	1.85

	Energy	Intensity	Δ
X-ray L	4.1	2.7 9	0.0002
X-ray Kα2	29.4580 10	6.68 22	0.0042
X-ray Kα1	29.7790 10	12.4 4	0.0079
X-ray Kβ	33.6	4.41 16	0.0032
γ 1	442.901 10	26.8 14	0.253
γ 2	526.557 14	2.41 12	0.0271
γ 3	613.493 13	0.351 21	0.0046
γ 4	969.458 20	0.63 4	0.0130
γ 5	1030.170 21	0.217 15	0.0048
γ 6	1140.079 23	1.17 6	0.0284
γ 7	1303.36 3	0.117 7	0.0032
γ 8	1629.07 4	0.135 7	0.0047
γ 9	1684.14 3	0.115 7	0.0041
γ 10	2155.68 5	0.161 14	0.0074

Maximum γ± intensity =137.16%

128BA EC DECAY (2.43 D 5)   I(MIN)= 0.10%

Radiation Type	Energy (keV)	Intensity (%)	Δ(g-rad/μCi-h)
Auger-L	3.55	78 4	0.0059
Auger-K	25.5	8.8 10	0.0048
ce-K-2	237.455 10	0.71 4	0.0036

(continued on next column)

avg	437.4 17	0.54 3	0.0050
β- 6 max	1471 4		
avg	545.0 18	89.3 9	1.04
Total β-			
avg	522.0 19	100.1 10	1.11

4 weak β's omitted (ΣIβ = 0.06%)

	Energy	Intensity	Δ
X-ray L	4	6.2 22	0.0005
γ 1	27.81 4	15.6 17	0.0092
γ 2	208.96 5	0.172 8	0.0008
γ 5	250.62 5	0.366 16	0.0020
γ 7	278.43 5	0.542 23	0.0032
γ 8	281.26 5	0.158 7	0.0009
γ 14	459.60 5	7.4 4	0.0721
γ 16	487.39 6	1.35 6	0.0141
γ 32	802.10 6	0.183 8	0.0031
γ 42	1083.85 6	0.471 21	0.0109
γ 43	1111.64 6	0.183 10	0.0043

37 weak γ's omitted (ΣIγ = 0.39%)

129TE IT DECAY (33.6 D 1)   I(MIN)= 0.10%
%IT=65 6
SEE ALSO 129TE B- DECAY (33.6 D)

Radiation Type	Energy (keV)	Intensity (%)	Δ(g-rad/μCi-h)
Auger-L	3.19	50 4	0.0034
Auger-K	22.7	4.1 10	0.0020
ce-K-1	73.69 5	33 3	0.0513

(continued on next page)

129TE IT DECAY (33.6 D 1) (CONTINUED)

Radiation Type	Energy (keV)	Intensity (%)	Δ(g-rad/μCi-h)
ce-L-1	100.56 5	24.9 24	0.0533
ce-MNO-1	104.49 5	7.3 7	0.0162
X-ray L	3.77	4.9 17	0.0004
X-ray Kα₂	27.20170 20	8.1 9	0.0047
X-ray Kα₁	27.47230 20	15.2 15	0.0089
X-ray Kβ	31	5.3 6	0.0035
γ	105.50 5	0.149 16	0.0003

129TE B- DECAY (33.6 D 1) I(MIN)= 0.10%
%B-=35 6
SEE ALSO 129TE IT DECAY (33.6 D)

Radiation Type	Energy (keV)	Intensity (%)	Δ(g-rad/μCi-h)
ce-L-1	22 3	0.110 19	≈0
β⁻ 1 max	203 4		
avg	55.8 12	0.156 17	0.0002
β⁻ 2 max	875 4		
avg	295.2 16	0.72 8	0.0045
β⁻ 3 max	909 4		
avg	308.7 16	3.1 4	0.0204
β⁻ 4 max	1605 4		
avg	603.3 18	31 6	0.398

129XE IT DECAY (8.0 D 2) I(MIN)= 0.10%

Radiation Type	Energy (keV)	Intensity (%)	Δ(g-rad/μCi-h)
Auger-L	3.43	147 8	0.0108
ce-K-1	5.02 3	79.10 20	0.0085
Auger-K	24.6	16 3	0.0083
ce-L-1	34.13 3	10.6 3	0.0077
ce-M-1	38.44 3	2.10 10	0.0017
ce-NOP-1	39.37 3	0.70 10	0.0006
ce-K-2	162.00 3	63.90 10	0.220
ce-L-2	191.11 3	24.4 7	0.0993
ce-M-2	195.42 3	5.50 20	0.0229
ce-NOP-2	196.35 3	1.50 10	0.0063
X-ray L	4.1	16 5	0.0014
X-ray Kα₂	29.4580 10	36.2 10	0.0227
X-ray Kα₁	29.7790 10	67.1 17	0.0426
X-ray Kβ	33.6	23.9 7	0.0171
γ 1	39.58 3	7.50 20	0.0063
γ 2	196.56 3	4.70 20	0.0197

129CS EC DECAY (32.06 H 6) I(MIN)= 0.10%

Radiation Type	Energy (keV)	Intensity (%)	Δ(g-rad/μCi-h)
Auger-L	3.43	110 9	0.0080
ce-K-1	5.020 15	31.4 22	0.0034
Auger-K	24.6	13 3	0.0068

	Energy	Intensity (%)	Δ(g-rad/μCi-h)
ce-L-1	34.128 15	4.2 3	0.0030
ce-MNO-1	38.439 15	1.02 7	0.0008
ce-K-3	58.768 4	0.57 4	0.0007
ce-K-14	337.3566 23	0.62 4	0.0044
ce-K-16	376.9286 23	0.359 22	0.0029
X-ray L	4.1	12 4	0.0011
X-ray Kα2	29.4580 10	29.7 21	0.0186
X-ray Kα1	29.7790 10	55 4	0.0350
X-ray Kβ	33.6	19.6 14	0.0140
γ 1	39.581 15	2.99 18	0.0025
γ 3	93.329 3	0.66 4	0.0013
γ 4	177.036 10	0.271 15	0.0010
γ 5	266.820 7	0.274 15	0.0016
γ 6	270.352 5	0.214 12	0.0012
γ 7	278.614 4	1.33 8	0.0079
γ 8	282.131 6	0.243 13	0.0015
γ 10	318.1800 20	2.46 14	0.0167
γ 14	371.9180 20	30.8 16	0.244
γ 16	411.4900 20	22.5 12	0.197
γ 21	548.945 8	3.42 18	0.0400
γ 27	588.549 8	0.61 4	0.0076
γ 32	906.425 6	0.221 12	0.0043

20 weak γ's omitted (ΣIγ = 0.37%)

Total β−
  avg   567.7 19   35 6   0.424

5 weak β's omitted (ΣIβ = 0.08%)

	Energy	Intensity (%)	Δ(g-rad/μCi-h)
γ 16	556.65 6	0.121 21	0.0014
γ 19	695.88 6	3.1 6	0.0453
γ 23	729.57 6	0.71 13	0.0111

36 weak γ's omitted (ΣIγ = 0.29%)

129I B− DECAY (1.57E7 Y 4)   I(MIN)= 0.10%

Radiation Type	Energy (keV)	Intensity (%)	Δ(g-rad/μCi-h)
Auger-L	3.43	74 4	0.0054
ce-K-1	5.02 3	79.10 20	0.0085
Auger-K	24.6	8.8 16	0.0046
ce-L-1	34.13 3	10.6 3	0.0077
ce-M-1	38.44 3	2.10 10	0.0017
ce-NOP-1	39.37 3	0.70 10	0.0006
β− 1 max	153 4		
avg	41.2 12	100	0.0878
X-ray L	4.1	8.2 25	0.0007
X-ray Kα2	29.4580 10	20.0 6	0.0126
X-ray Kα1	29.7790 10	37.1 9	0.0235
X-ray Kβ	33.6	13.2 4	0.0094
γ 1	39.58 3	7.50 20	0.0063

130I B- DECAY (12.36 H 1)   I(MIN)= 0.10%

Radiation Type	Energy (keV)	Intensity (%)	Δ(g-rad/μCi-h)
Auger-L	3.43	0.51 4	≈0
ce-K-13	501.529 20	0.624 23	0.0067
β⁻ 1 max	232 10		
avg	64 3	0.320 20	0.0004
β⁻ 2 max	355 10		
avg	103 4	0.330 20	0.0007
β⁻ 3 max	376 10		
avg	110 4	0.500 20	0.0012
β⁻ 4 max	557 10		
avg	173 4	0.185 6	0.0007
β⁻ 5 max	622 10		
avg	197 4	46.0 10	0.193
β⁻ 6 max	812 10		
avg	270 4	2.14 5	0.0123
β⁻ 7 max	902 10		
avg	306 4	0.175 16	0.0011
β⁻ 8 max	1040 10		
avg	361 5	47.7 12	0.367
β⁻ 9 max	1176 10		
avg	418 5	1.45 4	0.0129
Total β⁻ avg	279 5	99.1 16	0.590

6 weak β's omitted (ΣIβ = 0.27%)

X-ray Kα₂	29.4580 10	0.158 7	≈0
X-ray Kα₁	29.7790 10	0.293 13	0.0002
X-ray Kβ	33.6	0.104 5	≈0
γ 9	418.010 20	34.2 7	0.304
γ 11	457.720 20	0.237 16	0.0023
γ 12	510.350 20	0.85 3	0.0093
γ 13	536.090 20	99.0 20	1.13

131I B- DECAY (8.021 D 1) (CONTINUED)

Radiation Type	Energy (keV)	Intensity (%)	Δ(g-rad/μCi-h)
ce-K-14	329.919 11	1.54 6	0.0108
ce-L-14	359.027 11	0.244 8	0.0019
β⁻ 1 max	247.9 6		
avg	69.40 20	2.13 3	0.0031
β⁻ 2 max	303.9 6		
avg	87.00 20	0.620 20	0.0011
β⁻ 3 max	333.8 6		
avg	96.60 20	7.36 10	0.0151
β⁻ 4 max	606.3 6		
avg	191.6 3	89.4 10	0.365
β⁻ 5 max	806.9 6		
avg	283.2 3	0.420 20	0.0025
Total β⁻ avg	181.7 3	100.0 10	0.387

1 weak β's omitted (ΣIβ = 0.06%)

X-ray L	4.1	0.56 18	≈0
X-ray Kα₂	29.4580 10	1.37 5	0.0009
X-ray Kα₁	29.7790 10	2.54 9	0.0016
X-ray Kβ	33.6	0.90 4	0.0006
γ 1	80.183 10	2.62 5	0.0045
γ 4	177.210 10	0.265 4	0.0010
γ 7	284.298 11	6.06 9	0.0367
γ 12	325.781 11	0.251 6	0.0017
γ 14	364.480 11	81.2 12	0.631
γ 16	502.991 11	0.361 7	0.0039
γ 17	636.973 10	7.27 11	0.0986
γ 18	642.703 11	0.220 4	0.0030
γ 19	722.893 10	1.80 3	0.0278

10 weak γ's omitted (ΣIγ = 0.23%)

131XE IT DECAY (11.9 D 1)   I(MIN)= 0.10%
FEEDING OF 11.9-D 131XE IN 131I DECAY=1.086% 13

Radiation Type	Energy (keV)	Intensity (%)	Δ(g-rad/μCi-h)
Auger-L	3.43	75 4	0.0055
Auger-K	24.6	6.8 13	0.0036
ce-K-1	129.369 8	61.2 7	0.169
ce-L-1	158.477 8	28.6 6	0.0965
ce-M-1	162.788 8	6.50 20	0.0225
ce-NOP-1	163.722 6	1.78 6	0.0062
X-ray L	4.1	8 3	0.0007
X-ray Kα₂	29.4580 10	15.5 5	0.0097
X-ray Kα₁	29.7790 10	28.7 8	0.0182
X-ray Kβ	33.6	10.2 4	0.0073
γ 1	163.930 8	1.96 6	0.0068

131CS EC DECAY (9.69 D 1)   I(MIN)= 0.10%

Radiation Type	Energy (keV)	Intensity (%)	Δ(g-rad/μCi-h)
Auger-L	3.43	80 4	0.0058
Auger-K	24.6	9.3 17	0.0049
X-ray L	4.1	9 3	0.0008
X-ray Kα₂	29.4580 10	21.2 6	0.0133
X-ray Kα₁	29.7790 10	39.3 10	0.0249
X-ray Kβ	33.6	14.0 4	0.0100

γ 14	539.10 3	1.40 3	0.0160
γ 15	553.900 10	0.662 21	0.0078
γ 16	586.050 20	1.69 4	0.0211
γ 17	603.530 10	0.615 25	0.0079
γ 19	668.540 10	96.1 21	1.37
γ 20	685.990 10	1.07 3	0.0156
γ 22	739.480 20	82.3 18	1.30
γ 25	800.23 3	0.101 6	0.0017
γ 26	808.290 20	0.236 8	0.0041
γ 31	877.35 4	0.191 9	0.0036
γ 34	967.020 20	0.877 22	0.0181
γ 38	1096.48 3	0.552 15	0.0129
γ 39	1122.15 3	0.253 10	0.0061
γ 40	1157.470 10	11.31 25	0.279
γ 41	1222.56 3	0.179 7	0.0047
γ 42	1272.120 20	0.748 19	0.0203
γ 44	1403.900 20	0.345 14	0.0103

31 weak γ's omitted (ΣIγ = 0.81%)

131I B- DECAY (8.021 D 1)   I(MIN)= 0.10%
SEE ALSO 131XE IT DECAY

Radiation Type	Energy (keV)	Intensity (%)	Δ(g-rad/μCi-h)
Auger-L	3.43	5.1 3	0.0004
Auger-K	24.6	0.60 11	0.0003
ce-K-1	45.622 10	3.54 13	0.0034
ce-L-1	74.730 10	0.472 17	0.0008
ce-MNO-1	79.041 10	0.1310 24	0.0002
ce-K-7	249.737 11	0.248 9	0.0013

(continued on next column)

131BA EC DECAY (11.8 D 2)   I(MIN)= 0.10%

Radiation Type	Energy (keV)	Intensity (%)	Δ(g-rad/μCi-h)
Auger-L	3.55	104 5	0.0078
ce-K-1	18.98 11	0.59 7	0.0002
Auger-K	25.5	11.4 14	0.0062
ce-K-2	42.779 16	1.19 8	0.0011
ce-L-1	49.25 11	0.86 9	0.0009
ce-MNO-1	53.74 11	0.213 11	0.0002
ce-L-2	56.304 14	0.61 10	0.0007
ce-L-2	73.050 16	0.148 12	0.0002
ce-K-5	87.817 8	18.0 8	0.0337
ce-K-7	97.627 14	0.80 5	0.0017
ce-L-5	118.088 8	6.0 3	0.0150
ce-MNO-5	122.585 8	1.65 8	0.0043
ce-L-7	127.898 14	0.149 12	0.0004
ce-K-10	180.088 8	1.77 6	0.0068
ce-K-11	203.638 8	0.168 17	0.0007
ce-L-10	210.359 8	0.236 21	0.0011
ce-L-10	213.441 8	0.172 18	0.0008
ce-K-13	337.252 11	0.28 3	0.0020
ce-K-26	460.328 13	0.51 5	0.0050
X-ray L	4.29	13 4	0.0012
X-ray Kα2	30.6251 3	27.6 8	0.0180
X-ray Kα1	30.9728 3	51.2 13	0.0338
X-ray Kβ	35	18.4 6	0.0137
γ 2	78.764 16	0.729 25	0.0012
γ 4	92.289 14	0.64 9	0.0013
γ 5	123.802 8	28.9 9	0.0762
γ 7	133.612 14	2.15 7	0.0061
γ 9	157.147 9	0.190 6	0.0006
γ 10	216.073 8	19.7 5	0.0907
γ 11	239.623 8	2.40 5	0.0122
γ 12	246.879 12	0.640 20	0.0034
γ 13	249.426 8	2.82 7	0.0150

132TE B- DECAY (76.3 H 2)   I(MIN)= 0.10%   (CONTINUED)

Radiation Type	Energy (keV)	Intensity (%)	Δ(g-rad/μCi-h)
ce-K-2	78.59 8	0.86 13	0.0014
ce-K-3	83.13 8	0.90 14	0.0016
ce-L-2	106.57 8	0.24 4	0.0006
ce-L-3	111.11 8	0.19 4	0.0005
ce-K-4	194.99 6	7.1 3	0.0297
ce-L-4	222.97 6	1.34 5	0.0064
ce-MNO-4	227.09 6	0.273 6	0.0013
β- 1 max	215 4		
avg	59.4 12	100	0.127
X-ray L	4	8 3	0.0007
X-ray Kα2	28.3172 4	20.7 11	0.0125
X-ray Kα1	28.6120 3	38.5 19	0.0235
X-ray Kβ	32.3	13.5 7	0.0093
γ 1	49.720 10	15.0 4	0.0159
γ 2	111.76 8	1.75 5	0.0042
γ 3	116.30 8	1.97 6	0.0049
γ 4	228.16 6	88.2 18	0.429

132I B- DECAY (2.30 H 3)   I(MIN)= 0.10%

Radiation Type	Energy (keV)	Intensity (%)	Δ(g-rad/μCi-h)
Auger-L	3.43	0.89 7	≈0

Radiation Type	Energy (keV)	Intensity (%)	Δ(g-rad/μCi-h)
γ 14	294.508 20	0.166 5	0.0010
γ 16	351.188 24	0.102 7	0.0008
γ 18	373.237 11	14.0 4	0.111
γ 19	404.036 11	1.303 18	0.0112
γ 22	461.246 24	0.1027 9	0.0010
γ 23	462.0 3	0.1027 9	0.0010
γ 24	480.395 13	0.323 9	0.0033
γ 25	486.510 12	2.07 3	0.0214
γ 26	496.313 13	46.7 4	0.494
γ 30	572.672 15	0.155 5	0.0152
γ 32	585.026 15	1.219 11	0.0152
γ 33	620.095 17	1.36 9	0.0179
γ 34	674.415 20	0.133 4	0.0019
γ 35	696.470 20	0.148 5	0.0022
γ 39	831.60 3	0.231 7	0.0041
γ 42	923.846 22	0.729 20	0.0143
γ 45	1046.9	0.1401 12	0.0031
γ 46	1047.57 3	1.167 10	0.0261

23 weak γ's omitted (ΣIγ = 0.54%)

132TE B- DECAY (76.3 H 2)   I(MIN)= 0.10%

Radiation Type	Energy (keV)	Intensity (%)	Δ(g-rad/μCi-h)
Auger-L	3.3	78 6	0.0055
ce-K-1	16.551 10	73 3	0.0259
Auger-K	23.6	9.7 24	0.0049
ce-L-1	44.532 10	9.7 4	0.0092
ce-MNO-1	48.648 10	2.55 7	0.0026

(continued on next column)

Radiation Type	Energy (keV)	Intensity (%)	Δ(g-rad/μCi-h)
Auger-K	24.6	0.118 22	≈0
ce-K-24	488.09 9	0.127 14	0.0013
ce-K-34	633.157 4	0.35 3	0.0047
ce-K-41	738.04 8	0.206 24	0.0032
β⁻ 1 max	226 20		
avg	63 7	0.12 6	0.0002
β⁻ 2 max	320 20		
avg	92 7	0.261 3	0.0005
β⁻ 3 max	353 20		
avg	103 7	0.12 5	0.0003
β⁻ 4 max	366 20		
avg	107 7	0.19	0.0004
β⁻ 5 max	425 20		
avg	127 7	0.21 4	0.0006
β⁻ 6 max	504 20		
avg	154 8	0.24 6	0.0008
β⁻ 7 max	522 20		
avg	161 8	0.34 7	0.0012
β⁻ 8 max	740 20		
avg	242 8	1.81 10	0.0093
β⁻ 9 max	741 20		
avg	242 8	12.8 8	0.0660
β⁻ 10 max	826 20		
avg	275 8	0.37 6	0.0022
β⁻ 11 max	910 20		
avg	309 8	3.60 20	0.0237
β⁻ 12 max	967 20		
avg	331 9	8.1 4	0.0571
β⁻ 13 max	991 20		
avg	342 9	2.87 15	0.0209
β⁻ 14 max	996 20		
avg	344 9	3.46 16	0.0254
β⁻ 15 max	1155 20		
avg	409 9	2.57 7	0.0224
β⁻ 16 max	1185 20		
avg	422 9	19.0 7	0.171

(continued on next page)

132I B- DECAY (2.30 H 3) (CONTINUED)

Radiation Type	Energy (keV)	Intensity (%)	Δ(g-rad/μCi-h)
β⁻17 max	1413 20		
avg	519 9	1.7 6	0.0188
β⁻18 max	1468 20		
avg	543 9	2.0 8	0.0231
β⁻19 max	1470 20		
avg	543 9	9.9 10	0.115
β⁻20 max	1617 20		
avg	608 9	12.6 7	0.163
β⁻21 max	2140 20		
avg	841 9	17.0 22	0.305
Total β⁻			
avg	484 11	100 3	1.03

13 weak β's omitted (ΣIβ = 0.44%)

Radiation Type	Energy (keV)	Intensity (%)	Δ(g-rad/μCi-h)
X-ray Kα2	29.4580 10	0.269 15	0.0002
X-ray Kα1	29.7790 10	0.50 3	0.0003
X-ray Kβ	33.6	0.177 10	0.0001
γ 2	147.20 10	0.237 20	0.0007
γ 3	183.3	0.1579 4	0.0006
γ 4	254.80 20	0.19 3	0.0010
γ 5	262.70 10	1.44 9	0.0081
γ 7	284.80 10	0.79 7	0.0048
γ 10	306.6 4	0.11 4	0.0007
γ 12	316.5 4	0.16 4	0.0011
γ 15	363.5 4	0.49 10	0.0038
γ 16	387.8 4	0.17 3	0.0014
γ 17	416.8 4	0.46 9	0.0041
γ 18	431.9 4	0.45 9	0.0042
γ 19	446.0 4	0.67 8	0.0064
γ 20	473.4 7	0.27 5	0.0027
γ 22	487.5 7	0.18 5	0.0018
γ 23	505.90 15	5.03 20	0.0542

132I B- DECAY (2.30 H 3) (CONTINUED)

Radiation Type	Energy (keV)	Intensity (%)	Δ(g-rad/μCi-h)
γ 82	1476.80 20	0.138 20	0.0043
γ 94	1757.50 20	0.38 3	0.0140
γ 101	1921.08 12	1.18 9	0.0485
γ 103	2002.30 12	1.09 10	0.0463
γ 104	2086.82 15	0.25 4	0.0110
γ 105	2172.68 15	0.19 3	0.0087
γ 107	2223.17 15	0.118 20	0.0056
γ 110	2390.48 15	0.168 20	0.0085

60 weak γ's omitted (ΣIγ = 2.05%)

132CS B+ DECAY (6.475 D 10)   I(MIN)= 0.10%
%(EC+B+)=98.0 1
SEE ALSO 132CS B- DECAY

Radiation Type	Energy (keV)	Intensity (%)	Δ(g-rad/μCi-h)
Auger-L	3.43	78 4	0.0057
Auger-K	24.6	9.2 17	0.0048
β⁺ 1 max	411 21		
avg	193 10	0.33 8	0.0014
X-ray L	4.1	9 3	0.0008
X-ray Kα2	29.4580 10	21.0 6	0.0132
X-ray Kα1	29.7790 10	39.0 10	0.0248

		Energy (keV)	Intensity (%)	Δ(g-rad/μCi-h)
X-ray	Kβ	33.6	13.9 4	0.0099
γ	2	505.90 15	0.80 10	0.0086
γ	3	630.22 9	1.01 8	0.0136
γ	4	667.718 3	97.47 11	1.39
γ	6	1136.03 12	0.51 4	0.0123
γ	8	1317.927 7	0.58 5	0.0164

4 weak γ's omitted (ΣIγ = 0.27%)
Maximum γ± intensity = 0.66%

132CS B- DECAY (6.475 D 10)   I(MIN)= 0.10%
%B-=2.0 1
SEE ALSO 132CS B+ DECAY

Radiation Type	Energy (keV)	Intensity (%)	Δ(g-rad/μCi-h)
β⁻ 1 max	237 22		
avg	66 7	0.37 3	0.0005
β⁻ 2 max	804 22		
avg	266 9	1.57 12	0.0089
Total β⁻			
avg	222 11	2.00 13	0.0095

1 weak β's omitted (ΣIβ = 0.06%)

		Energy (keV)	Intensity (%)	Δ(g-rad/μCi-h)
γ	1	464.55 6	1.87 14	0.0185
γ	2	567.14 3	0.24 4	0.0029
γ	4	1031.70 3	0.122 12	0.0027

1 weak γ's omitted (ΣIγ = 0.06%)

γ	24	522.65 9	16.1 6	0.179
γ	25	535.5 4	0.52 8	0.0060
γ	26	547.10 20	1.25 9	0.0146
γ	30	620.8 10	0.3948 8	0.0052
γ	31	621.2 10	1.579 4	0.0209
γ	32	630.22 9	13.7 6	0.184
γ	33	650.60 20	2.66 20	0.0369
γ	34	667.718 3	98.70 20	1.40
γ	35	669.8 3	4.9 8	0.0704
γ	36	671.6 3	5.2 4	0.0748
γ	37	727	2.2 6	0.0336
γ	38	727.2 10	3.2 6	0.0489
γ	39	729.5 4	1.1 3	0.0169
γ	41	772.60 8	76.2 18	1.25
γ	42	780.2 3	1.23 6	0.0205
γ	43	784.5 4	0.42 5	0.0071
γ	45	809.80 20	2.9 3	0.0494
γ	46	812.20 20	5.6 5	0.0973
γ	47	863.30 20	0.59 5	0.0109
γ	48	876.80 20	1.08 5	0.0201
γ	50	910.30 20	0.92 5	0.0178
γ	51	927.6 3	0.44 8	0.0088
γ	53	954.55 9	18.1 6	0.367
γ	54	984.50 20	0.56 6	0.0118
γ	58	1034.70 20	0.57 5	0.0126
γ	64	1136.03 12	2.96 20	0.0716
γ	66	1143.40 20	1.38 10	0.0337
γ	67	1148.2 7	0.21 5	0.0051
γ	68	1173.20 20	1.09 10	0.0271
γ	71	1272.7 4	0.15 3	0.0040
γ	72	1290.7 3	1.14 6	0.0312
γ	73	1295.3 3	1.97 10	0.0545
γ	74	1298.2 5	0.89 10	0.0246
γ	76	1317.927 7	0.118 20	0.0033
γ	77	1372.07 13	2.47 10	0.0721
γ	78	1398.57 10	7.1 3	0.212
γ	80	1442.56 10	1.42 6	0.0437

(continued on next column)

133XE B- DECAY (5.243 D 1)   I(MIN)= 0.10%

Radiation Type	Energy (keV)	Intensity (%)	Δ(g-rad/μCi-h)
Auger-L	3.55	49.0 20	0.0037
Auger-K	25.5	5.5 7	0.0030
ce-K-1	43.636 11	0.30 6	0.0003
ce-K-2	45.0124 4	52.0 3	0.0498
ce-L-2	75.2827 4	8.49 20	0.0136
ce-MNO-2	79.7799 4	2.3 3	0.0039
β- 1 max	266 3		
avg	75.0 10	0.66 10	0.0011
β- 2 max	346 3		
avg	100.5 10	99.34 10	0.213
Total β-			
avg	100.3 10	100.01 15	0.214
X-ray L	4.29	6.1 17	0.0006
X-ray Kα2	30.6251 3	13.3 3	0.0087
X-ray Kα1	30.9728 3	24.6 5	0.0163
X-ray Kβ	35	8.84 20	0.0066
γ 1	79.621 11	0.22 6	0.0004
γ 2	81	37.1 4	0.0640

4 weak γ's omitted (ΣIγ = 0.07%)

133I B- DECAY (20.8 H 1)   I(MIN)= 0.10%
SEE ALSO 133XE IT DECAY (2.19 D)

Radiation Type	Energy (keV)	Intensity (%)	Δ(g-rad/μCi-h)
β- 1 max	170 30		
avg	46 9	0.414 14	0.0004
β- 2 max	370 30		
avg	110 10	1.24 4	0.0029
β- 3 max	410 30		
avg	122 11	0.396 10	0.0010
β- 4 max	460 30		
avg	140 11	3.74 6	0.0112
β- 5 max	520 30		
avg	162 11	3.12 6	0.0108
β- 6 max	710 30		
avg	230 12	0.541 19	0.0027
β- 7 max	880 30		
avg	299 12	4.15 11	0.0264
β- 8 max	1020 30		
avg	352 13	1.81 5	0.0136
β- 9 max	1230 30		
avg	441 13	83.5 18	0.784
β-10 max	1530 30		
avg	573 13	1.07 5	0.0131
Total β-			
avg	407 15	100.0 18	0.866
γ 5	262.702 12	0.356 10	0.0020
γ 6	267.173 22	0.117 6	0.0007
γ 7	345.43 5	0.104 18	0.0008
γ 8	361.09 6	0.11 4	0.0009
γ 12	417.556	0.153 11	0.0014
γ 13	422.910 16	0.309 9	0.0028
γ 15	510.530 11	1.81 5	0.0197
γ 18	529.872 11	86.3 18	0.974
γ 22	617.974 17	0.539 13	0.0071

133BA EC DECAY (10.5 Y 1) I(MIN)= 0.10%

Radiation Type	Energy (keV)	Intensity (%)	Δ(g-rad/μCi-h)
Auger-L	3.55	133 6	0.0101
ce-K-1	17.1764 11	10.6 4	0.0039
Auger-K	25.5	13.8 16	0.0075
ce-K-2	43.638 5	3.43 16	0.0032
ce-K-3	45.012 3	45.2 10	0.0434
ce-L-1	47.4467 11	1.45 20	0.0015
ce-MNO-1	51.9439 11	0.44 20	0.0005
ce-L-2	73.909 5	0.54 10	0.0008
ce-L-3	75.283 3	7.37 23	0.0118
ce-MNO-2	78.406 5	0.179 7	0.0003
ce-M-3	79.780 3	1.71 14	0.0029
ce-NOP-3	80.766 3	0.31 4	0.0005
ce-K-4	124.62 3	0.131 7	0.0003
ce-K-6	240.4134 21	0.331 10	0.0017
ce-K-7	266.8684 11	0.70 6	0.0040
ce-L-7	297.1387 11	0.103 15	0.0006
ce-K-8	320.0324 21	1.31 4	0.0089
ce-K-9	347.866 3	0.154 5	0.0011
ce-L-8	350.3027 21	0.218 7	0.0016
X-ray L	4.29	16 5	0.0015
X-ray Kα2	30.6251 3	33.5 8	0.0218
X-ray Kα1	30.9728 3	62.0 13	0.0409
X-ray Kβ	35	22.2 6	0.0166
γ 1	53.1610 10	2.19 4	0.0025
γ 2	79.623 5	2.45 9	0.0042
γ 3	80.997 3	32.3 5	0.0558
γ 4	160.609 25	0.653 13	0.0022
γ 5	223.12 4	0.451 8	0.0021
γ 6	276.3980 20	7.17 4	0.0422
γ 7	302.8530 20	18.32 9	0.118
γ 8	356.0170 20	62.1 3	0.471
γ 9	383.851 3	8.94 4	0.0731
γ 26	680.247 15	0.645 16	0.0093
γ 27	706.578 13	1.49 4	0.0225
γ 28	768.382 18	0.457 13	0.0075
γ 30	820.506 24	0.154 6	0.0027
γ 31	856.278 12	1.23 4	0.0225
γ 32	875.329 11	4.47 10	0.0834
γ 33	909.67 3	0.212 8	0.0041
γ 36	1052.296 21	0.552 13	0.0124
γ 37	1060.07 6	0.137 6	0.0031
γ 39	1236.411 12	1.49 4	0.0393
γ 40	1298.223 11	2.33 6	0.0644
γ 42	1350.38 3	0.148 5	0.0043

23 weak γ's omitted (ΣIγ = 0.62%)

133XE IT DECAY (2.19 D 1; I(MIN)= 0.10%
FEEDING OF 2.188-D 133XE IN 133I DECAY=2.883% 23

Radiation Type	Energy (keV)	Intensity (%)	Δ(g-rad/μCi-h)
Auger-L	3.43	70 4	0.0051
Auger-K	24.6	7.1 13	0.0037
ce-K-1	198.660 18	63.7 8	0.270
ce-L-1	227.768 18	20.8 6	0.101
ce-M-1	232.079 18	4.16 13	0.0206
ce-NOP-1	233.013 18	1.04 17	0.0052
X-ray L	4.1	7.8 24	0.0007
X-ray Kα2	29.4580 10	16.1 5	0.0101
X-ray Kα1	29.7790 10	29.9 8	0.0190
X-ray Kβ	33.6	10.6 4	0.0076
γ 1	233.221 18	10.3 3	0.0512

133BA IT DECAY (38.9 H 1) I(MIN)= 0.10%

Radiation Type	Energy (keV)	Intensity (%)	Δ(g-rad/μCi-h)
Auger-L	3.67	130 6	0.0102
ce-L-1	6.30 4	77.5916 5	0.0104
ce-M-1	11.00 4	15.9 5	0.0037
ce-NOP-1	12.04 4	5.20 20	0.0013
Auger-K	26.4	5.9 16	0.0033
ce-K-2	238.65 15	59.2 3	0.301
ce-L-2	270.10 15	18.0 6	0.104
ce-M-2	274.80 15	4.00 10	0.0234
ce-NOP-2	275.84 15	1.15 3	0.0068
X-ray L	4.47	18 5	0.0017
γ 1	12.29 4	1.39 4	0.0004
X-ray Kα₂	31.8171 3	15.2 5	0.0103
X-ray Kα₁	32.1936 3	28.0 9	0.0192
X-ray Kβ	36.4	10.2 4	0.0079
γ 2	276.09 15	18.0 5	0.106

134TE B- DECAY (41.8 M 8) I(MIN)= 0.10%

Radiation Type	Energy (keV)	Intensity (%)	Δ(g-rad/μCi-h)
Auger-L	3.3	32.3 22	0.0023
Auger-K	23.6	4.0 10	0.0020
ce-K-3	46.281 10	26.7 10	0.0263
ce-K-4	68.25 3	0.30 12	0.0004
ce-L-3	74.262 10	4.0 6	0.0063

134TE B- DECAY (41.8 M 8) (CONTINUED)

Radiation Type	Energy (keV)	Intensity (%)	Δ(g-rad/μCi-h)
γ 16	636.26 9	2.1 4	0.0291
γ 17	665.85	1.044 24	0.0148
γ 18	712.97 6	4.21 18	0.0639
γ 19	742.59 3	15.0 7	0.238
γ 20	767.20 3	30.7 10	0.502
γ 21	844.06 6	1.2 6	0.0221
γ 22	896.02 9	0.18 7	0.0035
γ 23	925.55 8	1.69 19	0.0333
γ 24	1027.00 10	0.46 13	0.0101

134I B- DECAY (52.6 M 5) I(MIN)= 0.10%

Radiation Type	Energy (keV)	Intensity (%)	Δ(g-rad/μCi-h)
Auger-L	3.43	1.14 18	≈0
Auger-K	24.6	0.14 4	≈0
ce-K-1	100.838 22	0.98 20	0.0021
ce-L-1	129.946 22	0.12 3	0.0003
ce-K-31	812.46 3	0.17 3	0.0030
ce-K-34	849.53 3	0.12 3	0.0022
β⁻ 1 max	770 60		
avg	255 24	1.48 9	0.0080
β⁻ 2 max	790 60		
avg	261 24	0.33 4	0.0018

	Energy	Intensity	
ce-M-3	78.378 10	0.79 11	0.0013
ce-NOP-3	79.264 10	0.20 3	0.0003
ce-K-6	147.72 3	3.21 20	0.0101
ce-K-8	168.07 3	0.98 20	0.0035
ce-L-6	175.70 3	0.69 5	0.0026
ce-K-9	177.301 20	2.1 3	0.0080
ce-MNO-6	179.82 3	0.175 10	0.0007
ce-K-11	244.781 20	0.8 3	0.0041
β^- 1 max	454 9		
avg	137 4	14.5 8	0.0423
β^- 2 max	637 9		
avg	203 4	42.6 17	0.184
β^- 3 max	713 9		
avg	232 4	43.5 14	0.215
Total β^-			
avg	206 5	100.6 24	0.441
X-ray L	4	3.4 12	0.0003
X-ray $K\alpha_2$	28.3172 4	8.6 4	0.0052
X-ray $K\alpha_1$	28.6120 3	16.0 8	0.0097
γ 1	29.55	0.184 5	0.0001
X-ray $K\beta$	32.3	5.6 3	0.0038
γ 2	76.230 20	0.29 3	0.0005
γ 3	79.450 10	20.9 8	0.0353
γ 4	101.42 3	0.34 7	0.0007
γ 5	131.05 20	0.15 3	0.0004
γ 6	180.89 3	18.4 11	0.0710
γ 7	183.05 13	0.6 3	0.0024
γ 8	201.24 3	8.9 7	0.0382
γ 9	210.470 20	22.4 11	0.100
γ 10	259.59 15	0.49 10	0.0027
γ 11	277.950 20	21.8 11	0.129
γ 12	435.06 5	19.0 13	0.176
γ 13	460.99 3	11.1 7	0.109
γ 14	464.64 6	5.22 25	0.0517
γ 15	565.99 3	19.3 11	0.233

(continued on next column)

	Energy	Intensity	
β^- 3 max	840 60		
avg	279 24	0.153 19	0.0009
β^- 4 max	1070 60		
avg	372 25	1.22 7	0.0097
β^- 5 max	1280 60		
avg	460 30	31.5 8	0.309
β^- 6 max	1380 60		
avg	500 30	0.53 4	0.0056
β^- 7 max	1500 60		
avg	550 30	7.6 5	0.0890
β^- 8 max	1560 60		
avg	580 30	16.2 5	0.200
β^- 9 max	1600 60		
avg	600 30	3.67 17	0.0469
β^- 10 max	1740 60		
avg	660 30	7.6 5	0.107
β^- 11 max	1800 60		
avg	690 30	11.2 7	0.165
β^- 12 max	1850 60		
avg	710 30	2.50 15	0.0378
β^- 13 max	1880 60		
avg	720 30	0.7 7	0.0107
β^- 14 max	2230 60		
avg	880 30	3.7 9	0.0694
β^- 15 max	2420 60		
avg	970 30	11.9 15	0.246
Total β^-			
avg	610 40	100.5 24	1.31
5 weak β's omitted ($\Sigma I_\beta = 0.23\%$)			
X-ray L	4.1	0.13 5	≈0
X-ray $K\alpha_2$	29.4580 10	0.32 6	0.0002
X-ray $K\alpha_1$	29.7790 10	0.60 10	0.0004
X-ray $K\beta$	33.6	0.21 4	0.0002
γ 1	135.399 22	3.76 22	0.0108
γ 2	139.03 3	0.69 5	0.0020

(continued on next page)

134I B- DECAY (52.6 M 5) (CONTINUED)

Radiation Type	Energy (keV)	Intensity (%)	A(g-rad/μCi-h)
γ 3	151.98 15	0.106 12	0.0003
γ 4	162.48 7	0.26 3	0.0009
γ 5	188.47 4	0.70 4	0.0028
γ 6	217.00 20	0.25 3	0.0011
γ 7	235.47 3	1.98 16	0.0100
γ 9	278.80 15	0.131 15	0.0008
γ 10	319.81 6	0.52 5	0.0035
γ 11	351.08 10	0.50 6	0.0037
γ 12	405.451 20	7.3 4	0.0634
γ 13	411.00 8	0.61 6	0.0053
γ 14	433.35 3	4.19 24	0.0387
γ 15	458.92 6	1.30 9	0.0127
γ 16	465.50 10	0.36 4	0.0036
γ 17	488.88 4	1.41 9	0.0147
γ 18	514.40 3	2.34 14	0.0256
γ 19	540.825 25	7.8 5	0.0901
γ 20	565.52 4	0.88 6	0.0106
γ 21	570.75 15	0.21 3	0.0026
γ 22	595.362 20	11.4 6	0.144
γ 23	621.790 25	10.6 6	0.140
γ 24	627.96 3	2.37 14	0.0316
γ 25	677.34 3	8.5 5	0.122
γ 26	706.65 10	0.83 6	0.0125
γ 27	730.74 4	1.91 12	0.0297
γ 28	739.18 8	0.76 8	0.0120
γ 29	766.68 4	4.1 3	0.0670
γ 30	816.38 7	0.52 5	0.0091
γ 31	847.025 25	95.4 3	1.72
γ 32	857.28 3	6.96 20	0.127
γ 33	864.0 3	0.19 3	0.0035
γ 34	884.09 3	65.3 10	1.23
γ 35	922.6 3	0.14 3	0.0028
γ 36	947.86 4	4.04 20	0.0815
γ 37	966.90 5	0.35 4	0.0073

134CS B- DECAY (2.065 Y 1)   I(MIN)= 0.10%

Radiation Type	Energy (keV)	Intensity (%)	A(g-rad/μCi-h)
Auger-L	3.67	0.84 6	≈0
ce-K-5	531.874 15	0.125 4	0.0014
ce-K-6	567.258 15	0.491 15	0.0059
ce-K-7	758.404 22	0.220 7	0.0036
β⁻ 1 max	88.6 4		
avg	23.08 11	27.40 12	0.0135
β⁻ 2 max	415.2 4		
avg	123.44 14	2.48 5	0.0065
β⁻ 3 max	658.0 4		
avg	210.14 15	70.1 4	0.314
β⁻ 4 max	890.6 4		
avg	299.92 16	0.45 15	0.0029
Total β⁻ avg	157.4 3	100.4 5	0.337
X-ray L	4.47	0.11 3	≈0
X-ray Kα₂	31.8171 3	0.238 9	0.0002
X-ray Kα₁	32.1936 3	0.440 16	0.0003
X-ray Kβ	36.4	0.160 6	0.0001
γ 3	475.35 5	1.46 4	0.0148
γ 4	563.227 15	8.37 3	0.100
γ 5	569.315 15	15.41 6	0.187
γ 6	604.699 15	97.6 4	1.26
γ 7	795.845 22	85.4 3	1.45
γ 8	801.932 22	8.68 3	0.148
γ 9	1038.57 3	0.984 5	0.0218
γ 10	1167.94 3	1.784 8	0.0444
γ 11	1365.15 4	3.001 13	0.0873

2 weak γ's omitted (ΣIγ = 0.04%)

135I β- DECAY (6.61 H 1)   I(MIN)= 0.10%

Radiation Type	Energy (keV)	Intensity (%)	Δ(g-rad/μCi-h)
Auger-L	3.43	2.46 15	0.0002
Auger-K	24.6	0.28 6	0.0001
ce-K-25	492.006 10	2.51 9	0.0263
ce-L-25	521.114 10	0.476 21	0.0053
β- 1 max	230 30		
β- 1 avg	65 9	0.140 20	0.0002
β- 2 max	240 30		
β- 2 avg	68 9	0.130 20	0.0002
β- 3 max	260 30		
β- 3 avg	74 9	0.14 3	0.0002
β- 4 max	300 30		
β- 4 avg	86 9	1.03 5	0.0019
β- 5 max	340 30		
β- 5 avg	98 10	0.91 4	0.0019
β- 6 max	350 30		
β- 6 avg	103 10	1.38 5	0.0030
β- 7 max	450 30		
β- 7 avg	137 10	4.73 11	0.0138
β- 8 max	480 30		
β- 8 avg	145 10	7.34 17	0.0227
β- 9 max	620 30		
β- 9 avg	196 11	1.57 7	0.0066
β-10 max	660 30		
β-10 avg	213 11	1.10 4	0.0050
β-11 max	740 30		
β-11 avg	243 11	7.90 20	0.0409
β-12 max	820 30		
β-12 avg	271 11	0.61 3	0.0035
β-13 max	920 30		
β-13 avg	312 12	8.70 20	0.0578

(continued on next page)

Radiation Type	Energy (keV)	Intensity (%)	Δ(g-rad/μCi-h)
γ 38	974.67 4	4.7 4	0.0970
γ 39	1040.25 10	1.91 19	0.0423
γ 43	1072.55 3	15.3 8	0.349
γ 45	1100.07 12	0.69 6	0.0161
γ 46	1103.18 12	0.73 6	0.0170
γ 47	1136.16 4	9.7 5	0.235
γ 48	1159.10 8	0.35 3	0.0087
γ 49	1164.0 3	0.13 3	0.0033
γ 51	1190.03 8	0.35 3	0.0089
γ 53	1239.0 3	0.21 6	0.0055
γ 55	1269.49 5	0.56 4	0.0152
γ 56	1322.4 3	0.10 4	0.0030
γ 57	1336.00 20	0.14 3	0.0041
γ 58	1352.62 8	0.45 4	0.0129
γ 61	1414.3 5	0.22 6	0.0066
γ 62	1428.2 3	0.17 4	0.0052
γ 63	1431.35 25	0.17 4	0.0052
γ 64	1455.24 5	2.29 15	0.0710
γ 65	1470.00 7	0.77 5	0.0242
γ 66	1505.5 4	0.11 4	0.0037
γ 67	1541.51 7	0.51 4	0.0166
γ 68	1613.80 5	4.36 24	0.150
γ 69	1629.24 8	0.26 4	0.0089
γ 70	1644.25 7	0.40 5	0.0140
γ 71	1655.19 10	0.23 3	0.0081
γ 72	1741.49 5	2.67 19	0.0991
γ 73	1806.84 4	5.7 4	0.220
γ 76	1925.88 10	0.181 19	0.0074
γ 78	2020.6 3	0.172 19	0.0074
γ 79	2119.9 3	0.21 3	0.0097
γ 82	2312.40 20	0.24 3	0.0117
γ 85	2467.4 3	0.153 19	0.0080

22 weak γ's omitted (ΣI$_γ$ = 1.33%)

135XE B- DECAY (9.09 H 1)   I(MIN)= 0.10%

Radiation Type	Energy (keV)	Intensity (%)	A(g-rad/μCi-h)
Auger-L	3.55	5.12 25	0.0004
Auger-K	25.5	0.59 7	0.0003
ce-K-3	213.809 15	5.63 17	0.0257
ce-L-3	244.080 15	0.739 23	0.0038
β⁻ 1 max	97 9		
avg	25.3 25	0.123 4	≈0
β⁻ 2 max	551 9		
avg	171 4	3.12 10	0.0114
β⁻ 3 max	751 9		
avg	246 4	0.68 3	0.0036
β⁻ 4 max	909 9		
avg	308 4	95.98 3	0.630
Total β⁻ avg	303 4	99.98 11	0.645

1 weak β's omitted (ΣIβ = 0.08%)

X-ray L	4.29	0.63 18	≈0
X-ray Kα₂	30.6251 3	1.43 6	0.0009
X-ray Kα₁	30.9728 3	2.66 9	0.0018
X-ray Kβ	35	0.95 4	0.0007
γ 1	158.197 18	0.289 10	0.0010
γ 3	249.794 15	90.13 8	0.480
γ 4	358.39 4	0.221 9	0.0017
γ 6	407.990 20	0.359 13	0.0031
γ 9	608.185 16	2.90 9	0.0376

8 weak γ's omitted (ΣIγ = 0.21%)

135I B- DECAY (6.61 H 1)   (CONTINUED)

Radiation Type	Energy (keV)	Intensity (%)	A(g-rad/μCi-h)
β⁻14 max	1030 30		
avg	358 12	21.7 4	0.165
β⁻15 max	1140 30		
avg	405 12	8.00 20	0.0690
β⁻16 max	1250 30		
avg	450 12	7.4 3	0.0709
β⁻17 max	1260 30		
avg	454 12	0.10 5	0.0010
β⁻18 max	1450 30		
avg	535 13	23.7 7	0.270
β⁻19 max	1580 30		
avg	591 13	1.2 6	0.0151
β⁻20 max	2180 30		
avg	857 13	1.9 6	0.0347
Total β⁻ avg	369 15	99.8 13	0.785

4 weak β's omitted (ΣIβ = 0.15%)

X-ray L	4.1	0.27 9	≈0
X-ray Kα₂	29.4580 10	0.64 3	0.0004
X-ray Kα₁	29.7790 10	1.18 5	0.0007
X-ray Kβ	33.6	0.419 19	0.0003
γ 6	220.502 11	1.75 6	0.0082
γ 7	229.720 23	0.241 6	0.0012
γ 10	264.26 9	0.183 6	0.0010
γ 11	288.451 11	3.09 9	0.0190
γ 12	290.27 3	0.304 21	0.0014
γ 18	361.85 13	0.186 23	0.0020
γ 19	403.03 3	0.232 6	0.0027
γ 20	414.83 3	0.301 18	0.0027
γ 21	417.633 19	3.52 10	0.0313
γ 22	429.926 21	0.304 21	0.0028

135XE IT DECAY (15.29 M 3)   I(MIN)= 0.10%

Radiation Type	Energy (keV)	Intensity (%)	Δ(g-rad/ μCi-h)
Auger-L	3.43	14.9 8	0.0011
Auger-K	24.6	1.7 3	0.0009
ce-K-1	492.006 10	15.2 3	0.159
ce-L-1	521.114 10	2.88 9	0.0320
ce-M-1	525.425 10	0.610 20	0.0068
ce-NOP-1	526.359 10	0.200 10	0.0022
X-ray L	4.1	1.7 5	0.0001
X-ray $K\alpha_2$	29.4580 10	3.84 13	0.0024
X-ray $K\alpha_1$	29.7790 10	7.13 23	0.0045
X-ray $K\beta$	33.6	2.54 9	0.0018
γ 1	526.567 10	81.2 6	0.911

135BA IT DECAY (28.7 H 2)   I(MIN)= 0.10%

Radiation Type	Energy (keV)	Intensity (%)	Δ(g-rad/ μCi-h)
Auger-L	3.67	63 4	0.0050
Auger-K	26.4	5.9 16	0.0033
ce-K-1	230.797 10	59.9 7	0.294
ce-L-1	262.249 10	18.7 5	0.104
ce-M-1	266.945 10	4.19 12	0.0238
ce-NOP-1	267.985 10	1.20 3	0.0068

(continued on next page)

#	Type	Energy (keV)	Intensity (%)	Δ(g-rad/μCi-h)
23	γ	433.741 16	0.553 21	0.0051
24	γ	451.631 20	0.315 18	0.0030
25	γ	526.567 10	13.5 5	0.151
27	γ	546.557 11	7.13 19	0.0830
28	γ	575.97 8	0.129 23	0.0016
31	γ	649.85 3	0.46 3	0.0063
35	γ	690.13 5	0.129 15	0.0019
36	γ	707.92 4	0.66 3	0.0099
37	γ	785.48 5	0.152 18	0.0025
39	γ	797.71 8	0.17 3	0.0029
41	γ	836.804 12	6.68 19	0.119
43	γ	961.46	0.15 3	0.0030
44	γ	972	0.89 3	0.0184
45	γ	972.6	1.20 3	0.0249
46	γ	995.09 10	0.15 3	0.0033
47	γ	1038.760 18	7.94 22	0.176
49	γ	1101.58 3	1.60 5	0.0376
50	γ	1124.00 3	3.61 10	0.0864
51	γ	1131.511 14	22.5 6	0.543
53	γ	1159.85 17	0.103 23	0.0025
54	γ	1169.04 4	0.87 3	0.0218
57	γ	1240.468 24	0.90 4	0.0238
59	γ	1260.409 13	28.6 7	0.769
65	γ	1367.89 3	0.61 3	0.0177
67	γ	1448.35 10	0.32 3	0.0097
68	γ	1457.555 23	8.65 22	0.269
69	γ	1502.79 3	1.07 4	0.0344
71	γ	1566.41 3	1.29 6	0.0430
72	γ	1678.027 18	9.5 3	0.341
73	γ	1706.459 18	4.10 15	0.149
74	γ	1791.196 18	7.71 19	0.294
75	γ	1830.69 4	0.579 24	0.0226
77	γ	1927.303 21	0.295 15	0.0121
79	γ	2045.88 3	0.87 4	0.0380
84	γ	2255.457 19	0.613 24	0.0295
86	γ	2408.65 3	0.95 3	0.0489

42 weak γ's omitted (ΣIγ = 1.46%)

135BA IT DECAY (28.7 H 2) (CONTINUED)

Radiation Type	Energy (keV)	Intensity (%)	Δ(g-rad/μCi-h)
X-ray L	4.47	8.6 22	0.0008
X-ray Kα₂	31.8171 3	15.3 6	0.0104
X-ray Kα₁	32.1936 3	28.3 10	0.0194
X-ray Kβ	36.4	10.3 4	0.0080
γ 1	268.238 10	16.0 5	0.0914

136CS B- DECAY (13.1 D 1)   I(MIN)= 0.10%

Radiation Type	Energy (keV)	Intensity (%)	Δ(g-rad/μCi-h)
Auger-L	3.67	20.8 13	0.0016
Auger-K	26.4	1.9 6	0.0011
ce-K-1	29.47 5	7.3 7	0.0046
ce-K-2	48.85 5	1.85 12	0.0019
ce-L-1	60.92 5	1.05 10	0.0014
ce-MNO-1	65.62 5	0.212 19	0.0003
ce-K-3	72.22 10	0.38 3	0.0006
ce-L-2	80.30 5	0.253 17	0.0004
ce-K-4	115.78 5	2.38 12	0.0059
ce-K-5	126.45 5	5.17 25	0.0139
ce-K-6	129.09 10	0.145 21	0.0004
ce-K-7	139.12 5	0.55 3	0.0016
ce-L-4	147.23 5	0.67 4	0.0021
ce-MNO-4	151.93 5	0.144 6	0.0005
ce-L-5	157.90 5	4.11 20	0.0138
ce-M-5	162.60 5	0.92 5	0.0032

136CS B- DECAY (13.1 D 1) (CONTINUED)

Radiation Type	Energy (keV)	Intensity (%)	Δ(g-rad/μCi-h)
γ 13	319.87 10	0.60 6	0.0041
γ 14	340.57 5	46.8 15	0.339
γ 16	507.21 10	0.98 6	0.0106
γ 18	818.50 4	100 3	1.74
γ 19	1048.07 7	79.7 25	1.78
γ 20	1235.34 5	19.8 7	0.520

8 weak γ's omitted (ΣIγ = 0.39%)

137CS B- DECAY (30.0 Y 2)   I(MIN) = 0.10%

Radiation Type	Energy (keV)	Intensity (%)	Δ(g-rad/μCi-h)
Auger-L	3.67	7.3 5	0.0006
Auger-K	26.4	0.77 21	0.0004
ce-K-1	624.219 3	7.78 5	0.103
ce-L-1	655.671 3	1.42 3	0.0198
ce-MNO-1	660.367 3	0.472 19	0.0066
β⁻ 1 max	511.5 9		
avg	173.5 3	94.43 7	0.349
β⁻ 2 max	1173.2 9		
avg	415.4 3	5.53 7	0.0489
Total β⁻ avg	186.9 4	99.96 10	0.398

ce-NOP-5	163.64	5	0.240	12	0.0008
ce-K-10	236.21	4	0.159	8	0.0008
ce-K-14	303.13	5	1.17	6	0.0075
ce-L-14	334.58	5	0.201	9	0.0014
ce-K-18	781.06	4	0.241	11	0.0040
ce-K-19	1010.63	7	0.112	5	0.0024
β^- 1 max	174.8	20			
avg	47.3	6	2.29	10	0.0023
β^- 2 max	192.0	20			
avg	52.4	6	0.21	3	0.0002
β^- 3 max	341.4	20			
avg	98.9	7	94	3	0.198
β^- 4 max	408.3	20			
avg	121.1	7	0.13	13	0.0003
β^- 5 max	494.6	20			
avg	150.9	7	0.8	8	0.0026
β^- 6 max	518.0	20			
avg	175.7	8	1.8	8	0.0067
β^- 7 max	682.0	20			
avg	219.1	8	2.0	20	0.0093
Total β^-					
avg	101.8	8	101	4	0.220
X-ray L	4.47		2.8	8	0.0003
X-ray Kα_2	31.8171	3	4.99	25	0.0034
X-ray Kα_1	32.1936	3	9.2	5	0.0063
X-ray Kβ	36.4		3.35	17	0.0026
γ 1	66.91	5	12.5	11	0.0178
γ 2	86.29	5	6.3	4	0.0116
γ 3	109.66	10	0.41	3	0.0010
γ 4	153.22	5	7.5	3	0.0244
γ 5	163.89	5	4.62	17	0.0161
γ 6	166.53	10	0.63	4	0.0022
γ 7	176.56	5	13.6	5	0.0511
γ 8	187.25	10	0.60	6	0.0024
γ 10	273.65	4	12.7	5	0.0740

(continued on next column)

X-ray L	4.47		1.0	3	≈0
X-ray Kα_2	31.8171	3	1.99	7	0.0014
X-ray Kα_1	32.1936	3	3.68	12	0.0025
X-ray Kβ	36.4		1.34	5	0.0010
γ 1	661.660	3	85.1	3	1.20

137BA IT DECAY (2.553 M 1)   I(MIN)= 0.10%

Radiation Type	Energy (keV)		Intensity (%)		Δ(g-rad/ μCi-h)
Auger-L	3.67		7.6	5	0.0006
Auger-K	26.4		0.79	21	0.0004
ce-K-1	624.219	3	8.02	7	0.107
ce-L-1	655.671	3	1.50	3	0.0209
ce-MNO-1	660.367	3	0.500	20	0.0070
X-ray L	4.47		1.0	3	≈0
X-ray Kα_2	31.8171	3	2.05	7	0.0014
X-ray Kα_1	32.1936	3	3.79	12	0.0026
X-ray Kβ	36.4		1.38	5	0.0011
γ 1	661.660	3	90.1	6	1.27

138XE B- DECAY (14.08 M 8)   I(MIN)= 0.10%

Radiation Type	Energy (keV)	Intensity (%)	Δ(g-rad/μCi-h)
Auger-L	3.55	49 3	0.0037
ce-M-1	3.63 5	32 5	0.0025
ce-NOP-1	4.62 5	10.6 17	0.0010
ce-L-2	5.14 5	51.0 24	0.0056
ce-M-2	9.63 5	10.4 5	0.0021
ce-NOP-2	10.62 5	3.41 16	0.0008
Auger-K	25.5	0.42 6	0.0002
ce-K-5	117.873 3	1.61 25	0.0040
ce-L-5	148.144 3	0.36 18	0.0011
ce-K-7	206.58 5	0.241 11	0.0011
ce-K-8	222.426 20	1.80 10	0.0085
ce-L-8	252.697 20	0.28 7	0.0015
ce-K-14	360.528 10	0.107 14	0.0008
ce-K-17	398.58 5	0.26 5	0.0022
β⁻ 1 max	230 50		
β⁻ 1 avg	64 16	0.23 17	0.0003
β⁻ 2 max	250 50		
β⁻ 2 avg	70 16	0.46 14	0.0007
β⁻ 3 max	400 50		
β⁻ 3 avg	119 17	3.07 13	0.0078
β⁻ 4 max	480 50		
β⁻ 4 avg	145 18	9.5 4	0.0293
β⁻ 5 max	710 50		
β⁻ 5 avg	231 20	32.7 13	0.161
β⁻ 6 max	720 50		
β⁻ 6 avg	233 20	0.28 3	0.0014
β⁻ 7 max	950 50		
β⁻ 7 avg	323 21	0.23 3	0.0016
β⁻ 8 max	1370 50		
β⁻ 8 avg	498 22	0.16 3	0.0017
β⁻ 9 max	1530 50		
β⁻ 9 avg	571 22	0.19 3	0.0023

138XE B- DECAY (14.08 M 8)   (CONTINUED)

Radiation Type	Energy (keV)	Intensity (%)	Δ(g-rad/μCi-h)
γ 27	588.84 8	0.123 11	0.0015
γ 30	654.08 8	0.145 14	0.0020
γ 47	865.82 7	0.296 22	0.0055
γ 49	869.35 6	0.62 4	0.0115
γ 50	896.87 12	0.132 14	0.0025
γ 52	912.51 7	0.328 22	0.0064
γ 53	917.13 6	0.92 4	0.0180
γ 54	936.36 11	0.135 14	0.0027
γ 55	941.25 8	0.230 18	0.0046
γ 60	1093.87 9	0.41 3	0.0095
γ 61	1098.77 11	0.214 18	0.0050
γ 62	1102.24 17	0.107 14	0.0025
γ 63	1114.29 10	1.47 9	0.0350
γ 64	1141.64 9	0.51 4	0.0125
γ 65	1145.44 18	0.132 20	0.0032
γ 80	1571.84 16	0.26 3	0.0089
γ 82	1614.57 18	0.24 3	0.0081
γ 84	1768.26 13	16.7 7	0.630
γ 87	1812.54 18	0.180 20	0.0069
γ 88	1850.86 13	1.42 7	0.0561
γ 90	1925.36 14	0.56 4	0.0231
γ 91	2004.75 14	5.35 23	0.229
γ 92	2015.82 14	12.3 5	0.526
γ 94	2079.17 14	1.44 7	0.0639
γ 95	2252.26 14	2.29 11	0.110
γ 97	2321.90 16	0.62 4	0.0307
γ 99	2475.26 16	0.312 20	0.0164
γ 101	2497.56 17	0.173 14	0.0092

57 weak γ's omitted (ΣIγ = 2.66%)

138CS B- DECAY (32.2 M 1)   I(MIN)= 0.10%

Radiation Type	Energy (keV)	Intensity (%)	Δ(g-rad/μCi-h)
Auger-L	3.67	0.65 5	≈0
ce-K-14	425.355 5	0.315 12	0.0029
β^- 1 max	745 22		
β^- 1 avg	243 9	0.26 3	0.0013
β^- 2 max	867 22		
β^- 2 avg	291 9	0.160 20	0.0010
β^- 3 max	1132 22		
β^- 3 avg	398 10	0.100 10	0.0008
β^- 4 max	1295 22		
β^- 4 avg	466 10	0.18 3	0.0018
β^- 5 max	1440 22		
β^- 5 avg	528 10	0.47 6	0.0053
β^- 6 max	1452 22		
β^- 6 avg	534 10	0.21 3	0.0024
β^- 7 max	1681 22		
β^- 7 avg	634 10	0.30 3	0.0041
β^- 8 max	1728 22		
β^- 8 avg	654 10	0.43 8	0.0060
β^- 9 max	2008 22		
β^- 9 avg	779 10	0.230 20	0.0038
β^- 10 max	2036 22		
β^- 10 avg	792 10	0.170 20	0.0029
β^- 11 max	2132 22		
β^- 11 avg	835 10	0.270 20	0.0048
β^- 12 max	2211 22		
β^- 12 avg	871 10	0.34 3	0.0063
β^- 13 max	2325 22		
β^- 13 avg	922 10	0.17 3	0.0033
β^- 14 max	2384 22		
β^- 14 avg	949 10	0.64 4	0.0129

(continued on next page)

Radiation Type	Energy (keV)	Intensity (%)	Δ(g-rad/μCi-h)
β^- 10 max	1790 50		
β^- 10 avg	682 23	0.27 4	0.0039
β^- 11 max	2290 50		
β^- 11 avg	908 23	20.1 7	0.389
β^- 12 max	2330 50		
β^- 12 avg	925 23	13.4 5	0.264
β^- 13 max	2480 50		
β^- 13 avg	996 23	5.6 21	0.119
β^- 14 max	2720 50		
β^- 14 avg	1107 23	9 6	0.212
β^- 15 max	2730 50		
β^- 15 avg	1109 23	5 5	0.118
Total β^- avg	610 40	100 9	1.32

5 weak β's omitted ($\Sigma I\beta = 0.24\%$)

Radiation Type	Energy (keV)	Intensity (%)	Δ(g-rad/μCi-h)
X-ray L	4.29	6.1 17	0.0006
γ 1	4.85 5	0.19 3	≈0
γ 2	10.85 5	0.696 25	0.0002
X-ray Kα_2	30.6251 3	1.02 8	0.0007
X-ray Kα_1	30.9728 3	1.89 13	0.0012
X-ray Kβ	35	0.68 5	0.0005
γ 5	153.858 3	5.95 25	0.0195
γ 7	242.56 5	3.50 14	0.0181
γ 8	258.411 20	31.5 13	0.173
γ 9	282.51 6	0.428 20	0.0026
γ 12	335.28 9	0.107 11	0.0008
γ 13	371.44 5	0.50 3	0.0040
γ 14	396.513 10	6.3 3	0.0532
γ 15	401.36 5	2.17 13	0.0186
γ 17	434.56 4	20.3 9	0.188
γ 18	500.22 6	0.362 18	0.0039
γ 19	530.07 7	0.252 16	0.0028
γ 21	537.76 13	0.117 17	0.0013
γ 23	555.95 9	0.117 14	0.0014
γ 24	568.53 6	0.306 19	0.0037

(continued on next column)

138CS B- DECAY (32.2 M 1) (CONTINUED)

Radiation Type	Energy (keV)	Intensity (%)	Δ(g-rad/μCi-h)
β-15 max	2443 22		
avg	976 10	0.21 4	0.0044
β-16 max	2494 22		
avg	999 10	0.37 20	0.0079
β-17 max	2523 22		
avg	1013 10	0.20 5	0.0043
β-18 max	2596 22		
avg	1046 10	1.59 7	0.0354
β-19 max	2735 22		
avg	1110 11	8.70 23	0.206
β-20 max	2792 22		
avg	1136 11	1.67 8	0.0404
β-21 max	2959 22		
avg	1200 10	0.63	0.0161
β-22 max	2929 22		
avg	1200 11	43.0 7	1.10
β-23 max	3067 22		
avg	1264 11	7.90 20	0.213
β-24 max	3157 22		
avg	1305 11	13.0 3	0.361
β-25 max	3476 22		
avg	1454 11	13.7 7	0.424
β-26 max	3939 22		
avg	1670 11	4.6 18	0.164
Total β-	1241 12	99.7 22	2.63

6 weak β's omitted (ΣIβ = 0.15%)

X-ray Kα2	31.8171 3	0.182 8	0.0001
X-ray Kα1	32.1936 3	0.337 14	0.0002
X-ray Kβ	36.4	0.122 6	≈0
γ 1	112.60 13	0.130 23	0.0003

138CS B- DECAY (32.2 M 1) (CONTINUED)

Radiation Type	Energy (keV)	Intensity (%)	Δ(g-rad/μCi-h)
γ 61	2062.34 17	0.111 12	0.0049
γ 64	2210.7 4	0.21 7	0.0101
γ 65	2218.00 10	15.2 4	0.717
γ 67	2499.4 3	0.17 5	0.0089
γ 69	2583.15 13	0.239 16	0.0131
γ 71	2639.59 13	7.63 23	0.429
γ 72	2731.12 15	0.120 8	0.0070
γ 78	3339.01 25	0.151 10	0.0107
γ 80	3366.98 25	0.227 13	0.0163

39 weak γ's omitted (ΣIγ = 1.62%)

139BA B- DECAY (82.8 M 2)   I(MIN)= 0.10%

Radiation Type	Energy (keV)	Intensity (%)	Δ(g-rad/μCi-h)
Auger-L	3.8	4.4 3	0.0004
Auger-K	27.4	0.48 14	0.0003
ce-K-1	126.928 7	5.06 16	0.0137
ce-L-1	159.587 7	0.653 21	0.0022
ce-MNO-1	164.492 7	0.1734 18	0.0006
β- 1 max	889 5		
avg	298.6 20	0.287 6	0.0018
β- 2 max	2143 5		
avg	838.4 23	29.7 3	0.530

β⁻	3 max	2309 5		
	avg	913.4 23	70.0 4	1.36
Total β⁻				
	avg	889.1 23	100.0 5	1.89

14 weak β's omitted (ΣIβ = 0.05%)

X-ray L		4.65	0.66 16	≈0
X-ray Kα₂		33.03410 20	1.30 6	0.0009
X-ray Kα₁		33.44180 20	2.40 11	0.0017
X-ray Kβ		37.8	0.88 4	0.0007
γ	1	165.853 7	23.76 25	0.0839
γ	12	1420.50 20	0.261 5	0.0079

26 weak γ's omitted (ΣIγ = 0.07%)

139CE EC DECAY (137.64 D 2)   I(MIN)= 0.10%

Radiation Type	Energy (keV)	Intensity (%)	Δ(g-rad/ μCi-h)
Auger-L	3.8	89 6	0.0072
Auger-K	27.4	8.3 23	0.0048
ce-K-1	126.928 7	17.9 5	0.0484
ce-L-1	159.587 7	2.40 7	0.0082
ce-MNO-1	164.492 7	0.640 20	0.0022
X-ray L	4.65	13 4	0.0013
X-ray Kα₂	33.03410 20	22.8 9	0.0160
X-ray Kα₁	33.44180 20	41.9 15	0.0299
X-ray Kβ	37.8	15.4 6	0.0124
γ 1	165.853 7	79.1 5	0.279

	Energy	Intensity	Δ
γ 2	138.10 6	1.49 9	0.0044
γ 3	191.96 6	0.50 4	0.0021
γ 4	193.89 8	0.328 23	0.0014
γ 5	212.32 8	0.175 14	0.0008
γ 6	227.76 6	1.51 4	0.0073
γ 7	324.90 8	0.290 19	0.0020
γ 9	363.93 8	0.244 23	0.0019
γ 10	365.29 13	0.191 23	0.0015
γ 12	408.98 6	4.66 10	0.0406
γ 13	421.59 7	0.427 23	0.0038
γ 14	462.796 5	30.7 7	0.303
γ 15	516.74 12	0.43 5	0.0047
γ 16	547.001 5	10.76 24	0.125
γ 19	683.59 15	0.108 14	0.0016
γ 23	766.10 12	0.146 15	0.0024
γ 24	773.31 10	0.233 19	0.0038
γ 25	782.08 9	0.33 3	0.0055
γ 31	871.80 8	5.11 14	0.0949
γ 32	880.8 3	0.11 3	0.0021
γ 33	935.03 12	0.181 16	0.0036
γ 36	1009.78 8	29.8 7	0.642
γ 38	1054.32 15	0.159 19	0.0036
γ 39	1147.22 9	1.24 7	0.0304
γ 40	1199.15 24	0.17 3	0.0043
γ 41	1203.69 13	0.40 4	0.0102
γ 42	1264.94 16	0.137 17	0.0037
γ 43	1343.59 9	1.14 6	0.0328
γ 46	1415.68 13	0.37 3	0.0110
γ 47	1435.86 9	76.3 16	2.33
γ 48	1445.04 25	0.97 19	0.0298
γ 49	1495.63 23	0.18 4	0.0058
γ 50	1555.31 10	0.366 23	0.0121
γ 51	1614.09 20	0.107 23	0.0047
γ 52	1717.1 9	0.107 23	0.0039
γ 53	1727.68 18	0.111 13	0.0041
γ 55	1778.25 23	0.137 23	0.0052
γ 60	2023.93 20	0.118 16	0.0051

(continued on next column)

140BA B- DECAY (12.76 D 1)   I(MIN)= 0.10%

Radiation Type	Energy (keV)	Intensity (%)	Δ(g-rad/μCi-h)
Auger-L	3.8	99 7	0.0080
ce-L-1	7.58 5	52 7	0.0083
ce-M-1	12.49 5	10.7 14	0.0028
ce-NOP-1	13.58 5	3.6 5	0.0010
ce-L-2	23.70 5	60.9 18	0.0308
Auger-K	27.4	0.14 4	≈0
ce-M-2	28.61 5	12.6 4	0.0077
ce-NOP-2	29.70 5	4.19 18	0.0026
ce-K-7	123.72 5	1.48 5	0.0039
ce-L-7	156.37 5	0.198 7	0.0007
β⁻ 1 max	453 10		
avg	136 4	24.69 23	0.0715
β⁻ 2 max	566 10		
avg	176 4	9.71 8	0.0364
β⁻ 3 max	871 10		
avg	305 4	3.86 12	0.0251
β⁻ 4 max	990 10		
avg	339 4	41 9	0.296
β⁻ 5 max	1004 10		
avg	357 4	21 9	0.160
Total β⁻ avg	276 5	100 13	0.589
X-ray L	4.65	15 4	0.0015
γ 1	13.85 5	1.20 15	0.0004
γ 2	29.97 5	14.1 4	0.0090
X-ray Kα₂	33.03410 20	0.391 18	0.0003
X-ray Kα₁	33.44180 20	0.72 4	0.0005
X-ray Kβ	37.8	0.264 13	0.0002
γ 6	132.84 12	0.220 20	0.0006
γ 7	162.64 5	6.21 8	0.0215
γ 8	304.840 20	4.30 5	0.0279

140LA B- DECAY (40.28 H 1)   (CONTINUED)

Radiation Type	Energy (keV)	Intensity (%)	Δ(g-rad/μCi-h)
β⁻ 7 avg	487.2 9	44.8 3	0.465
max	1410.7 21		
β⁻ 8 avg	513.7 9	0.530 20	0.0058
max	1412.6 21		
β⁻ 9 avg	514.5 9	5.13 5	0.0562
max	1677.3 21		
β⁻10 avg	629.3 9	20.7 5	0.277
max	2164.3 20		
avg	846.0 9	4.8 5	0.0865
Total β⁻ avg	525.8 10	99.9 8	1.12
3 weak β's omitted (ΣIβ = 0.10%)			
X-ray L	4.84	0.30 7	≈0
X-ray Kα₂	34.27890 20	0.493 19	0.0004
X-ray Kα₁	34.71970 20	0.90 4	0.0007
X-ray Kβ	39.3	0.337 14	0.0003
γ 4	109.419 7	0.190 20	0.0004
γ 5	131.119 8	0.430 10	0.0012
γ 6	173.541 9	0.110 10	0.0004
γ 7	241.961 12	0.430 10	0.0022
γ 8	266.546 9	0.500 10	0.0028
γ 10	328.761 8	20.74 18	0.145
γ 12	432.493 10	2.920 20	0.0269
γ 15	487.018 9	45.9 4	0.476
γ 17	751.83 8	4.41 4	0.0706
γ 18	815.766 16	23.64 17	0.411
γ 19	867.846 20	5.59 5	0.103
γ 20	919.550 23	2.73 3	0.0535
γ 21	925.189 21	7.05 8	0.139
γ 22	950.99 3	0.528 7	0.0107
γ 23	1596.182 20	95.40 8	3.24

γ			
γ 9	423.69 3	3.15 4	0.0284
γ 10	437.55 3	1.93 4	0.0180
γ 12	537.32 8	24.39 22	0.279

4 weak γ's omitted (ΣIγ = 0.09%)

γ			
γ 26	2347.85 4	0.850 10	0.0425
γ 28	2521.38 4	3.42 4	0.184
γ 29	2547.23 6	0.1010 20	0.0055

14 weak γ's omitted (ΣIγ = 0.35%)

141CE B- DECAY (32.50 D 1)   I(MIN)= 0.10%

Radiation Type	Energy (keV)	Intensity (%)	Δ(g-rad/μCi-h)
Auger-L	4	15.8 8	0.0014
Auger-K	29.4	1.5 5	0.0010
ce-K-1	103.449 3	18.2 4	0.0401
ce-L-1	138.605 3	2.50 11	0.0074
ce-MNO-1	143.929 4	0.68 15	0.0021
β- 1 max	434.8 13		
avg	129.7 5	69.6 4	0.192
β- 2 max	580.2 13		
avg	180.7 5	30.4 4	0.117
Total β- avg	145.2 6	100.0 6	0.309
X-ray L	5	2.6 4	0.0003
X-ray Kα₂	35.55020 20	4.74 18	0.0036
X-ray Kα₁	36.02630 20	8.7 3	0.0067
X-ray Kβ	40.7	3.26 13	0.0028
γ 1	145.440 3	48.44 14	0.150

140LA B- DECAY (40.28 H 1)   I(MIN)= 0.10%

Radiation Type	Energy (keV)	Intensity (%)	Δ(g-rad/μCi-h)
Auger-L	4	1.98 12	0.0002
ce-L-1	18.046 4	0.331 9	0.0001
Auger-K	28.4	0.17 5	0.0001
ce-K-3	28.473 6	0.227 20	0.0001
ce-K-4	68.976 7	0.150 17	0.0002
ce-K-10	288.318 8	0.81 3	0.0050
ce-L-10	322.212 8	0.108 4	0.0007
ce-K-15	446.575 9	0.443 14	0.0042
β- 1 max	1213.3 21		
avg	438.0 9	0.631 8	0.0059
β- 2 max	1239.0 21		
avg	440.9 9	11.02 10	0.103
β- 3 max	1244.7 21		
avg	443.3 9	5.72 5	0.0540
β- 4 max	1279.6 21		
avg	458.0 9	0.85 4	0.0083
β- 5 max	1296.5 21		
avg	465.1 9	5.61 5	0.0556
β- 6 max	1348.5 21		

(continued on next column)

142PR B- DECAY (19.13 H 4)   I(MIN)= 0.10%

Radiation Type	Energy (keV)	Intensity (%)	Δ(g-rad/μCi-h)
β⁻ 1 max	585 3		
avg	182.1 10	3.7 5	0.0144
β⁻ 2 max	2160.3 25		
avg	833.4 11	96.3 5	1.71
Total β⁻			
avg	809.1 12	100.0 7	1.72

1 weak β's omitted (ΣIβ = 0.02%)

γ 2	1575.6 5	3.7 5	0.124

1 weak γ's omitted (ΣIγ = 0.02%)

143CE B- DECAY (33.0 H 2)   I(MIN)= 0.10%

Radiation Type	Energy (keV)	Intensity (%)	Δ(g-rad/μCi-h)
Auger-L	4	59 4	0.0051
ce-K-1	15.3744 12	65.7 21	0.0215
Auger-K	29.4	5.8 17	0.0036
ce-L-1	50.5302 12	9.3 3	0.0100
ce-M-1	55.8540 13	1.95 7	0.0023
ce-NOP-1	57.0605 14	0.541 18	0.0007
ce-K-7	251.271 21	2.24 7	0.0120
ce-L-7	286.427 21	0.351 11	0.0021

143PR B- DECAY (13.58 D 3)   I(MIN)= 0.10%

Radiation Type	Energy (keV)	Intensity (%)	Δ(g-rad/μCi-h)
β⁻ 1 max	934.5 19		
avg	315.3 8	100	0.672

144CE B- DECAY (285.0 D 1)   I(MIN)= 0.10%

Radiation Type	Energy (keV)	Intensity (%)	Δ(g-rad/μCi-h)
Auger-L	4	10.7 5	0.0009
ce-K-3	11.439 10	0.70 4	0.0002
ce-K-4	17.04 3	0.467 16	0.0002
ce-L-1	26.785 10	0.99 11	0.0006
Auger-K	29.4	0.84 25	0.0005
ce-M-1	32.109 10	0.207 22	0.0001
ce-L-2	34.06 5	0.89 9	0.0006
ce-K-5	38.115 5	3.17 20	0.0026
ce-M-2	39.38 5	0.188 18	0.0002
ce-L-4	52.20 3	0.701 24	0.0008
ce-M-4	57.52 3	0.174 6	0.0002
ce-L-5	73.271 5	0.44 3	0.0007
ce-K-7	91.553 5	5.47 19	0.0107
ce-L-7	126.709 5	0.751 25	0.0020
ce-M-7	132.033 5	0.157 6	0.0004
β⁻ 1 max	184.7 20		
avg	50.2 6	20.0 4	0.0214

Radiation Type	Energy (keV)	Intensity (%)	Δ(g-rad/μCi-h)
β⁻ 1 max	63.9 20		
avg	16.4 6	0.39 4	0.0001
β⁻ 2 max	301.2 20		
avg	85.7 7	0.551 17	0.0010
β⁻ 3 max	523.8 19		
avg	160.6 7	1.51 4	0.0052
β⁻ 4 max	739.7 19		
avg	239.9 8	12.6 4	0.0644
β⁻ 5 max	1111.0 19		
avg	387.3 8	48.3 5	0.398
β⁻ 6 max	1404.2 19		
avg	510.3 9	36.2 4	0.393
Total β⁻			
avg	406.1 10	99.7 8	0.863

3 weak β's omitted (ΣIβ = 0.20%)

Radiation Type	Energy (keV)	Intensity (%)	Δ(g-rad/μCi-h)
X-ray L	5	9.6 15	0.0010
X-ray Kα₂	35.55020 20	17.7 8	0.0134
X-ray Kα₁	36.02630 20	32.3 14	0.0248
X-ray Kβ	40.7	12.1 6	0.0105
γ 1	57.3650 10	11.77 12	0.0144
γ 4	169.0 10	0.3 4	0.0011
γ 5	216.0 10	0.210 22	0.0010
γ 6	231.56 3	2.04 3	0.0101
γ 7	293.262 21	42.8 4	0.267
γ 8	338.0 10	0.29 3	0.0021
γ 9	350.59 5	3.27 4	0.0244
γ 15	433.02 7	0.166 6	0.0015
γ 16	439.0 10	0.120 13	0.0011
γ 18	490.36 7	2.127 24	0.0222
γ 22	587.28 15	0.257 18	0.0032
γ 24	664.55 10	5.60 6	0.0792
γ 26	721.96 11	5.32 6	0.0818
γ 31	880.39 13	1.014 13	0.0190
γ 38	1102.98 18	0.406 7	0.0095

26 weak γ's omitted (ΣIγ = 0.87%)

Radiation Type	Energy (keV)	Intensity (%)	Δ(g-rad/μCi-h)
β⁻ 2 max	238.1 20		
avg	66.1 6	4.3 3	0.0061
β⁻ 3 max	318.2 20		
avg	91.1 7	75.7 5	0.147
Total β⁻			
avg	81.8 7	100.0 7	0.174
X-ray L	5	1.7 3	0.0002
γ 1	33.620 10	0.26 3	0.0002
X-ray Kα₂	35.55020 20	2.56 11	0.0019
X-ray Kα₁	36.02630 20	4.69 19	0.0036
X-ray Kβ	40.7	1.76 8	0.0015
γ 2	40.89 5	0.41 4	0.0004
γ 3	53.430 10	0.102 5	0.0001
γ 5	80.106 5	1.50 8	0.0026
γ 7	133.544 5	11.09 16	0.0315

2 weak γ's omitted (ΣIγ = 0.04%)

144PR B- DECAY (17.28 M 5)   I(MIN)= 0.10%

Radiation Type	Energy (keV)	Intensity (%)	Δ(g-rad/μCi-h)
β⁻ 1 max	811 3		
avg	267.0 12	0.984 17	0.0056
β⁻ 2 max	2301 3		
avg	904.4 14	1.058 12	0.0204
β⁻ 3 max	2997 3		
avg	1221.8 14	97.950 25	2.55

(continued on next page)

144PR B- DECAY (17.28 M 5) (CONTINUED)

Radiation Type	Energy (keV)	Intensity (%)	Δ(g-rad/μCi-h)
Total β⁻			
avg	1209.0 15	100.00 4	2.58
γ 3	696.510 3	1.342 14	0.0199
γ 7	1489.160 5	0.272 5	0.0086
γ 9	2185.662 7	0.707 15	0.0329

7 weak γ's omitted (ΣIᵧ = 0.02%)

145PM EC DECAY (17.7 Y 4)   I(MIN)= 0.10%

Radiation Type	Energy (keV)	Intensity (%)	Δ(g-rad/μCi-h)
Auger-L	4.23	81 4	0.0073
ce-K-1	23.63 10	2.18 16	0.0011
ce-K-2	28.83 10	6.7 5	0.0041
Auger-K	30.5	6.6 20	0.0043
ce-L-1	60.07 10	3.13 23	0.0040
ce-L-2	65.27 10	0.94 7	0.0013
ce-M-1	65.62 10	0.71 6	0.0010
ce-NOP-1	66.88 10	0.189 14	0.0003
ce-M-2	70.82 10	0.198 14	0.0003
X-ray L	5.23	14.3 20	0.0016
X-ray Kα₂	36.8474 3	21.7 7	0.0170
X-ray Kα₁	37.3610 3	39.6 12	0.0315
X-ray Kβ	42.3	15.1 5	0.0136

147ND B- DECAY (10.98 D 1) (CONTINUED)

Radiation Type	Energy (keV)	Intensity (%)	Δ(g-rad/μCi-h)
X-ray L	5.43	8.0 11	0.0009
X-ray Kα₂	38.1712 5	13.1 7	0.0106
X-ray Kα₁	38.7247 5	23.7 13	0.0196
X-ray Kβ	43.8	9.1 5	0.0085
γ 3	91.109 4	28.2 13	0.0547
γ 4	120.490 10	0.369 17	0.0009
γ 8	196.640 20	0.170 9	0.0007
γ 10	275.400 13	0.75 3	0.0044
γ 11	319.413 10	1.87 8	0.0127
γ 12	398.130 16	0.82 4	0.0069
γ 14	410.49 4	0.116 8	0.0010
γ 15	439.871 14	1.13 5	0.0106
γ 16	489.260 20	0.134 7	0.0014
γ 17	531.0329 24	12.3 4	0.139
γ 20	594.800 20	0.236 13	0.0030
γ 22	685.90 3	0.75 3	0.0110

10 weak γ's omitted (ΣIᵧ = 0.11%)

147PM B- DECAY (2.6234 Y 2)   I(MIN)= 0.10%

Radiation Type	Energy (keV)	Intensity (%)	Δ(g-rad/μCi-h)
β⁻ 1 max	224.7 4		
avg	61.96 12	99.9940 10	0.132

Radiation Type	Energy (keV)	Intensity (%)	Δ(g-rad/μCi-h)
γ 1	67.20 10	0.65 5	0.0009
γ 2	72.40 10	2.16 13	0.0033

147ND B- DECAY (10.98 D 1)   I(MIN)= 0.10%

Radiation Type	Energy (keV)	Intensity (%)	Δ(g-rad/μCi-h)
Auger-L	4.38	42 3	0.0039
Auger-K	31.5	3.8 12	0.0025
ce-K- 3	45.925 4	49.0 22	0.0479
ce-K- 4	75.306 10	0.290 13	0.0005
ce-L- 3	83.681 4	7.2 4	0.0128
ce-M- 3	89.460 7	1.53 7	0.0029
ce-NOP- 3	90.779 7	0.434 20	0.0008
ce-K- 17	485.8489 25	0.171 6	0.0018
β- 1 max	209.6 9		
avg	57.5 3	2.23 9	0.0027
β- 2 max	215.0 9		
avg	59.1 3	0.18 3	0.0002
β- 3 max	364.5 9		
avg	106.0 3	15.3 6	0.0345
β- 4 max	406.3 9		
avg	119.8 3	0.80 4	0.0020
β- 5 max	485.0 9		
avg	146.6 4	0.57 3	0.0018
β- 6 max	804.4 9		
avg	263.9 4	81 4	0.455
Total β-			
avg	233.0 5	100 4	0.497

(continued on next column)

148PM B- DECAY (5.37 D 1)   I(MIN)= 0.10%

Radiation Type	Energy (keV)	Intensity (%)	Δ(g-rad/μCi-h)
Auger-L	4.53	0.142 10	≈0
ce-K- 1	503.27 20	0.182 7	0.0019
β- 1 max	415 14		
avg	122 5	1.33 4	0.0035
β- 2 max	1007 14		
avg	343 6	33.3 8	0.243
β- 3 max	1922 14		
avg	732 7	9.5 3	0.148
β- 4 max	2472 14		
avg	978 7	55.6 11	1.16
Total β-			
avg	731 9	99.7 14	1.55
γ 1	550.10 20	22.0 6	0.258
γ 2	592.83 3	0.353 11	0.0045
γ 3	611.26 3	1.02 3	0.0133
γ 4	874.18 3	0.235 9	0.0044
γ 5	896.42 3	0.981 24	0.0187
γ 6	914.85 3	11.5 3	0.223
γ 7	1465.10 20	22.2 5	0.693

148PM IT DECAY (41.3 D 1)   I(MIN)=0.10%
%IT=4.6 5
SEE ALSO 148PM B- DECAY (41.3 D)

Radiation Type	Energy (keV)	Intensity (%)	Δ(g-rad/µCi-h)
Auger-L	4.38	5.3 5	0.0005
ce-K-2	30.52 10	3.0 5	0.0020
Auger-K	31.5	0.23 8	0.0002
ce-L-1	54.07 10	3.4 4	0.0039
ce-M-1	59.85 10	0.96 11	0.0012
ce-NOP-1	61.17 10	0.28 3	0.0004
ce-L-2	68.27 10	0.42 7	0.0006
X-ray L	5.43	1.02 16	0.0001
X-ray Kα₂	38.1712 5	0.79 13	0.0006
X-ray Kα₁	38.7247 5	1.44 23	0.0012
X-ray Kβ	43.8	0.55 9	0.0005
γ 2	75.70 10	1.01 16	0.0016

148PM B- DECAY (41.3 D 1)   I(MIN)= 0.10%
%B-=95.4 5
SEE ALSO 148PM IT DECAY (41.3 D)

Radiation Type	Energy (keV)	Intensity (%)	Δ(g-rad/µCi-h)
Auger-L	4.53	4.3 5	0.0004
Auger-K	32.6	0.44 15	0.0003
ce-K-1	51.65 3	3.7 5	0.0041
ce-K-2	142.80 3	0.21 4	0.0006
ce-K-3	241.28 3	0.81 17	0.0041

149ND B- DECAY (1.73 H 1)   I(MIN)= 0.10%

Radiation Type	Energy (keV)	Intensity (%)	Δ(g-rad/µCi-h)
Auger-L	4.38	34.5 22	0.0032
ce-K-2	13.699 20	1.41 9	0.0004
ce-L-1	22.57 3	4.8 5	0.0023
ce-M-1	28.35 3	1.10 11	0.0007
ce-K-4	29.15 3	3.7 6	0.0023
ce-K-5	29.48 10	0.49 16	0.0003
ce-NOP-1	29.67 3	0.36 4	0.0002
ce-K-6	30.56 4	0.9 4	0.0006
Auger-K	31.5	2.6 9	0.0018
ce-L-2	51.455 20	0.223 13	0.0002
ce-K-8	51.823 15	2.21 20	0.0024
ce-L-4	66.90 3	1.7 4	0.0025
ce-L-6	68.31 4	0.42 15	0.0006
ce-K-10	69.137 14	17.0 19	0.0251
ce-M-4	72.68 3	0.39 8	0.0006
ce-NOP-4	74.00 3	0.105 22	0.0002
ce-L-8	89.579 15	0.35 5	0.0007
ce-K-15	94.026 14	0.23 3	0.0005
ce-L-10	106.893 14	2.6 3	0.0058
ce-K-16	110.692 10	0.41 5	0.0010
ce-M-10	112.672 15	0.55 6	0.0013
ce-NOP-10	113.991 15	0.157 17	0.0004
ce-K-19	143.456 9	0.36 5	0.0011
ce-K-21	153.744 9	0.28 3	0.0009
ce-K-22	162.964 10	0.44 8	0.0015
ce-K-23	166.123 8	4.3 3	0.0153
ce-L-19	181.212 9	0.107 13	0.0004
ce-K-27	195.034 8	2.17 22	0.0090
ce-L-23	203.879 8	0.66 5	0.0029
ce-MNO-23	209.658 10	0.182 14	0.0008
ce-K-32	222.508 9	0.49 6	0.0023
ce-K-33	224.981 8	0.169 18	0.0008
ce-L-27	232.790 8	0.39 4	0.0019

ce-K-7	367.24	3	0.112	19	0.0009
ce-K-11	503.44	3	0.81	8	0.0087
ce-K-18	678.87	3	0.138	17	0.0020
β⁻ 1 max	416	14			
avg	123	5	54.3	10	0.142
β⁻ 2 max	514	14			
avg	157	5	18.6	4	0.0622
β⁻ 3 max	704	14			
avg	225	6	21.9	5	0.105
β⁻ 4 max	1015	14			
avg	346	6	0.92	22	0.0068
Total β⁻					
avg	155	6	95.7	12	0.316
X-ray L	5.64		0.89	14	0.0001
X-ray Kα₂	39.5224	3	1.60	15	0.0014
X-ray Kα₁	40.1181	3	2.9	3	0.0025
X-ray Kβ	45.4		1.13	11	0.0011
γ 1	98.48	3	2.48	7	0.0052
γ 2	189.63	3	1.11	4	0.0045
γ 3	288.11	3	12.59	23	0.0772
γ 5	311.63	3	3.92	8	0.0261
γ 6	362.09	3	0.178	18	0.0014
γ 7	414.07	3	18.7	4	0.165
γ 8	432.78	3	5.36	11	0.0494
γ 9	460.57	3	0.419	20	0.0041
γ 10	501.26	3	6.76	14	0.0722
γ 11	550.27	3	95.1	18	1.11
γ 12	553.24	3	0.40	4	0.0047
γ 13	571.95	3	0.214	10	0.0026
γ 14	599.74	3	12.57	25	0.161
γ 15	611.26	3	5.49	13	0.0715
γ 16	629.97	3	89.2	15	1.20
γ 18	725.70	3	32.9	4	0.509
γ 19	915.33	3	17.2	4	0.335
γ 20	1013.81	3	20.3	4	0.439

3 weak γ's omitted (ΣIγ = 0.19%)

β⁻ 1 max	376	4			
avg	109.7	14	0.154	10	0.0004
β⁻ 2 max	454	4			
avg	135.9	14	0.92	5	0.0027
β⁻ 3 max	1033	4			
avg	354.3	17	19.5	13	0.147
β⁻ 4 max	1150	4			
avg	402.0	17	24.1	18	0.206
β⁻ 5 max	1226	4			
avg	433.4	17	0.6	10	0.0055
β⁻ 6 max	1263	4			
avg	448.8	17	0.91	8	0.0087
β⁻ 7 max	1291	4			
avg	460.7	17	3.8	5	0.0373
β⁻ 8 max	1300	4			
avg	464.6	17	0.5	5	0.0049
β⁻ 9 max	1328	4			
avg	476.2	17	0.53	13	0.0054
β⁻ 10 max	1418	4			
avg	514.2	17	19.1	16	0.209
β⁻ 11 max	1477	4			
avg	539.4	18	26.5	21	0.304
β⁻ 12 max	1499	4			
avg	549.1	18	2.9	13	0.0339
β⁻ 13 max	1574	4			
avg	581.1	18	1		0.0136
Total β⁻					
avg	457.0	18	101	4	0.980

3 weak β's omitted (ΣIβ = 0.02%)

X-ray L	5.43		6.6	9	0.0008
X-ray Kα₂	38.1712	5	9.1	6	0.0074
X-ray Kα₁	38.7247	5	16.5	11	0.0136
X-ray Kβ	43.8		6.4	5	0.0059
γ 2	58.883	20	1.51	8	0.0019
γ 4	74.33	3	1.25	20	0.0020

(continued on next page)

149PM B- DECAY (53.08 H 5) (CONTINUED)

Radiation Type	Energy (keV)	Intensity (%)	Δ(g-rad/μCi-h)
β- 2 max	787 4		
avg	256.9 16	3.1 3	0.0170
β- 3 max	1073 4		
avg	369.8 17	96.5 3	0.760
Total β-			
avg	365.3 17	100.0 5	0.778

11 weak β's omitted (ΣI$_\beta$ = 0.25%)

| γ 11 | 285.950 10 | 2.85 25 | 0.0174 |

49 weak γ's omitted (ΣI$_\gamma$ = 0.40%)

152EU EC DECAY (13.4 Y 1)   I(MIN) = 0.10%

%(EC+B+)=72.10 23
SEE ALSO 152EU B- DECAY (13.4 Y)

Radiation Type	Energy (keV)	Intensity (%)	Δ(g-rad/μCi-h)
Auger-L	4.53	74 4	0.0071
Auger-K	32.6	5.7 19	0.0040
ce-K-1	74.9482 12	19.4 7	0.0310
ce-L-1	114.0456 12	10.6 4	0.0258
ce-M-1	120.0596 14	2.43 8	0.0062
ce-NOP-1	121.4367 15	0.667 22	0.0017
ce-K-9	197.8647 12	0.612 21	0.0026
ce-L-9	236.9621 12	0.157 6	0.0008

149ND B- DECAY (1.73 H 1) (CONTINUED)

Radiation Type	Energy (keV)	Intensity (%)	Δ(g-rad/μCi-h)
γ 5	74.66 10	1.0 3	0.0016
γ 6	75.74 4	0.33 11	0.0005
γ 8	97.007 15	1.52 13	0.0031
γ 10	114.321 14	18.7 20	0.0456
γ 11	116.93 3	0.117 23	0.0003
γ 12	122.416 14	0.231 24	0.0006
γ 13	126.630 19	0.114 12	0.0003
γ 15	139.210 14	0.48 5	0.0014
γ 16	155.876 10	6.0 6	0.0200
γ 17	177.831 19	0.163 13	0.0006
γ 19	188.640 19	1.99 24	0.0080
γ 20	192.027 9	0.59 6	0.0024
γ 21	198.928 9	1.45 16	0.0062
γ 22	208.148 10	2.9 4	0.0129
γ 23	211.307 8	27.2 19	0.122
γ 24	213.946 17	0.41 5	0.0019
γ 25	226.846 21	0.16 4	0.0008
γ 27	240.218 8	3.9 4	0.0202
γ 28	245.699 8	1.03 11	0.0054
γ 31	258.064 14	0.38 4	0.0021
γ 32	267.692 9	6.0 6	0.0344
γ 33	270.165 8	10.7 11	0.0615
γ 34	273.25 4	0.23 5	0.0013
γ 35	275.445 13	0.59 12	0.0035
γ 36	276.960 19	0.32 7	0.0019
γ 37	282.455 11	0.61 6	0.0037
γ 38	288.192 12	0.67 8	0.0041
γ 39	294.807 12	0.58 6	0.0036
γ 40	301.133 16	0.38 4	0.0024
γ 41	310.982 14	0.52 5	0.0034
γ 42	326.556 11	4.7 5	0.0324
γ 44	347.833 23	0.18 9	0.0014
γ 45	349.233 10	1.47 16	0.0110

γ	Energy (keV)	Intensity (%)	Δ(g-rad/μCi-h)
48	360.055 20	0.163 13	0.0013
49	366.637 15	0.66 8	0.0052
51	384.691 18	0.33 4	0.0027
53	423.554 10	9.4 10	0.0849
55	443.550 12	1.50 16	0.0142
63	538.15 7	0.11 4	0.0012
64	540.510 10	7.7 8	0.0883
65	556.43 5	1.2 5	0.0139
70	630.238 21	0.220 24	0.0030
71	635.482 25	0.112 12	0.0015
72	654.831 14	7.3 8	0.102
77	686.933 25	0.103 12	0.0015
78	696.266 25	0.171 18	0.0025
92	808.834 22	0.169 18	0.0029
107	923.876 25	0.114 12	0.0022
116	979.02 4	0.112 12	0.0023
119	1022.78 3	0.120 12	0.0026
139	1234.12 5	0.29 3	0.0077

100 weak γ's omitted (ΣIγ = 1.58%)

149PM B- DECAY (53.08 H 5) I(MIN)= 0.10%

Radiation Type	Energy (keV)	Intensity (%)	Δ(g-rad/μCi-h)
Auger-L	4.53	0.259 20	≈0
ce-L-1	14.782 8	0.134 12	≈0
ce-K-11	239.116 10	0.208 20	0.0011
β⁻ 1 max	191 4		
avg	52.0 12	0.125 12	0.0001

(continued on next column)

---

2 weak β's omitted (ΣIβ = 0.03%)

	Energy (keV)	Intensity (%)	Δ(g-rad/μCi-h)
X-ray L	5.64	15.1 19	0.0018
X-ray Kα₂	39.5224 3	21.0 7	0.0177
X-ray Kα₁	40.1181 3	38.0 11	0.0325
X-ray Kβ	45.4	14.8 5	0.0143
γ 1	121.7824 11	28.4 3	0.0736
γ 9	244.6989 11	7.52 11	0.0392
γ 14	295.939 8	0.445 9	0.0028
γ 17	329.433 17	0.148 3	0.0010
γ 26	416.052 6	0.110 3	0.0010
γ 29	443.983 7	0.326 15	0.0031
γ 30	444	2.797 21	0.0264
γ 32	488.66 4	0.422 4	0.0044
γ 39	564.021 8	0.505 9	0.0061
γ 41	566.421 8	0.133 13	0.0016
γ 45	656.484 12	0.148 11	0.0021
γ 48	674.678 3	0.195 11	0.0028
γ 51	688.678 6	0.873 9	0.0128
γ 52	719.353	0.285 13	0.0044
γ 59	810.459 7	0.324 9	0.0056
γ 62	841.586 8	0.174 5	0.0031
γ 64	867.388 8	4.224 25	0.0780
γ 67	919.401 8	0.432 13	0.0085
γ 68	926.324 15	0.287 13	0.0057
γ 70	964.131 9	14.43 7	0.296
γ 71	964.13	0.143 5	0.0029
γ 74	1005.279 17	0.640 6	0.0137
γ 75	1084.0 10	0.23 3	0.0054
γ 76	1085.914 13	10.01 5	0.231
γ 77	1112.116 17	13.44 7	0.318
γ 80	1212.950 12	1.423 12	0.0368
γ 81	1249.946 13	0.182 5	0.0048
γ 85	1408.011 14	20.78 9	0.623
γ 86	1457.628 15	0.497 7	0.0154
γ 87	1528.115 19	0.280 3	0.0091

61 weak γ's omitted (ΣIγ = 1.02%)

152EU B- DECAY (13.4 Y 1)   I(MIN)= 0.10%
%B-=27.90 23
SEE ALSO 152EU EC DECAY (13.44 Y)

Radiation Type	Energy (keV)	Intensity (%)	Δ(g-rad/μCi-h)
Auger-L	4.84	0.78 4	≈0
ce-K-6	294.036 4	0.83 3	0.0052
ce-L-6	335.899 4	0.181 6	0.0013
β- 1 max	178.2 18		
avg	48.2 6	1.810 20	0.0019
β- 2 max	387.6 18		
avg	113.3 6	2.45 3	0.0059
β- 3 max	698.4 18		
avg	222.7 7	13.77 13	0.0653
β- 4 max	712.4 18		
avg	227.9 7	0.290 20	0.0014
β- 5 max	891.0 18		
avg	296.2 7	0.310 20	0.0020
β- 6 max	1066.2 18		
avg	365.8 8	0.930 20	0.0072
β- 7 max	1477.3 18		
avg	536.6 8	8.19 16	0.0936
Total β- avg	298.4 11	27.95 22	0.178

5 weak β's omitted (ΣIβ = 0.20%)

Radiation Type	Energy (keV)	Intensity (%)	Δ(g-rad/μCi-h)
X-ray L	6	0.184 21	≈0
X-ray Kα2	42.3089 3	0.241 10	0.0002
X-ray Kα1	42.9962 3	0.435 17	0.0004
X-ray Kβ	48.7	0.172 7	0.0002
γ 6	344.275 4	26.6 3	0.195
γ 8	367.788 4	0.864 11	0.0068
γ 9	411.115 5	2.265 22	0.0198
γ 12	503.387 5	0.160 4	0.0017

152EU EC DECAY (9.274 H 3)   (CONTINUED)

Radiation Type	Energy (keV)	Intensity (%)	Δ(g-rad/μCi-h)
γ 11	562.920 10	0.226 25	0.0027
γ 17	841.63 6	14.6 16	0.262
γ 19	961.06 22	0.204 25	0.0042
γ 20	963.34 6	12.0 13	0.246

20 weak γ's omitted (ΣIγ = 0.36%)

152EU B- DECAY (9.274 H 3)   I(MIN)= 0.10%
%B-=72 3
SEE ALSO 152EU EC DECAY (9.274 H)

Radiation Type	Energy (keV)	Intensity (%)	Δ(g-rad/μCi-h)
β- 1 max	552.5 18		
avg	169.7 7	1.60 20	0.0058
β- 2 max	1522.9 18		
avg	556.1 8	1.70 20	0.0201
β- 3 max	1867.2 18		
avg	705.2 8	69 3	1.04
Total β- avg	688.8 9	72 3	1.06

3 weak β's omitted (ΣIβ = 0.15%)

γ 5	344.273 9	2.4 3	0.0179

γ	17	586.294	6	0.455	17	0.0057
γ	20	678.623		0.474	11	0.0068
γ	23	764.905	8	0.197	11	0.0032
γ	24	778.910	10	12.92	12	0.214
γ	30	1089.700	15	1.735	17	0.0403
γ	31	1109.180	12	0.208	11	0.0049
γ	34	1299.124	12	1.621	17	0.0449

25 weak γ's omitted (ΣIγ = 0.68%)

152EU EC DECAY (9.274 H 3)   I(MIN)= 0.10%
%(EC+B+)=28 3
SEE ALSO 152EU B- DECAY (9.274 H)

Radiation Type	Energy (keV)	Intensity (%)	Δ(g-rad/μCi-h)
Auger-L	4.53	25.6 23	0.0025
Auger-K	32.6	2.1 7	0.0014
ce-K-1	74.943 4	4.9 6	0.0079
ce-L-1	114.040 4	2.7 4	0.0065
ce-M-1	120.054 4	0.61 7	0.0016
ce-NOP-1	121.431 4	0.166 20	0.0004

2 weak β's omitted (ΣIβ = 0.02%)

X-ray	L	5.64	5.2 8	0.0006
X-ray	Kα₂	39.5224 3	7.5 8	0.0063
X-ray	Kα₁	40.1181 3	13.6 13	0.0117
X-ray	Kβ	45.4	5.3 5	0.0051
γ	1	121.777 4	7.2 8	0.0187

(continued on next column)

γ	14	970.38	3	0.60	7	0.0125
γ	16	1314.670	10	0.96	11	0.0268

16 weak γ's omitted (ΣIγ = 0.31%)

153SM B- DECAY (46.7 H 1)   I(MIN)= 0.10%

Radiation Type	Energy (keV)	Intensity (%)	Δ(g-rad/μCi-h)
Auger-L	4.69	54 3	0.0054
ce-L-1	11.768 20	0.32	≈0
ce-K-4	21.1530 21	23.5 14	0.0106
Auger-K	33.7	4.5 10	0.0032
ce-K-6	34.848 3	0.50 8	0.0004
ce-K-7	40.965 4	0.37 5	0.0003
ce-K-9	48.911 3	0.190 8	0.0002
ce-K-10	54.6600 21	40.8 15	0.0474
ce-L-4	61.6200 21	3.83 22	0.0050
ce-M-4	67.8720 21	0.83 5	0.0012
ce-NOP-4	69.3118 22	0.278 16	0.0004
ce-L-6	75.315 3	0.24 4	0.0004
ce-L-10	95.1270 21	6.17 23	0.0125
ce-MNO-10	101.3790 21	1.78 4	0.0038

β⁻ 1	max	637 3		
	avg	200.3 10	34.7 16	0.148
β⁻ 2	max	707 3		
	avg	226.1 10	43.5 23	0.209
β⁻ 3	max	712 3		
	avg	228.2 10	0.59 5	0.0029

(continued on next page)

153SM B- DECAY (46.7 H 1) (CONTINUED)

Radiation Type	Energy (keV)	Intensity (%)	Δ(g-rad/μCi-h)
β- 4 max	809.8 25		
avg	265.2 10	20.9 17	0.118
Total β-			
avg	224.9 10	100 4	0.479

11 weak β's omitted (ΣIβ = 0.25%)

X-ray L	5.85	11.9 14	0.0015
γ 1	19.820 20	0.415	0.0002
X-ray Kα₂	40.9019 3	17.3 7	0.0151
X-ray Kα₁	41.5422 3	31.3 12	0.0277
X-ray Kβ	47	12.2 5	0.0123
γ 4	69.6720 20	5.25 25	0.0078
γ 5	75.421 3	0.190 20	0.0003
γ 6	83.367 3	0.21 3	0.0004
γ 7	89.484 4	0.170 20	0.0003
γ 9	97.430 3	0.730 20	0.0015
γ 10	103.1790 20	28.3 6	0.0622

58 weak γ's omitted (ΣIγ = 0.31%)

153GD EC DECAY (242 D 1)   I(MIN)= 0.10%

Radiation Type	Energy (keV)	Intensity (%)	Δ(g-rad/μCi-h)
Auger-L	4.69	110 5	0.0109

154EU B- DECAY (8.55 Y 6)   I(MIN)= 0.10%

Radiation Type	Energy (keV)	Intensity (%)	Δ(g-rad/μCi-h)
Auger-L	4.84	32.4 15	0.0033
Auger-K	34.9	1.8 6	0.0013
ce-K-1	72.831 4	26.7 11	0.0414
ce-L-1	114.694 4	16.7 7	0.0408
ce-M-1	121.189 4	3.87 15	0.0100
ce-NOP-1	122.694 4	1.09 5	0.0028
ce-K-5	197.700 8	0.542 21	0.0023
ce-L-5	239.563 8	0.154 6	0.0008
β- 1 max	248.7 16		
avg	69.1 5	29.0 5	0.0427
β- 2 max	307.7 16		
avg	87.4 5	0.780 20	0.0015
β- 3 max	322.7 16		
avg	92.2 5	0.15 4	0.0003
β- 4 max	350.7 16		
avg	101.2 6	1.61 4	0.0035
β- 5 max	408.7 16		
avg	120.2 6	0.15 3	0.0004
β- 6 max	436.7 16		
avg	129.6 6	0.228 14	0.0006
β- 7 max	570.7 16		
avg	176.2 6	36.5 7	0.137
β- 8 max	704.7 16		
avg	225.0 6	0.738 17	0.0035
β- 9 max	716.7 16		
avg	229.5 6	0.318 16	0.0016
β- 10 max	840.7 16		
avg	276.6 7	18.0 4	0.106
β- 11 max	920.7 16		
avg	307.8 7	0.10 4	0.0007
β- 12 max	971.7 16		
avg	327.9 7	2.56 12	0.0179

ce-L-1	6.0180 4	0.146 5	≈0
ce-L-2	11.768 20	0.298 9	≈0
ce-K-3	21.1544 5	10.8 7	0.0049
Auger-K	33.7	9.0 20	0.0065
ce-K-5	34.8486 5	0.49 6	0.0004
ce-K-6	40.9675 5	0.15 5	0.0001
ce-K-7	48.9126 5	7.6 4	0.0079
ce-K-8	54.6617 5	30.4 14	0.0354
ce-L-3	61.6214 5	1.77 10	0.0023
ce-MNO-3	67.8734 6	0.508 25	0.0007
ce-L-5	75.3156 5	0.23 3	0.0004
ce-L-7	89.3796 5	1.13 5	0.0022
ce-L-8	95.1287 5	4.60 22	0.0093
ce-MNO-7	95.6316 6	0.322 10	0.0007
ce-MNO-8	101.3807 6	1.33 5	0.0029
X-ray L	5.85	24 3	0.0030
X-ray Ka₂	40.9019 3	34.6 12	0.0301
X-ray Ka₁	41.5422 3	62.5 20	0.0553
X-ray Kβ	47	24.5 9	0.0245
γ 3	69.67340 20	2.42 12	0.0036
γ 5	83.3676 3	0.206 22	0.0004
γ 7	97.4316 3	29.5 9	0.0612
γ 8	103.1807 3	21.1 8	0.0464

5 weak γ's omitted (ΣIγ = 0.21%)

β⁻13 max	1152.7 16		
avg	400.9 7	0.31 3	0.0026
β⁻14 max	1844.7 16		
avg	695.3 7	9.6 16	0.142
Total β⁻			
avg	216.1 9	100.0 19	0.460
X-ray L	6	7.6 9	0.0010
X-ray Ka₂	42.3089 3	7.2 4	0.0065
X-ray Ka₁	42.9962 3	13.0 6	0.0119
X-ray Kβ	48.7	5.15 25	0.0053
γ 1	123.070 4	40.4 10	0.106
γ 3	188.22 5	0.237 9	0.0010
γ 5	247.939 8	6.69 16	0.0353
γ 8	401.30 5	0.173 15	0.0015
γ 9	444.39 4	0.581 15	0.0055
γ 11	478.26 5	0.222 11	0.0023
γ 14	557.56 5	0.268 10	0.0032
γ 15	582.00 5	0.924 19	0.0115
γ 16	591.76 3	5.08 9	0.0640
γ 18	625.22 5	0.328 10	0.0044
γ 20	676.59 5	0.166 18	0.0024
γ 21	692.42 4	1.84 4	0.0271
γ 22	715.76 5	0.170 8	0.0026
γ 23	723.30 3	20.6 4	0.317
γ 24	756.86 9	4.67 9	0.0752
γ 25	815.55 20	0.535 20	0.0093
γ 26	845.39 13	0.597 13	0.0108
γ 27	850.64 10	0.245 10	0.0044
γ 28	873.20 3	12.45 22	0.232
γ 30	892.73 13	0.530 13	0.0101
γ 31	904.05 19	0.927 19	0.0179
γ 33	996.30 3	10.65 19	0.226
γ 34	1004.76 4	18.4 4	0.394
γ 35	1118.50 10	0.103 8	0.0024
γ 36	1128.40 10	0.319 16	0.0077
γ 37	1140.90 10	0.238 7	0.0058

(continued on next page)

157DY EC DECAY (8.1 H 1) I(MIN)= 0.10%

Radiation Type	Energy (keV)	Intensity (%)	Δ(g-rad/μCi-h)
Auger-L	5	71 4	0.0076
ce-K-1	8.82 7	2.1 9	0.0004
ce-K-2	31.01 4	1.7 3	0.0011
Auger-K	36	5.5 20	0.0043
ce-L-1	52.11 7	0.35 14	0.0004
ce-L-2	74.30 4	0.25 4	0.0004
ce-K-7	274.16 20	1.09 4	0.0064
ce-L-7	317.45 20	0.151 5	0.0010
X-ray L	6.27	17.7 20	0.0024
X-ray Kα₂	43.7441 3	23.5 8	0.0219
X-ray Kα₁	44.4816 3	42.2 13	0.0400
X-ray Kβ	50.4	16.8 6	0.0180
γ 1	60.82 7	0.27 11	0.0003
γ 2	83.01 4	0.53 8	0.0009
γ 4	182.20 20	1.95 13	0.0076
γ 5	265.340 20	0.19 4	0.0011
γ 7	326.16 20	95.2 12	0.661

21 weak γ's omitted (ΣI_γ = 0.48%)

154EU B- DECAY (8.55 Y 6) (CONTINUED)

Radiation Type	Energy (keV)	Intensity (%)	Δ(g-rad/μCi-h)
γ 40	1241.60 20	0.135 18	0.0036
γ 41	1246.20 20	0.881 21	0.0234
γ 42	1274.52 6	35.4 6	0.961
γ 44	1494.08 7	0.728 14	0.0232
γ 47	1596.45 7	1.86 4	0.0632

16 weak γ's omitted (ΣI_γ = 0.84%)

155EU B- DECAY (4.72 Y 2) I(MIN)= 0.10%

Radiation Type	Energy (keV)	Intensity (%)	Δ(g-rad/μCi-h)
ce-L-1	2.024 20	1.21 14	≈0
Auger-L	4.84	34 3	0.0035
ce-M-1	8.519 20	0.26 3	≈0
ce-K-9	9.7709 21	8.5 9	0.0018
ce-L-3	10.40 4	14 3	0.0031
ce-L-4	12.644 20	0.95 8	0.0003
ce-M-3	16.90 4	3.1 7	0.0011
ce-L-5	18.137 21	0.49 4	0.0002
ce-NOP-3	18.40 4	1.03 20	0.0004
ce-M-4	19.139 20	0.220 19	≈0
ce-L-6	23.05 5	0.40 14	0.0002
ce-M-5	24.632 21	0.108 9	≈0
Auger-K	34.9	1.6 6	0.0012
ce-K-10	35.823 5	0.40 5	0.0003
ce-K-11	36.3039 21	11.2 8	0.0087

157TB EC DECAY (99 Y 10)   I(MIN)= 0.10%

Radiation Type	Energy (keV)	Intensity (%)	Δ(g-rad/μCi-h)
Auger-L	4.84	60.9 19	0.0063
Auger-K	34.9	1.4 5	0.0011
X-ray L	6	14.3 16	0.0018
X-ray Kα2	42.3089 3	5.7 3	0.0051
X-ray Kα1	42.9962 3	10.3 5	0.0094
X-ray Kβ	48.7	4.06 21	0.0042

166HO B- DECAY (1200 Y 180)   I(MIN)= 0.10%

Radiation Type	Energy (keV)	Intensity (%)	Δ(g-rad/μCi-h)
Auger-L	5.5	70 3	0.0083
ce-K-1	23.104 5	21.7 9	0.0107
ce-K-2	37.16 3	0.46 3	0.0004
Auger-K	39.7	2.2 9	0.0018
ce-K-3	61.55 3	0.18 7	0.0002
ce-K-4	63.67 3	0.166 9	0.0002
ce-L-1	70.838 5	50.4 20	0.0760
ce-M-1	78.382 5	12.2 5	0.0204
ce-NOP-1	80.140 5	3.32 13	0.0057
ce-L-4	111.41 3	0.149 8	0.0004
ce-K-9	126.930 6	14.9 5	0.0403
ce-K-11	158.39 3	0.336 12	0.0011
ce-L-7	36.9224 21	0.44 4	0.0003
ce-L-9	51.6344 21	1.71 18	0.0019
ce-K-12	55.069 3	4.4 4	0.0052
ce-M-9	58.1292 21	0.38 4	0.0005
ce-NOP-9	59.6342 22	0.107 12	0.0001
ce-L-11	78.1674 21	1.72 13	0.0029
ce-M-11	84.6622 21	0.37 3	0.0007
ce-NOP-11	86.1672 22	0.105 8	0.0002
ce-L-12	96.932 3	0.66 5	0.0014
ce-M-12	103.427 3	0.142 11	0.0003
β- 1 max	105.3 16		
avg	27.6 5	0.69 8	0.0004
β- 2 max	133.4 16		
avg	35.4 5	2.2 3	0.0017
β- 3 max	146.1 16		
avg	39.0 5	46 5	0.0382
β- 4 max	164.9 16		
avg	44.3 5	26 4	0.0245
β- 5 max	191.4 16		
avg	52.0 5	8.3 11	0.0092
β- 6 max	251.4 16		
avg	70.0 5	18 5	0.0268
Total β- avg	46.8 6	101 9	0.101
X-ray L	6	8.0 11	0.0010
γ 5	26.513 21	0.318 24	0.0002
X-ray Kα2	42.3089 3	6.5 4	0.0059
X-ray Kα1	42.9962 3	11.8 7	0.0108
γ 7	45.2980 20	1.28 10	0.0012
X-ray Kβ	48.7	4.6 3	0.0048
γ 9	60.0100 20	1.14 12	0.0015
γ 10	86.062 5	0.151 19	0.0003
γ 11	86.5430 20	30.9 20	0.0570
γ 12	105.308 3	20.6 14	0.0462

7 weak γ's omitted (ΣI_γ = 0.18%)

(continued on next page)

166HO B- DECAY (1200 Y 180) (CONTINUED)

Radiation Type	Energy (keV)	Intensity (%)	Δ(g-rad/μCi-h)
ce-L-9	174.664 6	6.99 23	0.0260
ce-M-9	182.208 6	1.67 6	0.0065
ce-NOP-9	183.966 6	0.461 15	0.0018
ce-K-14	222.964 8	1.81 6	0.0086
ce-K-15	243.258 20	0.187 6	0.0010
ce-L-14	270.699 8	0.549 18	0.0032
ce-M-14	278.243 8	0.128 4	0.0008
β^- 1 max	1314 3		
avg		3.31 16	≈0
β^- 2 max	948 3		
avg		1.19 3	≈0
β^- 3 max	643 3		
avg		2.07 7	≈0
β^- 4 max	483 3		
avg		0.62 10	≈0
β^- 5 max	304 3		
avg		0.408 15	≈0
β^- 6 max	72 3		
avg		74.2 7	≈0
β^- 7 max	32 3		
avg		17.42 17	≈0
Total β^- avg		99.2 8	
X-ray L	7	21.0 20	0.0031
X-ray Kα_2	48.2211 4	10.7 4	0.0110
X-ray Kα_1	49.1277 4	19.1 7	0.0199
X-ray Kβ	55.7	7.8 3	0.0092
γ 1	80.589 5	12.8 3	0.0219
γ 2	94.65 3	0.162 9	0.0003
γ 3	119.04 3	0.163 7	0.0004
γ 4	121.16 3	0.247 10	0.0006

169YB EC DECAY (32.03 D 1)   I(MIN)= 0.10%

Radiation Type	Energy (keV)	Intensity (%)	Δ(g-rad/μCi-h)
ce-K-3	3.7312 6	39.9 16	0.0032
Auger-L	5.67	160 9	0.0193
ce-M-1	6.1031 13	71 3	0.0093
ce-L-2	10.6280 9	8.5 7	0.0019
ce-M-2	18.4369 11	1.89 14	0.0007
ce-K-4	34.2255 7	8.38 20	0.0061
Auger-K	40.9	10 4	0.0089
ce-K-5	50.3906 6	34.9 7	0.0374
ce-L-3	53.0051 5	7.1 3	0.0080
ce-K-7	58.8005 12	1.33 5	0.0017
ce-M-3	60.8140 8	1.58 7	0.0021
ce-NOP-3	62.6491 10	0.429 17	0.0006
ce-K-8	71.1343 7	6.1 3	0.0092
ce-L-4	83.4994 6	1.44 4	0.0026
ce-M-4	91.3083 8	0.324 9	0.0006
ce-L-5	99.6645 5	5.62 11	0.0119
ce-M-5	107.4734 8	1.258 23	0.0029
ce-L-7	108.0744 11	1.37 5	0.0032
ce-NOP-5	109.3085 10	0.366 7	0.0009
ce-M-7	115.8833 13	0.332 12	0.0008
ce-K-10	117.8248 7	10.30 18	0.0259
ce-L-8	120.4082 6	5.2 3	0.0133
ce-M-8	128.2171 8	1.25 6	0.0034
ce-NOP-8	130.0522 10	0.349 17	0.0010
ce-K-11	138.5685 8	12.80 9	0.0378
ce-L-10	167.0987 7	1.84 3	0.0066
ce-M-10	174.9076 9	0.416 7	0.0016
ce-NOP-10	176.7427 11	0.119 4	0.0004
ce-L-11	187.8424 8	2.090 9	0.0084
ce-M-11	195.6513 10	0.468 3	0.0020
ce-NOP-11	197.4864 11	0.134 4	0.0006
ce-K-17	248.3486 10	0.523 17	0.0028
ce-L-17	297.6225 9	0.151 5	0.0010

γ 9	184.415 6	72.1 9	0.283
γ 10	190.711 25	0.221 9	0.0009
γ 11	215.88 3	2.57 5	0.0118
γ 12	231.28 4	0.208 8	0.0010
γ 13	259.716 20	1.073 20	0.0059
γ 14	280.450 20	29.7 3	0.177
γ 15	300.744 20	3.71 4	0.0238
γ 16	339.78 8	0.170 9	0.0012
γ 17	365.777 16	2.47 3	0.0192
γ 19	410.941 25	11.34 13	0.0992
γ 20	451.524 25	2.99 4	0.0287
γ 21	464.82 4	1.217 21	0.0121
γ 22	529.81 3	9.60 12	0.108
γ 23	571.00 3	5.50 7	0.0669
γ 24	594.37 3	0.577 18	0.0073
γ 25	611.52 7	1.437 22	0.0187
γ 27	644.45 10	0.158 11	0.0022
γ 28	670.51 4	5.44 6	0.0777
γ 29	691.21 5	1.343 15	0.0198
γ 30	711.69 4	54.9 5	0.832
γ 31	736.67 8	0.375 12	0.0059
γ 32	752.26 4	12.19 12	0.195
γ 33	778.82 4	3.06 3	0.0508
γ 34	810.31 4	57.7 6	0.996
γ 35	830.56 4	9.74 10	0.172
γ 36	875.64 5	0.729 9	0.0136
γ 37	950.94 6	2.73 3	0.0553
γ 39	1120.31 7	0.231 9	0.0055
γ 40	1146.82 7	0.203 3	0.0049
γ 41	1241.44 6	0.811 12	0.0215
γ 42	1282.12 7	0.178 4	0.0049
γ 43	1400.72 8	0.496 7	0.0148
γ 44	1427.05 8	0.493 9	0.0150

7 weak γ's omitted (ΣIγ = 0.51%)

X-ray L	7.18	51 7	0.0077
γ 1	8.4099 11	0.332 11	≈0
γ 2	20.7437 8	0.192 14	≈0
X-ray Kα₂	49.7726 4	53.0 17	0.0562
X-ray Kα₁	50.7416 4	94 3	0.101
X-ray Kβ	57.5	38.5 13	0.0471
γ 3	63.12080 20	43.7 12	0.0588
γ 4	93.6151 4	2.66 7	0.0053
γ 5	109.7802 3	17.4 4	0.0406
γ 7	118.1901 10	1.88 4	0.0047
γ 8	130.5239 4	11.1 5	0.0309
γ 10	177.2144 5	21.5 4	0.0810
γ 11	197.9581 6	34.94 5	0.147
γ 12	240.30 10	0.128 7	0.0007
γ 13	261.0788 7	1.897 25	0.0106
γ 17	307.7382 8	10.80 14	0.0708

23 weak γ's omitted (ΣIγ = 0.20%)

170TM B- DECAY (128.6 D 3)   I(MIN) = 0.10%
%B-=99.854 2

Radiation Type	Energy (keV)	Intensity (%)	Δ(g-rad/μCi-h)
Auger-L	5.84	12.1 8	0.0015
ce-K-1	22.9229 6	4.7 3	0.0023
Auger-K	42.2	0.23 10	0.0002
ce-L-1	73.7688 5	12.2 7	0.0192
ce-M-1	81.8571 5	3.00 18	0.0052
ce-NOP-1	83.7680 7	0.83 5	0.0015

(continued on next page)

170TM B- DECAY (128.6 D 3) (CONTINUED)

Radiation Type	Energy (keV)	Intensity (%)	Δ(g-rad/μCi-h)
β⁻ 1 max	883.7 10		
avg	290.5 4	24.0 10	0.149
β⁻ 2 max	968.0 10		
avg	323.1 4	76.0 10	0.523
Total β⁻			
avg	315.3 4	100.0 15	0.672
X-ray L	7.42	4.0 6	0.0006
X-ray $K\alpha_2$	51.3540 5	1.27 8	0.0014
X-ray $K\alpha_1$	52.3889 5	2.25 14	0.0025
X-ray $K\beta$	59.4	0.93 6	0.0012
γ 1	84.25523 23	3.26 16	0.0058

171ER B- DECAY (7.52 H 3) (CONTINUED)

Radiation Type	Energy (keV)	Intensity (%)	Δ(g-rad/μCi-h)
γ 6	124.017 4	9.1 4	0.0240
γ 10	210.60 3	0.64 3	0.0029
γ 11	237.14 4	0.302 14	0.0015
γ 13	277.43 5	0.58 3	0.0034
γ 15	295.901 14	28.9 12	0.182
γ 16	308.291 18	64 3	0.423
γ 18	371.96 9	0.257 13	0.0020
γ 34	670.70 20	0.252 9	0.0036
γ 36	676.1 3	0.285 11	0.0041
γ 42	784.10 20	0.240 9	0.0040
γ 43	796.60 20	0.640 24	0.0109
γ 48	907.7 4	0.635 23	0.0123

51 weak γ's omitted ($\Sigma I_\gamma = 1.09\%$)

171ER B- DECAY (7.52 H 3)   I(MIN)= 0.10%

Radiation Type	Energy (keV)	Intensity (%)	Δ(g-rad/μCi-h)
ce-L-2	2.269 8	6.3 10	0.0003
ce-M-1	2.718 6	70 3	0.0040
ce-NOP-1	4.553 6	23.3 11	0.0023
Auger-L	5.67	46 3	0.0055
ce-M-2	10.078 8	1.39 22	0.0003
ce-NOP-2	11.913 8	0.47 7	0.0001
Auger-K	40.9	2.5 10	0.0022
ce-K-4	52.231 4	39.2 23	0.0436
ce-K-5	57.266 6	1.69 9	0.0021

171TM B- DECAY (1.92 Y 1)   I(MIN)= 0.10%

Radiation Type	Energy (keV)	Intensity (%)	Δ(g-rad/μCi-h)
ce-K-1	5.386 7	1.02 20	0.0001
Auger-L	5.84	1.17 17	0.0001
ce-L-1	56.232 7	0.71 14	0.0008
ce-MNO-1	64.320 7	0.22 5	0.0003

Radiation Type	Energy (keV)	Intensity (%)	Δ(g-rad/µCi-h)
ce-K-6	64.627 4	5.7 3	0.0078
ce-L-4	101.505 4	6.4 4	0.0137
ce-L-5	106.540 6	1.78 9	0.0040
ce-M-4	109.314 4	1.41 9	0.0033
ce-NOP-4	111.149 4	0.410 24	0.0010
ce-L-6	113.901 4	5.3 3	0.0129
ce-M-5	114.349 6	0.430 22	0.0010
ce-NOP-5	116.184 6	0.120 6	0.0003
ce-M-6	121.710 4	1.29 7	0.0033
ce-NOP-6	123.545 4	0.364 20	0.0010
ce-K-15	236.511 14	0.60 3	0.0030
ce-K-16	248.901 18	1.04 5	0.0055
ce-L-16	298.175 18	0.149 7	0.0009
β^- 1 max	205.0 13		0.0004
avg	55.9 4	0.334 12	
β^- 2 max	491.4 13		0.0016
avg	147.5 5	0.52 3	
β^- 3 max	577.0 13		0.0080
avg	177.3 5	2.12 11	
β^- 4 max	814.1 13		0.0010
avg	264.3 5	0.177 18	
β^- 5 max	1065.1 13		0.733
avg	362.1 6	95 5	
β^- 6 max	1485.0 13		0.0262
avg	534.6 6	2.30 20	
Total β^- avg	359.5 6	101 5	0.771

6 weak β's omitted ($\Sigma I\beta$ = 0.18%)

Radiation Type	Energy (keV)	Intensity (%)	Δ(g-rad/µCi-h)
X-ray L	7.18	14.4 20	0.0022
X-ray $K\alpha_2$	49.7726 4	13.1 7	0.0139
X-ray $K\alpha_1$	50.7416 4	23.1 13	0.0250
X-ray $K\beta$	57.5	9.5 6	0.0116
γ 4	111.621 4	20.5 10	0.0487
γ 5	116.656 6	2.30 10	0.0057

(continued on next column)

Radiation Type	Energy (keV)	Intensity (%)	Δ(g-rad/µCi-h)
β^- 1 max	29.8 10		0.0003
avg	7.5 3	2.1 4	
β^- 2 max	96.5 10		0.0523
avg	25.1 3	97.9 4	
Total β^- avg	24.7 4	100.0 6	0.0527
X-ray L	7.42	0.39 8	≈0
X-ray $K\alpha_2$	51.3540 5	0.28 6	0.0003
X-ray $K\alpha_1$	52.3889 5	0.49 10	0.0005
X-ray $K\beta$	59.4	0.20 4	0.0003
γ 1	66.718 7	0.15 3	0.0002

175YB B- DECAY (4.19 D 1)   I(MIN)= 0.10%

Radiation Type	Energy (keV)	Intensity (%)	Δ(g-rad/µCi-h)
Auger-L	6	3.1 4	0.0004
Auger-K	43.5	0.19 9	0.0002
ce-K-1	50.489 4	3.7 5	0.0039
ce-K-2	74.342 6	0.131 19	0.0002
ce-L-1	102.933 4	0.88 12	0.0019
ce-M-1	111.312 4	0.20 3	0.0005
ce-K-6	333.008 20	0.24 4	0.0017
β^- 1 max	71.8 15		0.0041
avg	18.5 4	10.3 4	
β^- 2 max	216.6 15		
avg	68.3 5	0.80 12	0.0012
β^- 3 max	354.3 15		

(continued on next page)

178TA EC DECAY (2.45 H 5)   I(MIN)= 0.10%

Radiation Type	Energy (keV)	Intensity (%)	Δ(g-rad/μCi-h)
Auger-L	6.18	142 11	0.0187
ce-K-1	23.48 5	27 3	0.0134
ce-K-2	27.828 6	18.4 20	0.0109
Auger-K	44.8	7 3	0.0064
ce-L-1	77.56 5	4.6 6	0.0077
ce-L-2	81.908 6	46 5	0.0802
ce-M-1	86.23 5	1.05 12	0.0019
ce-NOP-1	88.29 5	0.30 4	0.0006
ce-M-2	90.578 6	11.4 13	0.0220
ce-NOP-2	92.641 6	3.3 4	0.0064
ce-K-3	148.083 6	10.7 12	0.0339
ce-L-3	202.163 6	5.4 6	0.0233
ce-M-3	210.833 6	1.32 15	0.0059
ce-NOP-3	212.896 6	0.37 4	0.0017
ce-K-4	260.203 8	4.0 5	0.0221
ce-K-5	266.31 5	3.7 5	0.0211
ce-L-4	314.283 8	1.25 13	0.0084
ce-L-5	320.39 5	0.57 7	0.0039
ce-M-4	322.953 8	0.30 4	0.0021
ce-M-5	329.06 5	0.128 16	0.0009
ce-K-6	361.004 10	2.13 23	0.0164
ce-L-6	415.084 10	0.53 6	0.0047
ce-M-6	423.754 10	0.126 14	0.0011
X-ray L	7.9	53 7	0.0088
X-ray Kα2	54.6114 8	40 4	0.0465
X-ray Kα1	55.7902 8	70 6	0.0832
X-ray Kβ	63.2	29.4 24	0.0395
γ 1	88.83 5	67 8	0.126
γ 2	93.179 6	16.6 17	0.0330
γ 3	213.434 6	76 8	0.348
γ 4	325.554 8	90 9	0.624
γ 5	331.66 5	31 4	0.216
γ 6	426.355 10	96 10	0.875

175YB B- DECAY (4.19 D 1)   (CONTINUED)

Radiation Type	Energy (keV)	Intensity (%)	Δ(g-rad/μCi-h)
β⁻ 4 avg	101.8 5	3.30 20	0.0072
max	468.1 15		
avg	139.2 5	86.5 17	0.256
Total β⁻ avg	125.1 7	100.9 18	0.269
X-ray L	7.66	1.10 18	0.0002
X-ray Kα2	52.9650 5	1.10 14	0.0012
X-ray Kα1	54.0698 5	1.93 24	0.0022
X-ray Kβ	61.3	0.80 10	0.0010
γ 1	113.803 4	1.91 25	0.0046
γ 2	137.656 6	0.117 16	0.0003
γ 3	144.861 5	0.33 5	0.0010
γ 5	282.517 14	3.1 4	0.0184
γ 6	396.322 20	6.5 8	0.0549

1 weak γ's omitted (ΣIγ = 0.08%)

177LU B- DECAY (6.71 D 1)   I(MIN)= 0.10%

Radiation Type	Energy (keV)	Intensity (%)	Δ(g-rad/μCi-h)
Auger-L	6.18	9.0 5	0.0012
ce-K-1	6.2952 21	0.114 6	≈0
Auger-K	44.8	0.28 12	0.0003
ce-K-2	47.601 3	5.3 3	0.0054

Radiation Type	Energy (keV)	Intensity (%)	Δ(g-rad/μCi-h)
ce-L-2	101.681 3	7.1 3	0.0155
ce-M-2	110.351 3	1.77 8	0.0042
ce-NOP-2	112.414 3	0.507 21	0.0012
ce-K-4	143.008 10	0.59 7	0.0018
ce-L-4	197.088 10	0.100 17	0.0004
β⁻ 1 max	176.0 10		
avg	47.3 3	12.2 7	0.0123
β⁻ 2 max	384.3 10		
avg	111.3 4	9.5 10	0.0225
β⁻ 3 max	497.3 10		
avg	148.9 4	78.2 8	0.248
Total β⁻ avg	132.9 5	99.9 15	0.283

1 weak β's omitted (ΣIβ = 0.05%)

Radiation Type	Energy (keV)	Intensity (%)	Δ(g-rad/μCi-h)
X-ray L	7.9	3.3 4	0.0006
X-ray Kα₂	54.6114 8	1.66 9	0.0019
X-ray Kα₁	55.7902 8	2.91 15	0.0035
X-ray Kβ	63.2	1.22 7	0.0016
γ 1	71.6460 20	0.159 5	0.0002
γ 2	112.952 3	6.50 20	0.0156
γ 4	208.359 10	11.0 4	0.0488
γ 5	249.686 25	0.208 9	0.0011
γ 6	321.313 9	0.227 9	0.0016

1 weak γ's omitted (ΣIγ = 0.06%)

180TA EC DECAY (8.152 H 6)   I(MIN)= 0.10%
%EC=81.8 9
SEE ALSO 180TA B- DECAY (8.15 H)

Radiation Type	Energy (keV)	Intensity (%)	Δ(g-rad/μCi-h)
Auger-L	6.18	60 4	0.0079
ce-K-1	27.9732 21	4.69 14	0.0028
Auger-K	44.8	3.3 14	0.0031
ce-L-1	82.0533 21	11.64 22	0.0203
ce-M-1	90.7231 21	2.89 8	0.0056
ce-NOP-1	92.7859 21	0.83 3	0.0016
X-ray L	7.9	22 3	0.0037
X-ray Kα₂	54.6114 8	19.4 6	0.0226
X-ray Kα₁	55.7902 8	34.0 9	0.0404
X-ray Kβ	63.2	14.3 5	0.0192
γ 1	93.3240 20	4.28 5	0.0085

180TA B- DECAY (8.152 H 6)   I(MIN)= 0.10%
%B-=18.2 9
SEE ALSO 180TA EC DECAY (8.15 H)

Radiation Type	Energy (keV)	Intensity (%)	Δ(g-rad/μCi-h)
Auger-L	6.53	1.43 9	0.0002
ce-K-1	34.04 9	0.65 4	0.0005
ce-L-1	91.46 9	1.52 8	0.0030

(continued on next page)

181W EC DECAY (121.2 D 3)   I(MIN)= 0.10%

Radiation Type	Energy (keV)	Intensity (%)	Δ(g-rad/μCi-h)
ce-M-1	3.50 3	68 16	0.0051
ce-NOP-1	5.64 3	22 6	0.0027
Auger-L	6.35	58 10	0.0078
Auger-K	46.2	3.0 15	0.0030
γ 1	6.21 3	1.05 3	0.0001
X-ray L	8.15	22 5	0.0039
X-ray Kα2	56.2770 10	19 4	0.0225
X-ray Kα1	57.5320 10	33 7	0.0401
X-ray Kβ	65.2	14 3	0.0192

2 weak γ's omitted (ΣIγ = 0.11%)

182TA B- DECAY (114.43 D 4)   I(MIN)= 0.10%

Radiation Type	Energy (keV)	Intensity (%)	Δ(g-rad/μCi-h)
Auger-L	6.53	58 3	0.0081
ce-K-5	15.1558 5	15.2 6	0.0049
ce-L-1	19.6381 7	1.17 7	0.0005
ce-M-1	28.9183 8	0.269 15	0.0002
ce-K-6	30.5815 5	12.6 5	0.0082
ce-L-2	30.6153 5	0.156 7	0.0001
ce-K-8	44.1473 5	5.01 19	0.0047
Auger-K	45.7	1.6 7	0.0015
ce-L-3	53.6227 5	7.0 3	0.0080

180TA B- DECAY (8.152 H 6)   (CONTINUED)

Radiation Type	Energy (keV)	Intensity (%)	Δ(g-rad/μCi-h)
ce-M-1	100.74 9	0.381 22	0.0008
ce-NOP-1	102.96 9	0.112 7	0.0002
β⁻ 1 max	604 4		
avg	185.8 15	3.50 10	0.0139
β⁻ 2 max	708 4		
avg	223.0 15	14.7 9	0.0698
Total β⁻			
avg	215.8 15	18.2 9	0.0837
X-ray L	8.4	0.61 7	0.0001
X-ray Kα2	57.9817 5	0.178 11	0.0002
X-ray Kα1	59.31820 10	0.310 19	0.0004
X-ray Kβ	67.2	0.132 9	0.0002
γ 1	103.56 9	0.79 4	0.0017

181HF B- DECAY (42.4 D 1)   I(MIN)= 0.10%

Radiation Type	Energy (keV)	Intensity (%)	Δ(g-rad/μCi-h)
ce-M-1	1.19 10	2.64 14	≈0
ce-NOP-1	3.33 10	1.36 7	≈0
ce-NOP-2	5.64 3	0.120 12	
Auger-L	6.35	40.3 24	0.0055
Auger-K	46.2	1.5 7	0.0015
ce-K-3	65.604 20	21.4 13	0.0299

ce-K-4	68.834 20	8.5 6	0.0124
ce-K-5	69.44 4	2.5 4	0.0037
ce-L-3	121.338 20	25.0 15	0.0647
ce-L-4	124.568 20	1.71 12	0.0045
ce-L-5	125.18 4	0.49 10	0.0013
ce-M-3	130.312 20	6.2 4	0.0173
ce-NOP-3	132.454 20	1.81 11	0.0051
ce-M-5	133.542 20	0.40 3	0.0011
ce-NOP-5	134.15 4	0.113 24	0.0003
ce-NOP-4	135.684 20	0.118 8	0.0003
ce-K-6	278.43 20	0.54 4	0.0032
ce-L-6	334.17 20	0.43 10	0.0012
ce-K-8	414.58 20	1.54 10	0.0136
ce-L-8	470.32 20	0.353 21	0.0035
β^- 1 max	408 3		
avg	119.0 10	7 3	0.0177
β^- 2 max	412 3		
avg	120.3 10	93 3	0.238
Total β^- avg	120.2 10	100 5	0.256
X-ray L	8.15	15.7 18	0.0027
X-ray Kα₂	56.2770 10	9.5 5	0.0113
X-ray Kα₁	57.5320 10	16.5 8	0.0202
X-ray Kβ	65.2	7.0 4	0.0097
γ 3	133.020 20	43.0 22	0.122
γ 4	136.250 20	6.1 3	0.0177
γ 5	136.86 4	1.80 9	0.0052
γ 6	345.85 20	14.0 7	0.103
γ 7	476.00 20	0.43 10	0.0044
γ 8	482.00 20	86 5	0.883
γ 9	615.5 5	0.250 13	0.0033

3 weak γ's omitted (ΣIγ = 0.05%)

ce-L-4	55.6502 4	6.5 3	0.0078
ce-M-3	62.9029 5	1.60 6	0.0021
ce-M-4	64.9304 5	1.49 6	0.0021
ce-NOP-3	65.1275 5	0.482 18	0.0007
ce-NOP-4	67.1550 5	0.422 17	0.0006
ce-L-5	72.5810 5	3.77 15	0.0058
ce-M-5	81.8612 5	0.89 4	0.0016
ce-NOP-5	82.9058 6	0.72 3	0.0013
ce-K-10	84.0858 6	0.267 11	0.0005
ce-K-11	86.8624 6	0.259 10	0.0005
ce-L-6	88.0067 5	32.6 12	0.0611
ce-M-6	97.2869 5	8.2 3	0.0170
ce-NOP-6	99.5115 5	2.40 9	0.0051
ce-L-8	101.5725 6	0.98 4	0.0021
ce-K-12	109.8698 6	1.57 10	0.0037
ce-M-8	110.8527 7	0.228 9	0.0005
ce-K-13	128.8280 7	0.245 10	0.0007
ce-L-10	140.3310 6	0.118 5	0.0004
ce-K-14	152.5848 7	0.298 11	0.0010
ce-K-15	159.7970 10	0.427 16	0.0015
ce-L-12	167.2950 6	0.434 14	0.0015
ce-M-12	176.5752 7	0.104 4	0.0004
ce-L-13	186.2532 7	0.156 6	0.0006
ce-K-16	194.2532 9	0.289 11	0.0012
ce-L-15	217.2222 10	0.221 8	0.0010
ce-L-16	251.9757 9	0.126 5	0.0007
ce-K-24	1051.776 5	0.105 4	0.0024
β^- 1 max	260.0 20		
avg	72.2 6	29.0 6	0.0446
β^- 2 max	325.7 20		
avg	92.4 7	1.04 21	0.0020
β^- 3 max	370.4 20		
avg	106.7 7	0.64 8	0.0015
β^- 4 max	439.4 20		
avg	129.2 7	19.6 6	0.0539
β^- 5 max	482.1 20		

(continued on next page)

182TA B- DECAY (114.43 D 4) (CONTINUED)

Radiation Type	Energy (keV)	Intensity (%)	Δ(g-rad/μCi-h)
β⁻ 6 avg	143.5 7	2.43 10	0.0074
	524.1 20		
β⁻ 6 max	157.8 7	41.8 12	0.140
β⁻ 7 avg	591.8 20		
β⁻ 7 max	181.4 7	5.0 9	0.0193
β⁻ 8 avg	1483.8 20		
β⁻ 8 max	528.6 9	0.17 14	0.0019
β⁻ 9 avg	1713.1 20		
β⁻ 9 max	624.7 9	0.8 8	0.0106
Total β⁻ avg	131.7 8	100.6 20	0.282

1 weak β's omitted (ΣIβ = 0.08%)

X-ray L	8.4	25 3	0.0045
γ 1	31.7379 6	0.91 5	0.0006
γ 2	42.7151 4	0.276 9	0.0003
X-ray Kα₂	57.9817 5	10.1 4	0.0125
X-ray Kα₁	59.31820 10	17.6 6	0.0222
γ 3	65.7225 3	3.00 7	0.0042
X-ray Kβ	67.2	7.5 3	0.0107
γ 4	67.75000 20	41.4 10	0.0598
γ 5	84.6808 3	2.54 7	0.0046
γ 6	100.1065 3	14.1 3	0.0301
γ 8	113.6723 4	1.90 4	0.0046
γ 9	116.4186 7	0.434 11	0.0011
γ 10	152.4308 5	6.88 13	0.0223
γ 11	156.3874 5	2.65 6	0.0088
γ 12	179.3948 5	3.01 7	0.0115
γ 13	198.3530 6	1.41 4	0.0059
γ 14	222.1098 6	7.44 14	0.0352
γ 15	229.3220 9	3.62 7	0.0177
γ 16	264.0755 8	3.59 8	0.0202

187W B- DECAY (23.9 H 1)   I(MIN)= 0.10%

Radiation Type	Energy (keV)	Intensity (%)	Δ(g-rad/μCi-h)
ce-K-7	0.326 4	7.7 4	≈0
ce-M-1	4.1683 4	0.9 5	≈0
Auger-L	6.7	17.4 12	0.0025
ce-K-13	42.070 8	0.227 13	0.0002
Auger-K	47	1.0 5	0.0010
ce-L-7	59.475 4	1.48 8	0.0019
ce-K-15	62.571 7	15.9 7	0.0212
ce-M-7	69.070 4	0.339 17	0.0005
ce-L-15	121.720 7	2.70 11	0.0070
ce-M-15	131.315 7	0.62 3	0.0017
ce-NOP-15	133.622 7	0.190 8	0.0005
ce-K-20	134.571 19	0.360 21	0.0010
ce-K-30	407.89 3	0.387 20	0.0034
ce-K-41	546.60 6	0.156 15	0.0018
β⁻ 1 max	432.9 18		
β⁻ 1 avg	126.9 6	0.52 6	0.0014
β⁻ 2 max	447.9 18		
β⁻ 2 avg	131.9 6	0.67 8	0.0019
β⁻ 3 max	539.6 18		
β⁻ 3 avg	163.0 7	4.2 4	0.0146
β⁻ 4 max	626.8 18		
β⁻ 4 avg	193.5 7	53.1 16	0.219
β⁻ 5 max	687.0 18		
β⁻ 5 avg	215.0 7	5.2 5	0.0238
β⁻ 6 max	694.2 18		
β⁻ 6 avg	217.7 7	5.2 5	0.0241
β⁻ 7 max	1178.3 18		
β⁻ 7 avg	401.7 7	0.7 3	0.0060
β⁻ 8 max	1312.5 18		
β⁻ 8 avg	457.1 8	30.4 16	0.296
Total β⁻ avg	275.3 9	100.0 25	0.587

	Energy (keV)	Intensity (%)	Δ (g-rad/μCi-h)
γ 19	927.98 4	0.637 21	0.0126
γ 20	959.72 4	0.360 13	0.0074
γ 21	1001.696 18	2.09 5	0.0446
γ 22	1044.43 9	0.238 12	0.0053
γ 23	1113.414 18	0.38 7	0.0091
γ 24	1121.301 5	35.0 7	0.836
γ 25	1157.5	0.35 11	0.0086
γ 26	1157.5	0.63 11	0.0155
γ 28	1189.050 5	16.3 3	0.413
γ 29	1221.408 5	27.2 5	0.707
γ 30	1231.016 5	11.55 21	0.303
γ 31	1257.418 5	1.49 3	0.0399
γ 32	1273.730 5	0.654 14	0.0178
γ 33	1289.156 5	1.351 25	0.0371
γ 34	1342.731 25	0.251 6	0.0072
γ 35	1373.836 5	0.220 6	0.0065

7 weak γ's omitted (ΣI_γ = 0.33%)

185W B- DECAY (75.1 D 3)   I(MIN) = 0.10%

Radiation Type	Energy (keV)	Intensity (%)	Δ (g-rad/μCi-h)
β⁻ 1 max	432.6 9		
avg	126.8 3	99.920 10	0.270

1 weak β's omitted (ΣI_β = 0.08%)

1 weak γ's omitted (ΣI_γ = 0.02%)

---

10 weak β's omitted (ΣI_β = 0.06%)

	Energy (keV)	Intensity (%)	Δ (g-rad/μCi-h)
X-ray L	8.65	7.8 9	0.0014
X-ray Kα2	59.7179 6	7.0 3	0.0088
X-ray Kα1	61.1403 6	12.0 5	0.0157
X-ray Kβ	69.3	5.15 21	0.0076
γ 7	72.002 4	10.8 4	0.0165
γ 15	134.247 7	8.6 4	0.0245
γ 20	206.247 19	0.138 7	0.0006
γ 23	246.276 22	0.115 6	0.0006
γ 30	479.57 3	21.1 9	0.216
γ 33	511.65 5	0.624 23	0.0068
γ 34	551.52 4	4.92 20	0.0578
γ 39	588.96 6	0.118 5	0.0015
γ 41	618.28 6	6.07 25	0.0799
γ 42	625.54 10	1.05 5	0.0140
γ 47	685.74 5	26.4 11	0.385
γ 50	745.29 10	0.288 12	0.0046
γ 52	772.91 6	3.98 17	0.0656
γ 57	864.71 18	0.325 14	0.0060
γ 58	879.56 19	0.137 6	0.0026

50 weak γ's omitted (ΣI_γ = 0.38%)

188W B- DECAY (69.4 D 5)   I(MIN) = 0.10%

Radiation Type	Energy (keV)	Intensity (%)	Δ (g-rad/μCi-h)
Auger-L	6.7	0.20 4	≈0
ce-L-1	51.05 3	0.29 5	0.0003

(continued on next page)

188W B- DECAY (69.4 D 5) (CONTINUED)

Radiation Type	Energy (keV)	Intensity (%)	Δ(g-rad/µCi-h)
β⁻ 1 max	58 3		
avg	14.9 8	0.85 5	0.0003
β⁻ 2 max	285 3		
avg	89.9 10	0.13 8	0.0002
β⁻ 3 max	349 3		
avg	99.7 10	99.02 10	0.210
Total β⁻			
avg	99.0 11	100.01 14	0.211
γ 1	63.58 3	0.108 18	0.0001
γ 6	227.090 20	0.221 14	0.0011
γ 7	290.669 13	0.401 25	0.0025

4 weak γ's omitted (ΣIγ = 0.02%)

188RE B- DECAY (16.98 H 2)   I(MIN)= 0.10%

Radiation Type	Energy (keV)	Intensity (%)	Δ(g-rad/µCi-h)
Auger-L	6.88	6.5 5	0.0009
Auger-K	48.3	0.19 9	0.0002
ce-K-1	81.169 20	4.9 3	0.0084
ce-L-1	142.072 20	5.5 4	0.0167
ce-M-1	151.991 20	1.40 9	0.0045
ce-NOP-1	154.386 20	0.420 25	0.0014

191OS B- DECAY (15.4 D 1)   I(MIN)= 0.10%

Radiation Type	Energy (keV)	Intensity (%)	Δ(g-rad/µCi-h)
ce-K-3	6.316 10	0.14 8	≈0
Auger-L	7	87 7	0.0131
ce-L-1	28.431 10	71 6	0.0430
ce-L-2	33.63 3	0.23 12	0.0002
ce-M-1	38.676 10	21.7 18	0.0179
ce-NOP-1	41.160 10	7.2 6	0.0063
Auger-K	49.6	2.2 11	0.0023
ce-K-4	53.320 5	58 6	0.0657
ce-L-3	69.009 10	0.10 6	0.0002
ce-L-4	116.012 5	12.0 12	0.0297
ce-M-4	126.257 5	2.8 3	0.0076
ce-NOP-4	128.741 5	0.88 9	0.0024
β⁻ 1 max	137 3		
avg	36.2 9	100	0.0771
X-ray L	9.18	43 5	0.0084
X-ray Kα2	63.2867 7	16.1 16	0.0217
X-ray Kα1	64.8956 7	28 3	0.0383
X-ray Kβ	73.6	12.0 12	0.0187
γ 4	129.431 5	25.7 24	0.0709

3 weak γ's omitted (ΣIγ = 0.03%)

191IR IT DECAY (4.94 S 3) I(MIN)= 0.10%

Radiation Type	Energy (keV)	Intensity (%)	Δ(g-rad/μCi-h)
ce-K-3	6.3104 15	0.14 7	≈0
Auger-L	7	88 5	0.0132
ce-L-1	28.431 10	71.1 7	0.0431
ce-L-2	33.63 3	0.22 12	0.0002
ce-M-1	38.676 10	21.7 6	0.0179
ce-NOP-1	41.160 10	7.10 20	0.0062
Auger-K	49.6	2.2 11	0.0023
ce-K-4	53.307 15	57.9 6	0.0658
ce-L-3	69.0029 15	0.10 5	0.0001
ce-L-4	115.999 15	12.2 3	0.0301
ce-M-4	126.244 15	2.87 9	0.0077
ce-NOP-4	128.728 15	0.9571 15	0.0026
X-ray L	9.18	43 4	0.0084
X-ray Kα2	63.2867 7	16.1 5	0.0217
X-ray Kα1	64.8956 7	27.8 7	0.0384
X-ray Kβ	73.6	12.0 4	0.0188
γ 4	129.418 15	25.8 6	0.0711

3 weak γ's omitted (ΣIγ = 0.03%)

Radiation Type	Energy	Intensity	Δ
β⁻ 1 max	178.6 9		
avg	48.1 3	0.107 7	0.0001
β⁻ 2 max	354.3 9		
avg	101.3 3	0.186 10	0.0004
β⁻ 3 max	657.2 9		
avg	204.0 4	0.52 3	0.0023
β⁻ 4 max	1033.3 9		
avg	345.1 4	0.64 3	0.0047
β⁻ 5 max	1486.7 9		
avg	527.5 4	1.60 14	0.0180
β⁻ 6 max	1964.7 9		
avg	728.6 4	25.1 13	0.390
β⁻ 7 max	2119.7 9		
avg	795.1 4	71.6 15	1.21
Total β⁻ avg	764.4 5	100.0 20	1.63

10 weak β's omitted (ΣIβ = 0.25%)

X-ray L	9	3.0 4	0.0006
X-ray Kα2	61.4867 7	1.35 9	0.0018
X-ray Kα1	63.0005 7	2.34 15	0.0031
X-ray Kβ	71.4	1.00 7	0.0015
γ 1	155.040 20	14.9 8	0.0491
γ 5	477.96 3	1.04 6	0.0106
γ 10	633.03 3	1.25 13	0.0168
γ 11	635.00 4	0.15 5	0.0020
γ 12	672.51 3	0.110 6	0.0016
γ 15	829.51 3	0.408 21	0.0072
γ 17	931.32 3	0.561 25	0.0111

34 weak γ's omitted (ΣIγ = 0.72%)

192IR EC DECAY (73.83 D 1) I(MIN)= 0.10%

%EC=4.74 6
SEE ALSO 192IR B- DECAY (74.02 D)

Radiation Type	Energy (keV)	Intensity (%)	Δ(g-rad/μCi-h)
Auger-L	6.88	3.24 19	0.0005
Auger-K	48.3	0.16 8	0.0002
ce-K-3	131.9247 8	0.645 21	0.0018
ce-L-3	192.8275 8	0.446 15	0.0018
ce-MNO-3	202.7470 8	0.1455 19	0.0006
X-ray L	9	1.52 16	0.0003
X-ray Kα₂	61.4867 7	1.17 4	0.0015
X-ray Kα₁	63.0005 7	2.02 5	0.0027
X-ray Kβ	71.4	0.87 3	0.0013
γ 2	201.3112 9	0.476 10	0.0020
γ 3	205.7955 6	3.31 5	0.0145
γ 4	283.255 20	0.268 5	0.0016
γ 6	374.4851 13	0.720 6	0.0057
γ 8	484.5780 13	3.17 7	0.0327
γ 9	489.06 3	0.423 13	0.0044

4 weak γ's omitted (ΣIγ = 0.09%)

192IR B- DECAY (73.83 D 18) (CONTINUED)

Radiation Type	Energy (keV)	Intensity (%)	Δ(g-rad/μCi-h)
X-ray L	9.44	4.0 5	0.0008
X-ray Kα₂	65.1220 20	2.63 7	0.0036
X-ray Kα₁	66.8320 20	4.51 10	0.0064
X-ray Kβ	75.7	1.96 6	0.0032
γ 1	136.3435 5	0.188 17	0.0005
γ 2	295.9583 8	28.68 15	0.181
γ 3	308.4569 8	29.70 16	0.195
γ 4	316.5079 9	82.86 16	0.559
γ 5	416.4714 13	0.664 6	0.0059
γ 6	468.0715 13	47.64 21	0.475
γ 8	588.5845 17	4.474 19	0.0561
γ 10	604.4146 16	8.11 5	0.104
γ 11	612.4656 15	5.26 3	0.0686
γ 12	884.5418 24	0.286 3	0.0054

5 weak γ's omitted (ΣIγ = 0.10%)

194IR B- DECAY (19.15 H 3) I(MIN)= 0.10%

Radiation Type	Energy (keV)	Intensity (%)	Δ(g-rad/μCi-h)
Auger-L	7.24	0.68 8	0.0001
ce-K-4	215.146 14	0.17 3	0.0008
ce-K-6	250.053 14	0.64 11	0.0034
ce-L-6	314.568 14	0.27 5	0.0018

192IR B- DECAY (73.83 D 18)   I(MIN)= 0.10%
%B-=95.26 6
SEE ALSO 192IR EC DECAY

Radiation Type	Energy (keV)	Intensity (%)	Δ(g-rad/µCi-h)
Auger-L	7.24	7.5 5	0.0012
Auger-K	51	0.35 13	0.0004
ce-K-1	57.9487 9	0.116 11	0.0001
ce-K-2	217.5635 11	1.901 11	0.0088
ce-K-3	230.0621 11	1.79 3	0.0088
ce-K-4	238.1131 12	4.47 14	0.0227
ce-L-2	282.0784 9	0.87 3	0.0052
ce-M-2	292.6623 12	0.217 7	0.0014
ce-L-3	294.5770 9	0.772 24	0.0048
ce-L-4	302.6280 10	1.95 6	0.0126
ce-M-3	305.1609 12	0.192 6	0.0012
ce-M-4	313.2119 13	0.484 15	0.0032
ce-NOP-4	315.7859 11	0.148 5	0.0010
ce-K-6	389.6767 15	1.01 3	0.0084
ce-L-6	454.1916 14	0.292 9	0.0028
ce-K-10	526.0198 18	0.165 5	0.0019
β- 1 max	256 4		
avg	70.8 12	5.528 21	0.0083
β- 2 max	536 4		
avg	161.2 14	41.33 19	0.142
β- 3 max	672 4		
avg	208.9 15	47.82 22	0.213
β- 4 max	1140 4		
avg	383.1 15	0.4 4	0.0033
Total β-			
avg	180.7 16	95.2 5	0.366

2 weak β's omitted (ΣIβ = 0.10%)

(continued on next column)

β- 1 max	450.5 18		
avg	132.4 6	0.330 20	0.0009
β- 2 max	625.7 18		
avg	192.3 7	0.160 20	0.0007
β- 3 max	735.9 18		
avg	231.8 7	0.55 3	0.0027
β- 4 max	768.6 18		
avg	243.8 7	0.63 4	0.0033
β- 5 max	980.7 18		
avg	323.5 7	1.76 8	0.0121
β- 6 max	1325.1 18		
avg	452.8 7	0.3	0.0029
β- 7 max	1625.8 18		
avg	582.6 8	1.34 13	0.0166
β- 8 max	1919.4 18		
avg	705.8 8	9.2 6	0.138
β- 9 max	2247.8 18		
avg	846.1 8	85.5 20	1.54
Total β-			
avg	806.8 9	100.0 21	1.72

14 weak β's omitted (ΣIβ = 0.28%)

X-ray L	9.44	0.37 6	≈0
X-ray Kα2	65.1220 20	0.24 3	0.0003
X-ray Kα1	66.8320 20	0.40 5	0.0006
X-ray Kβ	75.7	0.176 23	0.0003
γ 4	293.541 14	2.5 4	0.0158
γ 5	300.741 14	0.35 6	0.0022
γ 6	328.448 14	13.0 21	0.0909
γ 10	589.179 17	0.139 22	0.0017
γ 14	621.971 19	0.33 6	0.0044
γ 15	645.146 20	1.16 19	0.0160
γ 23	938.69 3	0.59 10	0.0119
γ 27	1150.75 5	0.59 10	0.0145
γ 30	1183.49 5	0.30 5	0.0076
γ 43	1468.91 7	0.19 3	0.0059

61 weak γ's omitted (ΣIγ = 0.84%)

195PT IT DECAY (4.02 D 1)   I(MIN)= 0.10%

Radiation Type	Energy (keV)	Intensity (%)	Δ(g-rad/ μCi-h)
ce-L-1	5.9 3	0.134 10	≈0
Auger-L	7.24	137 10	0.0211
ce-L-3	16.996 6	69 6	0.0251
ce-K-4	20.485 20	66 6	0.0288
ce-M-3	27.580 6	15.7 13	0.0092
ce-NOP-3	30.154 6	7.4 6	0.0048
Auger-K	51	3.0 11	0.0032
ce-K-5	51.11 20	13.0 10	0.0142
ce-K-6	51.362 20	1.33 11	0.0015
ce-L-4	85.000 20	11.6 10	0.0211
ce-M-4	95.584 20	2.74 22	0.0056
ce-NOP-4	98.158 20	0.85 7	0.0018
ce-L-5	115.62 20	61 5	0.151
ce-L-6	115.877 20	2.71 22	0.0067
ce-M-5	126.20 20	19.0 15	0.0512
ce-M-6	126.461 20	0.71 6	0.0019
ce-NOP-5	128.78 20	6.0 5	0.0165
ce-NOP-6	129.035 20	0.215 18	0.0006
X-ray L	9.44	74 9	0.0148
γ 3	30.876 6	2.28 12	0.0015
X-ray Kα2	65.1220 20	22.4 16	0.0310
X-ray Kα1	66.8320 20	38 3	0.0546
X-ray Kβ	75.7	16.7 12	0.0269
γ 4	98.880 20	11.4 9	0.0240
γ 6	129.757 20	2.83 21	0.0078

6 weak γ's omitted (ΣIγ = 0.21%)

195AU IT DECAY (30.5 S 2)   I(MIN)= 0.10%

Radiation Type	Energy (keV)	Intensity (%)	Δ(g-rad/ μCi-h)
Auger-L	7.42	62 4	0.0098
ce-L-1	42.45 3	71.61 5	0.0647
ce-L-2	47.11 3	1.64 12	0.0016
Auger-K	52.4	0.8 4	0.0009
ce-M-1	53.38 3	20.887 13	0.0237
ce-NOP-1	56.04 3	6.962 5	0.0083
ce-M-2	58.04 3	0.40 3	0.0005
ce-NOP-2	60.70 3	0.117 8	0.0002
ce-K-4	181.03 4	23.31 4	0.0899
ce-K-5	237.88 10	0.26 3	0.0013
ce-L-4	247.40 4	4.370 8	0.0230
ce-M-4	258.33 4	1.0683 18	0.0059
ce-NOP-4	260.99 4	0.3205 6	0.0018
ce-L-5	304.25 10	0.162 18	0.0010
X-ray L	9.7	35 4	0.0072
γ 2	61.46 3	0.173 13	0.0002
X-ray Kα2	66.9895 8	6.57 15	0.0094
X-ray Kα1	68.8037 8	11.24 23	0.0165
X-ray Kβ	78	4.92 12	0.0082
γ 3	200.38 4	1.70 12	0.0073
γ 4	261.75 4	67.9 7	0.379

2 weak γ's omitted (ΣIγ = 0.07%)

195AU EC DECAY (183 D 2) I(MIN)= 0.10%

Radiation Type	Energy (keV)	Intensity (%)	Δ(g-rad/μCi-h)
Auger-L	7.24	105 8	0.0162
ce-L-1	16.996 6	22.6 18	0.0082
ce-K-2	20.485 20	63 5	0.0275
ce-M-1	27.580 6	5.2 4	0.0031
ce-NOP-1	30.154 6	1.74 14	0.0011
Auger-K	51	3.9 14	0.0042
ce-K-3	51.362 20	0.39 3	0.0004
ce-L-2	85.000 20	11.1 9	0.0201
ce-M-2	95.584 20	2.62 21	0.0053
ce-NOP-2	98.158 20	0.82 7	0.0017
ce-L-3	115.877 20	0.78 6	0.0019
ce-M-3	126.461 20	0.202 14	0.0005
X-ray L	9.44	57 7	0.0114
γ 1	30.876 6	0.75 6	0.0005
X-ray Kα₂	65.1220 20	29.0 18	0.0403
X-ray Kα₁	66.8320 20	50 3	0.0709
X-ray Kβ	75.7	21.7 13	0.0350
γ 2	98.880 20	10.9 9	0.0230
γ 3	129.757 20	0.82 5	0.0023

2 weak γ's omitted (ΣIγ = 0.02%)

197PT B- DECAY (18.3 H 3) I(MIN)= 0.10%

Radiation Type	Energy (keV)	Intensity (%)	Δ(g-rad/μCi-h)
Auger-L	7.42	37 5	0.0059
Auger-K	52.4	0.13 6	0.0001
ce-L-1	63.00 5	55 6	0.0735
ce-M-1	73.93 5	13.3 15	0.0210
ce-NOP-1	76.59 5	4.2 5	0.0068
ce-K-2	110.712 10	3.53 15	0.0083
ce-L-2	177.084 10	0.600 25	0.0023
ce-M-2	188.012 10	0.139 6	0.0006
β⁻ 1 max	450.1 6		
avg	132.09 20	8.2 8	0.0231
β⁻ 2 max	641.6 6		
avg	197.63 22	81 3	0.342
β⁻ 3 max	718.9 6		
avg	225.28 22	11 3	0.0509
Total β⁻ avg	195.19 22	100 5	0.416
X-ray L	9.7	21 4	0.0043
X-ray Kα₂	66.9895 8	0.99 5	0.0014
X-ray Kα₁	68.8037 8	1.69 8	0.0025
γ 1	77.35 5	17.1 18	0.0281
X-ray Kβ	78	0.74 4	0.0012
γ 2	191.437 10	3.67 15	0.0150
γ 3	268.78 5	0.231 24	0.0013

197HG EC DECAY (64.1 H 1)   I(MIN)= 0.10%

Radiation Type	Energy (keV)	Intensity (%)	Δ(g-rad/μCi-h)
Auger-L	7.42	88 7	0.0140
Auger-K	52.4	2.7 13	0.0030
ce-L-1	62.997 10	58 4	0.0775
ce-M-1	73.925 10	14.0 9	0.0221
ce-NOP-1	76.591 10	4.3 3	0.0070
ce-K-2	110.639 15	0.59 4	0.0014
X-ray L	9.7	50 6	0.0103
X-ray Kα₂	66.9895 8	20.8 16	0.0296
X-ray Kα₁	68.8037 8	35 3	0.0520
γ 1	77.350 10	18.0 11	0.0297
X-ray Kβ	78	15.5 12	0.0258
γ 2	191.364 15	0.61 4	0.0025

1 weak γ's omitted (ΣIγ = 0.04%)

198AU B- DECAY (2.696 D 2)   I(MIN)= 0.10%

Radiation Type	Energy (keV)	Intensity (%)	Δ(g-rad/μCi-h)
Auger-L	7.6	2.08 16	0.0003
ce-K-1	328.7021 14	2.88 9	0.0201
ce-L-1	396.9651 15	1.02 3	0.0086
ce-M-1	408.2428 16	0.254 8	0.0022

199AU B- DECAY (3.139 D 7)   (CONTINUED)

Radiation Type	Energy (keV)	Intensity (%)	Δ(g-rad/μCi-h)
ce-NOP-2	157.5791 12	1.34 6	0.0045
ce-L-3	193.3667 12	1.02 4	0.0042
ce-M-3	204.6444 13	0.245 11	0.0011
β⁻ 1 max	244.2 7		
avg	67.15 21	18.9 6	0.0270
β⁻ 2 max	294.0 7		
avg	82.23 22	66.4 10	0.116
β⁻ 3 max	452.4 7		
avg	132.70 24	14.7 5	0.0415
Total β⁻ avg	86.80 23	100.0 13	0.185
X-ray L	10	12.8 14	0.0027
γ 1	49.82680 21	0.328 13	0.0003
X-ray Kα₂	68.8950 20	4.47 20	0.0066
X-ray Kα₁	70.8190 20	7.6 4	0.0115
X-ray Kβ	80.3	3.34 15	0.0057
γ 2	158.3794 5	36.9 11	0.124
γ 3	208.2060 6	8.4 3	0.0371

201TL EC DECAY (72.91 H 2)   I(MIN)= 0.10%

Radiation Type	Energy (keV)	Intensity (%)	Δ(g-rad/μCi-h)
ce-NOP-1	0.770 20	36 23	0.0006

Radiation Type	Energy (keV)	Intensity (%)	Δ(g-rad/μCi-h)
β⁻ 1 max	284.8 6		
avg	79.41 19	0.985 4	0.0017
β⁻ 2 max	960.7 6		
avg	314.64 23	98.990 7	0.663
Total β⁻			
avg	312.36 24	100.000 8	0.665

1 weak β's omitted (ΣIβ = 0.03%)

Radiation Type	Energy (keV)	Intensity (%)	Δ(g-rad/μCi-h)
X-ray L	10	1.27 15	0.0003
X-ray Kα₂	68.8950 20	0.81 4	0.0012
X-ray Kα₁	70.8190 20	1.37 6	0.0021
X-ray Kβ	80.3	0.603 25	0.0010
γ 1	411.8044 11	95.59 13	0.838
γ 2	675.8874 19	0.804 3	0.0116
γ 3	1087.6905 3	0.1591 21	0.0037

199AU B- DECAY (3.139 D 7)   I(MIN) = 0.10%

Radiation Type	Energy (keV)	Intensity (%)	Δ(g-rad/μCi-h)
Auger-L	7.6	20.9 15	0.0034
ce-L-1	34.9875 11	2.66 11	0.0020
ce-M-1	46.2652 12	0.75 4	0.0007
ce-NOP-1	49.0265 11	0.189 10	0.0002
Auger-K	53.8	0.5 4	0.0006
ce-K-2	75.2771 10	10.5 5	0.0168
ce-K-3	125.1037 10	5.5 3	0.0146
ce-L-2	143.5401 12	17.1 5	0.0522
ce-M-2	154.8178 12	4.5 4	0.0147

(continued on next column)

Radiation Type	Energy (keV)	Intensity (%)	Δ(g-rad/μCi-h)
Auger-L	7.6	76 6	0.0123
ce-L-3	15.76 3	10.1 5	0.0034
ce-L-4	17.35 3	8.8 5	0.0032
ce-M-3	27.04 3	2.35 11	0.0014
ce-M-4	28.63 3	2.05 11	0.0012
ce-K-5	52.24 4	7.9 4	0.0088
Auger-K	53.8	3.4 20	0.0039
ce-K-6	82.78 7	0.22 4	0.0004
ce-K-7	84.33 7	16.5 9	0.0295
ce-L-5	120.50 4	1.34 6	0.0035
ce-M-5	131.78 4	0.313 13	0.0009
ce-NOP-5	134.54 4	0.101 4	0.0003
ce-L-7	152.59 7	2.78 14	0.0090
ce-M-7	163.87 7	0.64 4	0.0022
ce-NOP-7	166.63 7	0.207 11	0.0007
γ 1	1.570 20	36 23	0.0012
X-ray L	10	46 6	0.0099
γ 3	30.60 3	0.272 12	0.0002
γ 4	32.19 3	0.276 15	0.0002
X-ray Kα₂	68.8950 20	27.7 10	0.0407
X-ray Kα₁	70.8190 20	47.2 16	0.0712
X-ray Kβ	80.3	20.8 8	0.0355
γ 5	135.34 4	2.80 11	0.0081
γ 6	165.88 7	0.159 22	0.0006
γ 7	167.43 7	10.6 5	0.0378

203HG B- DECAY (46.60 D 1) I(MIN)= 0.10%

Radiation Type	Energy (keV)	Intensity (%)	Δ(g-rad/μCi-h)
Auger-L	7.78	8.8 7	0.0015
Auger-K	55.2	0.44 20	0.0005
ce-K-1	193.6664 15	13.37 21	0.0551
ce-L-1	263.8501 14	3.91 17	0.0220
ce-MNO-1	275.4927 14	1.30 17	0.0077
β- 1 max	213.0 20		
avg	57.9 6	100	0.123
X-ray L	10.3	5.9 7	0.0013
X-ray Kα2	70.8319 9	3.75 10	0.0057
X-ray Kα1	72.8715 9	6.36 16	0.0099
X-ray Kβ	82.6	2.81 8	0.0049
γ 1	279.1968 13	81.5 8	0.485

203PB EC DECAY (51.88 H 1) I(MIN)= 0.10%

Radiation Type	Energy (keV)	Intensity (%)	Δ(g-rad/μCi-h)
Auger-L	7.78	57 5	0.0094
Auger-K	55.2	3.0 14	0.0035
ce-K-1	193.6664 15	13.22 23	0.0545
ce-L-1	263.8501 14	3.92 6	0.0220
ce-MNO-1	275.4927 14	0.951 12	0.0056
ce-K-2	315.785 12	0.50 4	0.0034
ce-L-2	385.968 12	0.166 10	0.0014

206BI EC DECAY (6.243 D 3) (CONTINUED)

Radiation Type	Energy (keV)	Intensity (%)	Δ(g-rad/μCi-h)
ce-K-27	449.45 4	2.32 8	0.0222
ce-L-25	481.20 4	0.2411 25	0.0025
ce-L-26	500.32 4	1.25 4	0.0133
ce-L-27	521.59 4	0.390 13	0.0043
ce-K-31	532.48 5	0.288 10	0.0033
ce-K-32	544.25 5	0.211 7	0.0024
ce-K-39	715.10 5	0.799 24	0.0122
ce-L-39	787.24 5	0.174 6	0.0029
ce-K-42	793.01 5	0.449 15	0.0076
ce-K-43	807.12 5	0.318 10	0.0055
ce-K-46	930.63 8	0.111 4	0.0022
X-ray L	10.6	50 5	0.0112
X-ray Kα2	72.8042 9	32.4 7	0.0502
X-ray Kα1	74.9694 9	54.7 11	0.0873
X-ray Kβ	84.9	24.3 6	0.0439
γ 6	184.02 3	15.8 3	0.0620
γ 10	234.26 7	0.241 12	0.0012
γ 11	262.71 5	3.02 5	0.0169
γ 13	313.67 7	0.359 10	0.0024
γ 15	343.51 3	23.4 3	0.171
γ 17	386.20 7	0.516 10	0.0042
γ 18	398.00 3	10.74 10	0.0910
γ 22	452.84 8	0.156 8	0.0015
γ 25	497.06 4	15.31 15	0.162
γ 26	516.18 4	40.7 4	0.448
γ 27	537.45 4	30.4 3	0.348
γ 29	576.36 10	0.112 10	0.0014
γ 30	581.97 8	0.485 25	0.0060
γ 31	620.48 5	5.76 6	0.0761
γ 32	632.25 5	4.47 5	0.0602
γ 33	657.16 5	1.91 3	0.0267
γ 35	739.24 8	0.157 8	0.0025

γ	Energy (keV)		Intensity (%)		Δ(g-rad/μCi-h)
36	754.96	7	0.527	10	0.0085
38	784.58	7	0.536	7	0.0090
39	803.10	5	98.89	3	1.69
41	841.28	7	0.186	9	0.0033
42	881.01	5	66.2	7	1.24
43	895.12	5	15.65	16	0.298
46	1018.63	8	7.59	8	0.165
50	1098.26	7	13.50	15	0.316
51	1142.37	10	0.111	5	0.0027
53	1194.69	8	0.277	15	0.0070
54	1202.58	10	0.105	6	0.0027
58	1332.33	10	0.282	15	0.0080
59	1405.01	8	1.434	25	0.0429
62	1496.18	8	0.176	10	0.0056
63	1560.30	8	0.378	20	0.0126
64	1565.34	8	0.304	15	0.0101
66	1595.27	8	5.01	6	0.170
67	1718.70	7	31.8	4	1.17
68	1844.49	10	0.569	25	0.0223
69	1878.65	8	2.01	4	0.0803
70	1903.56	10	0.349	15	0.0142
75	2599.60	20	0.130	10	0.0072

37 weak γ's omitted (ΣIγ = 1.24%)

207BI EC DECAY (32.2 Y 9)   I(MIN)= 0.10%

Radiation Type	Energy (keV)	Intensity (%)		Δ(g-rad/μCi-h)
Auger-L	8	47	4	0.0080
Auger-K	56.7	2.2	10	0.0027

(continued on next page)

X-ray L	10.3		38	4	0.0083
X-ray Kα2	70.8319	9	25.6	7	0.0387
X-ray Kα1	72.8715	9	43.4	10	0.0674
X-ray Kβ	82.6		19.2	5	0.0338
γ 1	279.1968	13	80.6	10	0.479
γ 2	401.315	12	3.47	17	0.0296
γ 3	680.502	15	0.70	9	0.0102

206BI EC DECAY (6.243 D 3)   I(MIN)= 0.10%

Radiation Type	Energy (keV)	Intensity (%)		Δ(g-rad/μCi-h)
Auger-L	8	71	5	0.0121
Auger-K	56.7	3.7	15	0.0044
ce-K-5	70.60 10	0.173	18	0.0003
ce-K-6	96.02 3	22.0	8	0.0451
ce-L-10	146.26 7	0.168	10	0.0005
ce-K-10	168.16 3	3.84	14	0.0138
ce-L-11	174.71 5	1.56	6	0.0058
ce-M-6	180.17 3	0.90	4	0.0035
ce-NOP-6	183.13 3	0.296	11	0.0012
ce-L-8	186.58 10	0.109	11	0.0004
ce-L-11	246.85 5	0.270	10	0.0014
ce-K-15	255.51 3	5.89	8	0.0321
ce-K-18	310.00 3	1.819	17	0.0120
ce-L-15	327.65 3	1.00	4	0.0070
ce-M-15	339.66 3	0.235	8	0.0017
ce-L-18	382.14 3	0.309	10	0.0025
ce-K-25	409.06 4	1.428	15	0.0124
ce-K-26	428.18 4	1.99	7	0.0182

(continued on next column)

218PO A DECAY (3.05 M) (RaA)

Radiation Type	Energy (keV)	Intensity (%)	Δ(g-rad/μCi-h)
α 1	6002.40 9	99.9789 20	12.8

214PB B- DECAY (26.8 M) (RaB)

Radiation Type	Energy (keV)	Intensity (%)	Δ(g-rad/μCi-h)
Auger-L	8.15	18.4 14	0.0032
ce-L-1	36.838 14	10.7 6	0.0084
ce-M-1	49.227 14	2.51 14	0.0026
ce-NOP-1	52.288 14	0.84 5	0.0009
Auger-K	58.2	0.7 4	0.0008
ce-K-2	151.38 3	5.19 21	0.0167
ce-K-3	168.26 6	0.302 15	0.0011
ce-K-5	204.644 20	7.1 3	0.0310
ce-L-2	225.52 3	0.90 4	0.0043
ce-M-2	237.91 3	0.211 9	0.0011
ce-K-7	261.37 3	9.1 3	0.0505
ce-L-5	278.782 20	1.277 25	0.0076
ce-M-5	291.171 20	0.301 6	0.0019
ce-NOP-5	294.232 20	0.1003 19	0.0006
ce-L-7	335.51 3	1.56 5	0.0112
ce-M-7	347.90 3	0.365 12	0.0027
ce-NOP-7	350.96 3	0.122 4	0.0009

207BI EC DECAY (32.2 Y 9) (CONTINUED)

Radiation Type	Energy (keV)	Intensity (%)	Δ(g-rad/μCi-h)
ce-K-2	481.6975 22	1.55 5	0.0159
ce-L-2	553.8412 21	0.430 13	0.0051
ce-K-4	975.657 4	7.19 22	0.149
ce-L-4	1047.801 4	1.78 6	0.0397
ce-MNO-4	1059.811 4	0.5927 20	0.0134

1 weak β's omitted (ΣIβ = 0.01%)

X-ray L	10.6	33 4	0.0074
X-ray Kα₂	72.8042 9	19.7 4	0.0306
X-ray Kα₁	74.9694 9	33.3 6	0.0532
X-ray Kβ	84.9	14.8 4	0.0268
γ 2	569.7020 20	97.7 4	1.19
γ 3	897.70 10	0.119 13	0.0023
γ 4	1063.662 4	74.08 25 @	1.68
γ 5	1442.20 20	0.129 5	0.0040
γ 6	1460.0 15	1.61 6	0.0502
γ 7	1770.237 10	6.87 3	0.259

226RA A DECAY (1600 Y 7) (Ra)

Radiation Type	Energy (keV)	Intensity (%)	Δ(g-rad/μCi-h)
Auger-L	8.7	0.97 10	0.0002
ce-K-1	87.59 4	0.677 23	0.0013

Radiation Type	Energy (keV)	Intensity (%)	Δ(g-rad/μCi-h)
ce-L-1	167.94 4	1.29 4	0.0046
ce-M-1	181.52 4	0.345 14	0.0013
ce-NOP-1	184.90 4	0.119 7	0.0005
α 1	4601.70 20	5.94 12	0.582
α 2	4784.38 25	94.05 12	9.58
X-ray L	11.7	0.86 10	0.0002
X-ray Kα2	81.070 20	0.193 8	0.0003
X-ray Kα1	83.780 20	0.320 12	0.0006
X-ray Kβ	94.9	0.145 6	0.0003
γ 1	185.99 4	3.51 5	0.0139

222RN A DECAY (3.8235 D 3) (Rn)

Radiation Type	Energy (keV)	Intensity (%)	Δ(g-rad/μCi-h)
α 1	5489.5 3	99.920 10	11.7

2 weak α's omitted (ΣIα = 0.08%)

1 weak γ's omitted (ΣIγ = 0.08%)

Radiation Type	Energy (keV)	Intensity (%)	Δ(g-rad/μCi-h)
β⁻ 1 max	178 12	2.43 7	0.0025
avg	48 4		
β⁻ 2 max	483 12	0.97 10	0.0030
avg	143 4		
β⁻ 3 max	665 12	46.1 4	0.201
avg	205 5		
β⁻ 4 max	722 12	40.8 6	0.196
avg	225 5		
β⁻ 5 max	1017 12	9.5 9	0.0676
avg	334 5		
Total β⁻			
avg	221 6	99.8 12	0.470
X-ray L	10.8	13.4 14	0.0031
γ 1	53.226 14	1.11 5	0.0013
X-ray Kα2	74.8148 10	6.11 18	0.0097
X-ray Kα1	77.1079 10	10.3 3	0.0169
X-ray Kβ	87.3	4.59 14	0.0085
γ 2	241.91 3	7.28 19	0.0375
γ 3	258.79 6	0.510 20	0.0028
γ 4	274.53 5	0.35 4	0.0020
γ 5	295.170 20	18.40 20	0.116
γ 7	351.90 3	35.4 3	0.265
γ 9	480.42 8	0.310 20	0.0032
γ 10	487.08 8	0.39 4	0.0040
γ 11	533.69 8	0.200 10	0.0023
γ 13	580.15 4	0.290 10	0.0036
γ 14	766.0 4	0.12 4	0.0020
γ 15	785.910 20	1.040 20	0.0174
γ 16	839.025 15	0.63 5	0.0113

3 weak γ's omitted (ΣIγ = 0.11%)

214BI B- DECAY (19.9 M 4) (RaC)

Radiation Type	Energy (keV)	Intensity (%)	Δ(g-rad/ μCi-h)
Auger-L	8.33	0.48 4	≈0
ce-K- 8	516.220 7	0.672 21	0.0074
ce-K-48	1322.700 5	0.390 14	0.0110
β⁻ 1 max	537 12		
avg	160 4	0.396 22	0.0013
β⁻ 2 max	547 12		
avg	164 4	0.194 11	0.0007
β⁻ 3 max	571 12		
avg	172 5	0.103 10	0.0004
β⁻ 4 max	760 12		
avg	238 5	0.163 20	0.0008
β⁻ 5 max	784 12		
avg	247 5	1.03 4	0.0054
β⁻ 6 max	818 12		
avg	259 5	2.35 7	0.0130
β⁻ 7 max	973 12		
avg	317 5	0.53 3	0.0036
β⁻ 8 max	1057 12		
avg	349 5	0.320 20	0.0024
β⁻ 9 max	1062 12		
avg	351 5	5.56 13	0.0416
β⁻10 max	1073 12		
avg	355 5	0.82 3	0.0062
β⁻11 max	1118 12		
avg	372 5	0.41 4	0.0032
β⁻12 max	1147 12		
avg	384 5	4.25 9	0.0348
β⁻13 max	1177 12		
avg	395 5	0.127 8	0.0011
β⁻14 max	1249 12		
avg	423 5	2.41 4	0.0217
β⁻15 max	1255 12		

214BI B- DECAY (19.9 M 4) (RaC) (CONTINUED)

Radiation Type	Energy (keV)	Intensity (%)	Δ(g-rad/ μCi-h)
γ 2	388.98 19	0.39 5	0.0032
γ 3	454.77 12	0.300 20	0.0029
γ 4	469.69 12	0.130 20	0.0013
γ 5	474.38 10	0.110 10	0.0011
γ 8	609.320 4	44.8 4	0.581
γ 12	665.453 22	1.51 5	0.0214
γ 13	703.11 4	0.460 20	0.0069
γ 15	719.86 3	0.390 20	0.0060
γ 18	752.84 3	0.130 10	0.0021
γ 19	768.373 10	4.76 7	0.0779
γ 21	806.191 10	1.19 3	0.0204
γ 23	821.18 3	0.150 10	0.0026
γ 24	826.2 3	0.110 20	0.0019
γ 26	934.052 20	3.07 4	0.0611
γ 27	964.08 3	0.370 20	0.0076
γ 29	1051.96 3	0.300 10	0.0067
γ 31	1069.96 8	0.25 4	0.0057
γ 33	1120.276 22	14.80 20	0.353
γ 35	1133.66 3	0.25 3	0.0060
γ 36	1155.190 20	1.63 5	0.0401
γ 38	1207.68 3	0.450 20	0.0116
γ 40	1238.11 3	5.83 7	0.154
γ 41	1280.960 20	1.42 3	0.0387
γ 42	1303.76 8	0.110 10	0.0031
γ 44	1377.65 3	3.90 10	0.114
γ 45	1385.31 3	0.80 4	0.0236
γ 46	1401.50 4	1.36 4	0.0406
γ 47	1408.005 8	2.40 4	0.0720
γ 50	1509.229 9	2.08 5	0.0669
γ 51	1538.50 6	0.40 4	0.0131
γ 52	1543.32 6	0.320 20	0.0105
γ 53	1583.22 4	0.690 20	0.0233
γ 54	1594.73 8	0.260 20	0.0088

Radiation Type		Energy (keV)	Intensity (%)	Δ(g-rad/μCi-h)
β⁻16	avg	426 5	1.46 4	0.0132
	max	1271 12		
β⁻17	avg	432 5	1.20 5	0.0110
	max	1376 12		
β⁻18	avg	473 5	1.52 4	0.0153
	max	1419 12		
β⁻19	avg	490 5	8.15 9	0.0851
	max	1501 12		
β⁻20	avg	524 5	16.9 3	0.189
	max	1523 12		
β⁻21	avg	532 5	0.25 3	0.0028
	max	1536 12		
β⁻22	avg	538 5	17.5 3	0.201
	max	1605 12		
β⁻23	avg	566 5	1.15 4	0.0139
	max	1723 12		
β⁻24	avg	614 5	3.27 5	0.0428
	max	1851 12		
β⁻25	avg	666 5	1.06 5	0.0150
	max	1888 12		
β⁻26	avg	682 5	7.56 14	0.110
	max	1991 12		
β⁻27	avg	725 5	0.23 7	0.0036
	max	2657 12		
β⁻28	avg	1006 6	0.7 5	0.0150
	max	3266 12		
	avg	1267 6	19.8 5	0.534
Total β⁻	avg	653 7	99.9 9	1.39

20 weak β's omitted (ΣIβ = 0.48%)

Radiation Type	Energy (keV)	Intensity (%)	Δ(g-rad/μCi-h)
X-ray L	11	0.37 4	≈0
X-ray Kα₂	76.862 5	0.301 10	0.0005
X-ray Kα₁	79.290 5	0.504 15	0.0009
X-ray Kβ	89.8	0.226 8	0.0004
γ 1	386.72 19	0.32 4	0.0026

(continued on next column)

Radiation Type	Energy (keV)	Intensity (%)	Δ(g-rad/μCi-h)
γ 55	1599.31 6	0.320 20	0.0109
γ 57	1661.321 15	1.12 3	0.0396
γ 58	1684.032 21	0.230 20	0.0082
γ 59	1729.645 14	2.92 6	0.108
γ 60	1764.543 16	15.30 20	0.575
γ 62	1838.36 5	0.350 20	0.0137
γ 63	1847.44 5	2.04 4	0.0803
γ 64	1873.16 6	0.22 3	0.0088
γ 66	1895.8 4	0.150 20	0.0061
γ 75	2118.551 9	1.17 3	0.0528
γ 78	2204.091 14	4.98 12	0.234
γ 80	2293.347 22	0.300 10	0.0147
γ 83	2447.695 16	1.55 4	0.0808

55 weak γ's omitted (ΣIγ = 1.68%)

214PO A DECAY (164.3 US 18) (RaC')

Radiation Type	Energy (keV)	Intensity (%)	Δ(g-rad/μCi-h)
α 1	7686.90 6	99.9895 6	16.4

2 weak α's omitted (ΣIα = 0.01%)

2 weak γ's omitted (ΣIγ = 0.01%)

210PB B- DECAY (22.3 Y 2) (RaD)

Radiation Type	Energy (keV)	Intensity (%)	Δ(g-rad/ μCi-h)
Auger-L	8.15	35 4	0.0060
ce-L-1	30.133 20	60 4	0.0383
ce-M-1	42.521 20	14.0 9	0.0127
ce-NOP-1	45.582 20	4.6 3	0.0045
β⁻ 1 max	16.5 5		
avg	4.14 13	82 3	0.0072
β⁻ 2 max	63.0 5		
avg	16.13 14	18 3	0.0062
Total β⁻ avg	6.30 18	100 5	0.0134
X-ray L	10.8	25 3	0.0058
γ 1	46.520 20	4.18 23	0.0041

210BI B- DECAY (5.013 D 5) (RaE)

Radiation Type	Energy (keV)	Intensity (%)	Δ(g-rad/ μCi-h)
β⁻ 1 max	1161.9 10		
avg	389.2 4	100	0.829

210TL B- DECAY (1.30 M 3) (RaC")

Radiation Type	Energy (keV)	Intensity (%)	Δ(g-rad/ μCi-h)
Auger-L	8	30 7	0.0052
ce-K-2	9 3	2.1 11	0.0004
Auger-K	56.7	0.29 13	0.0004
ce-L-1	67 3	21 5	0.0298
ce-M-1	79 3	5.5 11	0.0092
ce-L-2	81 3	20 10	0.0345
ce-NOP-1	82 3	1.8 4	0.0031
ce-M-2	93 3	5 3	0.0104
ce-NOP-2	96 3	1.7 9	0.0035
ce-K-3	210.0 10	5.3 7	0.0236
ce-K-4	268 10	0.9 5	0.0052
ce-L-3	282.1 10	3.1 4	0.0187
ce-M-3	294.1 10	0.80 11	0.0050
ce-NOP-3	297.1 10	0.26 4	0.0016
ce-L-4	340 10	0.15 8	0.0011
ce-K-8	711.70 10	0.807 25	0.0122
ce-L-8	783.84 10	0.176 6	0.0029
β⁻ 1 max	1382 13		
avg	478 6	2	0.0204
β⁻ 2 max	1602 13		
avg	567 6	7	0.0845
β⁻ 3 max	1858 13		
avg	673 6	24	0.344
β⁻ 4 max	2024 13		
avg	743 6	10	0.158
β⁻ 5 max	2412 13		
avg	907 6	10	0.193
β⁻ 6 max	4204 13		
avg	1683 6	30	1.08
β⁻ 7 max	4384 13		
avg	1762 6	20	0.751

	avg		
Total β⁻	1197 8	103	2.63
X-ray L	10.6	21 5	0.0048
X-ray Kα₂	72.8042 9	2.6 4	0.0040
X-ray Kα₁	74.9694 9	4.3 7	0.0069
γ 1	83 3	2.0 4	0.0035
X-ray Kβ	84.9	1.9 3	0.0035
γ 2	97 3	4.0 20	0.0082
γ 3	298.0 10	79 10	0.503
γ 4	356 10	4.0 20	0.0300
γ 5	382 10	3.0 20	0.0242
γ 6	480 20	2.0 10	0.0202
γ 7	670 20	2.0 10	0.0282
γ 8	799.70 10	98.96 3	1.69
γ 9	860 30	6.9 20	0.127
γ 10	910 30	3.0 20	0.0575
γ 11	1070 10	12 5	0.271
γ 12	1110 20	6.9 20	0.164
γ 13	1210 20	17 4	0.434
γ 14	1316 13	21 5	0.583
γ 15	1410 20	4.9 20	0.149
γ 16	1540 30	2.0 10	0.0649
γ 17	1590 30	2.0 10	0.0670
γ 18	1650 30	2.0 10	0.0696
γ 19	2010 30	6.9 20	0.297
γ 20	2090 30	4.9 20	0.220
γ 21	2270 13	3.0 20	0.144
γ 22	2360 30	8 3	0.398
γ 23	2430 30	9 3	0.461

210PO A DECAY (138.376 D 2) (RaF)

Radiation Type	Energy (keV)	Intensity (%)	Δ(g-rad/μCi-h)
α 1	5304.38 7	100	11.3

232TH A DECAY (1.405E10 Y 6) (Th)

Radiation Type	Energy (keV)	Intensity (%)	Δ(g-rad/μCi-h)
Auger-L	9	8.4 13	0.0016
ce-L-1	39.8 10	16.6 19	0.0141
ce-M-1	54.2 10	4.5 5	0.0052
ce-NOP-1	57.8 10	1.61 18	0.0020
ce-L-2	107 5	0.10 5	0.0002
α 1	3830	0.20 8	0.0163
α 2	3954 8	23.0 20	1.94
α 3	4013 3	77.0 20	6.58
X-ray L	12.3	8.4 13	0.0022
γ 1	59.0 10	0.190 20	0.0002

1 weak γ's omitted (ΣIγ = 0.04%)

228RA B- DECAY (5.75 Y 3) (MsTh1)

Radiation Type	Energy (keV)	Intensity (%)	Δ(g-rad/ μCi-h)
ce-M-2	1.671 5	45	0.0016
ce-NOP-2	5.401 5	15	0.0017
β⁻ 1 max	15.2 10		
avg	3.8 3	40	0.0032
β⁻ 2 max	39.0 10		
avg	9.9 3	60	0.0127
Total β⁻ avg	7.5 4	100	0.0159

228AC B- DECAY (6.13 H) (MsTh2)

Radiation Type	Energy (keV)	Intensity (%)	Δ(g-rad/ μCi-h)
Auger-L	9.48	35 5	0.0072
ce-K-3	19.4 3	0.60 5	0.0002
ce-L-1	37.29 6	58 3	0.0460
ce-M-1	52.58 6	15.8 7	0.0177
ce-NOP-1	56.43 6	5.8 3	0.0070
Auger-K	69.2	0.16 13	0.0002
ce-K-9	74.9 8	4 3	0.0072
ce-L-2	78.98 8	3.7 16	0.0062
ce-M-2	94.27 8	0.9 4	0.0018
ce-NOP-2	98.12 8	0.33 14	0.0007
ce-K-13	99.8 3	0.258 11	0.0005
ce-L-3	108.6 3	5.7 4	0.0133

228AC B- DECAY (6.13 H) (MsTh2) (CONTINUED)

Radiation Type	Energy (keV)	Intensity (%)	Δ(g-rad/ μCi-h)
β⁻14 avg	141.8 20	0.79 7	0.0024
max	490 6		
β⁻15 avg	143.4 20	1.06 10	0.0032
max	590 6		
β⁻16 avg	176.6 21	0.26 4	0.0010
max	597 6		
β⁻17 avg	179.0 21	8 3	0.0305
max	678 6		
β⁻18 avg	206.8 21	0.21 7	0.0009
max	696 6		
β⁻19 avg	213.3 21	3 3	0.0136
max	784 6		
β⁻20 avg	244.2 22	0.13 3	0.0007
max	901 6		
β⁻21 avg	286.6 22	0.68 11	0.0042
max	953 6		
β⁻22 avg	305.5 23	0.20 4	0.0013
max	960 6		
β⁻23 avg	308.0 23	3.06 17	0.0201
max	974 6		
β⁻24 avg	313.4 23	7 4	0.0467
max	1005 6		
β⁻25 avg	324.9 23	5.8 3	0.0401
max	1037 6		
β⁻26 avg	336.6 23	0.41 7	0.0029
max	1106 6		
β⁻27 avg	362.5 23	3.4 3	0.0263
max	1112 6		
β⁻28 avg	364.9 23	0.47 7	0.0037
max	1149 6		
β⁻29 avg	378.8 23	0.21 4	0.0017
max	1159 6		

Radiation	Energy	Intensity	
ce-M-3	123.9 3	1.57 11	0.0042
ce-NOP-3	127.8 3	0.59 4	0.0016
ce-K-18	160.6 3	0.130 6	0.0004
ce-L-9	164.1 8	1.1 8	0.0038
ce-K-19	169.4 10	0.18 7	0.0007
ce-K-25	228.7 3	0.261 11	0.0013
ce-K-26	231.2 12	0.11 9	0.0006
ce-K-34	353.3 3	0.142 6	0.0011
ce-K-69	685.1 3	0.14 9	0.0020
ce-K-82	801.42 3	0.242 10	0.0041
ce-K-89	859.2 5	0.132 6	0.0024
β^- 1 max	118 6		
β^- 1 avg	30.8 17	0.180 20	0.0001
β^- 2 max	183 6		
β^- 2 avg	49.0 18	0.22 3	0.0002
β^- 3 max	228 6		
β^- 3 avg	62.0 18	0.120 20	0.0002
β^- 4 max	234 6		
β^- 4 avg	63.8 18	0.140 20	0.0002
β^- 5 max	331 6		
β^- 5 avg	92.9 19	0.110 20	0.0002
β^- 6 max	368 6		
β^- 6 avg	104.3 19	0.21 3	0.0005
β^- 7 max	384 6		
β^- 7 avg	109.4 19	0.34 5	0.0008
β^- 8 max	393 6		
β^- 8 avg	112.1 20	0.150 20	0.0004
β^- 9 max	404 6		
β^- 9 avg	115.7 20	1.48 9	0.0036
β^- 10 max	440 6		
β^- 10 avg	127.1 20	1.96 11	0.0053
β^- 11 max	445 6		
β^- 11 avg	128.8 20	1.38 10	0.0038
β^- 12 max	482 6		
β^- 12 avg	140.9 20	4.32 12	0.0130
β^- 13 max	485 6		

(continued on next column)

Radiation	Energy	Intensity	
β^- 30 avg	382.8 23	29 4	0.236
β^- 30 max	1732 6		
β^- 31 avg	607.3 25	11.7 5	0.151
β^- 31 max	1941 6		
β^- 32 avg	691.8 25	0.9 7	0.0133
β^- 32 max	2070 6		
avg	744.5 25	14.6 12	0.232
Total β^- avg	398 3	102 8	0.860

4 weak β's omitted (ΣI_β = 0.03%)

Radiation	Energy	Intensity	
X-ray L	13	38 5	0.0106
γ 1	57.76 6	0.508 16	0.0006
X-ray Kα_2	89.9530 20	2.0 9	0.0037
X-ray Kα_1	93.3500 20	3.2 15	0.0064
γ 2	99.45 8	1.2 5	0.0025
X-ray Kβ	105	1.5 7	0.0033
γ 3	129.1 3	2.23 14	0.0061
γ 6	145.9 5	0.20 6	0.0006
γ 7	154.2 3	0.70 3	0.0023
γ 10	191.5 5	0.11 3	0.0004
γ 11	199.5 5	0.32 11	0.0014
γ 12	204.1 6	0.16 6	0.0007
γ 13	209.4 3	3.81 11	0.0170
γ 18	270.3 3	3.44 9	0.0198
γ 19	279.0 10	0.21 7	0.0012
γ 21	321.7 6	0.245 23	0.0017
γ 22	328.0 3	3.10 9	0.0217
γ 24	332.4 5	0.28 3	0.0020
γ 25	338.4 3	11.3 3	0.0814
γ 26	340.9 12	0.403 24	0.0029
γ 30	409.4 5	1.95 7	0.0170
γ 32	440.3 8	0.13 3	0.0012
γ 34	463.0 3	4.50 12	0.0444
γ 36	478.2 8	0.22 5	0.0022
γ 38	503.7 8	0.19 4	0.0020

(continued on next page)

228AC B- DECAY (6.13 H) (MsTh2) (CONTINUED)

Radiation Type	Energy (keV)	Intensity (%)	Δ(g-rad/μCi-h)
γ 39	509.6 10	0.45 11	0.0049
γ 42	523.0 10	0.11 3	0.0012
γ 43	546.3 3	0.20 4	0.0023
γ 44	562.3 5	0.89 5	0.0107
γ 45	570.7 10	0.21 3	0.0026
γ 46	572.1 10	0.15 5	0.0018
γ 47	583.20 20	0.14 4	0.0017
γ 60	701.5 8	0.18 4	0.0027
γ 61	707.3 10	0.14 4	0.0021
γ 62	727.0 8	0.73 21	0.0113
γ 64	755.2 5	1.04 6	0.0167
γ 65	772.1 5	1.45 6	0.0238
γ 67	782.0 5	0.48 6	0.0080
γ 69	794.8 3	4.34 11	0.0735
γ 73	830.4 5	0.50 5	0.0088
γ 74	835.6 5	1.53 9	0.0272
γ 75	840.2 5	0.97 5	0.0174
γ 81	904.2 5	0.79 4	0.0152
γ 82	911.07 3	26.6 7	0.516
γ 86	948.0 8	0.110 20	0.0022
γ 87	958.5 5	0.29 6	0.0059
γ 88	964.6 5	5.05 14	0.104
γ 89	968.9 5	16.2 4	0.334
γ 91	988.1 5	0.17 3	0.0036
γ 96	1033.1 5	0.21 3	0.0046
γ 100	1065.1 5	0.130 20	0.0029
γ 101	1095.7 5	0.121 18	0.0028
γ 103	1110.6 5	0.32 5	0.0076
γ 108	1153.6 5	0.145 21	0.0036
γ 113	1247.1 5	0.58 5	0.0154
γ 116	1287.5 5	0.108 16	0.0030
γ 126	1459.2 5	0.78 4	0.0242
γ 129	1495.8 5	0.84 7	0.0268

228TH A DECAY (1.913 Y 2) (RdTh) (CONTINUED)

Radiation Type	Energy (keV)	Intensity (%)	Δ(g-rad/μCi-h)
X-ray L	12.3	9.2 10	0.0024
γ 1	84.40 5	1.17 4	0.0021
γ 2	131.62 20	0.110 18	0.0003
γ 5	215.94 8	0.268 7	0.0012

2 weak γ's omitted (ΣIγ = 0.11%)

224RA A DECAY (3.66 D 4) (ThX)

Radiation Type	Energy (keV)	Intensity (%)	Δ(g-rad/μCi-h)
Auger-L	8.7	0.46 5	≈0
ce-K-1	142.60 10	0.454 17	0.0014
ce-L-1	222.95 10	0.502 18	0.0024
ce-M-1	236.53 10	0.133 5	0.0007
α 1	5448.7 12	4.9 4	0.569
α 2	5685.42 15	95.1 4	11.5

3 weak α's omitted (ΣIα = 0.02%)

X-ray L	11.7	0.41 5	0.0001
X-ray Kα2	81.070 20	0.129 6	0.0002
X-ray Kα1	83.780 20	0.215 9	0.0004
γ 1	241.00 10	4.05 8	0.0208

3 weak γ's omitted (ΣIγ = 0.02%)

220RN A DECAY (55.6 S 1) (Tn)

Radiation Type	Energy (keV)	Intensity (%)	Δ(g-rad/μCi-h)
α 1	6288.13 10	99.930 20	13.4

1 weak α's omitted (ΣIα = 0.07%)

1 weak γ's omitted (ΣIγ = 0.07%)

216PO A DECAY (0.145 S 2) (ThA)

Radiation Type	Energy (keV)	Intensity (%)	Δ(g-rad/μCi-h)
α 1	6778.3 5	99.9979 4	14.4

	Energy (keV)	Intensity (%)	Δ(g-rad/μCi-h)
γ 130	1501.5 5	0.43 5	0.0138
γ 134	1556.9 6	0.18 4	0.0060
γ 136	1580.2 5	0.52 9	0.0175
γ 137	1587.9 4	3.26 10	0.110
γ 139	1624.7 5	0.29 6	0.0100
γ 140	1630.4 4	1.53 9	0.0531
γ 141	1638.0 5	0.46 4	0.0160
γ 142	1666.4 6	0.19 4	0.0067
γ 154	1887.4 6	0.100 20	0.0040

94 weak γ's omitted (ΣIγ = 3.33%)

228TH A DECAY (1.913 Y 2) (RdTh)

Radiation Type	Energy (keV)	Intensity (%)	Δ(g-rad/μCi-h)
Auger-L	9	9.2 10	0.0018
ce-L-1	65.17 5	18.5 4	0.0257
ce-M-1	79.58 5	4.99 16	0.0085
ce-NOP-1	83.19 5	1.80 6	0.0032
α 1	5175	0.18	0.0198
α 2	5212	0.36	0.0400
α 3	5340.31 15	26.70 20	3.04
α 4	5423.20 22	72.7 4	8.40

1 weak α's omitted (ΣIα = 0.05%)

(continued on next column)

212PB B- DECAY (10.64 H 1) (ThB)

Radiation Type	Energy (keV)	Intensity (%)	Δ(g-rad/μCi-h)
Auger-L	8.15	21.0 17	0.0036
ce-K-1	24.650 7	3.43 12	0.0018
Auger-K	58.2	1.2 6	0.0014
ce-L-1	98.788 7	0.605 21	0.0013
ce-M-1	111.177 7	0.141 5	0.0003
ce-K-4	148.107 4	32.5 11	0.102
ce-K-5	209.561 10	1.22 6	0.0055
ce-L-4	222.245 4	5.62 19	0.0266
ce-M-4	234.634 4	1.33 5	0.0066
ce-NOP-4	237.695 4	0.443 15	0.0022
ce-L-5	283.699 10	0.216 6	0.0013
β- 1 max	156 4		
avg	41.4 12	4.94 10	0.0044
β- 2 max	332 4		
avg	93.9 13	83.6 12	0.167
β- 3 max	571 4		
avg	172.1 14	11.3 12	0.0414
Total β-			
avg	100.2 14	99.8 17	0.213
X-ray L	10.8	15.2 16	0.0035
X-ray Kα2	74.8148 10	10.5 4	0.0167
X-ray Kα1	77.1079 10	17.7 6	0.0290
X-ray Kβ	87.3	7.9 3	0.0147
γ 1	115.176 7	0.593 9	0.0015
γ 4	238.633 4	43.9 6	0.223
γ 5	300.087 10	3.24 5	0.0207

3 weak γ's omitted (ΣIγ = 0.08%)

212BI A DECAY (60.55 M 6) (ThC)
%A=35.94 6
SEE ALSO 212BI B- DECAY (60.55 M)

Radiation Type	Energy (keV)	Intensity (%)	Δ(g-rad/μCi-h)
Auger-L	7.78	12.3 10	0.0020
ce-L-1	24.511 4	20.5 8	0.0107
ce-M-1	36.154 4	4.77 18	0.0037
ce-NOP-1	39.012 4	1.59 6	0.0013
ce-K-5	202.54 7	0.115 14	0.0005
α 1	5607.1 3	0.4025 7	0.0481
α 2	5768.1 3	0.600 8	0.0737
α 3	6050.83 4	25.23 9	3.25
α 4	6089.94 4	9.63 8	1.25

4 weak α's omitted (ΣIα = 0.06%)

X-ray L	10.3	8.2 9	0.0018
γ 1	39.858 4	1.10 3	0.0009
γ 5	288.07 7	0.31 4	0.0019
γ 7	327.96 10	0.140 11	0.0010
γ 9	452.83 10	0.32 7	0.0031

9 weak γ's omitted (ΣIγ = 0.11%)

212BI B- DECAY (60.55 M 6) (ThC)
%B-=64.07 6
SEE ALSO 212BI A DECAY (60.55 M)

Radiation Type	Energy (keV)	Intensity (%)	Δ(g-rad/μCi-h)
β⁻ 1 max	440 4		
avg	128.1 13	0.70 5	0.0019
β⁻ 2 max	567 4		
avg	170.3 14	0.257 16	0.0009
β⁻ 3 max	625 4		
avg	190.6 14	1.74 9	0.0071
β⁻ 4 max	733 4		
avg	228.7 15	1.39 7	0.0068
β⁻ 5 max	1519 4		
avg	530.7 17	4.84 20	0.0547
β⁻ 6 max	2246 4		
avg	831.6 17	55.10 22	0.976
Total β⁻			
avg	767.7 19	64.1 4	1.05

1 weak β's omitted (Σiβ = 0.03%)

γ 4	727.18 6	7.00 18	0.108
γ 5	785.42 6	1.07 5	0.0179
γ 6	893.39 7	0.35 3	0.0067
γ 7	952.10 10	0.185 10	0.0038
γ 9	1078.62 10	0.58 4	0.0133
γ 11	1512.75 10	0.28 4	0.0090
γ 12	1620.56 7	1.38 8	0.0476
γ 15	1806.0 5	0.117 23	0.0045

8 weak γ's omitted (Σiγ = 0.09%)

212PO A DECAY (0.298 US 3) (ThC')

Radiation Type	Energy (keV)	Intensity (%)	Δ(g-rad/μCi-h)
α 1	8784.15 7	100	18.7

208TL B- DECAY (3.053 M 4) (ThC")

Radiation Type	Energy (keV)	Intensity (%)	Δ(g-rad/μCi-h)
Auger-L	8	4.1 3	0.0007
Auger-K	56.7	0.23 10	0.0003
ce-K-1	123.40 15	0.172 20	0.0005
ce-K-2	145.36 15	0.224 16	0.0007
ce-K-3	164.61 10	0.41 6	0.0014
ce-K-4	189.347 10	2.83 13	0.0114
ce-L-4	261.490 10	0.486 22	0.0027
ce-MNO-4	273.500 10	0.151 5	0.0009
ce-K-7	422.77 10	1.99 7	0.0179
ce-L-7	494.91 10	0.335 12	0.0035
ce-K-8	495.1865 22	1.29 4	0.0136
ce-MNO-7	506.92 10	0.1117 20	0.0012
ce-L-8	567.3302 21	0.353 11	0.0043
ce-K-16	772.559 5	0.282 9	0.0046
β⁻ 1 max	810 4		
avg	257.9 15	0.227 12	0.0012
β⁻ 2 max	865 4		
avg	278.2 15	0.168 11	0.0010

(continued on next page)

208TL B- DECAY (3.053 M 4) (ThC") (CONTINUED)

Radiation Type	Energy (keV)	Intensity (%)	Δ(g-rad/μCi-h)
β- 3 max	1029 4		
avg	340.2 16	3.09 17	0.0224
β- 4 max	1070 4		
avg	356.1 16	0.63 6	0.0048
β- 5 max	1282 4		
avg	438.8 16	24.5 6	0.229
β- 6 max	1515 4		
avg	532.6 17	21.8 4	0.247
β- 7 max	1792 4		
avg	646.6 17	48.7 12	0.671
Total β-			
avg	556.6 18	99.3 15	1.18

6 weak β's omitted (ΣIβ = 0.20%)

Radiation Type	Energy (keV)	Intensity (%)	Δ(g-rad/μCi-h)
X-ray L	10.6	2.9 3	0.0006
X-ray Kα2	72.8042 9	2.03 6	0.0031
X-ray Kα1	74.9694 9	3.42 10	0.0055
X-ray Kβ	84.9	1.52 5	0.0028
γ 1	211.40 15	0.180 20	0.0008
γ 2	233.36 15	0.310 20	0.0015
γ 3	252.61 10	0.71 9	0.0038
γ 4	277.351 10	6.30 20	0.0372
γ 7	510.77 10	22.8 4	0.248
γ 8	583.1910 20	84.8 8	1.05
γ 12	722.04 12	0.23 3	0.0035
γ 14	763.13 8	1.82 4	0.0296
γ 16	860.564 5	12.52 12	0.229
γ 18	927.60 20	0.132 9	0.0026
γ 19	982.70 20	0.205 11	0.0043
γ 21	1093.90 20	0.40 3	0.0093
γ 29	2614.533 13	100	5.57

16 weak γ's omitted (ΣIγ = 0.36%)

239NP B- DECAY (2.355 D 4) (CONTINUED)

Radiation Type	Energy (keV)	Intensity (%)	Δ(g-rad/μCi-h)
ce-M-3	38.717 20	1.6 4	0.0013
ce-NOP-3	43.091 20	0.44 11	0.0004
ce-M-4	43.477 20	2.6 6	0.0024
ce-L-8	44.78 3	6.6 17	0.0063
ce-NOP-4	47.851 20	0.81 18	0.0008
ce-M-5	51.327 20	4.8 7	0.0052
ce-MNO-7	55.527 3	0.106 16	0.0001
ce-NOP-5	55.701 20	1.76 25	0.0021
ce-K-15	59.92 6	0.34 7	0.0004
ce-M-8	61.95 3	1.8 5	0.0024
ce-NOP-8	66.32 3	0.64 16	0.0009
Auger-K	76	1.33 20	0.0022
ce-L-11	83.019 6	4.3 8	0.0077
ce-L-12	83.37 4	0.42 8	0.0007
ce-K-16	87.962 3	7.8 9	0.0147
ce-M-11	100.190 3	1.14 20	0.0024
ce-MNO-12	100.54 4	0.140 25	0.0003
ce-NOP-11	104.5640 23	0.39 8	0.0009
ce-K-17	104.63 8	0.6 4	0.0013
ce-K-18	106.3920 23	21.0 19	0.0477
ce-K-19	132.62 8	0.18 5	0.0005
ce-K-20	151.05 7	0.117 22	0.0004
ce-K-21	155.8080 23	17.3 6	0.0575
ce-L-16	186.649 6	1.63 18	0.0065
ce-L-17	203.32 8	0.140 19	0.0006
ce-L-18	205.079 5	4.4 4	0.0193
ce-MNO-18	222.2500 23	1.51 9	0.0072
ce-L-21	254.495 5	3.55 12	0.0192
ce-MNO-21	271.6660 23	1.136 16	0.0066
β- 1 max	210.2 20		
avg	57.0 10	1.80 20	0.0022
β- 2 max	330.4 20		

238PU A DECAY (87.74 Y 4) I(MIN)= 0.10%

Radiation Type	Energy (keV)	Intensity (%)	Δ(g-rad/μCi-h)
Auger-L	9.89	9.1 13	0.0019
ce-L-2	21.722 10	20.7 8	0.0096
ce-M-2	37.932 10	5.69 22	0.0046
ce-NOP-2	42.039 10	1.88 8	0.0017
α 1	5357.7	0.10 3	0.0114
α 2	5456.30 20	28.3 6	3.29
α 3	5499.07 20	71.6 6	8.39
X-ray L	13.6	11.6 14	0.0034

34 weak γ's omitted (ΣIγ = 0.05%)

239NP B- DECAY (2.355 D 4) I(MIN)= 0.10%

Radiation Type	Energy (keV)	Intensity (%)	Δ(g-rad/μCi-h)
ce-M-1	1.9170 20	6.6 6	0.0003
ce-NOP-1	6.2910 10	2.20 18	0.0003
Auger-L	10.3	37 6	0.0081
ce-L-3	21.546 21	7.1 18	0.0033
ce-L-4	26.306 21	9.6 21	0.0054
ce-L-5	34.156 21	17.0 24	0.0124
ce-L-7	38.356 6	0.33 5	0.0003

(continued on next column)

Radiation Type	Energy (keV)	Intensity (%)	Δ(g-rad/μCi-h)
β⁻ 3 avg	93.0 10	33.0 20	0.0654
β⁻ 3 max	391.9 20		
β⁻ 4 avg	112.0 10	7 3	0.0167
β⁻ 4 max	436.6 20		
β⁻ 5 avg	126.0 10	53 5	0.142
β⁻ 5 max	664.8 20		
β⁻ 6 avg	201.0 10	2.0 10	0.0086
β⁻ 6 max	714.2 20		
avg	219.0 10	4.0 20	0.0187
Total β⁻ avg	118.1 11	101 7	0.254

4 weak β's omitted (ΣIβ = 0.03%)

Radiation Type	Energy (keV)	Intensity (%)	Δ(g-rad/μCi-h)
X-ray L	14.3	56 7	0.0170
γ 4	49.410 20	0.101 22	0.0001
γ 5	57.260 20	0.104 15	0.0001
γ 7	61.4600 20	0.97 15	0.0013
X-ray Kα2	99.55 5	13.7 7	0.0292
X-ray Kα1	103.76 5	22.1 11	0.0488
γ 11	106.1230 20	22.9 12	0.0517
X-ray Kβ	117	10.4 5	0.0259
γ 15	181.71 6	0.112 15	0.0004
γ 16	209.7530 20	3.27 24	0.0146
γ 17	226.42 8	0.34 5	0.0016
γ 18	228.1830 10	10.8 6	0.0525
γ 19	254.41 8	0.101 17	0.0005
γ 21	277.5990 10	14.20 20	0.0840
γ 22	285.4600 20	0.78 8	0.0047
γ 23	315.880 3	1.60 11	0.0108
γ 24	334.3100 20	2.04 18	0.0146

21 weak γ's omitted (ΣIγ = 0.39%)

240PU A DECAY (6564 Y 11)   I(MIN)= 0.10%

Radiation Type	Energy (keV)	Intensity (%)	Δ(g-rad/μCi-h)
Auger-L	9.89	8.9 13	0.0019
ce-L-1	23.484 6	20.2 7	0.0101
ce-M-1	39.694 6	5.56 19	0.0047
α 1	5123.66 16	26.39 21	2.88
α 2	5168.17 15	73.5 4	8.09

6 weak α's omitted (ΣIα = 0.07%)

| X-ray L | 13.6 | 11.3 13 | 0.0033 |

13 weak γ's omitted (ΣIγ = 0.05%)

241AM A DECAY (432.2 Y 5)   I(MIN)= 0.10%

Radiation Type	Energy (keV)	Intensity (%)	Δ(g-rad/μCi-h)
ce-L-2	3.9180 23	14 5	0.0012
Auger-L	10	30 5	0.0066
ce-L-5	10.778 11	17 3	0.0040
ce-L-7	20.30 5	0.324 23	0.0001
ce-M-2	20.606 5	3.9 5	0.0017
ce-L-8	20.996 11	9.1 12	0.0040
ce-M-5	27.466 12	4.4 7	0.0026
ce-L-11	33.133 20	0.89 12	0.0006
ce-L-14	37.1100 23	30.2 22	0.0238
ce-M-8	37.684 12	2.4 3	0.0019
ce-M-11	49.821 21	0.24 3	0.0003
ce-M-14	53.798 5	8.1 3	0.0093

242PU A DECAY (3.76E+5 Y 2) (CONTINUED)

Radiation Type	Energy (keV)	Intensity (%)	Δ(g-rad/μCi-h)
ce-L-1	23.157 13	16.3 24	0.0080
ce-M-1	39.367 13	4.5 7	0.0038
ce-NOP-1	43.474 13	1.48 21	0.0014
α 1	4856.2 12	22.4 20	2.32
α 2	4900.5 12	78 3	8.09

2 weak α's omitted (ΣIα = 0.10%)

| X-ray L | 13.6 | 9.1 17 | 0.0026 |

3 weak γ's omitted (ΣIγ = 0.04%)

242CM A DECAY (162.8 D 4)   I(MIN)= 0.10%

Radiation Type	Energy (keV)	Intensity (%)	Δ(g-rad/μCi-h)
Auger-L	10.3	7.6 12	0.0017
ce-L-1	20.98 3	19.0 4	0.0085
ce-M-1	38.15 3	5.25 15	0.0043
ce-NOP-1	42.52 3	1.74 7	0.0016
α 1	6069.42 12	25.9 5	3.35
α 2	6112.77 8	74.1 5	9.65

6 weak α's omitted (ΣIα = 0.04%)

| X-ray L | 14.3 | 11.4 12 | 0.0035 |

9 weak γ's omitted (ΣIγ = 0.04%)

Radiation Type	Energy (keV)	Intensity (%)	Δ(g-rad/μCi-h)
ce-NOP-14	58.0360 15	34 4	0.0417
ce-L-22	76.543 20	0.229 8	0.0004
α 1	5388.0 10	1.40 20	0.161
α 2	5442.98 13	12.80 20	1.48
α 3	5485.74 12	85.2 8	9.96
α 4	5512.0 20	0.20 5	0.0235
α 5	5544.3 3	0.34 5	0.0402

25 weak α's omitted (ΣIα = 0.03%)

X-ray L	13.9	42 6	0.0125
γ 2	26.3450 10	2.41 6	0.0014
γ 5	33.205 10	0.1260 16	≈0
γ 14	59.5370 10	35.9 4	0.0455

162 weak γ's omitted (ΣIγ = 0.17%)

241PU B- DECAY (14.4 Y 2) I(MIN)= 0.10%

Radiation Type	Energy (keV)	Intensity (%)	Δ(g-rad/μCi-h)
β⁻ 1 max	20.82 20		
avg	5.23 5	100	0.0111

242PU A DECAY (3.76E+5 Y 2) I(MIN)= 0.10%

Radiation Type	Energy (keV)	Intensity (%)	Δ(g-rad/μCi-h)
Auger-L	9.89	7.2 15	0.0015

(continued on next column)

243AM A DECAY (7380 Y 40) I(MIN)= 0.10%

Radiation Type	Energy (keV)	Intensity (%)	Δ(g-rad/μCi-h)
ce-L-1	8.67 15	8.6 12	0.0016
Auger-L	10	13.7 22	0.0030
ce-L-2	20.6730 20	6.9 8	0.0030
ce-L-3	21.10 15	4.4 6	0.0020
ce-M-1	25.36 15	2.1 3	0.0011
ce-NOP-1	29.60 15	0.70 10	0.0004
ce-M-2	37.361 5	1.82 19	0.0014
ce-M-3	37.79 15	1.10 14	0.0009
ce-NOP-2	41.5990 10	0.61 7	0.0005
ce-NOP-3	42.03 15	0.36 5	0.0003
ce-L-6	52.24 15	12.6 14	0.0140
ce-M-6	68.93 15	3.1 4	0.0046
ce-NOP-6	73.17 15	1.10 12	0.0017
α 1	5181.0 10	1	0.121
α 2	5233.0 5	10.6	1.18
α 3	5275.3 5	87.9	9.88
α 4	5321.0 10	0.12	0.0136
α 5	5350.0 10	0.16	0.0182

10 weak α's omitted (ΣIα = 0.01%)

X-ray L	13.9	19.0 23	0.0056
γ 3	43.53 15	5.0 7	0.0047
γ 6	74.67 15	60 6	0.0954
γ 7	86.79 15	0.31 5	0.0006
γ 9	117.60 15	0.50 9	0.0013
γ 10	142.18 15	0.114 17	0.0003

12 weak γ's omitted (ΣIγ = 0.15%)

Appendix B

The Statistics of Radioactive Decay

B.1 The Experiments of Rutherford and Soddy

In the early experiments of Rutherford and Soddy, it was observed that, in the cases of all separated radioactive materials, the activity decreased in a geometric progression, which they expressed as

$$\frac{I_t}{I_0} = e^{-\lambda t}, \tag{B.1}$$

where I_0 was the ionization current, due to the radiations, observed at time $t = 0$, I_t was the ionization current observed at time t, and λ was a constant (Rutherford and Soddy, 1902).

On the assumption that each "ray or projected particle" would produce a "definite number of ions in its path," and that "each changing system gives rise to one ray," these authors concluded that

$$\frac{N_t}{N_0} = e^{-\lambda t}, \tag{B.2}$$

where N_t was the number of systems unchanged at time t and N_0 was the number initially present.

Treating N_t as a continuous function of t and differentiating, they deduced that "the rate of change of the system at any time," what would now be called the activity of a radionuclide at time t, "is always proportional to the amount remaining unchanged," namely

$$\frac{dN_t}{dt} = -\lambda N_t. \tag{B.3}$$

From this equation we see that λ, now known as the decay constant, is also the fractional rate (100λ is the percentage rate) at which N_t atoms at time t are decaying.

B.2 The Statistical Approach of von Schweidler

At the International Congress of Radiology held in Liège in 1905, von Schweidler (1906) pointed out that on the assumption that the probability of an atom disintegrating in time dt is λdt, then the probability of its surviving for time t would be $e^{-\lambda t}$, and its mean life would be $1/\lambda$. The proof of this result was given some years later in their book *Radioactivität*, by Stefan Meyer and E. von Schweidler (1927).

In this proof it was assumed that every atom was an entity on its own, that its decay was a statistical process governed by the laws of chance, and that its probability of decay is completely uninfluenced by the decay probability of any neighboring atoms. It was also assumed that the probability of decay of an individual atom was wholly independent of its previous history.

Meyer and von Schweidler then considered a small interval of time Δt and said that the probability of an atom decaying in this interval was proportional to the magnitude of that interval of time, and they set it equal to $\lambda \Delta t$, where λ was a constant characteristic of a given radioactive element. The probability p of an atom surviving this small interval of time is therefore

$$p = (1 - \lambda \Delta t).\tag{B.4}$$

Now consider the probability of an atom surviving the next interval of time Δt. As the atom knows nothing of its previous history, the probability of surviving the second interval is again equal to p and the probability of surviving both intervals is p^2. Considering k such intervals of time from $t = 0$ to time t such that $k\Delta t = t$, then in the limit, as k tends to become infinitely large, the probability of one atom surviving for a time t is given by

$$p = \lim_{k \to \infty} (1 - \lambda \Delta t)^k = \lim_{k \to \infty} \left(1 - \frac{\lambda t}{k}\right)^k = e^{-\lambda t}.\tag{B.5}$$

The probability of the complementary event of the atom disintegrating in the interval of time between $t = 0$ and time t is

$$w = 1 - e^{-\lambda t}.\tag{B.6}$$

If N_0 is the initial number of radioactive atoms at $t = 0$, then, because the probability, $e^{-\lambda t}$, of each atom surviving beyond time t is independent of that of all other atoms, then Eq. B.2, the empirically discovered law of Rutherford and Soddy, follows from the addition theorem for independent events. (By the addition theorem, the chance of getting either "heads" or "tails" is ½ + ½, in the toss of a coin.)

B.3 The Distribution of Observed Decays at a Given Time

B.3.1 *The Bernoulli or Binomial Distribution*

Suppose, now, that a series of experiments is conducted in which N_0 atoms of a radionuclide are initially present, and the number of decays n, in a fixed interval of time t, is determined. In actual practice such an experiment could be carried out by using a very long-lived substance such as ^{14}C, where the observation of n events, in a time t that is negligible compared with the half life of ^{14}C, leaves N_0 essentially unchanged during a series of measurements of n.

Because radioactivity is a random, or stochastic, process, the measured value of n will vary from one measurement to the other. If, however, a sufficiently large number of measurements of n is carried out, the mean value, $\bar{n}$, will approach the product of the total number of atoms and the probability of decay in time t, namely, $N_0(1 - e^{-\lambda t})$.

But because the value of n, determined for a fixed time t, varies from one measurement to another, it is of interest to know how n varies, and in what manner it is distributed around $\bar{n}$.

Consider the probability p_n of exactly n of N_0 atoms decaying in time t. The probability of a single atom decaying in the interval $t = 0$ to time t is $1 - e^{-\lambda t}$ and the probability of a single atom surviving beyond time t is $e^{-\lambda t}$.

By way of example, consider the probability of exactly one atom decaying ($n = 1$). Suppose that a measurement starts with N_0 atoms at time $t = 0$ and is "frozen" at time t. At that point the decayed atoms are placed into a box A and undecayed atoms into a box B. Suppose, further, that the atoms are identified $1, 2, 3 \cdots N_0$.

During the time interval from zero to time t of each measurement, the probability of one given atom decaying is w, and of $(N_0 - 1)$ other atoms surviving is $(1 - w)^{N_0-1}$ (see Eq. B.6).

Therefore, on stopping the measurement at time t, the probability of finding one given atom in box A, and the rest in box B, is

$$p_1 = w(1 - w)^{N_0-1} = (1 - e^{-\lambda t})(e^{-\lambda t})^{N_0-1}. \qquad (B.7)$$

But it is immaterial *which* of the N_0 atoms finds its way into box A, because we are interested only in the number distribution of atoms in box A and box B at time t, and not the order in which they are placed in the two boxes. In other words, the one atom in box A can be any one of $1, 2, 3, \cdots N_0$. Thus, the probability of finding *any* atom in box A is greater than that of finding a *given* atom in box A by a factor N_0, and is given by

$$P_1 = N_0 p_1 = N_0(1 - e^{-\lambda t})(e^{-\lambda t})^{N_0-1}. \tag{B.8}$$

Similarly, the probability of finding any two atoms in box A is

$$P_2 = \frac{N_0(N_0 - 1)}{2!}(1 - e^{-\lambda t})^2(e^{-\lambda t})^{N_0-2}, \tag{B.9}$$

for, again, we are only interested in the number of possible combinations of two atoms in box A $(1 + 2, 1 + 3, \cdots 2 + 3, 2 + 4, \cdots$ etc.) and not in the order in which they arrive; i.e., not in the number of permutations.

Thus, in the general case, the probability of exactly n decays occurring in time t, P_n, is given by the general term of the binomial expansion, namely

$$P_n = \frac{N_0(N_0 - 1) \cdots (N_0 - n + 1)}{n!}(1 - e^{-\lambda t})^n(e^{-\lambda t})^{N_0-n}. \tag{B.10}$$

This distribution is known as the Bernoulli or binomial distribution. It gives the distribution of the probabilities of decays, in time t, occurring for values of n equal to 1, 2, 3, $\cdots$ N_0. In general, the region of interest is in the values of n around average value of $\bar{n}$, e.g., $\bar{n} + 1, \bar{n} + 2, \cdots \bar{n} - 1, \bar{n} - 2, \cdots$. This distribution function was also derived by von Schweidler in 1903.

If the term $((1 - w) + w)^{N_0}$, where $w = 1 - e^{-\lambda t}$, is expanded by means of the binomial theorem, then

$$((1 - w) + w)^{N_0} = (1 - w)^{N_0} + N_0(1 - w)^{N_0-1}w + \cdots$$

$$+ \frac{N_0(N_0 - 1) \cdots (N_0 - n + 1)}{n!}(1 - w)^{N_0-n}w^n + \cdots$$

$$+ N_0(1 - w)w^{N_0-1} + w^{N_0}. \tag{B.11}$$

The first and last terms on the right-hand side of Eq. B.11 correspond respectively to the probability of finding no atoms in box A and all N_0 atoms in box A.

The left-hand side of this equation is identically equal to unity, which simply means that the sum of all probabilities of every possible decay occurring between $n = 0$ and $n = N_0$ is equal to unity; i.e.,

$$\sum_{n=0}^{n=N_0} P_n = 1. \tag{B.12}$$

Suppose that, in a large number of experiments each carried out for time t, ν_n is the frequency of n, i.e., the number of times *exactly* n decays occur in time t. Then as we increase the number of experiments,

$\nu_n/\sum_0^{N_0} \nu_n$ approaches closer and closer to the probability P_n, so that for a large number of experiments

$$\nu_n/\sum_0^{N_0} \nu_n = P_n. \tag{B.13}$$

The mean value, or the expectation value $E(n)$ of n is given by

$$E(n) = \frac{\sum_0^{N_0} \nu_n n}{\sum_0^{N_0} \nu_n} = \sum_0^{N_0} P_n n. \tag{B.14}$$

Using the value for P_n given in Eq. B.10, the expectation value has been calculated (Friedlander *et al.*, 1964; Stevenson, 1966) to be

$$E(n) = wN_0 = N_0(1 - e^{-\lambda t}). \tag{B.15}$$

Consideration of the decay of, say, one microcurie of ^{14}C, for which N_0 is of the order of 10^{16} atoms, will give some appreciation for the orders of magnitude involved. The decay rate is 3.7×10^4 s^{-1}, and, for $t = 10$ min, the expectation value for the number of disintegrations, $E(n)$, will be of the order of 10^7. An observation time of $t = 10$ min is of the order of 10^{-9} of the half life (5730 y), and the probability of surviving 10 min, $1 - w = e^{-\lambda t}$, is less than unity by only 2 in 10^9.

The binomial distribution has been derived for actual decay rates, but the same considerations apply to *count* rates observed when using radiation detectors having efficiencies $\epsilon < 1$. The probability of recording a decay will be scaled down by ϵ so that the expectation value for the number of counts in time t will be $\epsilon N_0 (1 - e^{-\lambda t})$.

B.3.2 *The Poisson Distribution*

If, as was the case with ^{14}C, $N_0 \gg 1$, $n \ll N_0$, and $\lambda t \ll 1$, then the binomial or Bernoulli distribution can be greatly simplified. Equation B.10 can be written in the form

$$P_n = \frac{N_0(N_0 - 1) \cdots (N_0 - n + 1)}{n!} w^n(1 - w)^{N_0 - n}. \tag{B.16}$$

But, for $N_0 \gg 1$ and $n \ll N_0$,

$$N_0 \approx N_0 - 1 \approx N_0 - 2 \cdots$$

and

$$N_0(N_0 - 1) \cdots (N_0 - n + 1) \approx N_0^n.$$

Also, for $\lambda t \ll 1$, w is a very small quantity, and then,

$$(1 - w)^{N_0 - n} \approx (1 - w)^{N_0}$$

$$= 1 - wN_0 + \frac{N_0(N_0 - 1)}{2!} w^2 - \cdots$$

$$\approx 1 - wN_0 + \frac{(wN_0)^2}{2!} - \cdots$$

$$\approx e^{-wN_0}.$$

Substituting these results in equation B.16 gives

$$P_n = \frac{(wN_0)^n e^{-wN_0}}{n!}, \tag{B.17}$$

which is the Poisson distribution. This distribution also gives a mean value, or expectation value, of $E(n) = wN_0$. The Poisson distribution is not symmetrical about the mean value as it follows that P_{wN_0} is equal to P_{wN_0-1}, i.e.,

$$P_{wN_0} = \frac{(wN_0)^{wN_0} e^{-wN_0}}{(wN_0)!}$$

$$= \frac{(wN_0)^{wN_0-1} e^{-wN_0}}{(wN_0 - 1)!}$$

$$= P_{wN_0-1}.$$

B3.3   *The Gaussian or Normal Distribution.*

For larger values of n, greater than, for example, 100 (Friedlander *et al.*, 1964), the Stirling formula can be used to give an approximate value of $E(n)!$, namely

$$E(n)! \approx E(n)^{E(n)} e^{-E(n)} \sqrt{2\pi E(n)}. \tag{B.18}$$

If we use $E(n)$ for wN_0 in Eq. B.17, then

$$P_n = \frac{[E(n)]^n e^{-E(n)}}{n!}. \tag{B.19}$$

Following Stevenson (1966), we can now transform the variable n to $x = [n - E(n)]$. This means that instead of taking n from 0 to N_0 we have x distributed, plus and minus, around the expected value of n, i.e., $E(n) \pm 1$, $E(n) \pm 2$, $\cdots$. Then,

$$P_x = \frac{[E(n)]^{E(n)+x}e^{-E(n)}}{[E(n) + x]!}$$

$$= \frac{[E(n)^{E(n)}[E(n)]^x e^{-E(n)}}{[E(n)]![E(n) + 1][E(n) + 2] \cdots [E(n) + x]} \qquad \text{(B.20)}$$

$$= \frac{[E(n)]^{E(n)}e^{-E(n)}}{[E(n)]!} \frac{[E(n)]^x}{[E(n) + 1][E(n) + 2] \cdots [E(n) + x]},$$

whence, substituting Eq. B.18,

$$P_x = \frac{1}{\sqrt{2\pi E(n)}} \frac{1}{\left[1 + \dfrac{1}{E(n)}\right]\left[1 + \dfrac{2}{E(n)}\right] \cdots \left[1 + \dfrac{x}{E(n)}\right]}. \qquad \text{(B.21)}$$

If now we restrict consideration of the distribution to a range of values, x, around the expectation value, $E(n)$, such that $|E(n) - n| \ll E(n)$, i.e., that $|x| \ll E(n)$, then the following approximate relations hold:

$$e^{1/E(n)} \approx 1 + 1/E(n),$$

and

$$e^{2/E(n)} \approx 1 + 2/E(n), \cdots, e^{x/E(n)} \approx 1 + x/E(n). \qquad \text{(B.22)}$$

Therefore,

$$[1 + 1/E(n)][1 + 2/E(n)] \cdots [1 + x/E(n)] \qquad \text{(B.23)}$$
$$\approx e^{(1+2+\cdots x)/E(n)} \approx e^{(x^2+x)/2E(n)}.$$

Substitution in Eq. B.21, neglecting x compared with x^2, gives

$$P_x = \frac{1}{\sqrt{2\pi E(n)}} e^{-x^2/2E(n)}, \qquad \text{(B.24)}$$

which is the ordinate of the Gaussian or normal distribution at point x, and which is seen to be symmetrical about $E(n)$.

B.3.4 *Deviations from the Mean*

In determining the expectation value of a distribution $E(n)$ (designated also as $\langle n \rangle$), it is also important to be able to obtain a measure of the deviations, or the scatter, of measured values on either side of the expectation value or mean.

One indication of the scatter would be given by a plot of the distribution of $(n - \bar{n})$, but the mean value of all $(n - \bar{n})$'s, the mean deviation, of course, vanishes as, in taking the sum, values that are larger or smaller than the mean must cancel out by the very nature of the concept of the mean. A measure of the scatter would, however, be given by the mean of the absolute deviations, namely $\langle |n - \bar{n}| \rangle$. Another measure is the mean of the squares of the deviations, $\langle (n - \bar{n})^2 \rangle$, known as the mean square deviation.

The mean square deviation is the mean value, for all values of n, of $(n - \bar{n})^2$, that is,

$$\langle (n - \bar{n})^2 \rangle = \frac{\sum\limits_{n} \nu_n (n - \bar{n})^2}{\sum\limits_{n} \nu_n}$$

$$= \sum_{n} P_n (n - \bar{n})^2$$

$$= \sum_{n} P_n n^2 - 2 \sum_{n} P_n n \bar{n} + \sum_{n} P_n \bar{n}^2 \qquad \text{(B.25)}$$

$$= \langle n^2 \rangle - \langle n \rangle^2,$$

or

$$\langle (n - \bar{n})^2 \rangle = E(n^2) - [E(n)]^2.$$

B.3.5   *Mean and Variance of "n" Based on a Poisson Distribution*

As an exercise, the expected value (mean) and the variance of n may be derived from Eq. B.17. The expected value $E(n)$, remembering that $0! = 1$ by definition, is given by summing nP_n over all values of n, i.e.,

$$E(n) = \sum_{n=0}^{N_0} n \frac{(wN_0)^n e^{-wN_0}}{n!}$$

$$= (wN_0) \sum_{n=1}^{N_0} \frac{(wN_0)^{n-1} e^{-wN_0}}{(n - 1)!} \qquad \text{(B.26)}$$

$$= wN_0,$$

since the sum term is equal to unity for large N_0.

For the variance, which is defined as

$$E(n^2) - [E(n)]^2,$$

the first term may be written as

$$E(n^2) = E[n(n-1) + n]$$

$$= \sum_{n=1}^{N_0} n(n-1) \frac{(wN_0)^n e^{-wN_0}}{n!} + E(n) \tag{B.27}$$

$$= (wN_0)^2 \sum_{n=2}^{N_0} \frac{(wN_0)^{n-2} e^{-wN_0}}{(n-2)!} + wN_0$$

$$= (wN_0)^2 + wN_0.$$

Hence, the variance of n, or

$$E(n^2) - [E(n)]^2 = (wN_0)^2 + wN_0 - (wN_0)^2 \tag{B.28}$$

$$= wN_0.$$

Thus, the mean and the variance of n are both equal to wN_0.

In actual measurements, the mean, $E(n)$, is estimated from k sets of data by

$$\bar{n} = \frac{1}{k} \sum_{i=1}^{k} n_i$$

and the variance, as a measure of scatter, can be estimated by the mean square deviation, or

$$\text{Var } (n) = \frac{1}{k-1} \sum_{i=1}^{k} (n_i - \bar{n})^2.$$

The square root of the variance, "the standard deviation of a single measurement," is the usual measure of the dispersion of the data and is in the same units as the data.

B.3.6 *Summary*

The Gaussian or normal distribution is symmetrical about $E(n)$, whereas for large n the Poisson distribution is nearly symmetrical about $E(n)$. The Bernoulli or binomial distribution is asymmetrical. As the Poisson and normal distributions are derived from the binomial by the assumption of large n, the binomial is more appropriate when considering the statistics of the measurement of activity at low or environmental levels. The derivations in this Appendix have been based on a consideration of actual numbers of disintegrations in a given time. In the counting of emitted radiations, in practice, the probability for detection is greatly reduced by the overall efficiency of the detecting system, but the statistics of counting will still conform to the distributions derived from a consideration of the decay of radioactive atoms. For a much fuller treatment of the subject matter in this Appendix, the reader should consult Evans (1955).

References

ADAMS, R. J. (1969). "Absolute standardization of ^{195}Au by β-γ coincidence counting," Int. J. Appl. Radiat. Isot. **20**, 808.

ADAMS, F. AND DAMS, R. (1970). *Applied Gamma-Ray Spectrometry*, 2nd ed. (Pergamon Press, Oxford and New York).

ADAMS, J. A. S. AND LOWDER, W. M., Eds. (1964). *The Natural Radiation Environment* (University of Chicago Press, Chicago).

ADAMS, R. J. AND BAERG, A. P. (1967). "Least squares polynomial fitting of coincidence counting data for the evaluation of absolute disintegration rates," page 123 in *Standardization of Radionuclides*, IAEA/STI/PUB/139 (International Atomic Energy Agency, Vienna).

ADAMS, R., JANSEN, C., GRAMES, G. M., AND ZIMMERMAN, C. D. (1974). "Deadtime of scintillation camera systems—definitions, measurement and applications," Med. Phys. **1**, 198.

AHLEN, S. P., PRICE, P. B., AND TARLE, G. (1981). "Track-recording solids," Phys. Today **34**, 32.

ALLEN, R. A. (1957). "The standardization of electron capture isotopes," Int. J. Appl. Radiat. Isot. **1**, 289.

ALLSWORTH, F. L., HOWES, J. H., AND WHITE, G. (1982). "Intrinsic germanium detectors used in a high counting rate system," Nucl. Instrum. Methods **193**, 57.

ALTSHULER, B. AND PASTERNACK, B. S. (1963). "Statistical measures of the lower limit of detection of a radioactivity counter," Health Phys. **9**, 293.

ALVAREZ, J. (1975). "Preparation of ^{113m}In radiopharmaceuticals," in *Radiopharmaceuticals*, Subramanian, G., Rhodes, B. A., Cooper, J. F. and Sodd, V. J., Eds. (Society of Nuclear Medicine, New York).

ANDAI, Á. AND JEDLOVSKY, R. (1983). "Pile-up rejection live-time correction unit for precision γ-ray spectrometry," Int. J. Appl. Radiat. Isot. **34**, 501.

ANDERSON, E. C., LIBBY, W. F., WEINHOUSE, S., REID, A. F., KIRSHENBAUM, A. D., AND GROSSE, A. V. (1947). "Natural radiocarbon from cosmic radiation." Phys. Rev. **72**, 931.

ANDREWS, G. A., KNISELEY, R. M., AND WAGNER, H. N., Eds. (1966). *Radioactive Pharmaceuticals*, CONF-651111 (National Technical Information Service, Springfield, Virginia).

ANSI (1971). American National Standards Institute. *Test Procedures for Germanium Gamma-Ray Detectors.* Report ANSI/IEEE 325–1971, Reaffirmed 1983. (American National Standards Institute, New York).

ANSI (1975). American National Standards Institute. *Calibration Techniques for the Calorimetric Assay of Plutonium-Bearing Solids Applied to Nuclear*

Materials Control. Report N15.22-1975 (American National Standards Institute, New York).

ANSI (1976a). American National Standards Institute. *Test Procedures for Amplifiers and Preamplifiers for Semiconductor Detectors for Ionizing Radiation,* Report ANSI/IEEE 301-1976, Reaffirmed 1983. (American National Standards Institute, New York).

ANSI (1976b). American National Standards Institute. *Glossary of Terms in Nuclear Science and Technology.* Report ANSI N1.1-1976/ANS-9 (American National Standards Institute, New York).

ANSI (1977). American National Standards Institute. *Test Procedures for High-Purity Germanium Detectors for Ionizing Radiation.* Report ANSI/IEEE 645-1977, Reaffirmed 1983. (American National Standards Institute, New York).

ANSI (1978a). American National Standards Institute. *Calibration and Usage of Germanium Detectors for Measurement of Gamma-Ray Emission Rates of Radionuclides.* Report ANSI N42.14-1978 (American National Standards Institute, New York).

ANSI (1978b). American National Standards Institute. *Calibration and Usage of "Dose Calibrator" Ionization Chambers for the Assay of Radionuclides.* Report ANSI N42.13-1978 (American National Standards Institute, New York).

ANSI (1978c). American National Standards Institute. *Germanium Semiconductor Detector Gamma-Ray Efficiency Determination Using a Standard Marinelli (Reentrant) Beaker Geometry.* ANSI/IEEE 680-1978 (American National Standards Institute, New York).

ANSI (1980a). American National Standards Institute. *Performance Verification of Liquid-Scintillation Counting Systems,* Report ANSI N42.15-1980 (American National Standards Institute, New York).

ANSI (1980b). American National Standards Institute. *Calibration and Usage of Sodium Iodide Detector Systems,* Report ANSI N42.12-1980 (American National Standards Institute, New York).

ANSI (1982a). American National Standards Institute. *Test Procedures for Semiconductor Charged-Particle Detectors,* Report ANSI/IEEE 300-1982 (American National Standards Institute, New York).

ANSI (1982b). American National Standards Institute. *Test Procedures for Semiconductor X-Ray Energy Spectrometers,* Report ANSI/IEEE 759-1982 (American National Standards Institute, New York).

ARMANTROUT, G. A., BRADLEY, A. E., AND PHELPS, P. L. (1972). "Sensitivity problems in biological and environmental counting," IEEE Trans. Nucl. Sci. **NS-19**, 107.

ATTIX, F. H., AND RITZ, V. H. (1957). "A determination of the gamma-ray emission of radium," J. Res. Nat. Bur. Stand. **59**, 293.

AXTON, E. J. AND RYVES, T. B. (1963). "Dead-time corrections in the measurement of short-lived radionuclides," Int. J. Appl. Radiat. Isot. **14**, 159.

AYRES, R. L. (1977). Private communication. (Radioactivity Section, National Bureau of Standards, Washington).

BAERG, A. P. (1965). "Variation on the paired source method of measuring dead time," Metrologia 1, 131.

BAERG, A. P. (1966). "Measurement of radioactive disintegration rate by the coincidence method," Metrologia 2, 23.

BAERG, A. P. (1967). "Absolute measurement of radioactivity," Metrologia 3, 105.

BAERG, A. P. (1971). Statistical Analysis of Coincidence Counting Data, Nat. Res. Counc. Can. Rep. No. NRC 12114 (National Research Council, Ottawa).

BAERG, A. P. (1973a). "Dead time, decay and background effects in counting," Int. J. Appl. Radiat. Isot. 24, 401.

BAERG, A. P. (1973b). "Pressurized proportional counters for coincidence measurements," Nucl. Instrum. Methods 112, 95.

BAERG, A. P. (1973c). "The efficiency extrapolation method in coincidence counting," Nucl. Instrum. Methods 112, 143.

BAERG, A. P. (1981). "Multiple channel $4\pi\beta$-γ anticoincidence counting," Nucl. Instrum. Methods 190, 345.

BAERG, A. P. AND BOWES, G. C. (1971). "Standardization of ^{63}Ni by efficiency tracing," Int. J. Appl. Radiat. Isot. 22, 781.

BAERG, A. P., MEGHIR, S., AND BOWES, G. C. (1964). "Extension of the efficiency tracing method for the calibration of pure β-emitters," Int. J. Appl. Radiat. Isot. 15, 279.

BAERG, A. P., BOWES, G. C., AND ADAMS, R. J. (1967). "Calibration of radionuclides with a coincidence system using a pressurized proportional $4\pi\beta$-detector," page 91 in Standardization of Radionuclides, IAEA/STI/PUB/139. (International Atomic Energy Agency, Vienna).

BAERG, A. P., MUNZENMAYER, K., AND BOWES, G. C. (1976). "Live-timed anticoincidence counting with extending dead-time circuitry," Metrologia 12, 77.

BAILLIE, L. A. (1960). "Determination of liquid scintillation counting efficiency by pulse height shift," Int. J. Appl. Radiat. Isot. 8, 1.

BAMBYNEK, W. (1967). "Precise solid angle counting," page 373 in Standardization of Radionuclides, IAEA/STI/PUB/139. (International Atomic Energy Agency, Vienna).

BAMBYNEK, W. (1973). "On selected problems in the field of proportional counters," Nucl. Instrum. Methods 112, 103.

BAMBYNEK, W., LERCH, O., AND SPERNOL, A. (1966). "Eine auf 1% genau absolute Zählung von X-Strahlung geringer Energie," Nucl. Instrum. Methods 39, 104.

BAMBYNEK, W., CRASEMANN, B., FINK, R. W., FREUND, H. U., MARK, H., SWIFT, C. D., PRICE, R. E., AND RAO, P. V. (1972). "X-ray fluorescence yields, Auger, and Coster-Kronig transition probabilities," Rev. Mod. Phys. 44, 716.

BARNES, I. L., GARFINKEL, S. B., AND MANN, W. B. (1971). "Nickel-63: standardization, half-life, and neutron-capture cross-section," Int. J. Appl. Radiat. Isot. 22, 777.

BARRALL, R. C. (1972). "Reduction of radioactive impurities in radiopharmaceuticals," J. Nucl. Med. **13**, 570.

BASSON, J. K. AND STEYN, J. (1954). "Absolute alpha standardization with liquid scintillators," Proc. Phys. Soc., London, Sect. A. **67**, 297.

BATEMAN, H. (1910). "Solution of a system of differential equations occurring in the theory of radioactive transformations," Proc. Cambridge Philos. Soc. **15**, 423.

BAY, Z., MANN, W. B., SELIGER, H. H., AND WYCKOFF, H. O. (1957). "Absolute measurement of W_{air} for sulfur-35 beta rays," Radiat. Res. **7**, 558.

BEARDEN, J. A. (1967). "X-ray wavelengths," Rev. Mod. Phys. **39**, 78.

BECKER, K. (1969). "Alpha particle registration in plastics and its applications for radon and neutron personnel dosimetry," Health Phys. **16**, 113.

BELCHER, E. H. AND VETTER, H., EDS. (1971). *Radioisotopes in Medical Diagnosis* (Butterworth and Co., London).

BELLEMARE, M., LACHANCE, Y., AND ROY, J. C. (1972). "The preparation of ^{241}Am sources by surface adsorption on different backings," Nucl. Instrum. Methods **104**, 615.

BENAKIS, A. (1971). "A new gelifying agent in liquid scintillation counting," page 735 in *Organic Scintillators and Liquid Scintillation Counting*, Horrocks, D. L. and Peng, C. T., Eds. (Academic Press, New York).

BENSON, R. H. (1976). "Characteristics of solgel scintillators for liquid scintillation counting of aqueous and non-aqueous samples," Int. J. Appl. Radiat. Isot. **27**, 667.

BERLMAN, I. B. (1971). *Handbook of Fluorescent Spectra of Aromatic Molecules*, 2nd. ed. (Academic Press, New York).

BERTOLINI, G. AND COCHE, A., EDS. (1968). *Semiconductor Detectors* (North Holland Publishing Company, Amsterdam).

BILLINGHURST, M. W., GROOTHEDDE, M., AND PALSER, R. (1974). "Radiochemical purity of ^{99m}Tc-pertechnetate," J. Nucl. Med. **15**, 266.

BIPM (1973). Bureau International des Poids et Mesures. *Le Système International d'Unités (SI)*, 2e ed. (Bureau International des Poids et Mesures, Sèvres).

BIPM (1976). Bureau International des Poids et Mesures. *The Detection and Estimation of Spurious Pulses*, Campion, P. J., Ed., Monographie BIPM-2 (Bureau International des Poids et Mesures, Sèvres).

BIPM (1980). Bureau International des Poids et Mesures. *The Application of Liquid-Scintillation Counting to Radionuclide Metrology*, Mann, W. B. and Taylor, J. G. V., Eds., Monographie BIPM-3 (Bureau International des Poids et Mesures, Sèvres).

BIPM (1981a). Bureau International des Poids et Mesures. *Bibliography on Pulse Pile-up Effects*, Gostely, J. J., Ed., Rapport BIPM-81/10 (Bureau International des Poids et Mesures, Sèvres).

BIPM (1981b). Bureau International des Poids et Mesures. *Bibliography on Dead-time Effects*, Müller, J. W. Ed., Rapport BIPM 81/11 (Bureau International des Poids et Mesures, Sèvres).

BIRKS, J. B. (1964). *The Theory and Practice of Scintillation Counting* (Pergamon Press, Oxford).

BIRKS, J. B. (1970). "Physics of the liquid scintillation process," page 3 in *Current Status of Liquid Scintillation Counting*, Bransome, E. D., Ed. (Grune & Stratton, New York).

BIRKS, J. B. (1975). *Introduction to Liquid Scintillation Counting* (Koch-Light Laboratories, Ltd., Colnbrook, Bucks, England).

BOAG, J. W. (1966). "Ionization chambers," page 1 in *Radiation Dosimetry*, Vol. 2, Attix, F. H., Roesch, W. C., and Tochilin, E., Eds. (Academic Press, New York and London).

BOLOTIN, H. H., STRAUSS, M. G., AND McCLURE, D. A. (1970). "Simple technique for precise determinations of counting losses in nuclear pulse processing systems," Nucl. Instrum. Methods **83**, 1.

BORTNER, T. E. AND HURST, G. S. (1954). "Ionization of pure gases and mixtures of gases by 5-MeV alpha particles," Phys. Rev. **93**, 1236.

BOUCHARD, J. AND VATIN, R. (1977). "Développement de la technique de coincidences $4\pi\beta$-γ," Bull. Inf. Bur. Nat. Metrol. **8**, 30.

BOWEN, B. M. AND WOOD, D. E. (1972). Authors' reply, J. Nucl. Med. **13**, 570.

BOWES, G. C. AND BAERG, A. P. (1970). "Sampling and storage of radioactive solutions," Nat. Res. Counc. Can. Rep. No. NRC-11513 (National Research Council, Ottawa).

BRANSOME, E. D., Ed. (1970). *The Current Status of Liquid Scintillation Counting* (Grune & Stratton, New York).

BRAUER, F. P. AND KAYE, J. H. (1974). "Detection systems for the low level radiochemical analysis of iodine-131, iodine-129, and natural iodine in environmental samples," IEEE Trans. Nucl. Sci. **NS-21**, 496.

BRAUER, F. P., KAYE, J. H., AND CONNALLY, R. E. (1970). "X-ray and β-γ coincidence spectrometry applied to radiochemical analysis of environmental samples," page 231 in *Radionuclides in the Environment* (American Chemical Society, Washington).

BRAUER, F. P., FAGER, J. E., AND MITZLAFF, W. A. (1973). "Evaluation of a large Ge(Li) detector for low-level radionuclide anlaysis," IEEE Trans. Nucl. Sci. **NS-20**, 57.

BRINER, W. H. AND HARRIS, C. C. (1974). "Radionuclidic contamination of eluates from fission-product molybdenum-technetium generators," J. Nucl. Med. **15**, 466.

BRINKMAN, G. A. AND ATEN, A. H. W. (1963). "Absolute standardization with a NaI(Tl) crystal—III," Int. J. Appl. Radiat. Isot. **14**, 503.

BRINKMAN, G. A. AND ATEN, A. H. W. (1965). "Absolute standardization with a NaI(Tl) crystal—V," Int. J. Appl. Radiat. Isot. **16**, 177.

BRINKMAN, G. A., ATEN, A. H. W., AND VEENBOER, J. TH. (1963a). "Absolute standardization with a NaI(Tl) crystal—I," Int. J. Appl. Radiat. Isot. **14**, 153.

BRINKMAN, G. A., ATEN, A. H. W., AND VEENBOER, J. TH. (1963b). "Absolute standardization with a NaI(Tl) crystal—II," Int. J. Appl. Radiat. Isot. **14**, 433.

BRINKMAN, G. A., ATEN, A. H. W., AND VEENBOER, J. TH. (1965). "Absolute standardization with a NaI(Tl) crystal—IV," Int. J. Appl. Radiat. Isot. **16**, 15.

BROUNS, R. J. (1968). *Absolute Measurement of Alpha Emission and Spontaneous Fission*, Monograph NAS-NS 3112 (National Technical Information Service, Springfield, Virginia).

BROWNLEE, K. A. (1960). *Statistical Theory and Methodology in Science and Engineering* (John Wiley & Sons, New York).

BRUCER, M. (1970). "Standards and the calibration of a well counter," *Vignettes in Nuclear Medicine*, **2** [47], 1 (Mallinckrodt, Inc., St. Louis, Missouri).

BRYANT, J. (1962). "Anticoincidence counting method for standardizing radioactive materials," Int. J. Appl. Radiat. Isot. **13**, 273.

BRYANT, J. (1963). "Coincidence counting corrections for dead-time loss and accidental coincidences," Int. J. Appl. Radiat. Isot. **14**, 143.

BRYANT, J. (1967). "Advantage of anticoincidence counting for standardizing radionuclides emitting delayed gamma rays," page 129 in *Standardization of Radionuclides*, IAEA/STI/PUB/139 (International Atomic Energy Agency, Vienna).

BRYANT, J. AND MICHAELIS, M. (1952). "The measurement of micro-microcurie amounts of radon by a scintillation counting method," Radiochemical Centre Report R26 (The Radiochemical Centre, Amersham, United Kingdom).

BUDINGER, T. F. (1974). "Quantitative nuclear medicine imaging: application of computers to the gamma camera and whole-body scanner," in *Recent Advances in Nuclear Medicine*, Lawrence, J. H., Ed. (Grune & Stratton, New York).

BUNEMANN, O., CRANSHAW, T. E., AND HARVEY, J. A. (1949). "Design of grid ionisation chambers," Can. J. Res. **A27**, 191.

BUNTING, R. L. AND KRAUSHARR, J. J. (1974). "Short-lived radioactivity induced in Ge(Li) gamma-ray detectors by neutrons," Nucl. Instrum. Methods **118**, 565.

CAMERON, J. F. (1967). "Survey of systems for concentration and low background counting of tritium in water," page 543 in *Radioactive Dating and Methods of Low-Level Counting*, IAEA/STI/PUB/152 (International Atomic Energy Agency, Vienna).

CAMPION, P. J. (1959). "The standardization of radioisotopes by the beta-gamma coincidence method using high efficiency detectors," Int. J. Appl. Radiat. Isot. **4**, 232.

CAMPION, P. J. (1973). "Spurious pulses in proportional counters: a review," Nucl. Instrum. Methods **112**, 75.

CAMPION, P. J., TAYLOR, J. G. V., AND MERRITT, J. S. (1960). "The efficiency tracing technique for eliminating self-absorption errors in $4\pi\beta$-counting," Int. J. Appl. Radiat. Isot. **8**, 8.

CAMPION, P. J., DALE, J. W. G., AND WILLIAMS, A. (1964). "A study of weighing techniques used in radionuclide standardization," Nucl. Instrum. Methods **31**, 253.

CARSWELL, D. J. AND MILSTED, J. (1957). "A new method for the preparation of thin films of radioactive material," J. Nucl. Energy **4**, 51.

CAVALLO, L. M., COURSEY, B. M., GARFINKEL, S. B., HUTCHINSON, J. M. R., AND MANN, W. B. (1973). "Needs for radioactivity standards and measurements in different fields," Nucl. Instrum. Methods **112**, 5.

CEMBER, H. (1983). *Introduction to Health Physics*-Second Edition, Revised and Enlarged (Pergamon Press, Oxford and New York).

CHARALAMBUS, S. AND GOEBEL, K. (1963). "Low-level proportional counter for tritium," Nucl. Instrum. Methods **25**, 109.

CHASE, R. L. (1961). *Nuclear Pulse Spectrometry* (McGraw-Hill, New York).

CHASE, R. L. AND POULO, L. R. (1967). "High precision D. C. restorer," IEEE Trans. Nucl. Sci. **NS-14**, 83.

CHERRY, R. D. (1963). "The determination of thorium and uranium in geological samples by an alpha-counting technique," Geochim. Cosmochim. Acta **27**, 183.

CHEW, W. M., McGEORGE, J. C., AND FINK, R. W. (1973). "A multiwire proportional counter technique for standardization of low energy photon sources," Nucl. Instrum. Methods **106**, 499.

CHEW, W. M., XENOULIS, A. C., FINK, R. W., AND PINAJIAN, J. J. (1974a). "Nuclear matrix elements from L/K electron capture ratios in ^{59}Ni decay," Nucl. Phys. **A218**, 372.

CHEW, W. M., XENOULIS, A. C., FINK, R. W., SCHIMA, F. J., AND MANN, W. B. (1974b). "The L/K electron capture ratio in first-forbidden ^{81s}Kr decay," Nucl. Phys. **A229**, 79.

CHYLINSKI, A., PASIEWICZ, K., RADOSZEWSKI, T., WOLSKI, D., AND WOJTOWICZ, S. (1972). "$4\pi\beta$-γ anticoincidence method and apparatus for radioactive concentration measurements of radioactive solutions," Nucl. Instrum. Methods **98**, 109.

CLARKE, R. K. (1950). "Absolute determination of the emission rate of beta rays," Rev. Sci. Instrum. **21**, 753.

CLINE, J. E. (1971). "Gamma rays emitted by the fissionable nuclides and associated isotopes," Idaho Nuclear Corporation Report IN-1448 (National Technical Information Service, Springfield, Virginia).

CLINE, J. E., PUTNAM, M. H., AND HELMER, R. G. (1973). "Gauss VI, a computer program for automatic batch analysis of gamma-ray spectra from Ge(Li) spectrometers," AEC Report No. ANCR-1113 (National Technical Information Service, Springfield, Virginia).

CLULEY, H. J. (1962). "Suspension scintillation counting of carbon-14 barium carbonate," Analyst **87**, 170.

COCHRAN, J. A., SMITH, D. G., MAGNO, P. G., AND SHLEIEN, B. C. (1970). *An Investigation of Airborne Radioactive Effluent from an Operating Nuclear Fuel Reprocessing Plant* BRH/NERHL 70-3 (National Technical Information Service, Springfield, Virginia).

CODATA (1973). Committee on Data for Science and Technology. "Report of Task group on fundamental constants," CODATA Bull. 11.

COHEN, E. J. (1974). "Live time and pile-up corrections for multichannel analyzer spectra," Nucl. Instrum. Methods **121**, 25.

COHEN, E. R. AND DuMOND, J. W. M. (1965). "Our knowledge of the fundamental constants of physics and chemistry in 1965," Rev. Mod. Phys. **37**, 537.

COHEN, E. R. AND TAYLOR, B. N. (1973). "The 1973 least-squares adjustment of the fundamental constants," J. Phys. Chem. Ref. Data **2**, No. 4, 663.

COHN, C. E. (1966). "The effect of deadtime on counting errors," Nucl. Instrum. Methods **41**, 338.

COLEY, R. F. AND FRIGERIO, N. A. (1975). "Self absorption and geometric correction factors for reactor off-gas samples relative to NBS standards," page 666 in *Proceedings of the Noble Gases Symposium,* ERDA CONF-730915, Stanley, R. E. and Moghissi, A. A., Eds. (National Technical Information Service, Springfield, Virginia).

COLLÉ, R. (1976). "AIF-NBS radioactivity measurements assurance program for the radiopharmaceutical industry," page 71 in *Proceedings of the NBS 75th Anniversary Symposium,* Nat. Bur. Stand. (U.S.) Spec. Publ. 456 (National Bureau of Standards, Washington).

COLLÉ, R., RUBIN, R. J., KNAB, L. I., AND HUTCHINSON, J. M. R. (1981). "Radon transport through and exhalation from building materials: A review and assessment," Nat. Bur. Stand. Tech. Note 1139.

COLLINSON, A. J. L., DEMETSOPOULLOS, I. C., DENNIS, J. A., AND ZARZYCKI, J. M. (1960). "Self-quenching Geiger counters containing mixtures of permanent gases," Nature **185**, 369.

CONDE, C. A. N., SANTOS, M. C. M., FERREIRA, M. F. A., AND SOUSA, C. A. (1975). "An argon gas scintillation counter with uniform electric field," IEEE Trans. Nucl. Sci. **NS-22**, 104.

CONNOR, W. S. AND YOUDEN, W. J. (1954). "Comparison of four national radium standards. Part 2. Statistical procedures and survey," J. Res. Nat. Bur. Stand. **53**, 273.

COOPER, J. A. (1970). "Factors determining the ultimate detection sensitivity of Ge(Li) gamma-ray spectrometers," Nucl. Instrum. Methods **82**, 273.

COOPER, J. A. AND PERKINS, R. W. (1971). *A Versatile Ge(Li)-NaI(Tl) Coincidence-Anti-Coincidence Gamma-Ray Spectrometer for Environmental and Biological Problems,* BNWL-SA-3972 (Battelle Pacific Northwest Laboratories, Richland, Washington).

COOPER, J. A., HOLLANDER, J. M., AND RASMUSSEN, J. O. (1965). "Effect of the chemical state on the lifetime of the 24-second isomer of Nb^{90*}," Phys. Rev. Letters **15**, 680.

CORNET, A., REGAL, R., PONPON, J. P., AND SIFFERT, P. (1975). "New generation of individual safeguards system," page 169 in *Proceedings of the Second Ispra Nuclear Electronics Symposium,* Report EUR 5370 (Commission of the European Communities, Luxembourg).

CORSON, D. R. AND WILSON, R. R. (1948). "Particle and quantum counters," Rev. Sci. Instrum. **19**, 207.

COURSEY, B. M. (1976). "Use of NBS mixed-radionuclide gamma-ray standards for calibration of Ge(Li) detectors used in the assay of environmental radioactivity," page 173 in *Proceedings of a Symposium on Measurements for the Safe Use of Radiation*, Nat. Bur. Stand. (U.S.) Spec. Publ. 456, Fivozinsky, S. P., Ed. (National Bureau of Standards, Washington).

COURSEY, B. M., NOYCE, J. R., AND HUTCHINSON, J. M. R. (1975). *Interlaboratory Intercomparisons of Radioactivity Measurements Using National Bureau of Standards Mixed Radionuclide Test Solutions*, Nat. Bur. Stand. (U.S.) Tech. Note 875 (National Bureau of Standards, Washington).

COURSEY, B. M., GIBSON, J. A. B., HEITZMANN, M. W., AND LEAKE, J. C. (1984). "Standardization of technetium-99 by liquid scintillation counting." Int. J. Appl. Radiat. Isot. (in press).

COX, D. R. AND ISHAM, V. (1977). "A bivariate point process connected with electronic counters," Proc. Roy. Soc. **A356**, 149.

CRADDUCK, T. D. (1973). "Fundamentals of scintillation counting," Semin. Nucl. Med. **3**, 205.

CRAMER, H. (1946). *Mathematical Methods of Statistics* (Princeton University Press, Princeton, New Jersey).

CRASEMANN, B. (1973). "Some aspects of atomic effects in nuclear transformations," Nucl. Instrum. Methods **112**, 33.

CRAWFORD, J. A. (1949). "Theoretical calculations concerning back-scattering of alpha particles," page 1307 in *The Transuranium Elements*, Pt. II, Seaborg, G. T., Katz, J. J., and Manning, W. M., Eds. (McGraw-Hill, New York).

CROOK, M. A. AND JOHNSON, P., Eds. (1974). *Liquid Scintillation Counting, Vol. 3. Proceedings of a Symposium on Liquid Scintillation Counting, Brighton, U.K., 1973.* (John Wiley & Sons, Inc., New York).

CROOK, M. A. AND JOHNSON, P., Eds. (1977). *Liquid Scintillation Counting, Vol. 4. Proceedings of a Symposium on Liquid Scintillation Counting, Bath, U.K., 1975.* (Heyden, London).

CROOK, M. A. AND JOHNSON, P., Eds. (1978). *Liquid Scintillation Counting, Vol. 5. Proceedings of a Symposium on Liquid Scintillation Counting, Bath, U.K., 1977.* (John Wiley & Sons, Inc., New York).

CROOK, M. A., JOHNSON, P., AND SCALES, B., Eds. (1972). *Liquid Scintillation Counting, Vol. 2. Proceedings of a Symposium on Liquid Scintillation Counting, Brighton, U.K., 1971.* (Heyden and Sons, Limited, London).

CUDDIHY, R. G., MCCLELLAN, R. O., CHAPMAN, L. D., WAYLAND, J. R., AND DUGAN, V. L. (1976). "Simulation modelling of environmental transport and health consequences of radioactive effluents from nuclear power systems," page 657 in *Transuranium Nuclides in the Environment*, IAEA/STI/PUB/410 (International Atomic Energy Agency, Vienna).

CULHANE, J. L. AND FABIAN, A. C. (1972). "Circuits for pulse rise time discrimination in proportional counters," IEEE Trans. Nucl. Sci. **NS-19**, 569.

CURRAN, S. C. (1950). "Counting and spectrometry with proportional tubes," Atomics **1**, 221.

CURRAN, S. C. AND WILSON, H. W. (1965). "Proportional counters and pulse ion chambers," page 303 in *Alpha-, Beta-, and Gamma-Ray Spectroscopy*,

Vol. I., Siegbahn, K., Ed. (North-Holland Publishing Co., Amsterdam).

CURRIE, L. A. (1968). "Limits for qualitative detection and quantitative determination. Applications to radiochemistry," Anal. Chem. **40**, 586.

CURRIE, L. A. (1972). "The measurement of environmental levels of rare gas nuclides and the treatment of very low-level counting data," IEEE Trans. Nucl. Sci. **NS-19**, 119.

CURRIE, L. A., Ed. (1982). *Nuclear and Chemical Dating Techniques: Interpreting the Environmental Record*, ACS Symposium Series No. 176 (American Chemical Society, Washington).

CURRIE, L. A. AND LINDSTROM, R. M. (1975). "The NBS measurement system for natural argon-37," page 40 in *Proceedings of the Noble Gases Symposium*, ERDA CONF-730915, Stanley, R. E. and Moghissi, A. A., Eds. (National Technical Information Service, Springfield, Virginia).

CURTIS, M. L. (1972). "Detection and measurement of tritium by bremsstrahlung counting," Int. J. Appl. Radiat. Isot. **23**, 17.

CURTISS, L. F. AND DAVIS, F. J. (1943). "A counting method for the determination of small amounts of radium and of radon," J. Res. Nat. Bur. Stand. **31**, 181.

DALE, J. W. G. (1961). "A beta-gamma ionization chamber for substandards of radioactivity—II," Int. J. Appl. Radiat. Isot. **10**, 72.

DALE, J. W. G., PERRY, W. E., AND PULFER, R. F. (1961). "A beta-gamma ionization chamber for substandards of radioactivity—I," Int. J. Appl. Radiat. Isot. **10**, 65.

DAMON, P. E. AND WINTERS, P. N. (1954). "Resolution losses in counters and trigger circuits," Nucleonics **12**, No. 12, 36.

DARRALL, K. G., RICHARDSON, P. J., AND TYLER, J. F. C. (1973). "An emanation method for determining radium using liquid scintillation counting," Analyst **98**, 610.

DAVENPORT, T. I., MANN, W. B., McCRAVEN, C. C., AND SMITH, C. C. (1954). "Comparison of four national radium standards. Part 1. Experimental procedures and results," J. Res. Nat. Bur. Stand. **53**, 267.

DAVIS, D. M. AND GUPTON, E. D. (1949). *Health Physics Instrument Manual*, ORNL-332, (TID-4500) (Oak Ridge National Laboratory, Oak Ridge, Tennessee).

DEBERTIN, K. AND MANN, W. B., Eds. (1983). *Gamma- and X-ray-Spectrometry Techniques and Applications*, Proceedings of a seminar of the International Committee for Radionuclide Metrology (ICRM), Int. J. Appl. Radiat. Isot. **34**, 477.

DEBERTIN, K., SCHOTZIG, U., WALZ, K. F., AND WEISS, H. M. (1976). "Efficiency calibration of semiconductor spectrometers—techniques and accuracies," page 59 in *Proceedings of ERDA Symposium on X- and Gamma-Ray Sources and Applications*, Report CONF-760539 (National Technical Information Service, Springfield, Virginia).

DE BIÈVRE, P., GALLET, M., AND WERZ, R. (1983). "Determination of the ^{241}Pu half life by isotope mass spectrometry," Int. J. Mass Spectrom. Ion Phys. **51**, 111.

DE BRUIN, M., THEN, S. S., BODE, P., AND KORTHOVEN, P. J. M. (1974). "A simple deadtime stabilizer for gamma-ray spectrometers," Nucl. Instrum. Methods 121, 611.

DERUYTTER, A. J. (1962). "Evaluation of the absolute activity of alpha emitters and of the number of nuclei in thin alpha active layers," Nucl. Instrum. Methods 15, 164.

DESLATTES, R. D. (1980). "The Avogadro constant," Ann. Rev. Phys. Chem. 31, 435.

DESLATTES, R. D., HENINS, A., BOWMAN, H. A., SCHOONOVER, R. M., CARROLL, C. L., BARNES, I. L., MACHLAN, L. A., MOORE, L. J., AND SHIELDS, W. R. (1974). "Determination of the Avogadro constant," Phys. Rev. Letters 33, No. 8, 463.

DESLATTES, R. D., HENINS, A., SCHOONOVER, R. M., CARROLL, C. L., AND BOWMAN, H. A. (1976). "Avogadro's constant-Corrections to an earlier report," Phys. Rev. Letters 36, 898.

DEVOE, J. R. (1961). Radioactive Contamination of Materials Used in Scientific Research, NAS-NRC Publ. 895 (National Academy of Sciences, Washington).

DIAMOND, H., FIELDS, P. R., STEVENS, C. S., STUDIER, M. H., FIELD, S. M., INGRAHAM, M. G., HESS, D. C., PYLE, G. L., MECH, J. F., AND MANNING, W. M. (1960). "Heavy isotope abundances in Mike thermonuclear device," Phys. Rev. 119, 2000.

DIETHORN, W. S. (1974). "Counter deadtime: a question of poor textbook advice," Int. J. Appl. Radiat. Isot. 25, 55.

DIETZ, L. A. AND PACHUCKI, C. F. (1973). "^{137}Cs and ^{134}Cs half-lives determined by mass spectrometry," J. Inorg. Nucl. Chem. 35, 1769.

DITMARS, D. A. (1976). "Measurement of the average total decay power of two plutonium heat sources in a Bunsen ice calorimeter," Int. J. Appl. Radiat. Isot. 27, 469.

DIXON, W. J. AND MASSEY, F. J. (1957). Introduction to Statistical Analysis (McGraw-Hill, New York).

DOBRILOVIČ, L. AND SIMOVIČ, M. (1973). "Preparation of mono-molecular radioactive sources," Nucl. Instrum. Methods 112, 359.

DONN, J. J. AND WOLKE, R. L. (1977). "The statistical interpretaton of counting data from measurements of low-level radioactivity," Health Phys. 32, 1.

DRAPER, N. H. AND SMITH, H. (1966). Applied Regression Analysis (John Wiley & Sons, New York).

DREVER, R. W. P., MOLJK, A., AND CURRAN, S. C. (1957). "A proportional counter system with small wall effect," Nucl. Instrum. Methods 1, 41.

DUNWORTH, J. V. (1940). "The application of the method of coincidence counting to experiments in nuclear physics," Rev. Sci. Instrum. 11, 167.

DUVAL, C. (1963). Inorganic Thermogravimetric Analysis (Elsevier, Amsterdam).

DYER, A., Ed. (1971). Liquid Scintillation Counting, Vol. 1. Proceedings of a Symposium on Liquid Scintillation Counting, Salford, U.K., 1970. (Heyden and Sons, Limited, London)

EASTHAM, J. F., WESTBROOK, H. L., AND GONZALES, D. (1962). "Liquid scintillation detection of tritium and other radioisotopes in insoluble or quenching organic samples," page 203 in *Tritium and the Physical and Biological Sciences*, Vol. 1, IAEA/STI/PUB/39 (International Atomic Energy Agency, Vienna).

EASTWOOD, T. A., BROWN, F., AND CROCKER, I. H. (1964). "A krypton-81 half-life determination using a mass separator," Nucl. Phys. **58**, 328.

ECKELMAN, W. C. AND COURSEY, B. M., Eds. (1982). *Technetium-99m Generators, Chemistry and Preparation of Radiopharmaceuticals*, Int. J. Appl. Radiat. Isot. **33**, 793.

EDWARDS, R. R. (1962). "Iodine-129: its occurrence in nature and its utility as a tracer," Science **137**, 851.

EICHELBERGER, J. F., JORDAN, K. C., ORR, S. R., AND PARKS, J. R. (1954). "Calorimetric determination of the half-life of polonium-210," Phys. Rev. **96**, 719 (U.S. Government Printing Office, Washington).

EISENHART, C. (1963). "Realistic evaluation of the precision and accuracy of instrument calibration systems," J. Res. Nat. Bur. Stand. **67C**, 161. (Reprinted in Ku, 1969.)

EISENHART, C. (1968). "Expression of the uncertainties of final results," Science **160**, 1201. (Reprinted in Ku, 1969.)

EISENHART, C., KU, H. H., AND COLLÉ, R. (1983). "Expression of the uncertainties of final measurement results: Reprints," Nat. Bur. Stand. (U.S.) Spec. Pub. 644.

ELDRIDGE, J. S. AND CROWTHER, P. (1964). "Absolute determination of ^{125}I in clinical applications," Nucleonics, **22** No. 6, 56.

ELDRIDGE, J. S., O'KELLEY, G. D., NORTHCUTT, K. J., AND SCHONFELD, E. (1973). "Nondestructive determination of radionuclides in lunar samples using a large low-background gamma-ray spectrometer and a novel application of least squares fitting," Nucl. Instrum. Methods **112**, 319.

EMANUEL, C. F. (1973). "Delivery precision of micropipets," Anal. Chem. **45**, 1568.

ENGELKE, C. E. (1960). "High pressure gas scintillators," IRE Trans. Nucl. Sci. **NS-7**, 32.

ENGELKEMEIR, A. G. AND LIBBY, W. F. (1950). "End and wall corrections for absolute beta-counting in gas counters," Rev. Sci. Instrum. **21**, 550.

EPA (1973). Environmental Protection Agency. "Tritium surveillance system, April–June 1973," Radiat. Data Rep. **14**, 601.

ERDA (1977). Energy Research and Development Administration (presently Department of Energy). *A Guide for Environmental Radiological Surveillance at ERDA Installations* ERDA77-24 (National Technical Information Service, Springfield, Virginia).

EVANS, R. D. (1935). "Apparatus for the determination of minute quantities of radium, radon and thoron in solids, liquids and gases," Rev. Sci. Instrum. **6**, 99.

EVANS, R. D. (1937). "Radium poisoning: II. The quantitative determination of the radium content and radium elimination rate of living persons," Am. J. Roentgenol. Radium Ther. **37**, 368.

EVANS, R. D. (1955). *The Atomic Nucleus* (McGraw-Hill, New York).

EVANS, R. D. AND EVANS, R. O. (1948). "Studies of self-absorption in gamma-ray sources," Rev. Mod. Phys. **20**, 305.

FAIRSTEIN, E. AND HAHN, J. (1965). "Nuclear pulse amplifiers—fundamentals and design practice," Nucleonics **23** No. 7, 56; **23** No. 9, 81; **23** No. 11, 50; **24** No. 1, 54; **24** No. 3, 68.

FANO, U. (1947). "Ionization yield of radiations, II. Fluctuations of the number of ions," Phys. Rev. **72**, 26.

FARUKHI, M. R. (1983). "$Bi_4Ge_3O_{12}$ (BGO)—A scintillator replacement for NaI(Tl)," page 115 in *Nuclear Radiation Detector Materials*, Haller, E. E., Kraner, H. W. and Higginbotham, W. A., Eds. (Elsevier Science and North-Holland Publishing Companies, New York and Amsterdam).

FEINGOLD, Q. M. AND FRANKEL, S. (1955). "Geometrical corrections in angular correlation instruments," Phys. Rev. **97**, 1025.

FELLER, W. (1948). "On probability problems in the theory of counters," page 105 in *Studies and Essays*, presented to R. Courant on his 60th Birthday, Jan. 8, 1948, Friedrichs, K. O., Neugebaur, O. E., and Stoker, J. J., Eds. (Interscience Publishers, Inc., New York).

FELLER, W. (1968). *An Introduction to Probability Theory and Its Applications*, Vol. I, 3rd ed. (John Wiley & Sons, New York).

FICKEN, V., ANDERSON, D., COX, P., AND COOTS, H. (1973). "Radionuclide contamination of fission generator eluate," J. Nucl. Med. Technol. **1**, 11.

FIELDS, P. F., DIAMOND, H., METTA, D. N., ROKOP, D. J., AND STEVENS, C. M. (1972). "^{237}Np, ^{236}U, and other actinides on the moon," Geochim. Cosmochim. Acta, Suppl. 3, **2**, 1637.

FINK, R. W. (1973). "Radioactive x-ray standards," Nucl. Instrum. Methods **112**, 243.

FLEISCHER, R. L., ALTER, H. W., FURMAN, S. C., PRICE, P. B., AND WALKER, R. M. (1972). "Particle track etching," Science **178**, 255.

FLYNN, K. F., GLENDENIN, L. E., AND PRODI, V. (1971). "Absolute counting of low energy beta emitters using liquid scintillation counting techniques," page 687 in *Organic Scintillators and Liquid Scintillation Counting*, Horrocks, D. L. and Peng, C. T., Eds. (Academic Press, New York).

FOGLIO PARA, A. AND BETTONI, M. MANDELLI (1970). "Statistical determination of half lives," Nucl. Instrum. Methods **79**, 125.

FOLSOM, T. R., YOUNG, D. R., JOHNSON, J. N., AND PILLAI, K. C. (1963). "Manganese-54 and zinc-65 in coastal organisms of California," Nature **200**, 327.

FOLSOM, T. R., PILLAI, K. C., AND FINNIN, L. E. (1965). "Assay of small traces of caesium-137 by internal conversion electrons," page 699 in *Radioisotopic Sample Measurement Techniques in Medicine and Biology.* IAEA/STI/PUB/106 (International Atomic Energy Agency, Vienna).

FOX, B. W. (1968). "The application of Triton X-100 colloid scintillation counting in biochemistry," Int. J. Appl. Radiat. Isot. **19**, 717.

FRANÇOIS B. (1973). "Detection of hard beta-emitting radionuclides in aqueous solution using Čerenkov radiation," Int. J. Nucl. Med. Biol. **1**, 1.

FREEDMAN, G. S., KINSELLA, T., AND DWYER, A. (1972). "A correction method for high-count-rate quantitative radionuclide angiography," Radiology **104**, 713.

FRIEDLÄNDER, E. M. (1964). "Correlation counting of coincidence events," Nucl. Instrum. Methods **31**, 293.

FRIEDLANDER, G., KENNEDY, J. W., AND MILLER, J. M. (1964). *Nuclear and Radiochemistry*, 2nd ed. (John Wiley & Sons, New York).

FRIEDMAN, H. (1949). "Geiger counter tubes," Proceedings IRE **37**, 791.

GALE, H. J. (1972). "Approximations to the pulse height distribution at the output of a scintillation counter," J. Phys. E. **7**, 258.

GALE, H. J. AND GIBSON, J. A. B. (1966). "Methods of calculating the pulse height distribution at the output of a scintillation counter," J. Sci. Instrum. **43**, 224.

GANDY, A. (1961). "Mesure absolue de l'activité des radionuclides par la méthode des coïncidence béta-gamma à l'aide de détecteurs de grand efficacité. Étude des coïncidences instrumentales," Int. J. Appl. Radiat. Isot. **11**, 75.

GANDY, A. (1962). "Mesure absolue de l'activité des radionuclides par la méthode des coïncidences béta-gamma à l'aide de detécteurs de grande efficacité: Corrections de temps morts," Int. J. Appl. Radiat. Isot. **13**, 501.

GARDNER, D. G. AND MEINKE, W. W. (1958). "β-ray spectroscopy using a hollow plastic scintillator," Int. J. Appl. Radiat. Isot. **3**, 232.

GARFINKEL, S. B. (1959). "Semiautomatic Townsend balance system," Rev. Sci. Instrum. **30**, 439.

GARFINKEL, S. B. AND HUTCHINSON, J. M. R. (1962). "Determination of source self-absorption in the standardization of electron-capturing radionuclides," Int. J. Appl. Radiat. Isot. **13**, 629.

GARFINKEL, S. B. AND HUTCHINSON, J. M. R. (1966). "The standardization of Cobalt-57," Int. J. Appl. Radiat. Isot. **17**, 587.

GARFINKEL, S. B. AND HUTCHINSON, J. M. R. (1970). "Recent activities of the NBS Radioactivity Section, with special reference to the standardization of thallium-208 in Radioactivity Calibration Standards," Nat. Bur. Stand. (U.S.) Spec. Publ. 331, Mann, W. B. and Garfinkel, S. B., Eds. (U.S. Government Printing Office, Washington).

GARFINKEL, S. B. AND HINE, G. J. (1973). *Dose Calibrator Pilot Study*, Nat. Bur. Stand. (U.S.) Tech. Note 791 (U.S. Government Printing Office, Washington).

GARFINKEL, S. B., MANN, W. B., MEDLOCK, R. W., AND YURA, O. (1965). "The calibration of the National Bureau of Standards tritiated-toluene standard of radioactivity," Int. J. Appl. Radiat. Isot. **16**, 27.

GARFINKEL, S. B., MANN, W. B., AND YOUDEN, W. J. (1966). "Design and statistical procedures for the evaluation of an automatic gamma-ray point-source calibration," J. Res. Nat. Bur. Stand. **70C**, 503.

GARFINKEL, S. B., MANN, W. B., AND PARARAS, J. L. (1973a). "The National

Bureau of Standards $4\pi\beta$-γ coincidence-counting and γ-ray comparator automatic sample changers," Nucl. Instrum. Methods **112**, 213.

GARFINKEL, S. B., MANN, W. B., SCHIMA, F. J., AND UNTERWEGER, M. P. (1973b). "Present status in the field of internal gas counting," Nucl. Instrum. Methods **112**, 59.

GEIGER, E. L., TSCHAECHA, A. N., AND WHITTAKER, E. L. (1961). "Simplified autoradiography technique for α-emitters," Health Phys. **11**, 1092.

GEIGER, H. AND WERNER, A. (1924). "Die Zahl der von Radium ausgesandten α-Teilchen," Z. Phys. **21**, 187.

GENERAL ELECTRIC COMPANY (1962). *Tables of the Individual and Cumulative Terms of Poisson Distribution* (D. Van Nostrand Co., Inc., Princeton, New Jersey).

GENNA, S., WEBSTER, E. W., BRANTLEY, J. C., BROWNELL, G. L., CARDARELLI, J. A., CASTRONOVO, F. P., CHANDLER, H. L., AND MASSE, F. X. (1972). "A nuclear medicine quality control program," J. Nucl. Med. **13**, 285.

GERRARD, M. (1971). "Survey of the use of radionuclides in medicine—Phase II," Isot. Radiat. Technol. **8**, 236.

GIACOMO, P. (1981). "News from BIPM," Metrologia **17**, 73.

GIBSON, J. A. B. (1972). "Liquid scintillation counting as an absolute method" page 23 in *Liquid Scintillation Counting*, Vol. 2. Crook, M. A., Johnson, P., and Scales, B., Eds. (Heyden and Sons, Limited, London).

GIBSON, J. A. B. (1976). "Liquid scintillation counting of novel radionuclides" page 153 in *Liquid Scintillation: Science and Technology*, Noujaim, A. A., Ediss, C. and Weibe, L. I., Eds. (Academic Press, New York).

GIBSON, J. A. B. (1980). "The statistics of the scintillation process and determination of the counting efficiency" page 36 in *The Application of Liquid Scintillation Counting to Radionuclide Metrology*, Monographie BIPM-3 (Bureau International des Poids et Mesures, Sèvres)

GIBSON, J. A. B. AND GALE, H. J. (1968). "Absolute standardization with liquid scintillation counters," J. Phys. E. **1**, Ser. 2, 99.

GIBSON, J. A. B. AND MARSHALL, M. (1972). "The counting efficiency for ^{55}Fe and other E. C. nuclides in liquid scintillator solutions," Int. J. Appl. Radiat. Isot. **23**, 321.

GIBSON, W. M., MILLER, G. L., AND DONOVAN, P. F. (1965). "Semiconductor particle spectrometers," page 345 in *Alpha-, Beta-, and Gamma-Ray Spectroscopy*, Vol. 1, Siegbahn, K., Ed. (North Holland Publishing Company, Amsterdam).

GOLAS, D. B. AND CALHOUN, J. M. (1983). "U.S. National Bureau of Standards/Atomic Industrial Forum radioactivity measurements," Int. J. Nucl. Med. Biol. **10**, 163.

GOLD, R., ARMANI, R. J., AND ROBERTS, J. H. (1968). "Absolute fission rate measurements with solid-state track recorders," Nucl. Sci. Eng. **34**, 13.

GOLD, S., BARKHAU, H. W., SHLEIEN, B., AND KAHN, B. (1964). "Measurement of naturally occurring radionuclides in air," page 369 in *The Natural Radiation Environment*, Adams, J. A. S. and Lowder, W. M., Eds. (University of Chicago Press, Chicago).

GOLDANSKY, V. I. AND PODGORETSKY, M. I. (1955). "Possible use of a correlation function for the study of nuclear disintegration," Dok. Akad. Nauk. SSSR **100**, 237.

GOODIER, I. W., HUGHES, F. H., AND WILLIAMS, A. (1965). "Precise measurement of the activity of low-energy beta-emitters with calibrated ionization chambers," page 613 in *Radioisotope Sample Measurement Techniques in Medicine and Biology* IAEA/STI/PUB/106 (International Atomic Energy Agency, Vienna).

GOULDING, F. S. (1966). "Semiconductor detectors for nuclear spectrometry," Nucl. Instrum. Methods **43**, 1.

GOULDING, F. S. (1978). "Semiconductor detectors—an introduction," IEEE Trans. Nucl. Sci. **NS-25**, 916.

GRAU MALONDA, A. (1982). "Counting efficiency for electron-capturing nuclides in liquid scintillator solutions," Int. J. Appl. Radiat. Isot. **33**, 371.

GRAU MALONDA, A. AND GARCIA TORAÑO, E. (1982). "Evaluation of counting efficiency in liquid scintillation counting of pure β-ray emitters," Int. J. Appl. Radiat. Isot. **33**, 249.

GRAU MALONDA, A. AND LOS-ARCOS, J. M. (1983). "Un nuevo procedimiento para la calibracion del ^{90}Sr y del ^{90}Y mediante centelleo en fase liquida," An. Fis. **B79**, 5.

GRAY, L. H. (1949). "The experimental determination by ionization methods of the rate of emission of beta- and gamma-ray energy by radioactive substances," Br. J. Radiol. **22**, 677.

GREENE, R. E., PRESSLY, R. S., AND CASE, F. N. (1972). *A Review of Alpha Source Preparation Methods and Applications,* ORNL-4819 (Oak Ridge National Laboratory, Oak Ridge, Tennessee).

GREENWOOD, R. C., HELMER, R. G., AND GEHRKE, R. J. (1979). "Precise x-ray energies for calibration of Ge semiconductor spectrometer" Nucl. Instrum. Methods, **159**, 465.

GRODSTEIN, G. W. (1957). *X-Ray Attenuation Coefficients from 10 keV to 100 MeV,* Nat. Bur. Stand. Circ. 583 (U.S. Government Printing Office, Washington).

GROSS, W. AND FAILLA, G. (1950). "A new method of radioactivity measurement," Phys. Rev. **79**, 209.

GROSS, W., WINGATE, C., AND FAILLA, G. (1957a). "The average energy lost by S^{35} beta rays per ion pair produced in air," Radiat. Res. **7**, 570.

GROSS, W., WINGATE, C., AND FAILLA, G. (1957b). "Determination of disintegration rate for gamma-emitting isotopes," Radiology **69**, 699.

GROSSE, G. AND BAMBYNEK, W. (1983). *Physics Data: International Directory of Certified Radioactive Sources,* (Commission of the European Communities, Joint Nuclear Research Center, Geel, Belgium).

GRUMMITT, W. E., BROWN, R. M., CRUIKSHANK, A. J., AND FOWLER, I. L. (1956). "Recent developments in low-background Geiger-Müller counters," Can. J. Chem. **34**, 206.

GRUN, A. E. AND SCHOPPER, E. (1951). "Über die Fluoreszenz von Gasen bei

Anregung dürch α-Teilchen," Z. Naturf. **6a**, 698.

GUNN, S. R. (1964). "Radiometric calorimetry: a review," Nucl. Instrum. Methods **29**, 1.

GUNN, S. R. (1970). "Radiometric calorimetry: a review (1970 supplement)," Nucl. Instrum. Methods **85**, 285.

GUNNINK, R. AND NIDAY, J. B. (1972). *Computerized Quantitative Analysis by Gamma-Ray Spectrometry* Volumes I–IV, UCRL-51061 (National Technical Information Service, Springfield, Virginia)

GUNNINK, R. AND NIDAY, J. B. (1976). "A working model for Ge(Li) detector counting efficiencies," page 55 in *Proceedings of ERDA Symposium on X- and Gamma-Ray Sources and Applications,* CONF-760539 (National Technical Information Service, Springfield, Virginia).

GUNNINK, R., COLBY, L. J., JR., AND COBBLE, J. W. (1959). "Absolute beta standardization using 4π beta-gamma coincidence techniques," Anal. Chem. **31**, 796.

HAIGHT, F. A. (1967). *Handbook of the Poisson Distribution.* (John Wiley & Sons, New York).

HALLDEN, N. A. AND HARLEY, J. H. (1960). "An improved alpha-counting technique," Anal. Chem. **32**, 1861.

HALLIDAY, D. (1955). *Introductory Nuclear Physics,* 2nd Ed. (John Wiley & Sons, New York).

HANDLER, L. H. AND ROMBERGER, J. A. (1973). "The use of a solubilizer-scintillator mixture for counting aqueous systems containing high- and low-energy radionuclides," Int. J. Appl. Radiat. Isot. **24**, 129.

HANSEN, J. S. MCGEORGE, J. C., AND FINK, R. W. (1973). "Efficiency calibration of semiconductor detectors in the X-ray region," Nucl. Instrum. Methods **112**, 239.

HANSEN, W. L. AND HALLER, E. E. (1974). "A view of the present status and future prospects of high purity germanium," IEEE Trans. Nucl. Sci. **NS-21**, 251.

HARDING, J. E., SCHWEBEL, A., AND STOCKMANN, L. L. (1958). "Report on the determination of small amounts of radium in solution," J. Assoc. Off. Agri. Chem. **41**, 311.

HARE, D. L., HENDEE, W. R., WHITNEY, W. P., AND CHANCY, E. L. (1974). "Accuracy of well ionization chamber isotope calibrators," J. Nucl. Med. **15**, 1138.

HARLEY, J. H., Ed. (1972). HASL Procedures Manual, USAEC Report HASL-300 (National Technical Information Service, Springfield, Virginia).

HARLEY, J. H., HALLDEN, N. A., AND FISENNE, I. M. (1962). "Beta scintillation counting with thin plastic phosphors," Nucleonics **20** No.1, 59.

HARMS, J. (1967). "Automatic dead-time correction for multichannel pulse-height analyzers at variable counting rates," Nucl. Instrum. Methods **53**, 192.

HARPER, P. V., SIEMENS, W. D., LATHROP, K. A., AND ENDLICH H. (1963). "Production and use of Iodine-125," J. Nucl. Med. **4**, 277.

HASL (1969). "Mechanical and electrical drawings for low level beta scintillation counters," HASL Technical Memorandum 62-7 (Internal Publication, ERDA Health and Safety Laboratory, New York).

HASL (1973). Health and Safety Laboratory. *Fallout Program Quarterly Summary Report*, Hardy, E. P. Ed., U.S. Atomic Energy Commission Report HASL-274 (National Technical Information Service, Springfield, Virginia).

HAUSER, W. (1974). "The 1972 nuclear medicine survey—radionuclide identification and activity measurement," Am. J. Clin. Pathol. **61**, 943.

HAUSER, W. AND CAVALLO, L. M. (1975). "Measurement and quality assurance of the amount of administered tracer," page 154 in *Quality Control in Nuclear Medicine* (C. V. Mosby Company, St. Louis).

HÄUSSER, O. LONE, M. A., ALEXANDER, T. K., KUSHNERIUK, S. A., AND GASCON, J. (1983). "The prompt response of bismuth germanate and NaI(Tl) scintillation detectors to fast neutrons," Nucl. Instrum. Methods **213**, 301.

HAWKINGS, R. C., MERRITT, W. F., AND CRAVEN, J. H. (1952). "The maintenance of radioactive standards with the 4π proportional counter," page 35 in *Developments and Techniques in the Maintenance of Standards*, National Physical Laboratory, May 1951. (H.M. Stationery Office, London).

HAYWARD, R. W. (1961). "On the determination of disintegration rates by the coincidence method using high efficiency detectors," Int. J. Appl. Radiat. Isot. **12**, 148.

HAYWARD, R. W., HOPPES, D. D., AND MANN, W. B. (1955). "Branching ratio in the decay of polonium-210," J. Res. Nat. Bur. Stand. **54**, 47.

HEATH, R. L. (1957). *Scintillation Spectrometry Gamma-Ray Spectrum Catalogue*, page 13, USAEC Report, IDO-16408 (National Technical Information Service, Springfield, Virginia).

HEATH, R. L. (1964). *Scintillation Spectrometry, Gamma-Ray Spectrum Catalogue*, 2nd. ed., USAEC Report IDO-16880-1 (National Technical Information Service, Springfield, Virginia).

HEATH, R. L. (1966). "Computer techniques for the analysis of gamma-ray spectra obtained with NaI and lithium-ion drifted germanium detectors," Nucl. Instrum. Methods **43**, 209.

HEATH, R. L. (1969). "Gamma-ray spectrometry and automated data systems for activation analysis," page 959 in *Modern Trends in Activation Analysis*, Vol. II Nat. Bur. Stand. Spec. Publ. 312, DeVoe, J. R., Ed. (U.S. Government Printing Office, Washington).

HEATH, R. L. (1974). *Gamma-Ray Spectrum Catalogue, Ge(Li) and Si(Li) Spectrometry*, Vol. 2, Aerojet Nuclear Co. Report ANCR-1000-2 (National Technical Information Service, Springfield, Virginia).

HEATH, R. L. (1983). "Status of photon counting using solid scintillators," page 156 in *Advances in Scintillation Counting*, McQuarrie, S. A., Ediss, C. and Wiebe, L. I., Eds. (University of Alberta Press, Edmonton, Alberta, Canada).

HEATH, R. L., HELMER, R. G., SCHMITTROTH, L. A., AND CAZIER, G. A. (1967). "A method for generating single gamma-ray shapes for the analysis of spectra," Nucl. Instrum. Methods **47**, 281.

HEIMBUCH, A. H. (1961). "Scintillating ion exchange resins," Trans. Am. Nucl. Soc. **4,** 251.

HELF, S., WHITE, C. G., AND SHELLEY, R. N. (1960). "Radioassay of finely divided solids by suspension in a gel scintillator," Anal. Chem. **32,** 238.

HELMER, R. G. AND PUTNAM, M. H. (1972). "Gauss V, a computer program for the analysis of gamma-ray spectra from Ge(Li) spectrometers," USAEC Report No. ANCR-1043 (National Technical Information Service, Springfield, Virginia).

HELMER, R. G., HEATH, R. L., PUTNAM, M., AND GIPSON, D. H. (1967a). "Photopeak analysis program for photon energy and intensity determinations (Ge(Li) and NaI(Tl) spectrometers)," Nucl. Instrum. Methods **57,** 46.

HELMER, R. G., HEATH, R. L., SCHMITTROTH, L. A., JAYNE, G. A., AND WAGNER, L. M. (1967b). "Analysis of gamma-ray spectra from NaI(Tl) and Ge(Li) spectrometers. Computer programs," Nucl. Instrum. Methods **47,** 305.

HELMER, R. G., CLINE, J. E., AND GREENWOOD, R. C. (1975). "Gamma-ray energy and intensity measurements with Ge(Li) spectrometers," page 775 in *The Electromagnetic Interaction in Nuclear Spectroscopy,* Hamilton, W. D., Ed. (North Holland and American Elsevier Publishing Companies, Amsterdam and New York).

HELMER, R. G., GREENWOOD, R. C., AND GEHRKE, R. J. (1978). "Reevaluation of precise gamma-ray energies for calibration of Ge(Li) spectrometers," Nucl. Instrum. Methods **166,** 189.

HELMER, R. G., VAN ASSCHE, R. H. M., AND VAN DER LEUN, C. (1979). "Recommended standards for gamma-ray energy calibration (1979)," At. Data Nuc. Data Tables **24,** 39.

HENDEE, W. R. (1973). *Radioactive Isotopes in Biological Research* (John Wiley & Sons, New York).

HENDRICKS, R. W. (1972). "Gas amplification factor in xenon filled proportional counters," Nucl. Instrum. Methods **102,** 309.

HENSLEY, W. K., BASSETT, W. A., AND HIUZENGA, J. R. (1973). "Pressure dependence of the radioactive decay constant of beryllium-7," Science **181,** 1164.

HERCEG NOVI (1973). "Proceedings of the first international summer school on radionuclide metrology," Nucl. Instrum. Methods **112,** 1–398.

HERCULES, D. M. (1970). "Physical basis of chemiluminescence," page 315 in *The Current Status of Liquid Scintillation Counting,* Bransome, E. D., Ed. (Grune & Stratton, New York).

HERRERA, N. E. AND PARAS, P. (1983). "Q-series, The College of American Pathologists surveillance of activity (dose) calibrators," Int. J. Nucl. Med. Biol. **10,** 159.

HIGASHIMURA, T., YAMADA, O., NOHARA, N., AND SHIDEI, T. (1962). "External standard method for the determination of the efficiency in liquid scintillation counting," Int. J. Appl. Radiat. Isot. **13,** 308.

HILL, C. R. (1961). "A method of alpha particle spectroscopy for materials of very low specific activity," Nucl. Instrum. Methods **12,** 299.

HINE, G. J. (1956). "Appendix," in *Radiation Dosimetry*, Hine, G. J. and Brownell, G. L., Eds. (Academic Press, New York).

HINE, G. J. (1967). "γ-ray sample counting," page 275 in *Instrumentation in Nuclear Medicine*, Vol. 1, Hine, G. J., Ed. (Academic Press, New York).

HINE, G. L. AND SORENSON, J. A., Eds. (1974). *Instrumentation in Nuclear Medicine*, Vol. 2 (Academic Press, New York).

HIRSHFELD, A. T., HOPPES, D. D., AND SCHIMA, R. J. (1976). "Germanium detector efficiency calibration with NBS standards," page 90 in *Proceedings of ERDA Symposium on X- and Gamma-Ray Sources and Applications*, Report CONF-760539 (National Technical Information Service, Springfield, Virginia).

HOFFER, P. B., BECK, R. N., AND GOTTSCHALK, A., Eds. (1971). *Semiconductor Detectors in the Future of Nuclear Medicine* (Society of Nuclear Medicine, New York).

HOFSTADTER, R. (1948). "Alkali halide scintillation counters," Phys. Rev. **74**, 100.

HOLM, E. AND PERSSON, R. B. R. (1976). "Transfer of fall-out plutonium in the food-chain lichen→reindeer→man," page 435 in *Transuranium Nuclides in the Environment*, IAEA/STI/PUB/410 (International Atomic Energy Agency, Vienna).

HORROCKS, D. L. (1966a). "Low-level alpha disintegration rate determinations with a one-multiplier phototube liquid scintillation spectrometer," Int. J. Appl. Radiat. Isot. **17**, 441.

HORROCKS, D. L., ED. (1966b). *Organic Scintillators.* (Gordon and Breach Science Publishers, New York).

HORROCKS, D. L. (1968). "Direct measurement of $^{14}CO_2$ in a liquid scintillation counter," Int. J. Appl. Radiat. Isot. **19**, 859.

HORROCKS, D. L. (1971). "Obtaining the possible maximum 90 percent efficiency for counting of ^{55}Fe in liquid scintillator solutions," Int. J. Appl. Radiat. Isot. **22**, 258.

HORROCKS, D. L. (1974). *Applications of Liquid Scintillation Counting.* (Academic Press, New York).

HORROCKS, D. L. (1975). "Liquid scintillation counting in high-density polyethylene vials," Int. J. Appl. Radiat. Isot. **26**, 243.

HORROCKS, D. L. (1977). U.S. Patent No. 4,060,728. November 29, 1977.

HORROCKS, D. L. (1978). "A new method of quench monitoring in liquid scintillation counting: the H-number concept" page 145 in *Liquid Scintillation Counting, Vol. 5*, Crook, M. A. and Johnson, P. Eds. (Heyden, London).

HORROCKS, D. L. AND STUDIER, M. H. (1964). "Determination of radioactive noble gases with a liquid scintillator," Anal. Chem. **36**, 2077.

HORROCKS, D. L. AND PENG, C. T., EDS. (1971). *Organic Scintillators and Liquid Scintillation Counting* (Academic Press, New York).

HOUTERMANS, F. G. AND OESCHGER, H. (1958). "Proportionalzählrohr zur Messung schwacher Aktivitäten weicher β-Strahlung," Helv. Phys. Acta **31**, 117.

HOUTERMANS, H. (1973). "Probability of non-detection in liquid scintillation counting," Nucl. Instrum. Methods **112**, 121.

HOUTERMANS, H. AND MIGUEL, M. (1962). "4π-β-γ coincidence counting for the calibration of nuclides with complex decay schemes," Int. J. Appl. Radiat. Isot. **13**, 137.

HUBBARD, G. S. (1983). "Materials aspects of germanium radiation detector fabrication," page 161 in *Nuclear Radiation Detector Materials* Haller, E. E., Kraner, H. W. and Higginbotham, W. A., Eds. (Elsevier Science and North-Holland Publishing Companies, New York and Amsterdam).

HUBBELL, J. H. (1969). *Photon Cross Sections, Attenuation Coefficients, and Energy Absorption Coefficients from 10 keV to 100 GeV.* Report NSRDS-NBS 29 (U.S. Government Printing Office, Washington).

HUBBELL, J. H. (1977). "Photon mass attenuation and mass energy-absorption coefficients for H, C, N, O, Ar, and seven mixtures from 0.1 keV to 20 MeV," Radiat. Res. **70**, 58.

HUBBELL, J. H. (1982). "Photon mass attenuation and energy-absorption coefficients from 1 keV to 20 MeV" Int. J. Appl. Radiat. Isot. **33**, 1269.

HUBBELL, J. H., VEIGELE, W. J., BRIGGS, E. A., BROWN, R. T., CROMER, D. T., AND HOWERTON, R. J. (1975). "Atomic form factors, incoherent scattering functions, and photon scattering cross sections," J. Phys. Chem. Ref. Data **4**, 471.

HUGHES, D. J. (1953). *Pile Neutron Research.* (Addison-Wesley, Cambridge, Massachusetts).

HUGHES, E. E. AND MANN, W. B. (1964). "The half-life of carbon-14; comments on the mass-spectrometric method," Int. J. Appl. Radiat. Isot. **15**, 97.

HURST, G. S., PAYNE, M. G., KRAMER, S. D., AND CHEN, C. H. (1980). "Counting the atoms," Physics Today **33**, 24.

HUTCHINSON, J. M. R. AND MULLEN, P. A. (1983a). "Calibration of K-x-ray-emission rates in the decay of ^{49}V," Int. J. Appl. Radiat. Isot. **34**, 539.

HUTCHINSON, J. M. R. AND MULLEN, P. A. (1983b). "Pin-well-NaI(Tl) counting of 59.5 keV γ rays in the decay of ^{241}Am," Int. J. Appl. Radiat. Isotop. **34**, 543.

HUTCHINSON, J. M. R., NAAS, C. R., WALKER, D. H., AND MANN, W. B. (1968). "Backscattering of alpha particles from thick metal backings as a function of atomic weight," Int. J. Appl. Radiat. Isot. **19**, 517.

HUTCHINSON, J. M. R., LANTZ, J. L., MANN, W. B., MULLEN, P. A., AND RODRIGUEZ-PASQUÉS, R. H. (1972). "An anti-coincidence shielded NaI(Tl) system at NBS," IEEE Trans. Nucl. Sci. **NS-19**, 117.

HUTCHINSON, J. M. R., MANN, W. B., AND MULLEN, P. A. (1973a). "Sum-peak counting with two crystals," Nucl. Instrum. Methods **112**, 187.

HUTCHINSON, J. M. R., MANN, W. B., AND PERKINS, R. W. (1973b). "Low-level radioactivity measurements," Nucl. Instrum. Methods **112**, 305.

HUTCHINSON, J. M. R., LUCAS, L. L., AND MULLEN, P. A. (1976a). "Study of the scattering correction for thick uranium-oxide and other alpha-particle sources—Part II: Experimental," Int. J. Appl. Radiat. Isot. **27**, 43.

HUTCHINSON, J. M. R., MANN, W. B., AND MULLEN, P. A. (1976b). "Development of the National Bureau of Standards low-energy-photon-emission-rate radioactivity standards," page 25 in *Proceedings of ERDA Symposium on X- and Gamma-Ray Sources and Applications,* CONF 760539 (National Technical Information Service, Springfield, Virginia).

HUTCHINSON, J. M. R., MULLEN, P. A., AND COLLÉ, R. (1984). "Development of a regenerative radon-in-water radioactivity standard," Nucl. Instrum. Methods **223,** 451.

IAEA (1965). International Atomic Energy Agency. *Radioisotope Sample Measurement Techniques in Medicine and Biology,* IAEA/STI/PUB/106 (International Atomic Energy Agency, Vienna).

IAEA (1967a). International Atomic Energy Agency. *Standardization of Radionuclides,* IAEA/STI/PUB/139 (International Atomic Energy Agency, Vienna).

IAEA (1967b). International Atomic Energy Agency. *Radioactive Dating and Methods of Low-Level Counting,* IAEA/STI/PUB/152 (International Atomic Energy Agency, Vienna).

IAEA (1970). International Atomic Energy Agency. *Analytical Control of Radiopharmaceuticals,* IAEA/STI/PUB/253 (International Atomic Energy Agency, Vienna).

IAEA (1971). International Atomic Energy Agency. *Radioisotope Production and Quality Control,* IAEA/STI/DOC/10/128 (International Atomic Energy Agency, Vienna).

IAEA (1973). International Atomic Energy Agency. *New Developments in Radiopharmaceuticals and Labelled Compounds,* IAEA/STI/PUB/344 (International Atomic Energy Agency, Vienna).

IAEA (1981). International Atomic Energy Agency. *Methods of Low-Level Counting and Spectrometry* IAEA/STI/PUB/592 (International Atomic Energy Agency, Vienna).

ICRM (1978). International Committee for Radionuclide Metrology. *Metrology Needs in the Measurement of Environmental Radioactivity,* Proceedings of a seminar, Hutchinson, J. M. R. and Mann, W. B., Eds., Env. Int. **1,** 7.

ICRM (1980). International Committee for Radionuclide Metrology. *Traceability and Quality Control in the Measurement of Environmental Activity,* Proceedings of a seminar, Hutchinson, J. M. R. and Mann, W. B., Eds., Env. Int. **3,** 361.

ICRM (1983a). International Committee for Radionuclide Metrology. *Gamma- and X-Ray Spectrometry Techniques and Applications,* Proceedings of a seminar, Debertin, K. and Mann, W. B., Eds., Int. J. Appl. Radiat. Isot. **34,** 477.

ICRM (1983b). International Committee for Radionuclide Metrology. *Applied Radionuclide Metrology,* Proceedings of a seminar, Mann, W. B., Lowenthal, G. C. and Taylor, J. G. V., Eds., Int. J. Appl. Radiat. Isot. **34,** 1039.

ICRM (1983c). International Committee for Radionuclide Metrology. *Radionuclide Metrology in the Life Sciences,* Proceedings of a seminar, Woods, M. J., Ed., Int. J. Nucl. Med. Biol. **10,** 65.

ICRM (1984a). International Committee for Radionuclide Metrology. *Alpha*

Particle Spectrometry and Low Level Measurements, Proceedings of a seminar, Glover, K., Lally, A. E. and Ivanovich, M., Eds., Nucl. Instrum. Methods **223**, 1.

ICRM (1984b). International Committee for Radionuclide Metrology. *Alpha Particle Spectrometry Techniques and Applications* Proceedings of a seminar, Bortels, G., Brinkman, G. A. and Mann, W. B., Eds., Int. J. Radiat. Isot. **35**, 4.

ICRP (1975). International Commission on Radiological Protection. *Report of the Task Group on Reference Man,* ICRP Publication 23 (Pergamon Press, New York).

ICRU (1961). International Commission on Radiological Units and Measurements (1959), published as NBS Handbook 78 (International Commission on Radiation Units and Measurements, Bethesda, Maryland).

ICRU (1963). International Commission on Radiological Units and Measurements. *Radioactivity,* ICRU Report 10c, issued as National Bureau of Standards Handbook 86 (International Commission on Radiation Units and Measurements, Bethesda, Maryland).

ICRU (1964). International Commission on Radiological Units and Measurement. *Physical Aspects of Irradiation,* ICRU Report 10b published as National Bureau of Standards Handbook 85 (International Commission on Radiation Units and Measurements, Bethesda, Maryland).

ICRU (1968). International Commission on Radiation Units and Measurements. *Certification of Standardized Radioactive Sources,* ICRU Report 12 (International Commission on Radiation Units and Measurements, Bethesda, Maryland).

ICRU (1970). International Commission on Radiation Units and Measurements. *Radiation Dosimetry: X Rays Generated at Potentials of 5 to 150 kV,* ICRU Report 17 (International Commission on Radiation Units and Measurements, Bethesda, Maryland).

ICRU (1972). International Commission on Radiation Units and Measurements. *Measurement of Low-Level Radioactivity,* ICRU Report 22 (International Commission on Radiation Units and Measurements, Bethesda, Maryland).

ICRU (1979a). International Commission on Radiation Units and Measurements. *Average Energy Required to Produce an Ion Pair,* ICRU Report 31 (International Commission on Radiation Units and Measurements, Bethesda, Maryland).

ICRU (1979b). International Commission on Radiation Units and Measurement. *Methods of Assessment of Absorbed Dose in Clinical Use of Radionuclides,* ICRU Report 32 (International Commission on Radiation Units and Measurements, Bethesda, Maryland).

ICRU (1980). International Commission on Radiation Units and Measurements, *Radiation Quantities and Units,* ICRU Report 33 (International Commission on Radiation Units and Measurements, Bethesda, Maryland).

INN, K. W. G., LIGGETT, W. S., AND HUTCHINSON, J. M. R. (1984). "The National Bureau of Standards Rocky Flats soil standard reference material,"

Nucl. Instrum. Methods Phys. Res. **223**, 443.

ISRAEL, H. I., LIER, D. W., AND STORM, E. (1971). "Comparison of detectors used in measurement of 10 to 300 keV x-ray spectra," Nucl. Instrum. Methods **91**, 141.

IVANOV, V. B. AND SHIPILOV, V. I. (1974). "A summing Ge(Li)-Compton spectrometer," Nucl. Instrum. Methods **119**, 313.

JAFFEY, A. H. (1954). "Radiochemical assay by alpha and fission measurements," page 596 in *The Actinide Elements*, Seaborg, G. T. and Katz, J. J., Eds. (McGraw-Hill, New York, Toronto, and London).

JARRETT, A. A. (1946). *Statistical Methods Used in the Measurement of Radioactivity with Some Useful Graphs and Nomographs*, USAEC Report AECU-262 (U.S. Department of Energy, Technical Information Center, Oak Ridge, Tennessee).

JEFFAY, H. AND ALVAREZ, J. (1961). "Liquid scintillation counting of carbon 14; use of ethanolamine-ethylene glycol monomethyl ether-toluene," Anal. Chem. **33**, 612.

JELLEY, J. V. (1958). *Čerenkov Radiation and Its Applications* (Pergamon Press, Oxford and New York).

JESSE, W. P. AND SADAUSKIS, J. (1953). "Alpha-particle ionization in pure gases and the average energy to make an ion pair," Phys. Rev. **90**, 1120.

JESSE, W. P. AND SADAUSKIS, J. (1955). "Ionization in pure gases and the average energy to make an ion pair for alpha and beta particles," Phys. Rev. **99**, 1668.

JOHNS, F. B., ED. (1975). *Handbook of Radiochemical Analytical Methods.* EPA-680/4-75-001 (National Technical Information Service, Springfield, Virginia).

JOHNS, H. E. AND CUNNINGHAM, J. R. (1969). Table A-8 in *The Physics of Radiology*, 3rd Ed. (C. C Thomas, Springfield).

JOHNSON, N. R., EICHLER, E., AND O'KELLEY, G. D. (1963). *Nuclear Chemistry*, Vol. 2 of *Techniques of Inorganic Chemistry*, Jonassen, H. B., Ed. (Interscience Publishers, New York).

JONES, D. G. AND MCNAIR, A. (1970). "Standardization of radionuclides by efficiency-tracing methods using a liquid-scintillation counter," page 37 in *Radioactivity Calibration Standards*, Nat. Bur. Stand. Spec. Publ. 331, Mann, W. B. and Garfinkel, S. B., Eds. (U.S. Government Printing Office, Washington).

JONES, W. M. (1955). "Half-life of tritium," Phys. Rev. **100**, 124.

JORDAN, K. C. (1973). Private communication. (Monsanto Mound Laboratory, Miamisburg, Ohio).

JORDAN, K. C. AND BLANKE, B. C. (1967). "Half-lives of actinium-227 and thorium-227 measured calorimetrically," page 567 in *Standardization of Radionuclides*, IAEA/STI/PUB/139 (International Atomic Energy Agency, Vienna).

KAHN, B., BLANCHARD, R. L., KRIEGER, H. Z., KOLDE, H. E., SMITH, D. B., MARTIN, A., GOLD, S., AVERETT, W. J., BRINCK, W. L., AND KARCHES, G. J. (1971a). *Radiological Surveillance Studies at a Boiling Water Nuclear*

Power Reactor, PHS Report BRH/DER 70-1. (National Technical Information Service, Springfield, Virginia).

KAHN, B., BLANCHARD, R. L., KOLDE, H. E., KRIEGER, H. L., GOLD, S., BRINCK, W. L., AVERETT, W. J., SMITH, D. B., AND MARTIN, A. (1971b). *Radiological Surveillance Studies at a Pressurized Water Nuclear Power Reactor*, EPA Report RD-71-1 (National Technical Information Service, Springfield, Virginia).

KALBHEN, D. A. (1970). "Chemiluminescence as a problem in liquid scintillation counting," page 337 in *The Current Status of Liquid Scintillation Counting*, Bransome, E. D., Ed. (Grune & Stratton, New York).

KALLMANN, H. (1947). "Der elementar Prozess der Lichtanregung in Leuchtstoffen. Über eine neue Methode der Zählung und Energiemessung geladener Teilchen," Natur. Tech. (July).

KATCOFF, S., SCHAEFFER, O. A., AND HASTINGS, J. M. (1951). "Half-life of I^{129} and the age of the elements," Phys. Rev. **82**, 688.

KEENE, J. P. (1950). "An absolute method for measuring the activity of radioactive isotopes," Nature **166**, 601.

KEISCH, B., KOCH, R. C., AND LEVINE, A. S. (1965). "Determination of biospheric levels of I-129 by neutron activation analysis," page 284 in *Modern Trends in Activation Analysis*. (Center for Trace Characterization, Texas A & M University, College Station, Texas).

KEITH, H. D. AND LOOMIS, T. C. (1976). "Calibration and use of lithium-drifted silicon detector for accurate analysis of x-ray spectra," X-Ray Spectrometry, **5**, 93.

KEITH, R. L. G. AND WATT, D. E. (1966). "A routine procedure for 4π proportional counting weak beta emitters from solid sources," Nucl. Instrum. Methods **42**, 68.

KEITHLEY (1977). *Electrometer Measurements* Revised 2nd Ed. (Keithley Instruments, Inc., Cleveland, Ohio).

KLOTS, C. E. (1968). "Energy deposition mechanisms," page 1 in *Fundamental Processes in Radiation Chemistry*, Ausloos, P. J., Ed. (John Wiley & Sons, New York).

KNOLL, G. F. (1979). *Radiation Detection and Measurement* (John Wiley and Sons, New York)

KOBAYASHI, Y. AND MAUDSLEY, D. V. (1970). "Practical aspects of double isotope counting," page 76 in *The Current Status of Liquid Scintillation Counting*, Bransome, E. D., Ed. (Grune & Stratton, New York).

KOCH, L. (1958). "La scintillation dans les gaz nobles et leurs mélanges," page 151 in *Nuclear Electronics*, IAEA/STI/PUB/2, Vol. 1 (International Atomic Energy Agency, Vienna).

KOCH, R. C. AND KEISCH, B. (1964). "Neutron activation analysis studies of ^{129}I in the biosphere," page 47 in *L'Analyse Par Radioactivation* (Presses Universitaires de France, Vendome).

KOCHER, D. C. (1981). *Radioactive Decay Data Tables. A Handbook of Decay Data for Application to Radiation Dosimetry and Radiological Assessments*, U.S. Department of Energy Report DOE/TIC-11026 (National Technical

Information Service, Springfield, Virginia).

KOHMAN, T. P. (1945). "A general method for determining coincidence corrections of counting instruments," page 1655 in *The Transuranium Elements, Part II*, Seaborg, G. T., Katz, J. J., and Manning, W. M., Eds. (McGraw-Hill, New York).

KOLAROV, V., LEGALLIC, Y., AND VATIN, R. (1970). "Absolute direct measurement of the activity of pure β emitters by liquid scintillation" Int. J. Appl. Radiat. Isot. **21**, 443. (In French).

KONDO, S. AND RANDOLPH, M. L. (1960). "Effect of finite size of ionization chambers on measurements of small photon sources," Radiat. Res. **13**, 37.

KORENMAN, I. M. (1966). *Analytical Chemistry of Low Concentrations*. (Israel Program for Scientific Translations Limited, Jerusalem).

KOSTROUN, V. O., CHEN, M. S., AND CRASEMANN, B. (1971). "Atomic radiation transition probabilities to the 1s state and theoretical K-shell fluorescence yields," Phys. Rev. **A3**, 533.

KOWALSKI, E. (1965). "Absolutmessung der radioactiven Quellstarke nach der allgemeinen Koinzidenmethode; Einfluss der Richtungskorrelation, Streukoinzidenzen und Summenimpulse," Nucl. Instrum. Methods **33**, 29.

KOWALSKI, E., ANLIKER, R., AND SCHMID, K. (1967). "Criteria for the selection of solutes in liquid scintillation counting: new efficient solutes with high solubility," Int. J. Appl. Radiat. Isot. **18**, 307.

KRANER, H. W., SCHROEDER, G. L., LEWIS, A. R., AND EVANS, R. D. (1964). "Large volume scintillation chamber for radon counting," Rev. Sci. Instrum. **35**, 1259.

KREY, P. W. AND HARDY, E. P. (1970). "Plutonium in soil around the Rocky Flats plant," USAEC Report HASL-235 (National Technical Information Service, Springfield, Virginia).

KRUGERS, J. (1973). *Instrumentation in Applied Nuclear Chemistry*. (Plenum Press, New York).

KU, H. H. (1966). "Notes on the use of propagation of error formulas," J. Res. Nat. Bur. Stand. **70C**, 263. (Reprinted in Ku, 1969).

KU, H. H., Ed. (1969). *Precision Measurement and Calibration—Statistical Concepts and Procedures*, Nat. Bur. Stand. (U.S.) Publ. 300, Vol. 1 (U.S. Government Printing Office, Washington).

KUCHTA, D. R., RUNDO, J., AND HOLTZMAN, R. B. (1976). "Low-level determination of skeletal ^{228}Ra and ^{228}Th in the presence of gross amounts of ^{226}Ra," page 183 in *Biological and Environmental Effects of Low-Level Radiation*, Vol. II, IAEA/STI/PUB/409 (International Atomic Energy Agency, Vienna).

LACHANCE, Y. AND ROY, J. C. (1972). "Method for vacuum evaporation of SiO on thin plastic films," Rev. Sci. Instrum. **43**, 1382.

LAICHTER, Y., NOTEA, A., AND SHAFRIR, N. H. (1973). "A general approach to the calibration of a Ge(Li) gamma spectrometry system for measurement of environmental and biological samples," Nucl. Instrum. Methods **113**, 61.

LAPP, R. E. AND ANDREWS, H. L. (1964). *Nuclear Radiation Physics* (Prentice-Hall, Englewood Cliffs, New Jersey).

LATNER, N. AND SANDERSON, C. G. (1972). "The HASL Ge(Li)-NaI(Tl) low-level counting system," IEEE Trans. Nucl. Sci. **NS-19,** 141.

LAVI, N. (1973). "^{166m}Ho as an efficiency calibration standard for Ge(Li) detectors," Nucl. Instrum. Methods **109,** 265.

LEDERER, C. M., HOLLANDER, J. M., AND PERLMAN, I. (1967). *Table of Isotopes,* 6th Ed. (John Wiley & Sons, New York).

LEGRAND, J. (1973). "Calibration of γ-spectrometers," Nucl. Instrum. Methods **112,** 229.

LEGRAND, J., BLONDEL, M., AND MAGNIER, P. (1973). "High-pressure 4π proportional counter for internal conversion electron measurements (^{139}Ce, ^{109}Cd, ^{99m}Tc)," Nucl. Instrum. Methods **112,** 101.

LEWIS, W. B. (1948). *Electrical Counting* (Cambridge University Press, Cambridge).

LEWIS, V. E., WOODS, M. J., AND GOODIER, I. W. (1972). "The calibration of the 1383A ionization chamber for ^{67}Ga," Int. J. Appl. Radiat. Isot. **23,** 279.

LEWIS, V. E., SMITH, D., AND WILLIAMS, A. (1973). "Correlation counting applied to the determination of absolute disintegration rates for nuclides with delayed states," Metrologia **9,** 14.

LIEBERMAN, R. AND MOGHISSI, A. A. (1970). "Low-level counting by liquid scintillation, II. Applications of emulsions in tritium counting," Int. J. Appl. Radiat. Isot. **21,** 319.

LIEBSON, S. H. AND FRIEDMAN, H. (1948). "Self-quenching halogen-filled cavities," Rev. Sci. Instrum. **19,** 303.

LINDENBAUM, A. AND SMYTH, M. A. (1971). "Determination of ^{239}Pu and ^{241}Am in animal tissues by liquid scintillation spectrometry," page 951 in *Organic Scintillators and Liquid Scintillation Counting,* Horrocks, D. L. and Peng, C. T., Eds. (Academic Press, New York).

LLOYD, R. D., MAYS, C. W., AND ATHERTON, D. R. (1967). "Knothole, a new side-well gamma ray detector," Nucl. Instrum. Methods **49,** 109.

LMRI (1975). Laboratoire de Metrologie de Rayonnements Ionisants. *Annual Report of the Laboratoire de Metrologie des Rayonnements Ionisants.* (Centre d'Études Nucleaires, Saclay).

LOEVINGER, R. AND BERMAN, M. (1951). "Efficiency criteria in radioactivity counting," Nucleonics **9,** No. 1, 26.

LOEVINGER, R. AND BERMAN, M. (1976). *A Revised Schema for Calculating the Absorbed Dose from Biologically Distributed Radionuclides,* MIRD Pamphlet No. 1, Revised (Society of Nuclear Medicine, New York).

LOFTUS, T. P. AND WEAVER, J. T. (1974). "Standardization of ^{60}Co and ^{137}Cs gamma-ray beams in terms of exposure," J. Res. Nat. Bur. Stand. **78A,** 465.

LOOSLI, H., OESCHGER, H., STUDER, R., WAHLEN, M., AND WIEST, W. (1975). "Argon-37 activity in air samples and its atmospheric mixing," page 240 in *Proceedings of the Noble Gases Symposium,* Stanley, R. E. and Moghissi, A. A., Eds. ERDA CONF-730915 (National Technical Information Service, Springfield, Virginia).

LOWENTHAL, G. C. (1969). "Secondary standard instruments for the activity measurement of pure β emitters—a review," Int. J. Appl. Radiat. Isot. **20**, 559.

LOWENTHAL, G. C. AND SMITH, A. M. (1964). "Use of Au-20% Pd for metallising thin source supports for 4π proportional gas flow counters," Nucl. Instrum. Methods **30**, 363.

LOWENTHAL, G. C. AND HICKS, R. M. (1966). "An isothermal microcalorimeter operated in a water triple point cell and designed for radioactivity measurements," J. Sci. Instrum. **43**, 36.

LOWENTHAL, G. C. AND PAGE, V. (1970). "A syringe pycnometer for the accurate weighing of milligram quantities of aqueous solutions," Anal. Chem. **42**, 815.

LOWENTHAL, G. C. AND WYLLIE, H. A. (1973a). "Special methods of source preparation," Nucl. Instrum. Methods **112**, 353.

LOWENTHAL, G. C. AND WYLLIE, H. A. (1973b). "Thin 4π sources on electrosprayed ion exchange resin for radioactivity standardization," Int. J. Appl. Radiat. Isot. **24**, 415.

LOWENTHAL, G. C., PAGE, V., AND WYLLIE, H. A. (1973). "The use of sources made on electrosprayed pads of ion exchange resins for efficiency tracer measurements," Nucl. Instrum. Methods **112**, 197.

LOWENTHAL, G. C., WYLLIE, H. A., AND PAGE, V. (1982). "A method for measuring the activity of ^{55}Fe using a 4π proportional counter," Int. J. Appl. Radiat. Isot. **33**, 755.

LUCAS, H. F. (1957). "Improved low-level alpha-scintillation counter for radon," Rev. Sci. Instrum. **28**, 680.

LUCAS, H. F. (1964). *The Low-Level Gamma Counting Room: Radon Removal and Control,* ANL Report ANL-6938 (Argonne National Laboratory, Argonne, Illinois).

LUCAS, H. F. (1965). *A Radon Removal System for the NASA Lunar Sample Laboratory: Design and Discussion,* ANL Report ANL-7060 (Argonne National Laboratory, Argonne, Illinois).

LUCAS, L. L. AND HUTCHINSON, J. M. R. (1976). "Study of the scattering correction for thick uranium-oxide and other alpha-particle sources—Part I; Theoretical," Int. J. Appl. Radiat. Isot. **27**, 35.

LUCAS, L. L. AND MANN, W. B., Eds. (1978). *The Half Life of ^{239}Pu: Measurements Organized by the ERDA Half-Life Evaluation Committee,* Int. J. Appl. Radiat. Isot. **29**, 479.

LUCAS, L. L. AND MANN, W. B., Eds. (1984). *The Half Life of Plutonium-240,* Int. J. Appl. Radiat. Isot. **35**, 153.

LUKENS, H. R. (1961). "The relationship between fluorescence intensity and counting efficiency with liquid scintillators," Int. J. Appl. Radiat. Isot. **12**, 134.

MAGNO, P., REAVEY, T., AND APIDANAKIS, J. (1970). *Liquid Waste Effluents from a Nuclear Fuel Reprocessing Plant,* PHS Report BRH/NERHL 70-2 (National Technical Information Service, Springfield, Virginia).

MAKOWER, W. (1909). "On the number and the absorption by matter of the

β particles emitted by radium," Philos. Mag. 17 [6], 171.

MALM, H. L., RAUDORF, T. W., MARTINI, M., AND ZANIO, K. R. (1973). "Gamma-ray efficiency comparisons for Si(Li), Ge, CdTe, and HgI$_2$ detectors," IEEE Trans. Nucl. Sci. NS-20, 500.

MANLEY, J. H., AGNEW, H. M., BARSCHALL, H. H., BRIGHT, W. C., COON, J. H., GRAVES, E. R., JORENSON, T., AND WALDMAN, B. (1946). "Elastic backscattering of d-d neutrons," Phys. Rev. 70, 602.

MANN, H. M., BILGER, H. R., AND SHERMAN, I. S. (1966). "Observations on the energy resolution of germanium detectors for 0.1 to 10 MeV gamma rays," IEEE Trans. Nucl. Sci. NS-13, 252.

MANN, W. B. (1954a). "Use of Callendar's 'Radio-Balance' for the measurement of the energy emission from radioactive sources," J. Res. Nat. Bur. Stand. 52, 177.

MANN, W. B. (1954b). "A radiation balance for the calorimetric comparison of four national radium standards," J. Res. Nat. Bur. Stand. 53, 277.

MANN, W. B. (1962). "Calorimetric measurements of radioactivity," page 538 in Encyclopaedic Dictionary of Physics, Vol. 1, Thewlis, J., Ed-in-Chief (Pergamon Press, Oxford and MacMillan Co., New York).

MANN, W. B. (1973). "Radioactive calorimetry: a review of the work at the National Bureau of Standards," Nucl. Instrum. Methods 112, 273.

MANN, W. B. (1980). "Nuclear-decay data: the statement of uncertainties," Int. J. Appl. Radiat. Isot. 31, 387 (and in a number of other international journals).

MANN, W. B. (1983). "An international reference material for radiocarbon dating," Radiocarbon 25, 519.

MANN, W. B. AND PARKINSON, G. B. (1949). "A Geiger-Müller counting unit and external quenching equipment for the estimation of C^{14} in carbon dioxide," Rev. Sci. Instrum. 20, 41.

MANN, W. B. AND SELIGER, H. H. (1958). "Preparation, maintenance, and application of standards of radioactivity," Nat. Bur. Stand. (U.S.) Circ. 594 (U.S. Government Printing Office, Washington).

MANN, W. B. AND GARFINKEL, S. B. (1966). Radioactivity and Its Measurement (D. Van Nostrand Co., Inc., Princeton, New Jersey). (Revised as Mann, Ayres, and Garfinkel, 1980).

MANN, W. B. AND HUTCHINSON, J. M. R. (1976). "The standardization of iodine-129 by beta-photon coincidence counting," Int. J. Appl. Radiat. Isot. 27, 187.

MANN, W. B., STOCKMANN, L. L., YOUDEN, W. J., SCHWEBEL, A., MULLEN, P. A., AND GARFINKEL, S. B. (1959). "The preparation of new solution standards of radium," J. Res. Nat. Bur. Stand. 62, 27.

MANN, W. B., SELIGER, H. H., MARLOW, W. F., AND MEDLOCK, R. W. (1960). "Recalibration of the NBS carbon-14 standard by Geiger-Müller and proportional gas counting," Rev. Sci. Instrum. 31, 690.

MANN, W. B., MARLOW, W. F., AND HUGHES, E. E. (1961). "The half-life of carbon-14," Int. J. Appl. Radiat. Isot. 11, 57.

MANN, W. B., MEDLOCK, R. W., AND YURA, O. (1964). "A recalibration of the National Bureau of Standards tritiated water standards by gas counting," Int. J. Appl. Radiat. Isot. **15**, 351.

MANN, W. B., AYRES, R. L., AND GARFINKEL, S. B. (1980). *Radioactivity and Its Measurement*, 2nd Ed. (Pergamon Press, Oxford and New York).

MANN, W. B., HUTCHINSON, J. M. R., AND EDGERLY, D. E. (1981). "National and international traceability in radioactivity measurements," p. 173 in *Methods of Low-Level Counting and Spectrometry*, IAEA/STI/PUB/592 (International Atomic Energy Agency, Vienna).

MANN, W. B., UNTERWEGER, M. P., AND COURSEY, B. M. (1982). "Comments on the NBS tritiated-water standards and their use," Int. J. Appl. Radiat. Isot. **33**, 383.

MANN, W. B., UNTERWEGER, M. P., AND COURSEY, B. M. (1983a). "Further, Comments on tritium measurement scales for water," Int. J. Appl. Radiat. Isot. **34**, 1410.

MANN, W. B., LOWENTHAL, G. C., AND TAYLOR, J. G. V., Eds. (1983b). *Applied Radionuclide Metrology*, Int. J. Appl. Radiat. Isot. **34**, 1039.

MARINENKO, G. AND TAYLOR, J. K. (1966). "Coulometric calibration of microvolumetric apparatus," J. Res. Nat. Bur. Stand. **70C**, 1.

MARLOW, W. F. AND MEDLOCK, R. W. (1960). "Preparation and standardization of a carbon-14 beta-ray standard: benzoic-acid-7-C^{14} in toluene," J. Res. Nat. Bur. Stand. **64A**, 143.

MARTIN, G. R. (1961). "The estimation of the resolving time of a counting apparatus," Nucl. Instrum. Methods **13**, 263.

MARTIN, M. J. (1973). *Radioactive Atoms—Supplement I.* Report ORNL-4923 (Oak Ridge National Laboratory, Oak Ridge, Tennessee).

MARTIN, M. J. (1976). *Nuclear Decay Data for Selected Radionuclides*, Report ORNL-5114 (Oak Ridge National Laboratory, Oak Ridge, Tennessee).

MARTIN, M. J. (1978). Personal communication. (Oak Ridge National Laboratory, Oak Ridge, Tennessee).

MARTIN, M. J. (1982). "Radioisotopes," page 682 in *Encyclopedia of Chemical Technology*, Vol. 19, 3rd Ed. (John Wiley & Sons, Inc., New York).

MARTIN, M. J. AND BLICHERT-TOFT, P. H. (1970). *Radioactive Atoms*, Nucl. Data Tables, **8**, No. 1-2. (Oak Ridge National Laboratory, Oak Ridge, Tennessee).

MASTERS, C. F. AND EAST, L. V. (1970). "An investigation into the Harms dead time correction procedure using Monte Carlo modelling techniques," IEEE Trans. Nucl. Sci. **NS-17**, 383.

MAY, H. A. AND MARINELLI, L. D. (1960). "Factors determining the ultimate detector sensitivity of gamma scintillation spectrometers," page 71 in Semi-Annual Report, Jan-June, Argonne National Laboratory Report ANL-6199 (National Technical Information Service, Springfield, Virginia).

MAYHUGH, M. R., LUCAS, A. C., AND UTTS, B. K. (1978). "Low background beta counting with CaF_2 (Eu) in a phoswich configuration," IEEE Trans. Nucl. Sci. **NS-25**, 569.

MAZAKI, H., NAGATOMO, T., AND SHIMIZU, S. (1972). "Effect of pressure on the decay constant of ^{99m}Tc," Phys. Rev. **C5**, 1718.

McBeth, G. W., Winyard, R. A., and Lutkin, J. E. (1971). *Pulse Shape Discrimination with Organic Scintillators* (Koch-Light Laboratories, Ltd., Colnbrook, Bucks, England).

McCallum, G. J. and Coote, G. E. (1975). "Influence of source-detector distance on relative intensity and angular correlation measurements with Ge(Li) spectrometers," Nucl. Instrum. Methods **30**, 189.

McCurdy, D. E. and Mellor, R. A. (1981). "Determination of radium-224, -226, and -228 by coincidence spectrometry," Anal. Chem. **53**, 2212.

McDaniel, E. W., Schaefer, H. J., and Colehour, J. K. (1956). "Dual proportional counter for low-level measurement of alpha activity of biological materials," Rev. Sci. Instrum. **27**, 864.

McDowell, W. J. (1971). "Liquid scintillation counting techniques for the higher actinides," page 937 in *Organic Scintillators and Liquid Scintillation Counting*, Horrocks, D. L. and Peng, C. T., Eds. (Academic Press, New York).

McDowell, W. J. (1975). "High resolution liquid scintillation method for the analytical determination of alpha emitters in environmental samples," IEEE Trans. Nucl. Sci. **NS-22**, 649.

McKay, K. (1951). "Electron-hole production in germanium by alpha particles," Phys. Rev. **84**, 829.

McLaughlin, W. L., Ed. (1982). *Trends in Radiation Dosimetry*, Int. J. Appl. Radiat. Isot. **33**, 953.

McNelles, L. A. and Campbell, J. L. (1973). "Absolute efficiency calibration of coaxial Ge(Li) detectors for the energy range 160 to 1330 keV," Nucl. Instrum. Methods **109**, 241.

McNish, A. G. (1958). "Classification and nomenclature for standards of measurement," IRE Trans. Instrum. **I 7**, 371.

McNish, A. G. (1960). "Nomenclature for standards of radioactivity," Int. J. Appl. Radiat. Isot. **8**, 145.

Merritt, J. S. (1973). "Present status in quantitative source preparation," Nucl. Instrum. Methods **112**, 325.

Merritt, J. S. and Taylor, J. G. V. (1967a). *Gravimetric Sampling in the Standardization of Solutions of Radionuclides*. Report No. AECL-2679 (Atomic Energy of Canada, Limited, Chalk River).

Merritt, J. S. and Taylor, J. G. V. (1967b). "Response of 4π proportional counters to γ rays," page 147 in *Standardization of Radionuclides*, IAEA/STI/PUB/139 (International Atomic Energy Agency, Vienna).

Merritt, J. S. and Taylor, J. G. V. (1971). "The standardization of ^{63}Ni and the decay of ^{65}Ni," Int. J. Appl. Radiat. Isot. **22**, 783.

Merritt, J. S., Taylor, J. G. V., and Campion, P. J. (1959). "Self-absorption in sources prepared for 4π beta counting," Can. J. Chem. **37**, 1109.

Merritt, J. S., Taylor, J. G. V., Merritt, W. F., and Campion, P. J. (1960). "The absolute counting of sulfur-35," Anal. Chem. **32**, 310.

Meyer, K. P., Schmid, P., and Huber, P. (1959). "Absolut-Messung radioaktiver quellstarken mit hilfe einer neugestaltung der koinzidenzmethode,"

Helv. Phys. Acta **32**, 423.

MEYER, R. A., TIRSELL, K. G., AND ARMANTROUT, G. A. (1976). "Current research relevant to the improvement of γ-ray spectroscopy as an analytical tool," page 40 in *Proceedings of ERDA Symposium on X- and Gamma-Ray Sources and Applications*, CONF-760539 (National Technical Information Service, Springfield, Virginia).

MEYER-SCHÜTZMEISTER, L. AND VINCENT, D. H. (1952). "Absoluteichungen Energiearmer β Strahler mit dem 4π-Zählrohr," Z. Phys. **134**, 9.

MEYER, S. AND VON SCHWEIDLER, E. (1927). *Radioaktivitat*, ed. 2 (B. G. Teubner, Berlin).

MILLER, W. W. (1947). "High-efficiency counting of long-lived radioactive carbon as CO_2," Science **105**, 123.

MOGHISSI, A. A. (1970). "Low-level liquid scintillation counting of α- and β-emitting nuclides," page 86 in *The Current Status of Liquid Scintillation Counting*, Bransome, E. D., Ed. (Grune & Stratton, New York).

MOGHISSI, A. A. AND CARTER, M. W. (1968). "Internal standard with identical system properties for determination of liquid scintillation efficiency," Anal. Chem. **40**, 812.

MOGHISSI, A. A. AND CARTER, M. W., Ed. (1973). *Tritium* (Messenger Graphics, Phoenix, Arizona).

MOGHISSI, A. A., MCNELIS, D. N., PLOTT, W. F., AND CARTER, M. W. (1971). "A rapid liquid scintillation technique for monitoring ^{14}C in air," page 391 in *Rapid Methods for Measuring Radioactivity in the Environment*, IAEA/STI/PUB/289 (International Atomic Energy Agency, Vienna).

MORAN, J. J. (1967). "Measurement of volume: liquid," page 4279 in *Treatise on Analytical Chemistry*, Part 1, vol. 7, Kolthoff, I. M., Elving, P. J., and Sandell, E. B., Eds. (John Wiley, New York).

MOSELEY, H. G. J. (1912). "The number of β particles emitted in the transformation of radium," Proc. R. Soc. London Ser. **A 87**, 230.

MOSZYNSKI, M., GRESSET, C., VACHER, J., AND ODRU, R. (1981). "Timing properties of BGO scintillator," Nucl. Instrum. Methods **188**, 403.

MUGHABGHAB, S. F. AND GARBER, D. I. (1973). "Neutron cross sections," in *Vol. 1, Resonance Parameters*, Report BNL 325, 3rd Ed. (also as Report TID-4500) (National Technical Information Service, Springfield, Virginia).

MÜLLER, J. W. (1973). "Dead-time problems," Nucl. Instrum. Methods **112**, 47.

MÜLLER, J. W. (1974). "Some formulae for a dead time distorted Poisson process," Nucl. Instrum. Methods **117**, 401.

MÜLLER, J. W. (1975). "Principles of correlation counting," Report BIPM-75/5 (Bureau International des Poids et Mesures, Sèvres).

MÜLLER, J. W. (1976a). "On the evaluation of the correction factor $\mu(\rho',\tau)$ for the periodic pulse method," Report BIPM-76/3 (Bureau International des Poids et Mesures, Sèvres).

MÜLLER, J. W. (1976b). "The source-pulser method revisited," Report BIPM-76/5 (Bureau International des Poids et Mesures, Sèvres).

MÜLLER, J. W. (1979). "Some second thoughts on error statements," Nucl.

Instrum. Methods **163,** 241.

MÜLLER, J. W. (1981). "Selective sampling—An alternative to coincidence counting," Nucl. Instrum. Methods **189,** 449.

MULLER, R. A. (1979). "Radioisotope dating with accelerators," Physics Today, **32,** 23.

MUNZENMAYER, K. AND BAERG, A. P. (1969). "Delay matching of beta-gamma coincident pulses," Nucl. Instrum. Methods **70,** 157.

NAS-NRC (1974). National Academy of Sciences-National Research Council. *Users' Guides for Radioactivity Standards,* National Academy of Sciences Nuclear Science Series Report No. 3115 (National Academy of Sciences, Washington).

NATRELLA, M. G. (1963). *Experimental Statistics,* Nat. Bur. Stand. (U.S.) Handb. 91 (U.S. Government Printing Office, Washington). (Reprinted with corrections 1966).

NBS (1963). National Bureau of Standards. "General physical constants— New values for the physical constants. Recommended by NAS-NRC," Nat. Bur. Stand. (U.S.) Tech. News Bull. **47,** No. 10, 175.

NBS (1977). National Bureau of Standards. *The International System of Units (SI),* Nat. Bur. Stand. (U.S.) Spec. Publ. **330,** 3rd Ed. (U.S. Government Printing Office, Washington).

NCRP (1961). National Committee on Radiation Protection and Measurements. *A Manual of Radioactivity Procedures,* NCRP Report No. 28, published as Nat. Bur. Stand. Handbook 80 (National Council on Radiation Protection and Measurements, Bethesda, Maryland).

NCRP (1975). National Council on Radiation Protection and Measurements. *Natural Background Radiation in the United States,* NCRP Report No. 45 (National Council on Radiation Protection and Measurements, Bethesda, Maryland).

NCRP (1976a). National Council on Radiation Protection and Measurements. *Tritium Measurement Techniques for Laboratory and Environmental Use,* NCRP Report No. 47 (National Council on Radiation Protection and Measurements, Bethesda, Maryland).

NCRP (1976b). National Council on Radiation Protection and Measurements. *Environmental Radiation Measurements,* NCRP Report No. 50 (National Council on Radiation Protection and Measurements, Bethesda, Maryland).

NCRP (1983). National Council on Radiation Protection and Measurements. *Iodine-129: Evaluation of Releases From Nuclear Power Generation,* NCRP Report No. 75 (National Council on Radiation Protection and Measurements, Bethesda, Maryland).

NCRP (1984). National Council on Radiation Protection and Measurements. *Carbon-14 in the Environment,* NCRP Report No. 81 (National Council on Radiation Protection and Measurements, Bethesda, Maryland).

NERVIK, W. E. AND STEVENSON, P. C. (1952). "Self-scattering and self-absorption of betas by moderately thick samples," Nucleonics **10** No. 3, 18.

NESTOR, O. H. (1983). "Scintillator material growth," page 77 in *Nuclear Radiation Detector Materials,* Haller, E. E., Kraner, H. W. and Higginbotham, W. A., Eds. (Elsevier Science and North-Holland Publishing Companies, New York and Amsterdam).

NESTOR, O. H. AND HUANG, C. Y. (1975). "Bismuth germanate: A high-Z gamma-ray and charged-particle detector, "IEEE Trans. Nucl. Sci. **NS-22,** 68.

NEYMAN, J. AND PEARSON, E. S. (1936). "Contribution to the theory of testing statistical hypotheses," page 1 in *Statistical Research Memoirs,* Vol. 1, Neyman, J. and Pearson, E. S., Eds. (The University Press, Cambridge, England).

NICHOLSON, P. W. (1974). *Nuclear Electronics* (John Wiley & Sons, New York).

NIELSEN, J. M. (1972). "Gamma-ray spectrometry," page 678 in *Physical Methods of Chemistry,* Vol. 1, Weissberger, A. and Rossiter, B. W., Eds. (John Wiley & Sons, New York).

NIELSEN, J. M. AND PERKINS, R. W. (1967). "Anti-coincidence shielded multidimensional gamma-ray spectrometers for low-level counting," page 687 in *Radioactive Dating and Methods of Low-Level Counting,* IAEA/STI/ PUB/152 (International Atomic Energy Agency, Vienna).

NIH (1973). National Institutes of Health. *The National Institutes of Health Radiation Safety Guide,* DHEW Publication NIH-73-18 (Department of Health, Education and Welfare, Washington).

NOAKES, J. E., NEARY, M. P., AND SPAULDING, J. D. (1973). "Tritium measurements with a new liquid scintillation counter," Nucl. Instrum. Methods **109,** 177.

NOBLES, R. A. (1956). "Detection of charged particles with gas scintillation counters," Rev. Sci. Instrum. **27,** 280.

NORTHRUP, J. A. AND NOBLES, R. A. (1956). "Some aspects of gas scintillation counters," IRE Trans. Nucl. Sci. **NS-3,** 59.

NOUJAIM, A. A., EDISS, C., AND WEIBE, L. I. (1976). *Liquid Scintillation Science and Technology,* Proceedings of an International Conference, Banff, Canada, 1976. (Academic Press, New York).

NOWLIN, C. H. AND BLANKENSHIP, C. L. (1965). "Elimination of undesirable undershoot in the operation and testing of nuclear pulse amplifiers," Rev. Sci. Instrum. **36,** 1830.

NUCLEAR DATA INC. (1984). Advertisement, Phys. Today **37,** No. 1, 85.

O'KELLEY, G. D. (1962). *Detection and Measurement of Nuclear Radiation,* NAS-NS 3105 (National Technical Information Service, Springfield, Virginia).

OLLER, W. L. AND PLATO, P. (1972). "Beta spectrum analysis: a new method to analyze mixtures of beta-emitting radionuclides by liquid scintillation techniques," Int. J. Appl. Radiat. Isot. **23,** 481.

OLSSON, I. U., KARLEN, I., TURNBULL, A. H., AND PROSSER, N. J. D. (1962). "A determination of the half-life of C^{14} with a proportional counter," Ark. Fys. **22,** 237.

ORTEC (1983). *Radiation Detection-Measurement-Analysis,* 1983/84 Catalog, E. G. and G. Ortec (Oak Ridge, Tennessee).

OSBORNE, R. V. AND HILL, C. R. (1964). "High resolution alpha particle spectroscopy at very low specific activities," Nucl. Instrum. Methods **29**, 101.

OTT, D. G., RICHMOND, C. R., TRUJILLO, T. T., AND FOREMAN, H. (1959). "Cab-O-Sil suspensions for liquid-scintillation counting," Nucleonics **17** No. 9, 106.

OUSEPH, T. J. (1975). *Introduction to Nuclear Radiation Detectors* (Plenum Publishing Company, New York).

PAINTER, K. (1974). "Choice of counting vials for liquid scintillation: A review," page 431 in *Recent Developments in Liquid Scintillation Counting*, Stanley, P. and Scoggins, B., Eds. (Academic Press, New York).

PALMER, H. E. (1975). "Recent advances in gas scintillation proportional counters," IEEE Trans. Nucl. Sci. **NS-22**, 100.

PALMER, H. E. AND BEASLEY, T. M. (1967). "Iron-55 in man and the biosphere," Health Phys. **13**, 889.

PARKER, R. P. AND ELRICK, R. H. (1970). "Čerenkov counting as a means of assaying β-emitting radionuclides," page 110 in *The Current Status of Liquid Scintillation Counting*, Bransome, E. D., Ed. (Grune & Stratton, New York).

PARMENTIER, J. H. AND TEN HAAF, F. E. L. (1969). "Developments in liquid scintillation counting since 1968," Int. J. Appl. Radiat. Isot. **20**, 305.

PARZEN, E. (1960). *Modern Probability Theory and Its Applications* (John Wiley & Sons, New York).

PASTERNACK, B. S. AND HARLEY, N. H. (1971). "Detection limits for radionuclides in the analysis of multi-component gamma ray spectrometer data," Nucl. Instrum. Methods **91**, 533.

PATE, B. D. AND YAFFE, L. (1955a). "A new material and technique for the fabrication and measurement of very thin films for use in 4π-counting," Can. J. Chem. **33**, 15.

PATE, B. D. AND YAFFE, L. (1955b). "Disintegration-rate determination by 4π-counting, Pt. I," Can. J. Chem. **33**, 610.

PATE, B. D. AND YAFFE, L. (1955c). "Disintegration-rate determination by 4π-counting, Pt. II. Source-mount absorption correction," Can. J. Chem. **33**, 929.

PATE, B. D. AND YAFFE, L. (1955d). "Disintegration-rate determination by 4π-counting, Pt. III. Absorption and scattering of radiation," Can. J. Chem. **33**, 1656.

PATE, B. D. AND YAFFE, L. (1956). "Disintegration-rate determination by 4π-counting, Pt. IV. Self-absorption correction: general method and application to Ni^{63} β-radiation," Can. J. Chem. **34**, 265.

PATTERSON, M. S. AND GREENE, R. C. (1965). "Measurement of low energy beta emitters in aqueous solution by liquid scintillation counting of emulsions," Anal. Chem. **37**, 854.

PAULE, R. C. AND MANDEL, J. (1982). "Consensus values and weighting factors," J. Res. Nat. Bur. Stand., **87**, 377.

PEELLE, R. W. AND MAIENSCHEIN, F. C. (1960). "Gamma-gamma coincidence standardization of sources, with application to the B^{11} (p,γ) C^{12} reaction," page ‘96 in *Neutron Physics Annual Progress Report*, ORNL-3016 (Oak Ridge National Laboratory, Oak Ridge, Tennessee).

PEHL, R. H. (1977). "Germanium gamma-ray detectors," Phys. Today, **30**, No. 11, 50.

PEHL, R. H., MADDEN, N. W., ELLIOTT, J. H., RAUDORF, T. W., TRAMELL, R. C., AND DARKEN, L. S. (1979). "Radiation damage resistance of reverse electrode Ge coaxial detectors," IEEE Trans. Nucl. Sci. **NS-26**, 321.

PELL, E. M. (1960). "Ion drift in an n-p junction," J. Appl. Phys. **31**, 291.

PENG, C. T. (1975). "Liquid scintillation and Čerenkov counting," page 79 in *Radiochemical Methods in Analysis*, Coomber, D. I., Ed. (Plenum Press, New York).

PENG, C. T. (1977). *Sample Preparation in Liquid Scintillation Counting*, RCC Review 17 (The Radiochemical Centre, Ltd., Amersham, England).

PERKINS, R. W. (1965). "An anticoincidence shielded-multidimensional gamma-ray spectrometer," Nucl. Instrum. Methods **33**, 71.

PERKINS, R. W., NIELSEN, J. M., AND DIEBEL, R. N. (1960). "Total absorption gamma-ray spectrometers utilizing anticoincidence shielding." Rev. Sci. Instrum. **31**, 1344.

PERKINS, R. W., RANCITELLI, L. A., COOPER, J. A., KAYE, J. H., AND WOGMAN, N. A. (1970). "Cosmogenic and primordial radionuclides in lunar samples by nondestructive gamma-ray spectrometry," Science **167**, 577.

PETEL, M. AND HOUTERMANS, H. (1967). "Methodes d'indicateur pour l'etalonnage du fer-55," page 301 in *Standardization of Radionuclides*, IAEA/STI/PUB/139 (International Atomic Energy Agency, Vienna).

PETREE, B., HUMPHREYS, J. C., LAMPERTI, P. J., LOFTUS, T. P., AND WEAVER, J. T. (1975). "Comparison of cobalt-60 exposure determinations by calorimetry and by ionization chamber techniques," Metrologia **11**, 11.

PETROFF, M. D. AND DOGGETT, W. O. (1956). "Application of a statistical correlation function for determining daughter half-lives in the range 10^{-4} to 10^{-1} second," Rev. Sci. Instrum. **27**, 838.

PHELPS, P. L. AND HAMBY, K. O. (1972). "Experience in the use of an anticoincidence shielded Ge(Li) gamma-ray spectrometer for low-level environmental radionuclide analysis," IEEE Trans. Nucl. Sci. **NS-19**, 155.

PHILLIPS, G. W. AND MARLOW, K. W. (1976). "Automatic analysis of gamma-ray spectra from germanium detectors," Nucl. Instrum. Methods, **137**, 525.

PICCIOTTO, E. AND WILGAIN, S. (1954). "Thorium determination in deep-sea sediments," Nature **173**, 632.

PILTINGSRUD, H. V. AND STENCEL, J. R. (1972). "Determination of ^{90}Y, ^{90}Sr, and ^{89}Sr in samples by use of liquid scintillation beta spectroscopy," Health Phys. **23**, 121.

POCHWALSKI, K., BRODA, R., RADOSZEWSKI, T., AND ZELAZNY, P. (1981). "Standardization of low-level beta-emitter solutions by using the triple-to-double coincidence ratio (TDCR) method," page 487 in *Methods of Low-Level Counting and Spectrometry*, IAEA Publication STI-PUB-592 (International Atomic Energy Agency, Vienna).

POLICARPO, A. J. P. L., CONDE, C. A. N., AND ALVES, M. A. F. (1968). "Large light output gas proportional scintillation counters," Nucl. Instrum. Methods **58**, 151.

POLICARPO, A. J. P. L., ALVES, M. A. F., SANTOS, M. C. M., AND CARVALHO, M. J. T. (1972). "Improved resolution for low energies with gas proportional scintillation counters," Nucl. Instrum. Methods **102**, 337.

PONTIUS, P. E. (1966). *The Measurement Philosophy of the Pilot Program for Mass Calibration*, Nat. Bur. Stand. (U.S.) Tech. Note 288 (Superintendent of Documents, U.S. Government Printing Office, Washington, D.C.).

POVINEC, P., Ed. (1981). "Simultaneous activity and background measurement by the same detector," Int. J. Appl. Radiat. Isot. **32**, 729.

POVINEC, P., Ed. (1982). *Low Radioactivities '80*, Physics and Applications, **8**.

PRICE, W. J. (1958). *Nuclear Radiation Detection* (McGraw-Hill, New York, Toronto, London).

PRICE, W. J. (1964). *Nuclear Radiation Detection*, 2nd Ed. (McGraw-Hill, New York).

PUTMAN, J. L. (1953). "Limitations and extensions of the coincidence method for measuring the activity of beta-gamma emitters," in AERE Report 1/M-26 (Atomic Energy Research Establishment, Harwell, United Kingdom).

RAGAINI, R. C. (1976). "Methods in environmental sampling for radionuclides," IEEE Trans. Nucl. Sci. **NS-23**, 1202.

RAMTHUN, H. (1973). "Recent developments in calorimetric measurements of radioactivity," Nucl. Instrum. Methods **112**, 265.

RANDOLPH, R. B. (1975). "Determination of strontium-90 and strontium-89 by Čerenkov and liquid scintillation counting," Int. J. Appl. Radiat. Isot. **26**, 9.

RAO, P. V., CHEN, M. H., AND CRASEMANN, B. (1972). "Atomic vacancy distributions produced by inner-shell ionization," Phys. Rev. **A5**, 997.

RAPKIN, E. (1967). "Preparation of samples for liquid scintillation counting," page 181 in *Instrumentation in Nuclear Medicine*, Vol. 1, Hine, G. J., Ed. (Academic Press, New York).

RAUDORF, T. W., TRAMMELL, R. C., AND DARKEN, L. S. (1979). "N-type high purity germanium coaxial detectors," IEEE Trans. Nucl. Sci. **NS-26**, 297.

REIFF, G., SQUILLANTE, M. R., SERREZE, H. B., ENTINE, G., AND HUTH, G. C. (1983). "Large area silicon avalanche photodiodes: Photo-multiplier tube alternate," page 131 in *Nuclear Radiation Detector Materials*, Haller, E. E., Kraner, H. W., and Higginbotham, W. A., Eds. (Elsevier Science and North-Holland Publishing Companies, New York and Amsterdam).

RIECK, H. G., MYERS, I. T., AND PALMER, R. F. (1956). "Tritiated water standard," Radiat. Res. **4**, 451.

ROBERTS, P. B. AND DAVIES, B. L. (1966). "A transistorized radon measuring equipment," J. Sci. Instrum. **43**, 32.

ROBINSON, B. M. W. (1960). "Routine standardization of radionuclides in the UK," page 41 in *Metrology of Radionuclides*, IAEA/STI/PUB/6 (International Atomic Energy Agency, Vienna).

ROBINSON, C. V. (1967). "Geiger-Müller and proportional counters," page 57 in *Instrumentation in Nuclear Medicine*, Vol. 1, Hine, G. J., Ed. (Academic

Press, New York).

ROBINSON, H. P. (1960). "A new design for a high precision, high geometry alpha counter," page 147 in *Metrology of Radionuclides*, IAEA/STI/PUB/6 (International Atomic Energy Agency, Vienna).

RODRIGUEZ-PASQUÉS, R. H. (1971). "Microactivity radiochemistry," page 278 in *Proceedings of the Secunda Conferencia Interamericana de Radioquimica*, CONF-680423 (Organization of American States, Washington).

RODRIGUEZ-PASQUÉS, R. H., STEINBERG, H. L., HARDING, J. E., MULLEN, P. A., HUTCHINSON, J. M. R., AND MANN, W. B. (1972). "Low-level radioactivity measurements on aluminum, steel, and copper," Int. J. Appl. Radiat. Isot. **23**, 445.

ROLLE, R. (1970). "Determination of radon daughters on filters by simple liquid scintillation technique," Am. Ind. Hyg. Assoc. J. **31**, 718.

ROSE, M. E. AND KORFF, S. A. (1941). "An investigation of proportional counters I," Phys. Rev. **59**, 850.

ROSS, H. H. (1970). "Čerenkov radiation: photon yield application to carbon-14 assay," page 123 in *Current Status of Liquid Scintillation Counting*, Bransome, E. D., Ed. (Grune & Stratton, New York).

ROSS, H. H. (1971). "Performance parameters of selected waveshifting compounds for Čerenkov counting," page 757 in *Organic Scintillators and Liquid Scintillation Counting*, Horrocks, D. L. and Peng, C. T., Eds. (Academic Press, New York).

ROSSI, B. (1952). *High Energy Particles* (Prentice Hall, Inc., Englewood Cliffs, New Jersey).

ROUTTI, J. T. AND PRUSSIN, S. G. (1969). "Photopeak method for the computer analysis of gamma-ray spectra from semiconductor detectors," Nucl. Instrum. Methods **72**, 125.

ROUX, A. M. (1974). "Comparison of activity and power measurements of a 1 Ci ^{60}Co source," Metrologia **10**, 103.

ROY, J. C. AND RYTZ, A. (1965). "Analyse de resultats de la comparaison internationale de ^{35}S, Juin 1962," BIPM Report (Bureau International des Poids et Mesures, Sèvres, France).

ROY, J. C. AND TURCOTTE, J. (1967). "Use of ionic bombardment to increase the conductivity of metalised thin plastic films for making source mounts," Can. J. Chem. **65**, 1379.

RUTHERFORD, E. (1905). "Charge carried by the α and β rays of radium," Philos. Mag. **10** [6], 193.

RUTHERFORD, E. AND SODDY, F. (1902). "The cause and nature of radioactivity: Part I Growth and decay," Philos. Mag. **4** [6], 370.

RUTHERFORD, E. AND SODDY, F. (1903). "Radioactive change," Philos. Mag. **5** [6], 576.

RUTHERFORD, E. AND BARNES, H. T. (1904). "Heating effects of the radium emanation," Philos. Mag. **7** [6], 202.

RUTHERFORD, E. AND GEIGER, H. (1908). "An electrical method of counting the number of α-particles from radioactive substances," Proc. Roy. Soc. **81**, 141.

RUTHERFORD, E. AND ROBINSON, H. (1913). "Heating effect of radium and its emanation," Philos. Mag. **25** [6], 312.

RUTHERFORD, E., CHADWICK, J., AND ELLIS, C. D. (1931). *Radiations From Radioactive Substances* (Cambridge University Press, Cambridge).

RYTZ, A. (1964). *Rapport Sur La Comparaison International de* [241] *Am, Juillet 1963*, (Bureau International des Poids et Mesures Report, Sèvres).

RYTZ, A. (1977). "Report on the international comparison of activity measurements of a solution of [139]Ce (March, 1976)," BIPM Rapport 77/4. This report summarized by same author in Nucl. Instrum. Methods **157**, 131.

RYTZ, A. (1978). "International coherence of activity measurements," Env. Int. **1**, 15.

RYTZ, A. (1980). "[134]Cs, report of an international comparison of activity measurements (October, 1978), Rapport BIPM-80/2 (Bureau International des Poides et Mesures, Sèvres).

RYTZ, A. (1983). "The international reference system for activity measurements of γ-ray emitting nuclides," Int. J. Appl. Radiat. Isot. **34**, 1047.

SALMON, L. (1961). "Analysis of gamma-ray scintillation spectra by the method of least squares," Nucl. Instrum. Methods **14**, 193.

SAYRES, A. AND WU, C. S. (1957). "Gas scintillation counter," Rev. Sci. Instrum. **28**, 758.

SCHELL, W. R. (1970). "An internal gas proportional counter for measuring low level environmental radionuclides," page 173 in *Radionuclides in the Environment* (American Chemical Society, Washington).

SCHONFELD, E., KIBBY, A. H., AND DAVIS, W. (1966). "Determination of nuclide concentrations in solutions containing low levels of radioactivity by least-squares resolution of the gamma-ray spectra," Nucl. Instrum. Methods **45**, 1.

SCHOPPER, H. (1966). *Weak Interactions and Nuclear Beta Decay* (North-Holland Publishing Co., Amsterdam).

SCHWEBEL, A., ISBELL, H. S., AND KARABINOS, J. V. (1951). "A rapid method for measurement of carbon-14 in formamide solution," Science **113**, 465.

SCHWEBEL, A., ISBELL, H. S., AND MOYER, J. D. (1954). "Determination of carbon-14 in solutions of carbon-14 labeled materials by means of a proportional counter," J. Res. Nat. Bur. Stand. **53**, 221.

SELIGER, H. H. (1960). "Liquid scintillation counting of α particles and energy resolution of the liquid scintillator for α- and β-particles," Int. J. Appl. Radiat. Isot. **8**, 29.

SELIGER, H. H. AND SCHWEBEL, A. (1954). "Standardization of beta-emitting nuclides," Nucleonics **12**, No. 7, 54.

SELIGER, H. H., MANN, W. B., AND CAVALLO, L. M. (1958). "The average energy of sulfur-35 beta decay," J. Res. Nat. Bur. Stand. **60**, 447.

SHALEK, R. J. AND STOVAL, M. (1969). "Dosimetry in implant therapy," page 743 in *Radiation Dosimetry*, Vol. 3, 2nd Ed., Attix, F. H. and Tochilin, E., Eds. (Academic Press, New York).

SHAPIRO, J. (1972). *Radiation Protection* (Harvard University Press, Cambridge, Massachusetts).

SHARPE, J. AND WADE, F. (1951). "T.P.A. Mk II ionization chamber," AERE EL/R806 (U.K. Atomic Energy Research Establishment, Harwell).

SHER, D. W., KAARTINEN, N., EVERETT, L. J., AND JUSTES, V. (1971). "Preparing samples for liquid scintillation counting with the Packard sample oxidizer," page 849 in *Organic Scintillators and Liquid Scintillation Counting,* Horrocks, D. L. and Peng, C. T., Eds. (Academic Press, New York).

SHONKA, F. R. AND STEPHENSON, R. J. (1949). *Chicago Pressure Ionization Chamber for Gamma Ray Measurement,* Univ. Chicago Metallurgical Laboratory Report MUC-RJS-2, AECD-2463 (National Technical Information Service, Springfield, Virginia).

SHORE, L. G. (1949). "Long-lived self-quenching counter filling," Rev. Sci. Instrum. **20,** 956.

SHUPING, R. E., PHILLIPS, C. R., AND MOGHISSI, A. A. (1969). "Low-level counting of environmental krypton-85 by liquid scintillation," Anal. Chem. **41,** 2082.

SIEGBAHN, K., Ed. (1965). *Alpha-, Beta-, and Gamma-Ray Spectroscopy* (North-Holland Publishing Co., Amsterdam).

SIFFERT, P. (1983). "Cadmium telluride detectors and applications," page 87 in *Nuclear Radiation Detector Materials,* Haller, E. E., Kraner, H. W. and Higginbotham, W. A., Eds. (Elsevier Science and North-Holland Publishing Companies, New York).

SILL, C. W. AND OLSON, D. G. (1970). "Sources and prevention of recoil contamination of solid-state alpha detectors," Anal. Chem. **42,** 1596.

SIMMONNET, G. AND ORIA, M. (1977). *Radioactivity Measurement Using Liquid Scintillation Counters* (Eyrolles, Paris) (In French).

SIMMONS, G. H. (1970). *A Training Manual for Nuclear Medicine Technologists,* BRH/DMRE 70-3 (National Technical Information Service, Springfield, Virginia).

SINCLAIR, W. K., TROTT, N. G., AND BELCHER, E. H. (1954). "The measurement of radioactive samples for clinical use," Br. J. Radiol. **27,** 565.

SMITH, C. M. H. (1965). *Textbook of Nuclear Physics,* Chapter 11 (Pergamon Press, Oxford; distributed by MacMillan, New York).

SMITH, D. (1975). "An improved method of data collection for $4\pi\beta$-γ coincidence measurements," Metrologia **11,** 73.

SMITH, D. (1978). "Improved correction formula for coincidence counting," Nucl. Instrum. Methods **152,** 505.

SMITH, D. (1980). "The problem of after pulses," page 65 in *The Application of Liquid Scintillation Counting to Radionuclide Metrology* Monographic BIPM-3 (Bureau International des Poids et Mèsures, Sevres, France).

SMITH, D. AND WILLIAMS, A. (1971). "A $4\pi\beta$-γ coincidence method for mixed electron-capture/positron emitters: The absolute measurement of ^{68}Ga," Int. J. Appl. Radiat. Isot. **22,** 615.

SMITH, D. AND STUART, L. E. H. (1975). "An extension of the $4\pi\beta$-γ coincidence technique: Two dimensional extrapolation," Metrologia **11,** 67.

SMITH, K. F. AND CLINE, J. E. (1966). "Low-noise charge sensitive preamplifier for semiconductor detectors using paralleled field-effect-transistors,"

IEEE Trans. Nucl. Sci. **NS-13,** 468.

SNELL, A. H., Ed. (1962). *Nuclear Instruments and Their Uses,* Vol. 1 (John Wiley & Sons, New York).

SONG, L., WU, X. Z., AND WANG, Z. E. (1980). Liquid scintillation counting for the standardization of β-particle emitters," Nucl. Instrum. Methods **175,** 503.

SPERNOL, A. AND DENECKE, B. (1964a). "High-precision absolute gas counting of tritium. I—preparation of hydrogen and counting gas," Int. J. Appl. Radiat. Isot. **15,** 139.

SPERNOL, A. AND DENECKE, B. (1964b). "High-precision absolute gas counting of tritium. II—characteristics and construction of internal gas counters, especially for hydrogen-methane mixtures," Int. J. Appl. Radiat. Isot. **15,** 195.

SPERNOL, A. AND DENECKE, B. (1964c). "High precision absolute gas counting of tritium. III—absolute counting of tritium in the internal gas counter," Int. J. Appl. Radiat. Isot. **15,** 241.

SPERNOL, A. AND LERCH, O. (1965). "Eine auf 0.2% genau Zählung von Alphateilchen mit Plastikdetektoren," Nucl. Instrum. Methods, **32,** 293.

SPERNOL, A., DeROOST, E., AND MUTTERER, M. (1973). "Problems and possibilities of bremsstrahlung counting," Nucl. Instrum. Methods **112,** 169.

SQUILLANTE, M. R., LIS, S., HAZLETT, T., AND ENTINE, G. (1983). "State-of-the-art x-ray detectors fabricated on PCG grown mercuric iodide," page 199 in *Nuclear Radiation Detector Materials,* Haller, E. E., Kraner, H. W., and Higginbotham, W. A., Eds. (Elsevier Science and North-Holland Publishing Companies, New York and Amsterdam).

STANLEY, R. E. AND MOGHISSI, A. A., Eds. (1975). *Noble Gases,* ERDA Report CONF-730915 (National Technical Information Service, Springfield, Virginia).

STAUB, H. H. (1953). "Detection methods," page 1 in *Experimental Nuclear Physics,* Segré, E., Ed. (John Wiley, New York and Chapman and Hall, London).

STEIDLEY, J., NAKASHIAN, M., AND ENTINE, G. (1980). "Build a personal radiation monitor," Popular Electronics, **January,** 39.

STEVENSON, P. C. (1966). *Processing of Counting Data,* National Academy of Sciences-National Research Council Nuclear Science Series, Report NAS-NS 3109 (National Techical Information Service, Springfield, Virginia).

STEYN, J. (1967). "Internal liquid scintillation counting applied to the absolute disintegration-rate measurement of electron capture nuclides," page 35 in *Standardization of Radionuclides,* IAEA/STI/PUB/139 (International Atomic Energy Agency, Vienna).

STEYN, J. (1973a). "Special problems involved in liquid scintillation counting," Nucl. Instrum. Methods **112,** 137.

STEYN, J. (1973b). "Tracer method for pure beta emitter measurement," Nucl. Instrum. Methods **112,** 157.

STEYN, J. AND HAASBROEK, F. J. (1958). "The application of internal liquid scintillation counting to a 4π beta-gamma coincidence method for the absolute standardization of radioactive nuclides," page 95 in *Peaceful Uses of Atomic Energy*, Vol. 21 (United Nations, Geneva).

STEYN, J. AND HAHNE, F. J. W. (1963). "Absolute disintegration rate measurement of beta-emitters by application of efficiency tracing to 4π-liquid scintillation counting," page 30 in *National Conference on Nuclear Energy; Application of Isotopes and Radiation* (Atomic Energy Board, Pelindaba, South Africa).

STEYN, J. AND DEJESUS, A. S. M. (1966). *Coincidence Measurements on Nuclides Decaying by Electron Capture*, National Physical Research Lab. Report NP-15912 (Council for Scientific and Industrial Research, Pretoria, South Africa).

STOCKMANN, L. L. (1955). "Radium assay of ores and sludges by the radon method," Unpublished manuscript (Radioactivity Section, National Bureau of Standards, Washington).

STROMSWOLD, D. C. (1981). "Comparison of sodium iodide, cesium iodide, and bismuth germanate scintillation detectors for borehole gamma-ray logging," IEEE Trans. Nucl. Sci. **NS-28**, 290.

SUBRAMANIAN, G., RHODES, B. A., COOPER, J. F., AND SODD, V. J., Eds. (1975). *Radiopharmaceuticals* (The Society of Nuclear Medicine, Inc., New York).

SUDAR, S., VAS, L., AND BIRO, T. (1973). "Pulse-shape discrimination in the proportional counting of tritium betas," Nucl. Instrum. Methods **112**, 399.

SURIN, V. M. AND FORMISHKIN, EH. F. (1983). "Half-lives of long-lived isotopes of transactinium elements from ^{228}Th to ^{257}Fm," International Nuclear Data Committee Report INDC (CCP)-196/GE (International Atomic Energy Agency, Vienna).

SUSCHNY, O., DUBCZYNSKI, R., AND TUGSAVUL, A. (1978). "The analytical quality control programme of the International Atomic Energy Agency—areas of common interests with the ICRM," Environ. Int. **1**, 19.

SUTHERLAND, L. G. AND BUCHANAN, J. D. (1967). "Error in the absolute determination of disintegration rates of extended sources by coincidence counting with a single detector—Application to I-125 and Co-60," Int. J. Appl. Radiat. Isot. **18**, 786.

SWIERKOWSKI, S. P., ARMANTROUT, G. A., AND WICHNER, R. (1974). "Recent advances with HgI_2 x-ray detectors," IEEE Trans. Nucl. Sci. **NS-21**, 302.

TAMERS, M., BIRBRON, R., AND DELIBRIAS, G. (1962). "A new method for measuring low-level tritium using a benzene liquid scintillator," page 303 in *Tritium in the Physical and Biological Sciences*, Vol. 1 IAEA/STI/PUB/39 (International Atomic Energy Agency, Vienna).

TANAKA, E. (1961). "A low background beta-ray scintillation spectrometer using a coincidence method with a Geiger counter," Nucl. Instrum. Methods **13**, 43.

TANAKA, E., ITOH. S., HIRAMOTO, T., AND IINUMA, T. A. (1965). "Cosmic ray contribution to the background of NaI scintillation spectrometers," Jpn. J. Appl. Phys. **4**, 785.

WOELLER, F. H. (1961). "Liquid scintillation counting of $C^{14}O_2$ with phenethylamine," Anal. Biochem. **2**, 508.

WOGMAN, N. A. (1974). "Comparison of Ge(Li) and anti-Compton systems for measurement of environmental samples," IEEE Trans. Nucl. Sci. **NS-21**, 526.

WOGMAN, N. A. AND BRODZINSKI, R. L. (1973). "The development and application of a beta-gamma-gamma multidimensional spectrometer," IEEE Trans. Nucl. Sci. **NS-20**, 73.

WOGMAN, N. A., PERKINS, R. W., AND KAYE, J. H. (1969). "An all sodium iodide anticoincidence shielded multidimensional gamma-ray spectrometer for low-activity samples," Nucl. Instrum. Methods **74**, 197.

WOLF, M. A., WAECHTER, D. A., AND UMBARGER, C. J. (1982). "A new tiny computerized radiation dosimeter," IEEE Trans. Nucl. Sci. **NS-29**, 773.

WOOD, D. E. AND BOWEN, B. M. (1971). "^{95}Zr and ^{124}Sb in ^{99m}Tc generators," J. Nucl. Med. **12**, 307.

WOOD, R. E. AND PALMS, J. M. (1974). "A gamma-ray spectrum analysis technique for low-level environmental radionuclides," IEEE Trans. Nucl. Sci. **NS-21**, 536.

WOODS, M. J. (1970). "Calibration figures for the type 1383A ionization chamber," Int. J. Appl. Radiat. Isot. **21**, 752.

WOODS, M. J. AND LUCAS, S. E. M. (1975). "Calibration of the 1383A ionization chamber for ^{125}I," Int. J. Appl. Radiat. Isot. **26**, 488.

WOODS, M. J., GOODIER, I. W., AND LUCAS, S. E. M. (1975). "The half-life of ^{133}Xe and its calibration factor for the 1383A ionization chamber," Int. J. Appl. Radiat. Isot. **26**, 485.

WRIGHT, G. T. (1954). "Low energy measurements with the photomultiplier scintillation counter," J. Sci. Instr. **31**, 462.

YAFFE, L. (1962). "Preparation of thin films, sources, and targets," Annu. Rev. Nucl. Sci. **12**, 153.

YAFFE, L. AND FISHMAN, J. B. (1960). "Self-absorption studies with a $4\pi\beta$-proportional flow counter," page 185 in *Metrology of Radionuclides* IAEA/STI/PUB/6 (International Atomic Energy Agency, Vienna).

YAGODA, H. J. (1949). *Radioactive Measurements with Nuclear Emulsions* (John Wiley & Sons, New York).

YAMADA, S. (1962). "The detection of low activity in aqueous solution using Čerenkov radiation," J. Phys. Soc. Jpn. **17**, 865.

YAMASHITA, M., YURA, O., AND KAWADA, Y. (1983). "Utilization of high-gain first dynode PMT's for measuring the average numbers of photoelectrons in weak light and scintillations," Bul. Electrotech. Lab. **47**, 116.

YOUDEN, W. J. (1951). *Statistical Methods for Chemists* (John Wiley & Sons, New York).

YOUDEN, W. J. (1963). "Statistics in chemical analysis," page 14-1 in *Handbook of Analytical Chemistry*, Meites, L., Ed., (McGraw-Hill, New York).

YOUNG, J. Z. AND ROBERTS, F. (1951). "A flying-spot microscope," Nature **167**, 231.

YURA, O. (1983). "4$\pi\beta$(LS)-γ coincidence counting in the electrotechnical laboratory," Bul. Electrotech. Lab. **47,** Nos. 9, 10, 21.

ZANIO, K. (1978). *Semiconductors and Semimetals: Cadmium Telluride,* Vol. 13, Willardson, R. K. and Beer, A. C., Eds. (Academic Press, New York).

ZSDANSZKY, K. (1973). "Precise measurement of small currents," Nucl. Instrum. Methods **112,** 299.

ZUMWALT, L. R. (1950). *Absolute Beta Counting Using End-Window Geiger-Müller Counters and Experimental Data on Beta-Particle Scattering Effects,* AECU-567 (Oak Ridge National Laboratory, Oak Ridge, Tennessee).

The NCRP

The National Council on Radiation Protection and Measurements is a nonprofit corporation chartered by Congress in 1964 to:

1. Collect, analyze, develop, and disseminate in the public interest information and recommendations about (a) protection against radiation and (b) radiation measurements, quantities, and units, particularly those concerned with radiation protection;
2. Provide a means by which organizations concerned with the scientific and related aspects of radiation protection and of radiation quantities, units, and measurements may cooperate for effective utilization of their combined resources, and to stimulate the work of such organizations;
3. Develop basic concepts about radiation quantities, units, and measurements, about the application of these concepts, and about radiation protection;
4. Cooperate with the International Commission on Radiological Protection, the International Commission on Radiation Units and Measurements, and other national and international organizations, governmental and private, concerned with radiation quantities, units, and measurements and with radiation protection.

The Council is the successor to the unincorporated association of scientists known as the National Committee on Radiation Protection and Measurements and was formed to carry on the work begun by the Committee.

The Council is made up of the members and the participants who serve on the eighty-one scientific committees of the Council. The scientific committees, composed of experts having detailed knowledge and competence in the particular area of the committee's interest, draft proposed recommendations. These are then submitted to the full membership of the Council for careful review and approval before being published.

The following comprise the current officers and membership of the Council:

Currently, the following subgroups are actively engaged in formulating recommendations:

SC-1:   Basic Radiation Protection Criteria
SC-3:   Medical X-Ray, Electron Beam and Gamma-Ray Protection for Energies Up to 50 MeV (Equipment Performance and Use)
SC-16:   X-Ray Protection in Dental Offices
SC-18:   Standards and Measurements of Radioactivity for Radiological Use
SC-38:   Waste Disposal
             Task Group on Krypton-85
             Task Group on Carbon-14
             Task Group on Disposal of Accident Generated Waste Water
             Task Group on Disposal of Low-Level Waste
             Task Group on the Actinides
             Task Group on Xenon
             Task Group on Definitions of Radioactive Waste
SC-40:   Biological Aspects of Radiation Protection Criteria
             Task Group on Atomic Bomb Survivor Dosimetry
             Subgroup on Biological Aspects of Dosimetry of Atomic Bomb Survivors
SC-42:   Industrial Applications of X Rays and Sealed Sources
SC-44:   Radiation Associated with Medical Examinations
SC-45:   Radiation Received by Radiation Employees
SC-46:   Operational Radiation Safety
             Task Group 1 on Warning and Personnel Security Systems
             Task Group 2 on Uranium Mining and Milling—Radiation Safety Programs
             Task Group 3 on ALARA for Occupationally Exposed Individuals in Clinical Radiology
             Task Group 4 on Calibration of Survey Instrumentation
             Task Group 5 on Maintaining Radiation Protection Records
             Task Group 6 on Radiation Protection for Allied Health Personnel
SC-47:   Instrumentation for the Determination of Dose Equivalent
SC-48:   Apportionment of Radiation Exposure
SC-52:   Conceptual Basis of Calculations of Dose Distributions
SC-53:   Biological Effects and Exposure Criteria for Radiofrequency Electromagnetic Radiation
SC-54:   Bioassay for Assessment of Control of Intake of Radionuclides

SC-55: Experimental Verification of Internal Dosimetry Calculations
SC-57: Internal Emitter Standards
Task Group 2 on Respiratory Tract Model
Task Group 3 on General Metabolic Models
Task Group 6 on Bone Problems
Task Group 7 on Thyroid Cancer Risk
Task Group 8 on Leukemia Risk
Task Group 9 on Lung Cancer Risk
Task Group 10 on Liver Cancer Risk
Task Group 11 on Genetic Risk
Task Group 12 on Strontium
Task Group 13 on Neptunium
SC-59: Human Radiation Exposure Experience
SC-60: Dosimetry of Neutrons from Medical Accelerators
SC-61: Radon Measurements
SC-63: Control of Exposure to Ionizing Radiation from Accident or Attack
SC-64: Radionuclides in the Environment
Task Group 5 on Public Exposure to Nuclear Power
Task Group 6 on Screening Models
Task Group 7 on Contaminated Soil as a Source of Radiation Exposure
SC-65: Quality Assurance and Accuracy in Radiation Protection Measurements
SC-66: Biological Effects and Exposure Criteria for Ultrasound
SC-67: Biological Effects of Magnetic Fields
SC-68: Microprocessors in Dosimetry
SC-69: Efficacy Studies
SC-70: Quality Assurance and Measurement in Diagnostic Radiology
SC-71: Radiation Exposure and Potentially Related Injury
SC-72: Radiation Protection in Mammography
SC-74: Radiation Received in the Decontamination of Nuclear Facilities
SC-75: Guidance on Radiation Received in Space Activities
SC-76: Effects of Radiation on the Embryo-Fetus
SC-77: Guidance on Occupational and Public Exposure Resulting from Diagnostic Nuclear Medicine Procedures
SC-78: Practical Guidance on the Evaluation of Human Exposures to Radiofrequency Radiation
SC-79: Extremely Low-Frequency Electric and Magnetic Fields
SC-80: Radiation Biology of the Skin (Beta-Ray Dosimetry)
SC-81: Assessment of Exposure from Therapy

Committee on Public Education
Committee on Public Relations
Ad Hoc Committee on Policy in Regard to the International System of Units
Ad Hoc Committee on Comparison of Radiation Exposures
Study Group on Comparative Risk
  Task Group on Comparative Carcinogenicity of Pollutant Chemicals
Task Force on Occupational Exposure Levels
Ad Hoc Group on Model Used for Assessing Transport of Low-Level Radioactive Waste
Ad Hoc Group on Medical Evaluation of Radiation Workers

In recognition of its responsibility to facilitate and stimulate cooperation among organizations concerned with the scientific and related aspects of radiation protection and measurement, the Council has created a category of NCRP Collaborating Organizations. Organizations or groups of organizations that are national or international in scope and are concerned with scientific problems involving radiation quantities, units, measurements and effects, or radiation protection may be admitted to collaborating status by the Council. The present Collaborating Organizations with which the NCRP maintains liaison are as follows:

American Academy of Dermatology
American Association of Physicists in Medicine
American College of Nuclear Physicians
American College of Radiology
American Dental Association
American Industrial Hygiene Association
American Institute of Ultrasound in Medicine
American Insurance Association
American Medical Association
American Nuclear Society
American Occupational Medical Association
American Podiatric Medical Association
American Public Health Association
American Radium Society
American Roentgen Ray Society
American Society of Radiologic Technologists
American Society of Therapeutic Radiology and Oncology
Association of University Radiologists
Atomic Industrial Forum
Bioelectromagnetics Society
College of American Pathologists
Federal Communications Commission
Federal Emergency Management Agency
Genetics Society of America
Health Physics Society
National Bureau of Standards
National Electrical Manufacturers Association
Radiation Research Society
Radiological Society of North America
Society of Nuclear Medicine
United States Air Force
United States Army
United States Department of Energy
United States Department of Housing and Urban Development
United States Department of Labor
United States Environmental Protection Agency
United States Navy
United States Nuclear Regulatory Commission
United States Public Health Service

The NCRP has found its relationships with these organizations to be extremely valuable to continued progress in its program.

Another aspect of the cooperative efforts of the NCRP relates to the special liaison relationships established with various governmental organizations that have an interest in radiation protection and measurements. This liaison relationship provides: (1) an opportunity for participating organizations to designate an individual to provide liaison between the organization and the NCRP; (2) that the individual designated will receive copies of draft NCRP reports (at the time that these are submitted to the members of the Council) with an invitation to comment, but not vote; and (3) that new NCRP efforts might be discussed with liaison individuals as appropriate, so that they might have an opportunity to make suggestions on new studies and related matters. The following organizations participate in the special liaison program:

Defense Nuclear Agency
Federal Emergency Management Agency
National Bureau of Standards
Office of Science and Technology Policy
Office of Technology Assessment
United States Air Force
United States Army
United States Coast Guard
United States Department of Energy
United States Department of Health and Human Services
United States Department of Labor
United States Department of Transportation
United States Environmental Protection Agency
United States Navy
United States Nuclear Regulatory Commission

The NCRP values highly the participation of these organizations in the liaison program.

The Council's activities are made possible by the voluntary contribution of time and effort by its members and participants and the generous support of the following organizations:

Alfred P. Sloan Foundation
Alliance of American Insurers
American Academy of Dental Radiology
American Academy of Dermatology
American Association of Physicists in Medicine
American College of Radiology
American College of Radiology Foundation
American Dental Association
American Industrial Hygiene Association
American Insurance Association
American Medical Association

American Nuclear Society
American Occupational Medical Association
American Osteopathic College of Radiology
American Podiatric Medical Association
American Public Health Association
American Radium Society
American Roentgen Ray Society
American Society of Radiologic Technologists
American Society of Therapeutic Radiology and Oncology
American Veterinary Medical Association
American Veterinary Radiology Society
Association of University Radiologists
Atomic Industrial Forum
Battelle Memorial Institute
Center for Devices and Radiological Health
College of American Pathologists
Commonwealth of Pennsylvania
Defense Nuclear Agency
Edison Electric Institute
Edward Mallinckrodt, Jr. Foundation
Electric Power Research Institute
Federal Emergency Management Agency
Florida Institute of Phosphate Research
Genetics Society of America
Health Physics Society
James Picker Foundation
National Association of Photographic Manufacturers
National Bureau of Standards
National Cancer Institute
National Electrical Manufacturers Association
Radiation Research Society
Radiological Society of North America
Society of Nuclear Medicine
United States Department of Energy
United States Department of Labor
United States Environmental Protection Agency
United States Navy
United States Nuclear Regulatory Commission

To all of these organizations the Council expresses its profound appreciation for their support.

Initial funds for publication of NCRP reports were provided by a grant from the James Picker Foundation and for this the Council wishes to express its deep appreciation.

The NCRP seeks to promulgate information and recommendations based on leading scientific judgment on matters of radiation protection and measurement and to foster cooperation among organizations concerned with these matters. These efforts are intended to serve the public interest and the Council welcomes comments and suggestions on its reports or activities from those interested in its work.

NCRP Publications

NCRP publications are distributed by the NCRP Publications Office. Information on prices and how to order may be obtained by directing an inquiry to:

NCRP Publications
7910 Woodmont Avenue
Suite 800
Bethesda, MD 20814-3095

The currently available publications are listed below.

573

101   *Exposure of the U.S. Population from Occupational Radiation* (1989)
102   *Medical X-Ray, Electron Beam and Gamma-Ray Protection for Energies Up to 50 MeV (Equipment Design, Performance and Use)* (1989)
103   *Control of Radon in Houses* (1989)
104   *The Relative Biological Effectiveness of Radiations of Different Quality* (1990)
105   *Radiation Protection for Medical and Allied Health Personnel* (1989)
106   *Limit for Exposure to "Hot Particles" on the Skin* (1989)
107   *Implementation of the Principle of As Low As Reasonably Achievable (ALARA) for Medical and Dental Personnel* (1990)
108   *Conceptual Basis for Calculations of Absorbed-Dose Distributions* (1991)
109   *Effects of Ionizing Radiation on Aquatic Organisms* (1991)
110   *Some Aspects of Strontium Radiobiology* (1991)
111   *Developing Radiation Emergency Plans for Academic, Medical or Industrial Facilities* (1991)
112   *Calibration of Survey Instruments Used in Radiation Protection for the Assessment of Ionizing Radiation Fields and Radioactive Surface Contamination* (1991)
113   *Exposure Criteria for Medical Diagnostic Ultrasound: I. Criteria Based on Thermal Mechanisms* (1992)
114   *Maintaining Radiation Protection Records* (1992)
115   *Risk Estimates for Radiation Protection* (1993)
116   *Limitation of Exposure to Ionizing Radiation* (1993)
117   *Research Needs for Radiation Protection* (1993)
118   *Radiation Protection in the Mineral Extraction Industry* (1993)

Binders for NCRP reports are available. Two sizes make it possible to collect into small binders the "old series" of reports (NCRP Reports Nos. 8-30) and into large binders the more recent publications (NCRP Reports Nos. 32-118). Each binder will accommodate from five to seven reports. The binders carry the identification "NCRP Reports" and come with label holders which permit the user to attach labels showing the reports contained in each binder.

The following bound sets of NCRP reports are also available:

Volume I.   NCRP Reports Nos. 8, 22
Volume II. NCRP Reports Nos. 23, 25, 27, 30
Volume III. NCRP Reports Nos. 32, 35, 36, 37
Volume IV. NCRP Reports Nos. 38, 40, 41

(Titles of the individual reports contained in each volume are given above.)

NCRP Commentaries

No.   Title

1   *Krypton-85 in the Atmosphere—With Specific Reference to the Public Health Significance of the Proposed Controlled Release at Three Mile Island* (1980)

2   *Preliminary Evaluation of Criteria for the Disposal of Transuranic Contaminated Waste* (1982)

3   *Screening Techniques for Determining Compliance with Environmental Standards—Releases of Radionuclides to the Atmosphere* (1986), Revised (1989)

4   *Guidelines for the Release of Waste Water from Nuclear Facilities with Special Reference to the Public Health Significance of the Proposed Release of Treated Waste Waters at Three Mile Island* (1987)

5   *Review of the Publication, Living Without Landfills* (1989)

6   *Radon Exposure of the U.S. Population—Status of the Problem* (1991)

7   *Misadministration of Radioactive Material in Medicine—Scientific Background* (1991)

8   *Uncertainty in NCRP Screening Models Relating to Atmospheric Transport, Deposition and Uptake by Humans* (1993)

Proceedings of the Annual Meeting

14 *Radiation Protection in Medicine,* Proceedings of the Twenty-eighth Annual Meeting held on April 1-2, 1992 (including Taylor Lecture No. 16) (1993)

Lauriston S. Taylor Lectures

No. Title

1 *The Squares of the Natural Numbers in Radiation Protection* by Herbert M. Parker (1977)

2 *Why be Quantitative about Radiation Risk Estimates?* by Sir Edward Pochin (1978)

3 *Radiation Protection—Concepts and Trade Offs* by Hymer L. Friedell (1979) [Available also in *Perceptions of Risk,* see above]

4 *From "Quantity of Radiation" and "Dose" to "Exposure" and "Absorbed Dose"—An Historical Review* by Harold O. Wyckoff (1980)

5 *How Well Can We Assess Genetic Risk? Not Very* by James F. Crow (1981) [Available also in *Critical Issues in Setting Radiation Dose Limits,* see above]

6 *Ethics, Trade-offs and Medical Radiation* by Eugene L. Saenger (1982) [Available also in *Radiation Protection and New Medical Diagnostic Approaches,* see above]

7 *The Human Environment—Past, Present and Future* by Merril Eisenbud (1983) [Available also in *Environmental Radioactivity,* see above]

8 *Limitation and Assessment in Radiation Protection* by Harald H. Rossi (1984) [Available also in *Some Issues Important in Developing Basic Radiation Protection Recommendations,* see above]

9 *Truth (and Beauty) in Radiation Measurement* by John H. Harley (1985) [Available also in *Radioactive Waste,* see above]

10 *Biological Effects of Non-ionizing Radiations: Cellular Properties and Interactions* by Herman P. Schwan (1987) [Available also in *Nonionizing Electromagnetic Radiations and Ultrasound,* see above]

11 *How to be Quantitative about Radiation Risk Estimates* by Seymour Jablon (1988) [Available also in *New Dosimetry at Hiroshima and Nagasaki and its Implications for Risk Estimates,* see above]

12 *How Safe is Safe Enough?* by Bo Lindell (1988) [Available also in *Radon,* see above]

Symposium Proceedings

The Control of Exposure of the Public to Ionizing Radiation in the Event of Accident or Attack, Proceedings of a Symposium held April 27-29, 1981 (1982)

NCRP Statements

Other Documents

The following documents of the NCRP were published outside of the NCRP Report, Commentary and Statement series:

Somatic Radiation Dose for the General Population, Report of the Ad Hoc Committee of the National Council on Radiation Protection and Measurements, 6 May 1959, Science, February 19, 1960, Vol. 131, No. 3399, pages 482–486

Dose Effect Modifying Factors In Radiation Protection, Report of Subcommittee M-4 (Relative Biological Effectiveness) of the National Council on Radiation Protection and Measurements, Report BNL 50073 (T-471) (1967) Brookhaven National Laboratory (National Technical Information Service Springfield, Virginia)

The following documents are now superseded and/or out of print:

NCRP Reports

No.   Title

1   *X-Ray Protection* (1931) [Superseded by NCRP Report No. 3]

2   *Radium Protection* (1934) [Superseded by NCRP Report No. 4]

3   *X-Ray Protection* (1936) [Superseded by NCRP Report No. 6]

4   *Radium Protection (1938)* [Superseded by NCRP Report No. 13]

5   *Safe Handling of Radioactive Luminous Compound* (1941) [Out of Print]

6   *Medical X-Ray Protection Up to Two Million Volts* (1949) [Superseded by NCRP Report No. 18]

7   *Safe Handling of Radioactive Isotopes* (1949) [Superseded by NCRP Report No. 30]

9   *Recommendations for Waste Disposal of Phosphorus-32 and Iodine-131 for Medical Users* (1951) [Out of Print]

10   *Radiological Monitoring Methods and Instruments* (1952) [Superseded by NCRP Report No. 57]

11   *Maximum Permissible Amounts of Radioisotopes in the Human Body and Maximum Permissible Concentrations in Air and Water* (1953) [Superseded by NCRP Report No. 22]

12   *Recommendations for the Disposal of Carbon-14 Wastes* (1953) [Superseded by NCRP Report No. 81]

NCRP Proceedings

Index

Note: For individual radionuclides, see List of Radionuclides (Appendix A.2) for which nuclear-decay data are given in Appendix A.3

Note: For individual radionuclides, see List of Radionuclides (Appendix A.2) for which nuclear-decay data are given in Appendix A.3

Note: For individual radionuclides, see List of Radionuclides (Appendix A.2) for which nuclear-decay data are given in Appendix A.3

Note: For individual radionuclides, see List of Radionuclides (Appendix A.2) for which nuclear-decay data are given in Appendix A.3

Note: For individual radionuclides, see List of Radionuclides (Appendix A.2) for which
nuclear-decay data are given in Appendix A.3

Note: For individual radionuclides, see List of Radionuclides (Appendix A.2) for which nuclear-decay data are given in Appendix A.3

Note: For individual radionuclides, see List of Radionuclides (Appendix A.2) for which nuclear-decay data are given in Appendix A.3

Note: For individual radionuclides, see List of Radionuclides (Appendix A.2) for which nuclear-decay data are given in Appendix A.3

Note: For individual radionuclides, see List of Radionuclides (Appendix A.2) for which nuclear-decay data are given in Appendix A.3

Note: For individual radionuclides, see List of Radionuclides (Appendix A.2) for which nuclear-decay data are given in Appendix A.3

Note: For individual radionuclides, see List of Radionuclides (Appendix A.2) for which nuclear-decay data are given in Appendix A.3

Note: For individual radionuclides, see List of Radionuclides (Appendix A.2) for which nuclear-decay data are given in Appendix A.3